Exploring Medicinal Fruits and Berries

Exploring Medicinal Fruits and Berries is a clear, authoritative guide to fruits long valued in traditional medical systems and increasingly examined by modern science. It explains how naturally occurring bioactive compounds—antioxidants, polyphenols, flavonoids, vitamins and minerals—contribute to pathways relevant to health risk and well-being. Bridging traditional knowledge with state-of-the-art research, the book surveys fruits and berries with evidence across immunity, inflammation management, cardiovascular and metabolic health, healthy ageing, skin health and gastrointestinal function. It also examines underutilized species, emerging nutraceutical applications and cultivation practices that support sustainability and biodiversity.

Part of the *Exploring Medicinal Plants* series, this book is designed for students, researchers, practitioners and growers, as well as health-conscious readers. It provides a rigorous resource for understanding medicinal fruits and berries in contemporary health and wellness, with an emphasis on scientific significance and practical applications.

Exploring Medicinal Plants

Series Editor: Azamal Husen, *Wolaita Sodo University, Ethiopia*

Medicinal plants render a rich source of bioactive compounds used in drug formulation and development; they play a key role in traditional or indigenous health systems. As the demand for herbal medicines increases worldwide, supply is declining as most of the harvest is derived from naturally growing vegetation. Considering global interests and covering several important aspects associated with medicinal plants, the Exploring Medicinal Plants series comprises volumes valuable to academia, practitioners, and researchers interested in medicinal plants. Topics provide information on a range of subjects including diversity, conservation, propagation, cultivation, physiology, molecular biology, growth response under extreme environment, handling, storage, bioactive compounds, secondary metabolites, extraction, therapeutics, mode of action, and healthcare practices.

Led by Azamal Husen, PhD, this series is directed to a broad range of researchers and professionals consisting of topical books exploring information related to medicinal plants. It includes edited volumes, references, and textbooks available for individual print and electronic purchases.

Wild Edibles of Deserts: Food Security and Sustainability
Salman Majeed, Muhammad Zafar, Mushtaq Ahmad, Mohd Kafeel Ahmad Ansari, Shazia Sultana, and Andrea Pieroni

Impact of Climate Change on Medicinal and Herbal Plant micro-RNA
Kanchanlata Tungare, Parul Johri, Sachidanand Singh, and Surojeet Das

Exploring Medicinal Bamboos
Edited by Manu Pant and Azamal Husen

Exploring Herbal Synergies for Optimal Human Health
Edited by Art Devi, Nancy George, Zaved Ahmed Khan, and Vagish Dwibedi

Exploring Medicinal and Commercial Uses of Flowers
Edited by Rakesh Kumar Bachheti, Archana Bachheti, and Azamal Husen

Phenolic Compounds from Medicinal Plants: Pharmaceutical and Health Benefits
Edited by Rakesh Kumar Bachheti, Archana Bachheti, and Azamal Husen

Therapeutic Potential of Vegetables
Edited by Azamal Husen and Megha Barot

Molecular Marker-Assisted Improvement and Production of Medicinal Plan
Edited by Arpan Modi, Megha Barot, Sachidanand Singh, and Azamal Husen

Exploring Medicinal Fruits and Berries
Edited by Azamal Husen

For more information about this series, please visit: www.routledge.com/Exploring-Medicinal-Plants/book-series/CRCEMP

Exploring Medicinal Fruits and Berries

Edited by
Azamal Husen

CRC Press
Taylor & Francis Group
Boca Raton London New York

CRC Press is an imprint of the
Taylor & Francis Group, an **informa** business

Designed cover image: Shutterstock

First edition published 2027
by CRC Press
2385 NW Executive Center Drive, Suite 320, Boca Raton FL 33431

and by CRC Press
4 Park Square, Milton Park, Abingdon, Oxon, OX14 4RN

CRC Press is an imprint of Taylor & Francis Group, LLC

ISBN: 978-1-041-16384-8 (hbk)
ISBN: 978-1-041-17159-1 (pbk)
ISBN: 978-1-003-68825-9 (ebk)

DOI: 10.1201/9781003688259

Typeset in Times
by SPi Technologies India Pvt Ltd (Straive)

Contents

Preface

Fruits and berries have been the main aspects in sustaining health and the cure of diseases since the earliest stages of human civilization. In addition to being the main carriers of nutritional value, the natural resources have also been shown to have the ability to cure diseases. In the majority of the historic traditions practised in the continents of Asia, Europe and Africa, fruits have remained the main natural resource in the cure of diseases. Among the fruits used by the Ayurvedics in ancient days, the Traditional Chinese, the Romans, the Native Africans in the continents of Africa, the Native Americans, the aborigines in the landmass of Europe and the Native Australians to cure diseases are the choice of fruit to be used in restoring health. The fruits and berries used in herbal medicine contain a high concentration of bioactive constituents, which play key roles as antioxidants, vitamin and mineral stores and dietary fibre, among other active biochemicals that play a role together to account for the medical and therapeutic benefits conferred by natural medications to the user. In contemporary society, there is a marked increase in lifestyle and other diseases ranging from heart diseases, diabetes, cancer, neurodegenerative diseases and other inflammatory diseases, which largely seek intense and novel preventive and medical approaches and solutions to combat and manage population health and wellness, and where fruits and berries have been recognized and elevated to new status as a source and package towards a long and healthy life and living well.

The advances of modern scientific research have made significant contributions to the better understanding of various medicinal properties displayed by different types of medicinal fruits and berries over the past several decades. In this regard, significant advancements were made with regard to phytochemical analysis, as well as modern molecular biological techniques, through which it can be proved that medicinal properties displayed by different varieties of medicinal fruits and berries play significant roles with regard to various biological activities, including enhancement of cellular functions, better management of cellular responses, improvement of antioxidants, management of gene expressions, influence on cellular signals, maintenance of metabolism homeostasis, etc. The increasing consumption of plant-based diets and natural health products has further raised the importance of this group of medicinal fruits and berries. This category of fruits and berries is increasingly being used in nutraceuticals, functional foods, herbal preparations and cosmetic products. However, despite their extensive traditional usage and commercial development, there is a continuing demand for well-documented and scientifically valid books that systematically outline their properties in an understandable form. This book covers this gap by gathering together traditional knowledge, experimental studies and applied aspects into a single, coherent text. The various chapters included in the book have been designed in an appropriate way to offer readers a stepwise understanding of the concept of medicinal fruits and berries. The initial chapters of the book have focused on the diversity and distribution of these fruits worldwide, together with their compounds constituting their medicinal importance. Later in the book, the importance of these fruits in augmenting the immune system, reducing inflammation, supporting the cardiovascular system, improving cognitive functions, stabilizing blood glucose and maintaining gastrointestinal health is explained in detail. Later in the book, the roles of these fruits are explained in the following aspects, namely their use in skin care, anti-ageing methods, detoxification mechanisms and cancer management. One of the defining features of this book has been its focus on bioactive fruits and berries in medicine, which exhibit synergistic effects, not individual bioactive molecules, in their bioactivity. Also, unlike other fruits, these fruits and berries exhibit their bioactivity by means of synergies between bioactive molecules, which, in most cases, increases not only their therapeutic values but also their bioactivity. The co-use of fruits and berries with herbal medicine has also been examined in this book, which is similar to other holistic systems.

Besides the known edible fruits, the present book introduces the reader to the knowledge of the underutilized medicinal fruits and berries with remarkable therapeutic potential. Many of the plant species used by the population belong to the category of medicinally underexplored species, despite the long period for which they have been used in the indigenous health practices of the region. At the same time, the mention of the above fruits emphasizes the need to conserve ethnobotanical knowledge inherited by the population. Sustainability is yet another important subject running as a thread throughout the entire content of this book. In the face of growing global demand for natural health products, medicinal plants are under increasing pressure. Factors such as unsustainable harvesting, habitat destruction and global warming have created serious sustainability issues surrounding the availability of fruit-bearing medicinal plants. Discussed in this book are the ways to implement sustainability in the cultivation, harvesting and production of medicinal fruits and berries in an eco-friendly manner to cater to the requirements of the coming generations.

The readers of this book will include a wide variety of people. The readers of this book will include undergraduate students, postgraduate students, researchers and professionals across different fields of botany, horticulture, pharmacognosy, nutrition sciences, among many other fields. The practitioners of traditional sciences will get valuable insights into cultivation practices. The readers will include people affiliated with different sections of nutraceuticals, pharmaceuticals and functional foods, among many other fields. The health-conscious readers of this book will find valuable information on natural wellness. *Exploring Medicinal Fruits and Berries* is intended to bridge the gap between traditional wisdom and modern science. It epitomizes the shifting notion that natural foods function not only as purveyors of nutrition but also as agents of health promotion and disease prevention. The juxtaposition of evidence-based knowledge with traditional practices in this book emphasizes an integrated approach to health—one that honours cultural heritage while embracing scientific innovation.

It is hoped that the information provided in this book will encourage informed use, further research and responsible stewardship of these medicinal fruits and berries. By raising greater awareness of their therapeutic importance and ecological roles, this text aspires to support healthier lives and a more sustainable relationship with nature. This book acts as a testament to the endurance that fruits and berries have played throughout history as fundamental connectors between nature, medicine and human health.

Azamal Husen

About the Editor

Azamal Husen has served as Full Professor and Head of the Department of Biology at the University of Gondar, Ethiopia. He is currently affiliated as a Foreign Delegate with Wolaita Sodo University, Ethiopia, and holds an Adjunct Professorship at the Department of Biotechnology, Graphic Era (Deemed to be University) in Dehradun, Uttarakhand, India. His earlier academic roles include Full Professor at Sankalchand Patel University in Visnagar, India, and Visiting Faculty at the Forest Research Institute and Doon College of Agriculture and Forestry in Dehradun, India. With more than 25 years of experience in academia and research, he is highly regarded for his expertise in nanobiotechnology, plant physiology, environmental botany, medicinal plants and plant propagation. His research has been supported by prominent national and international organizations, including the World Bank, the National Agricultural Technology Project, the Indian Council of Agricultural Research, the Indian Council of Forestry Research and Education and the Japan Bank for International Cooperation. He is a prolific scholar, having published more than 350 academic works, including research articles, books and book chapters. His editorial contributions are equally noteworthy; he serves on the editorial boards and peer-review panels of numerous prestigious journals published by Elsevier, Frontiers Media, Springer Nature, Taylor & Francis, RSC, Oxford University Press, the Royal Society, CSIRO, MDPI, PLOS, Sciendo, John Wiley & Sons and UPM Journals. He is also a member of the advisory board of Cambridge Scholars Publishing, United Kingdom. He is a fellow of the Plantae group under the American Society of Plant Biologists and holds membership in several scientific organizations, including the International Society of Root Research, the Asian Council of Science Editors and the International Natural Product Sciences Taskforce. Throughout his career, he has delivered invited lectures, keynote addresses and technical presentations at numerous international conferences, symposia and workshops. In the field of academic development, he has made significant contributions through the design and delivery of orientation programmes for university faculty, the development of educational modules and the formulation of curricula at the undergraduate, postgraduate and doctoral levels. In recognition of his contributions to teaching, research and community service, he was honoured with a special award from the University of Gondar. He currently serves as Editor-in-Chief of the American Journal of Plant Physiology and is Series Editor for several major academic collections, including *Exploring Medicinal Plants* (Taylor & Francis, USA), *Plant Biology, Sustainability and Climate Change* (Elsevier, USA) and *Smart Nanomaterials Technology* (Springer Nature, Singapore). He also leads the Handbook Series on *Nanotechnology and Its Applications* (Springer Nature, Singapore). His outstanding contributions to science have earned him international acclaim, including the prestigious distinction of being named among the "World's Top 2% Scientists" for three consecutive years (2020 onwards) by Stanford University, a recognition published in the Elsevier Data Repository. In addition to his regular academic responsibilities, he has served as Guest Editor for several high-impact special issues, including Advances in Plant Metabolomics: Insights, Applications, and Integration *(Plant Physiology and Biochemistry)*, Exploring Nanotechnology in Medicinal Plants *(Journal of Herbal Medicine)*, *Integrating Nanotechnology with Plant Compounds: Enhancing Abiotic Stress Resilience and Crop Improvement via Signaling Molecules (Frontiers in Plant Science)*, and Nanobiotechnology for Sustainable Agriculture and Crop Improvement *(Discover Applied Sciences)*, among others. His work continues to shape contemporary research at the intersection of plant science, biotechnology and sustainable development. His academic leadership, research excellence and international collaborations have positioned him as a prominent figure in global scientific circles.

Contributors

N. Zaheer Ahmed
Central Council for Research in Unani Medicine (CCRUM)
New Delhi, India

Jamal Akhtar
Central Council for Research in Unani Medicine (CCRUM)
New Delhi, India

Shah Alam
Regional Research Institute of Unani Medicine (RRIUM)
Mumbai, India

Ahmad Ali
Department of Life Science
University of Mumbai
Mumbai, India

Hina Alim
Department of Life Science
University of Mumbai
Mumbai, India

Tanfidz Alishlah
Faculty of Pharmacy
University of Jember
Indonesia

Nighat Anjum
Central Council for Research in Unani Medicine (CCRUM)
New Delhi, India

Anoja Priyadarshani Attanayake
Department of Biochemistry
Faculty of Medicine
University of Ruhuna
Galle, Sri Lanka

Sandeep K. Behera
Department of Life Science
University of Mumbai
Mumbai, India

Noureddine Chaachouay
Agri-Food and Health Laboratory (AFHL), ESEF of Berrechid
Hassan First University
Settat, Morocco

Kamonwan Chaikhong
Center of Excellence on Natural Products for Neuroprotection and Anti-Ageing
Chulalongkorn University
Bangkok, Thailand

Siriporn Chuchawankul
Immunomodulation of Natural Products Research Unit
Chulalongkorn University
Bangkok, Thailand
Department of Transfusion Medicine and Clinical Microbiology
Faculty of Allied Health Sciences
Chulalongkorn University
Bangkok, Thailand

Nayomi Deshani De Silva
Department of Biochemistry
Faculty of Medicine
University of Ruhuna
Sri Lanka

Dris Djemaa
Department of Applied Biology
Faculty of Natural and Life Sciences
Echahid Cheikh Larbi Tebessi University
Tebessa, Algeria

Hery Diar Febryanto
Faculty of Pharmacy
University of Jember
Indonesia

Patrick Valere Tsouh Fokou
Department of Biochemistry
Faculty of Science
University of Bamenda
Bambili, Cameroon
Advanced Research and Health Innovation Hub (ARHIH)
Yaounde, Cameroon

Sarfaraz Ali Ghumro
College of Science and Engineering
James Cook University
Cairns, Australia

Nazim Husain
Regional Research Institute of Unani Medicine
Silchar, India

Azamal Husen
Wolaita Sodo University
Wolaita, Ethiopia
Department of Biotechnology
Graphic Era (Deemed to be University)
Dehradun, India

Sajad Khan
Department of Biotechnology
Abdul Wali Khan University Mardan
Mardan, Pakistan

Uswatun Khasanah
College of Science and Engineering
James Cook University
Cairns, Australia
Faculty of Medicine
Brawijaya University
Malang, Indonesia

Balamurugan Krishnaswamy
Immunomodulation of Natural Products Research Unit
Chulalongkorn University
Bangkok, Thailand
Department of Biotechnology
Alagappa University
Karaikudi, India

Catherine M. Miller
College of Medicine and Dentistry
James Cook University
Cairns, Australia

Arpan Modi
Axenic Agritech Private Limited
Ahmedabad, India

Bynedi Seshadhri Chinna Mounish
Department of Biotechnology
Alagappa University
Karaikudi, India

Felix Wambua Muema
College of Science and Engineering
James Cook University
Cairns, Australia

Pandar Nartsawee
Program in Clinical Biochemistry and Molecular Medicine
Department of Clinical Chemistry
Faculty of Allied Health Sciences
Chulalongkorn University
Bangkok, Thailand

Malik Nasibullah
Department of Chemistry
Integral University
Lucknow, India

Azizi Nassima
Department of Applied Biology
Faculty of Natural and Life Sciences
Echahid Cheikh Larbi Tebessi University
Tebessa, Algeria

Y. P. Naveen
Department of Biochemistry
Adichunchanagiri School of Natural Sciences
Adichunchanagiri University
BG Nagara, India

H. G. Navyashree
Centre for Research and Innovation
Adichunchanagiri University
BG Nagara, India

Sumaira Naz
Department of Biochemistry
University of Malakand
Chakdara, Pakistan

Chaachouay Noureddine
Agri-Food and Health Laboratory (AFHL), ESEF of Berrechid
Hassan First University
Settat, Morocco

Ari Satia Nugraha
Drug Utilisation and Discovery Research Group
Faculty of Pharmacy
Universitas Jember
East Java, Indonesia

Nimisha Patel
Department of Life Science
J. C. Bose University of Science and Technology, YMCA
Faridabad, India

Anchalee Prasansuklab
Center of Excellence on Natural Products for Neuroprotection and Anti-Ageing
Chulalongkorn University
Bangkok, Thailand
College of Public Health Sciences
Chulalongkorn University
Bangkok, Thailand

Deepika B. Prashar
Department of Plant Sciences
School of Life Sciences
Central University of Himachal Pradesh
Kangra, India

Abdellatif Rahim
Laboratory of Biochemistry, Neurosciences, Natural Resources and Environment
Faculty of Sciences and Techniques
Hassan First University of Settat
Morocco

Sharika Rajasekharan
Immunomodulation of Natural Products Research Unit
Chulalongkorn University
Bangkok, Thailand
Department of Transfusion Medicine and Clinical Microbiology
Faculty of Allied Health Sciences
Chulalongkorn University
Bangkok, Thailand

Bendjazia Rania
Department of Applied Biology
Faculty of Natural and Life Sciences
Echahid Cheikh Larbi Tebessi University
Tebessa, Algeria

Patcharaporn Saengratwatchara
Center of Excellence on Natural Products for Neuroprotection and Anti-Ageing
Chulalongkorn University
Bangkok, Thailand
Department of Pre-Pharmacy Education
Faculty of Pharmacy
Payap University
Chiang Mai, Thailand

Thushara Indika Sampath
Department of Natural Products and Nutraceuticals
Sri Lanka Institute of Nanotechnology
Homagama, Sri Lanka

Munish Sharma
Department of Plant Sciences
School of Life Sciences
Central University of Himachal Pradesh
Kangra, India

Pankaj Sharma
Department of Microbiology
CCS Haryana Agricultural University
Hisar, India
Department of Ophthalmology and Visual Sciences
University of Illinois Chicago
USA

Shilpa
Department of Plant Sciences
School of Life Sciences
Central University of Himachal Pradesh
Kangra, India

Gazala Tabassum
Department of Life Science
University of Mumbai
Mumbai, India

Simplice Joël Ndendoung Tatsimo
Department of Chemistry
Higher Teachers' Training College
University of Maroua
Maroua, Cameroon

Tewin Tencomnao
Department of Clinical Chemistry
Faculty of Allied Health Sciences
Chulalongkorn University
Bangkok, Thailand
Center of Excellence on Natural Products for Neuroprotection and Anti-Ageing
Chulalongkorn University
Bangkok, Thailand

Shilpi Vashisth
Department of Plant Sciences
School of Life Sciences
Central University of Himachal Pradesh
Kangra, India

Phurpa Wangchuk
College of Science and Engineering
James Cook University
Cairns, Australia
Australian Institute of Health and Tropical Medicine (AITHM)
James Cook University
Cairns, Australia

Firdha Aprillia Wardhani
Faculty of Pharmacy
University of Jember
Indonesia

Weria Weisany
Department of Agriculture Science and Engineering
SRC, Islamic Azad University
Tehran, Iran

Muhammad Zahoor
Department of Biochemistry
University of Malakand
Chakdara, Pakistan

1 An Overview of Medicinal Fruits and Berries

Shilpa, Deepika B. Prashar, Shilpi Vashisth, and Munish Sharma

1.1 INTRODUCTION

Medicinal fruits and berries have been considered as an important source of nourishment and medicinal compounds in human diets since antiquity, and help in the prevention and treatment of human illness. Over the last several decades, the scientific interest and demand for fruits and berries as functional food have increased due to the presence of bioactive metabolites and advancements in the area of food science and phytochemistry, alongside nutritional epidemiology. These plant foods have been widely recognized to mediate oxidative stress, inflammation, metabolic dysfunction and cellular signalling pathways related to chronic disease, and have expanded the fields of nutrition, pharmacology and preventive medicine (Liu 2013; Husen 2023, 2024a, 2024b; Aguilera 2024). Berries constitute a significant group of medicinal fruits owing to their high phytochemical diversity, including polyphenolic compounds such as anthocyanins, flavonols, ellagitannins and phenolic acids, which are abundant in berries, especially in blueberries, strawberries, raspberries, blackberries and cranberries. Such compounds contribute to the characteristic pigmentation, sensory properties and biological behaviour of the berries to a great extent. A number of research studies have demonstrated the potent antioxidant and anti-inflammatory effects of phytochemicals present in the berries, aiding in the neutralization of reactive oxygen species, the regulation of inflammatory mediators and the prevention of oxidative damage in cells (Nile and Park 2014; Lamenza et al. 2024). The phytocompounds in berries have also been reported to be involved in various molecules and physiological processes such as regulation of gene expression, cell cycle progression, inducing apoptosis in aberrant cells and inhibiting angiogenesis, which justify their possible role in cancer chemoprevention. Moreover, the polyphenols of berries undergo a high degree of interaction with gut microbiota, biotransformation to bioactive products, which have the potential to impact metabolic health, immune response and inflammation. The existing evidence focuses on the fact that the combination of diverse phytochemicals, dietary fibre, vitamins and minerals collectively contributes towards the health advantages of berries (Williamson et al. 2018; Lamenza et al. 2024). Due to the varied phytochemical and nutritional profile, other medicinal fruits such as citrus fruits, pomegranates, tropical fruits and dragon fruit (*Hylocereus* spp.) containing phenolic, flavonoid, betalains, carotenoid, vitamins and dietary fibre have also made an immense contribution to human health. Dragon fruit is characterized by betalain pigments, phenolic compounds and vitamin C, all of which combine to form antioxidant activity and possible health benefits. It has also been used as a traditional medicine, and its medicinal benefits have been supported by recent scientific findings on its ability to stimulate or suppress the immune response, control metabolism and reduce oxidative stress (Sharma et al. 2024). The health significance of medicinal fruits and berries is also enhanced by the nutritional matrix. These foods also contain important micronutrients, including potassium, magnesium, folate and vitamin C, soluble and insoluble dietary fibre. Fibre is important in controlling gut, glycaemic, lipid metabolism and composition of microbes, which enhances the biological activities of the

DOI: 10.1201/9781003688259-1

related phytochemicals. Epidemiological research has consistently linked increased intake of fruit and berries with a decreased risk of cardiovascular disease, type 2 diabetes and some cancers, which supports the inclusion of fruit and berries in health promotion at the population level (Slavin and Lloyd 2012; Liu 2013). The preservation and processing of fruits and berries significantly affect their medicinal potential. Phytochemical stability, bioaccessibility and bioavailability can be modified through thermal and non-thermal processing methods such as drying, freezing, juicing, fermentation and high-pressure processing. Although traditional thermal treatment can result in deterioration of heat-sensitive compounds like anthocyanins and vitamin C, the emerging and optimized technologies can maintain or even improve functions. The freezing in most instances is actually for preserving polyphenolic content, but high-pressure processing can enhance extractability and stability of specific bioactive compounds. Therefore, processed fruit products, such as juices, purees and dried fruits, continue to be significant sources of dietary intake of medicinal phytochemicals (Aguilera 2024). Overall, the present chapter provides a systematic basis for understanding fruits and berries as functional and medicinal foods, incorporating evidence from recent reviews and experimental studies. Moreover, it identifies gaps in existing knowledge and research directions that should be pursued to enhance clinical translation, support health claims and efficiently explore medicinal fruits and berries in preventive and treatment policies. The present chapter promotes sustainable use of fruits and berries in agriculture, food technology and medicine by bridging traditional knowledge with modern science

1.2 SECONDARY METABOLITES PRESENT IN MEDICINAL FRUITS AND BERRIES

Phytochemical screening of fruits and berries reveals the various classes of bioactive compounds, including phenolic compounds (flavonoids, anthocyanins and flavonols), tannins, stilbenes, vitamins, organic acids and carotenoids. They also contain primary metabolites that contribute to their nutritional value and sensory properties; sugars and organic acids are examples of such compounds that vary with genetic background and environmental factors like temperature, light intensity and water supply, affecting taste and quality (Zheng et al. 2018). Some of the major secondary metabolite classes present in fruits and berries.

1.2.1 Polyphenols

The distinctive feature of medicinal fruits and berries is the presence of varied phytochemical composition, the most numerous and biologically important being polyphenols. Polyphenols include phenolic acids, flavonoids, stilbenes and tannins, which play significant roles in antioxidant activity, the suppression or regulation of inflammatory events and cellular signalling networks. They are produced by plants as defence systems and are not uniformly distributed throughout the tissue of fruits; they are commonly found in greater levels in the peel and seeds. Its presence in the fruits and berries supports the medicinal and functional value of their products (Aguilera 2024; Thorakkattu et al. 2025).

1.2.1.1 Phenolic Acids

Phenolic acids are widely found in medicinal berries and fruits, and are primarily categorized into hydroxybenzoic and hydroxycinnamic acid derivatives. Hydroxybenzoic acid compounds such as gallic, protocatechuic, vanillic and syringic acids, and hydroxycinnamic acid compounds such as caffeic, chlorogenic, ferulic, p-coumaric and sinapic acids, play a significant role in antioxidant activity, hydrolysing free radicals and chelating metals. They are especially found in berries like blueberries, blackberries, raspberries, mulberries, bilberries and chokeberries, and are important in the regulation of oxidative stress (Mujanović et al., 2024; Kim et al. 2025).

1.2.1.2 Flavonoids

The most diverse subclass of polyphenols in medicinal fruits and berries is flavonoids, which are highly diverse structurally and functionally. They belong to the following groups of anthocyanins, flavonols, flavan-3-ols, flavones and flavanones that have their own distinctive effects on biological activity and sensory properties (Aguilera 2024).

1.2.1.3 Tannins

Tannins are polyphenols with high molecular weight that cause astringency and strong biological actions in medicinal fruits and berries. They are either widely defined as condensed tannins or hydrolysable tannins.

1.2.1.4 Stilbenes

Stilbenes are minor but biologically significant polyphenols found in certain medicinal fruits, notably grape berries and certain wild ones. Resveratrol, pterostilbene and viniferins are antioxidants, cardioprotective, anti-inflammatory and anti-ageing compounds that complement the action of other polyphenols in fruit matrices (Aguilera 2024). Its production in grapes can be enhanced by various abiotic and biotic factors (Błaszczyk et al. 2018).

1.2.2 Anthocyanins

Anthocyanins are a group of water-soluble pigments of red, purple and blue colour present in many medicinal berries. They are found mostly as the glycosylated forms of anthocyanidins, including cyanidin, delphinidin, malvidin, pelargonidin, peonidin and petunidin. Anthocyanins are commonly the most significant source of total polyphenols in dark-coloured berries such as blueberries, bilberries, blackberries, mulberries, haskap berries and chokeberries. These substances are closely related to antioxidant, anti-inflammatory, vasoprotective and metabolic regulation (Zurek et al., 2024; Jurja et al. 2025).

1.2.3 Flavonols

Flavonols (quercetin, kaempferol, myricetin and isorhamnetin) compounds are commonly found in fruits and berries, which are mostly glycosides. They play a role in preventing lipid peroxidation, controlling inflammatory mediators and stimulating endogenous antioxidant defence. Anthocyanins often interact with flavonols in the berry matrices, increasing the overall functional capacity of the fruit (Kim et al. 2025).

1.2.4 Flavan-3-ols

The flavan-3-ols such as catechin, epicatechin, gallocatechin and epigallocatechin are found in monomeric and polymeric forms. These compounds are especially found in cranberries, strawberries, raspberries and blackberries and are related to cardiovascular protection, endothelial functionality and associations with the gut microbiota (Thorakkattu et al. 2025).

1.2.5 Condensed Tannins (Proanthocyanidin)

Proanthocyanidin (condensed tannins) are flavan-3-ol polymers that are found in cranberries, chokeberries, blueberries and blackberries. Their polymerization level also determines their bioavailability and antimicrobial action and can be associated with their functional significance (Kim et al. 2025).

1.2.6 Hydrolysable Tannins

The tannins that are hydrolysed can be of the ellagitannin type (sanguiin H-6 and lambertianin C) and gallotannin type, typically present in raspberries and strawberries. These are metabolized by the microbes in the gastrointestinal tract and yield urolithins that increase bioavailability and biological effects (Thorakkattu et al. 2025).

1.2.7 Carotenoids

Carotenoids are coloured pigments that impart yellow, orange and red colour to the fruits and berries. The carotenoids that are commonly found in medicinal fruits are 2-carotene, 5-carotene, lutein, zeaxanthin and lycopene. Although berries contain lower concentrations of carotenoids when compared to non-berry fruits, strawberries and raspberries have detectable carotenoids that synergize with polyphenols to increase antioxidant capacity (Aguilera 2024).

1.2.8 Betalains

Betalains are nitrogen-based pigments found in several medicinal fruits like dragon fruit (*Hylocereus* spp.). They are characterized by their ability to provide red violet (betacyanins) and yellow-orange (betaxanthins) colours and possess high antioxidant and anti-inflammatory effects. Betalains are different from anthocyanins and add a unique phytochemical dimension to pharmaceutical and functional fruits (Thorakkattu et al. 2025).

1.2.9 Volatile Compounds

Volatile compounds such as monoterpenes, sesquiterpenes, esters, aldehydes and alcohols are highly abundant in medicinal fruits and berries. These compounds affect the quality of sensory properties such as aroma, texture, flavour and colour, and could have antimicrobial and antioxidant effects (Martins et al. 2025).

1.2.10 Organic Acids

Organic acids such as citric, malic, tartaric and succinic acids constitute an important part of fruit phytochemistry. They affect acidity, stability, mineral bioavailability and phytochemical preservation in the processing and digestion process (Aguilera 2024).

Moreover, species, cultivar, geographical origin, environmental conditions and the maturity stage determine the phytochemical make-up of medicinal fruits and berries. Biological effects of these fruits are due to complicated synergistic actions of polyphenols, pigments, tannins, carotenoids, organic acids and volatile compounds instead of individual constituents. This combined chemical complexity forms the basis of medicinal and functional use of the whole fruits and berries (Jurja et al. 2025; Thorakkattu et al. 2025).

1.3 FACTORS AFFECTING PHYTOCHEMICAL COMPOSITION

The complex interrelationships among different environmental, developmental, genetic and agronomic factors, in a highly integrated manner, have a varied impact on the concentration, diversity and stability of phytochemical composition in medicinal fruits and berries.

1.3.1 Genetic Factors and Cultivar Variability

The composition of secondary metabolites in medicinal fruits and berries is strongly influenced by genetic factors and cultivar differences. These factors adversely affect the therapeutic potential of medicinal fruits and berries. For example, differences in stilbene content across developmental stages in berry cultivars have been studied, which might be due to the genetic differences in the regulation of biosynthetic pathways among cultivars (Błaszczyk et al., 2018). The genetic make-up has a significant role in determining the nutritional quality of the medicinal fruits and berries by controlling the production and metabolic pathways of essential phytochemicals, such as sugars, organic acids and polyphenols. Studies on grapes have demonstrated that genetic variation plays a key role in how the plant responds to environmental factors, ultimately leading to changes in phytochemical content (Zheng et al. 2018). In strawberries, breeding programmes exploit genetic variation to enhance the levels of health-promoting compounds, such as vitamin C and phenolics (Mezzetti et al. 2016). Similarly, hot pepper cultivars show genotype-driven differences in phytochemical traits (Tripodi et al. 2018). Polyphenols are abundant in medicinal berries; these compounds vary among cultivars, contributing to antioxidant capacity and potential health benefits in the prevention of chronic diseases (Rao and Snyder 2010; Mazzoni et al. 2015). Selective breeding strategies exploit these genetic variations to improve nutritional quality by increasing specific health-promoting phytochemicals (Battino et al. 2009).

1.3.2 Environmental Factors

The environmental conditions, such as light, temperature and water supply, critically influence the accumulation of phytochemicals in medicinal fruits and berries. The increased light intensity has been associated with higher sugar contents, whereas low water supply and high exposure tend to decrease malic acid contents, as observed in grapes (*Vitis vinifera*) (Zheng et al. 2018). In addition, interactions of the plant genotype, their developmental stage, growth period and growing season configure the metabolite profiles that ultimately determine fruit flavour and overall quality. The pharmacologically important compounds, including stilbenes in edible berries, are also affected by climate change, cultivation practices and post-harvest treatments. Abiotic and biotic conditions are known to stimulate stilbene formation, thereby emphasizing the role of controlled environmental signals in increasing the bioactivity of medicinal berries (Błaszczyk et al., 2018). Nutrient stress, such as iron and phosphorus deficiency, promotes the accumulation of phenolic compounds, including pelargonidin 3-glucoside and flavonols, under hydroponic conditions (Liu et al. 2019). The management of light intensity also alters the phytochemical composition of berries. Greenhouse shade enhances anthocyanin and total phenolic contents, along with improved antioxidant capacity in many berry species (Blando et al., 2018). Additionally, specific light qualities enhance the biosynthesis of anthocyanins and flavonoids in fruits such as blueberries and strawberries. Anthocyanin, flavonoid and phenolic biosynthesis in blueberries and strawberries have been stimulated by light quality, including UV and visible spectrums. Moreover, cultivation systems such as high tunnels may reduce specific phytochemicals in raspberries and tomatoes, suggesting that light-environment manipulation affects bioactive compound accumulation. Water deficit stress increases the levels of anthocyanins, flavonoids and phenolic acids in fruits such as grapes and pomegranates, demonstrating that controlled abiotic stress enhances phytochemical profiles (Xu et al. 2015). Secondary metabolite production in plants is highly regulated by the soil fertility, salinity and water retention. Since bioactive metabolites are generally produced in response to the plant's adaptation to stress conditions, optimization of cultivation conditions is essential for improving the phytochemicals of medicinal plants (Jamwal et al. 2018). Post-harvest processing helps preserve phytochemicals in berries, and freeze-drying provides better preservation of phenolic and antioxidant compounds. Moisture diffusivity improves the retention of bioactive compounds by pre-drying blanching techniques (Johnson et al. 2011).

1.3.3 Post-harvesting Handling

The nutritional quality and health benefits are influenced by post-harvesting management for enhancing the phytochemical content in medicinal fruits and berries. Phenolic compounds like anthocyanins, flavonols, flavan-3-ols and phenolic acids are sensitive to post-harvest storage conditions, such as temperature management or duration, that can preserve content through inhibition of enzymatic degradation during storage at room temperatures (Yan et al. 2021). In table grape berries, exogenous melatonin extends shelf life and maintains bioactive compounds by reducing oxidative stress markers and modulating antioxidant enzyme activities (Nasser et al. 2022). The initial phytochemical content depends on harvest timing; e.g. goji berries show differences in vitamin C, total phenolics, anthocyanins and antioxidants based on harvest month. Post-harvest storage conditions are key for final content in berries, e.g. in strawberries, both genotypes and post-harvest storage affect vitamin and phenolic contents, influencing antioxidant properties and health-promoting efficacy (Alvarez-Suarez et al. 2014). Moreover, phytochemical integrity has been maintained by use of controlled storage, freeze-drying and minimum processing methods. Bioaccessibility of some compounds can also be enhanced by processing and disrupting the cell walls, emphasizing the need for preservation and bioavailability (Folmer et al. 2014; Di Lorenzo et al. 2021).

1.3.4 Interactions between Phytochemical Synergies

The therapeutic effects of medicinal fruits and berries mostly depend on the integrated action of bioactive compounds within the whole fruit matrix, besides individual phytochemicals. Various metabolites, such as phenolics, carotenoids, dietary fibres and flavonoids, combine to confer anti-inflammatory, antidiabetic, chemoprotective and cardioprotective benefits. Phenols and flavonoids primarily help in retaining antioxidant capacity and modulating NF-κB and MAPK signalling pathways involved in inflammation. On the other hand, carotenoids and vitamins C and E protect cellular membranes and inhibit lipid peroxidation (Liu 2013). The whole fruit extract exhibits superiority in the production of antioxidant and anti-inflammatory activity as compared to purified phytochemicals (Jacobo-Velázquez and Cisneros-Zevallos, 2009). Therefore, the combined phytochemicals affect many processes, such as enzymatic mechanisms, gene expression and gut microbiota, and indicate health benefits beyond the simple antioxidant mechanism (Scalbert et al., 2005). The complexity synergy within fruits and berries reflects the challenges of standardizing marker isolation for therapeutic efficacy. Knowledge of dietary guidelines supports natural fruits and berries as a versatile food rather than their supplements.

1.4 SIGNIFICANCE OF MEDICINAL FRUITS AND BERRIES

The importance of medicinal fruits and berries is due to the presence of high phytochemical content, including phenolic acids and flavonoids (anthocyanins, flavan-3-ols), tannins, as well as vitamins, minerals and organic acids, which are the key for prevention of chronic diseases such as diabetes mellitus, cardiovascular disease, neurological disorders and cancer (Aly et al. 2018). Research on phytochemicals in berries such as blueberries, blackberries, raspberries and strawberries shows how digestion influences bioavailability, which is important for assessing their biological efficacy when consumed as functional foods or therapeutic agents. Polyphenols regulate lipid metabolism, glucose homeostasis and endothelial activity by regulating insulin metabolism, inhibiting carbohydrate digestive enzymes and promoting the bioavailability of nitric oxide (Wei et al. 2025). Specific compounds such as anthocyanins have been shown to provide protection against vision problems, cardiovascular health issues and antitumour activity, further demonstrating the medicinal value of fruits (Xue et al. 2022). Phenolic compounds containing fruits like *Lonicera caerulea*, possess chemopreventive and antioxidant activities that prevent chronic conditions such as diabetes mellitus or cancer (Svarcova et al. 2007). The traditional medicinal system in Mizoram suggests therapeutic effects of wild fruits and berries for a wide range of ailments (gastrointestinal disease, respiratory infections and dermatological issues), which signifies the need to preserve documentation and validate indigenous knowledge through scientific studies (Hazarika et al. 2012). Medicinal fruits, such as *Rubia cordifolia*, provide valuable models in ethnopharmacology as its parts contain bioactive compounds and other metabolites that exhibit antioxidation, anti-inflammation, antitumour and neuroprotective effects, showing that fruits can serve as molecular scaffolds for novel drug discovery (Shan et al. 2016).

The interaction of phytochemicals of fruit with the gut microbiota has been an emerging aspect of medicinal relevance. A large percentage of dietary polyphenols make it to the colon in an unmetabolized state, and they are subsequently biotransformed by the microbiota of the intestine into small-molecule metabolites with increased bioavailability and bioactivity. These substances, in turn, regulate the microbial flora, stimulating the development of the useful bacterial populations and inhibiting the pathogenic ones. This bidirectional relationship between fruits and gut microbiota is associated with the systemic anti-inflammatory, metabolic, immunomodulatory and health-promoting effects (Di Lorenzo et al. 2021; Wan et al. 2025). Wild and underutilized species of fruits have been found to have greater levels of bioactive compounds than their intensively grown varieties, thus further highlighting their potential as useful medicines. These factors are important to understand in order to formulate cultivation strategies that are intended to increase the therapeutic value of fruits and berries (Maurer et al. 2023). The nutritional and medicinal benefits of fruits and berries are depicted in Table 1.1.

TABLE 1.1
Nutritional Composition and Biological Properties of Some of the Medicinal Fruits and Berries

Fruit /Berries	Nutritional Composition	Biological Activities	Key References
Kiwifruit (*Actinidia deliciosa*)	Vitamin C, fibre, potassium, phenolics and flavonoids	Antioxidant, anti-inflammatory, gastrointestinal health, improved lipid profile	Kumar et al. (2024)
Pomegranate (*Punica granatum*)	Fibre, vitamins (C, K), minerals; low fat	Antioxidant, anti-inflammatory, antidiabetic, cardioprotective, anticancer	Laurindo et al. (2024); Noreen et al. (2025)
Guava (*Psidium guajava*)	Vitamin C, fibre, potassium, micronutrients	Antioxidant, antimicrobial, anti-inflammatory, antidiabetic, anticancer	Butt et al. (2025)
Banana (*Musa spp.*)	Carbohydrates, dietary fibre, potassium, vitamins B6 and C	Antidiabetic potential, antiatherosclerotic, antioxidant, cardiovascular support	Shinde et al. (2025)
Mango (*Mangifera indica*)	Vitamins (A, C), dietary fibre, minerals; supports overall diet quality	Cardiometabolic benefits, improved glycaemic control, anti-inflammatory, antioxidant	Swaroop et al. (2018)
Melon (*Cucumis melo*)	Vitamins (A, C), hydration, dietary fibre	Antioxidant, anti-inflammatory	Manchali et al. (2021)
Dragon Fruit (*Hylocereus polyrhizus*)	Vitamin C, fibre and phytonutrients	Anti-inflammatory, anticancer potential and blood sugar regulation effects	Singh et al. (2024)
Orange (*Citrus sinensis*)	Vitamin C, folate, potassium;	Antioxidant, anti-inflammatory, may reduce chronic disease risk	Manchali et al. (2021)
Papaya (*Carica papaya*)	Vitamins A, C, E, fibre and digestive enzymes	Antioxidant, anti-inflammatory, anticancer potential, digestive support	Lanjhiyana (2024)
Persimmon (*Diospyros kaki*)	Vitamins (A, C and K) and fibre with antioxidants	Cardiovascular support, anti-inflammatory, improved lipid profile	Pérez-Piñero et al. (2024)
Blueberry (*Vaccinium corymbosum*)	fibre, vitamin C, vitamin K, manganese; high antioxidant capacity	Neuroprotective, cardioprotective, anti-inflammatory, antidiabetic	Thorakkattu et al. (2025)

(*Continued*)

TABLE 1.1 (Continued)

Fruit /Berries	Nutritional Composition	Biological Activities	Key References
Strawberry (*Fragaria ananassa)*	Vitamin C, folate, fibre, polyphenols; low glycaemic load	Anti-inflammatory, improves insulin sensitivity	Jain et al. (2025)
Raspberry (*Rubus ideas*)	Fibre, vitamin C, manganese, ellagitannins	Antioxidant, antiobesity, antidiabetic	Sivapragasam et al. (2025a)
Blackberry (*Rubus fruticosus*)	Fibre, vitamin C, vitamin K, anthocyanins	Anti-inflammatory, anticancer, oxidative stress reduction	Zhang et al. (2025a)
Cranberry (*Vaccinium macrocarpon*)	Vitamin C, fibre, flavonoids	Antiadhesion activity for UTI pathogens; antioxidant	Perna et al. (2025a)
Goji berry (*Lyciumbarbarum*)	Carotenoids, vitamin A, polysaccharides, amino acids	Immunomodulatory, hepatoprotective, anti-ageing	Perna et al. (2025b)
Elderberry (*Sambucus nigra*)	Anthocyanins, vitamin C, flavonols	Antiviral, immune-boosting, anti-inflammatory	Chacón-Fuentes et al. (2025a)
Bilberry (*Vaccinium myrtillus*)	Anthocyanins, fibre, vitamins	Improves vision, vascular protection, antidiabetic	Chacón-Fuentes et al. (2025b)
Mulberry (*Morus alba*)	Iron, vitamin C, resveratrol-like compounds	Antihyperglycaemic, lipid-lowering, neuroprotective	Sivapragasam et al. (2025b)
Blackcurrant *(Ribes nigrum)*	Vitamin C, polyphenols, gamma-linoleic acid	Anti-inflammatory, eye health, immune modulation	Zhang et al. (2025b)

1.5 PHARMACOLOGICAL PROPERTIES OF FRUITS AND BERRIES

Owing to the presence of a diverse spectrum of bioactive metabolites, the medicinal fruits and berries have gained much attention for their health benefits and to combat a number of diseases such as obesity, diabetes, cancer, cardiovascular and other metabolic disorders. Biological effects of these fruits are due to complicated synergistic actions of polyphenols, pigments, tannins, carotenoids, organic acids and volatile compounds instead of individual constituents. This combined chemical complexity is the basis of medicinal and functional use of the whole fruits and berries (Jurja et al. 2025; Thorakkattu et al. 2025). Some of the biological activities possessed by the edible fruits and berries have been described and tabulated (Table 1.1).

1.5.1 Cardiovascular Protection

Epidemiologic data as well as interventional studies suggest that regular consumption of polyphenol-containing fruits and berries (grape stilbenes and black raspberry anthocyanins) is associated with favourable cardiovascular risk markers such as improved blood pressure, arterial stiffness, endothelial activity and lipid circulation. A randomized 8-week blind placebo-controlled trial in post-menopausal women has been attempted, which revealed that daily consumption of blueberries lowers systolic blood pressure and arterial stiffness plausibly via nitric-oxide-explained vasodilation (Woolf et al. 2023). Meta-analyses of grape products indicate a small decrease in systolic and diastolic blood pressure in randomized studies (Asbaghi et al. 2021).

1.5.2 Metabolic Regulation (Glucose and Lipid Homeostasis)

A number of clinical studies show that dietary supplementation with fruits rich in anthocyanins or purified preparations has positive effects on glycaemic control and dyslipidaemia. Randomized controlled trials using purified anthocyanins have demonstrated the lowering of serum triglycerides, LDL-C levels and improved insulin resistance markers (Mao et al. 2023). Clinical evidence is also available wherein the regular consumption of berries reflects the reduced risk for type 2 diabetes incidence, while short-term interventions show elevated post-prandial glycaemic response (PPGR) and modest improvement in their responsiveness to exogenous glucose (Kilt et al. 2020). Therefore, the fruit-derived polyphenols are encouraged as a dietary supplement intervention against metabolic risks.

1.5.3 Neuroprotective Function and Cognitive Abilities

Dietary consumption of berry fruits has been linked to better cognitive performance, enhanced memory retention and delayed age-related cognitive decline. Such effects can be mediated by means of better neuronal signalling, reduced neuroinflammatory responses and protective effects against oxidative neuronal damage, which highlights the applicability of fruits in promoting brain health across the lifespan (Giampieri et al. 2015; Kalt et al. 2020). Randomized clinical trials have been conducted using whole-fruit (blueberries), which showed that berries can enhance cognitive performance or prevent age-related decline, improvements in memory/executive function, outcomes/delayed ageing of the central nervous system (CNS). Following consumption of fruit berries, anthocyanin metabolites can pass into the brain to modulate neuroinflammatory responses, synaptic signalling and neuronal resilience, which might be the biological basis for the cognitive improvement (De Amicis et al. 2022).

1.5.4 Anti-Inflammatory and Immune Modulation

In many human studies and observational trials, fruits and berries have been shown to lower biomarkers for systemic inflammation, such as CRP and IL-6, in line with the suppression of NF-Kb/COX2 pathways by polyphenols of berries in vitro (Giampieri et al. 2015; Di Lorenzo et al. 2021). The translational use of the anti-inflammatory effects of concentrated berries has been suggested through evidence showing a reduction in pro-inflammatory/pro-survival biomarker expression using concentrated berry preparations in oral cancer patients undergoing clinical trials (Knobloch et al. 2016).

1.5.5 Anticancer Potential

Translational studies in human pilot trials have demonstrated that the berries' phytochemicals, such as black raspberry formulations, can modulate carcinogenesis-related pathways such as oxidative DNA damage, proliferation and apoptosis; biomarkers of premalignant/early-stage lesions suppression (Knobloch et al. 2016).

1.5.6 Dermatological and Cosmetic Applications

Polyphenols like ellagic acid and anthocyanins have proved to be effective in controlling skin oxidative stress, pigmentation pathways and inflammation. There are human studies and cosmetic trials, which show improvement of oxidative damage markers, photo-protection and alleviation of hyperpigmentation with topical interventions and dietary interventions (Draelos et al. 2013; Giampieri et al. 2018). These studies revealed nutraceutical and topical applications of fruit compounds in the field of dermatology. The phytochemical constituents found in medicinal fruits and berries have been illustrated schematically with major health benefits (Figure 1.1).

1.6 CLINICAL TRIALS APPLICATIONS

Health benefits of medicinal fruits and berries and their phytochemicals (polyphenols, flavones, flavonols, flavanols, tannins, phenolic acids, carotenoids, vitamins and minerals) on human health,

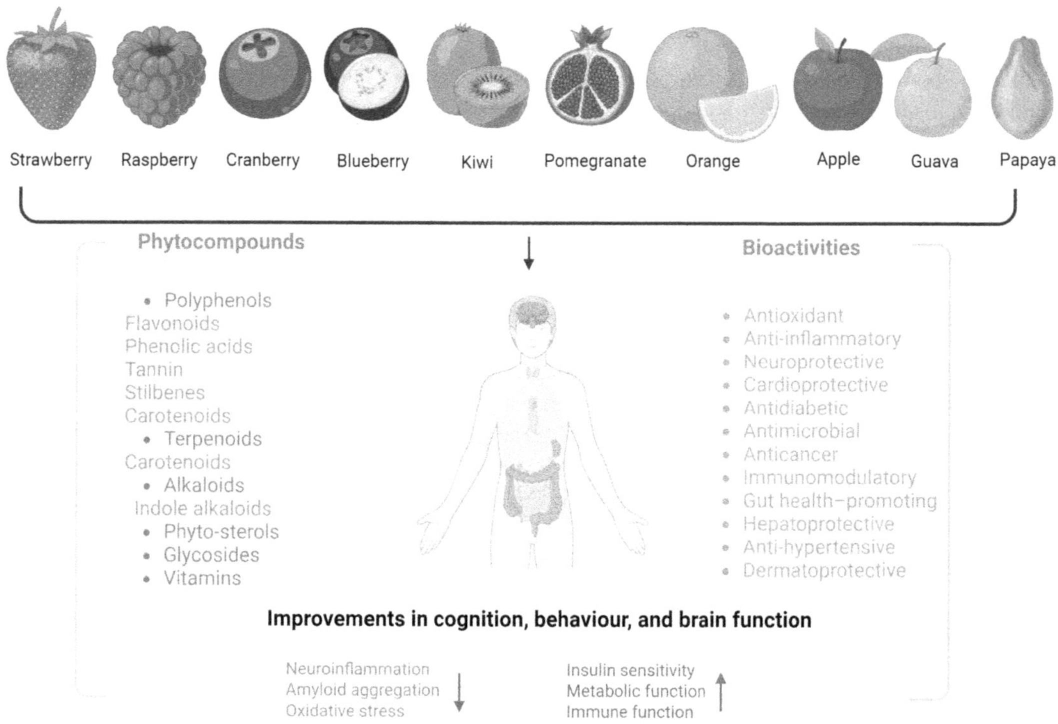

FIGURE 1.1 Medicinal potential of some of the fruits and berries.

including antioxidant activity, cardiovascular function, metabolic syndrome prevention and inflammation reduction, have been investigated in clinical trials. Studies have analysed extracts of *Aronia melanocarpa* fruits, like cyanidin-3-O-galactoside and chlorogenic acid, for cancer and COVID-19 (Ren et al. 2022). Sea buckthorn berry consumption has been examined in six human trials, mainly focusing on cardiovascular outcomes, but evidence was limited and inconsistent, requiring more extensive research (Sayegh et al. 2014). Acai and Jucara berries were examined in 23 trials, which are rich in anthocyanins and unsaturated fatty acids, suggesting increased antioxidant defences and reduced metabolic stress, though outcomes varied among trials (Baptista et al. 2021). Research studies indicate that long-term berry consumption shows improved hepatic function, increased insulin sensitivity and HDL cholesterol, decreased serum glucose and LDL cholesterol and reduced type 2 diabetes risk, indicating that berries can help prevent metabolic syndrome (Kowalska and Olejnik 2016). Continuous consumption of berries can reduce protein and lipid oxidation and improve antioxidant status in healthy individuals as well as those with cardiovascular risk factors, as shown in intervention studies (Del Bo' et al. 2015). Berry consumption improves vascular reactivity markers and arterial stiffness indicators, with outcomes varying by exposure in time, berry type, dose and biomarkers assessed, found in a systematic review of 22 randomized controlled trials (Martini et al. 2019). Berries such as blackberries, cranberries, blueberries, raspberries and strawberries contain phenolic compounds that help in treating chronic diseases through effects on oxidative stress and inflammation. Despite that, clinical trials are required to enhance bioavailability and validate their medicinal potential (Golovinskaia and Wang 2021).

1.7 PREPARATION, USES AND FORMULATIONS

Medicinal fruits and berries are widely used in traditional and modern forms. Bioactive extracts from fruits and berries are prepared by harvesting, drying, comminuting and extraction using ethanol, methanol or water. Novel green extraction methods like superfluid extraction (SFE) increase yield, reduce solvent use and protect thermolabile phytoconstituents like polyphenols and flavonoids with antioxidant and anti-inflammatory effects (Aguilera 2024). These extracts are useful for the treatment of chronic diseases (Bishayee et al. 2015; Golovinskaia and Wang 2021). Preparation of tea and decoctions of medicinal fruits and berries provides health benefits through their bioactive compounds. Goji berry tea is highly popular and is associated with antioxidant properties against oxidative stress (Ma et al. 2019). Rubus species decoctions contain ellagic acid and anthocyanins that help in treating gastrointestinal issues (Hummer 2010). In contemporary uses, these fruits are incorporated into functional foods, nutraceutical capsules, extracts, concentrates, teas and beverages, allowing their bioactive compounds to be delivered in standardized or convenient forms. Traditionally, they are processed into juices, decoctions, infusions, syrups, jams and fermented products. The composition of fresh and fairly dried fruits varies. Garden-fresh berries generally have very high moisture content with proteins; however, dried berries have a different composition of nutrients. In the case of cranberries, freeze-drying is often advised over commercial processing as it helps to preserve certain compounds (Grace et al. 2012). In comparison to fresh fruits, dried versions often carry high amounts of phenolics and flavonoids, supporting elevated antioxidant effects. Prunes are beneficial for overall bone health, and dried fruits help in improving gut microbiota (Alasalvar et al. 2023). Goji berries are a rich source of antioxidants both in fresh and dried form (Ma et al. 2019; Shi et al. 2024). Berries demonstrate significant cardiovascular benefits by improving lipid profiles, modulating blood pressure and reducing oxidative stress (Basu et al. 2010). Juicing of fruits can help with the concentration of bioactive compounds. However, dried fruits risk mycotoxin contamination during storage (González-Curbelo and Kabak 2023). Haskap berries exhibit significantly higher ascorbic acid content (Celli et al. 2014). Advanced formulations using microencapsulation, nanoemulsion and edible coatings protect bioactive agents from degradation and enhance their absorption. Berry extracts in skincare products help protect collagen and reduce oxidative stress for healthy skin and wound healing (Azevedo et al. 2025). Fruit extracts containing polyphenols and antioxidants are added to beverages, fortified foods and supplements for improved functionality and shelf life (Aguilera, 2024; Aaqil et al., 2024). The method of preparation can alter the medicinal value and overall efficacy. Infusions help in conserving heat-sensitive compounds, while decoctions are known to concentrate polysaccharides (Olennikov et al. 2015). Berry powder that is freeze-dried retains anthocyanins and polyphenols, and encapsulation elevates the stability of the compound (Casati et al., 2019).

1.8 STANDARDIZATION AND QUALITY CONTROL

The safety and efficacy of medicinal fruit berries as therapeutic agents are critical aspects to standardize the quality control parameters that include phytochemical composition, organoleptic properties and the presence or absence of markers for the toxicity of products (Golovinskaia and Wang 2021). Standardization and quality control processes include methodical assessment of botanical identity, physicochemical characteristics and bioactive compound profiles as a way of minimizing variance that arises due to geographical variations, harvesting conditions, post-harvest management and processing methods. However, there is a lack of standardized procedures related to material sourcing that includes drying techniques used during post-harvest handling, as well as dosage formulation and quality assurance, which limits global acceptance and commercial success potential associated with herbal medicines derived from fruits and berries. Standardization requires quality control parameters that measure the phytochemical composition, organoleptic properties, e.g. taste, colour of fruit extracts by advanced analytical techniques, including screening for bioactive compounds through spectroscopy methods, e.g. Fourier transform infrared (FT-IR) (Aly et al. 2018). Metabolomic studies have been conducted for guava fruit for optimization of chemical composition and shelf life using advanced methods, including LC-MS (liquid chromatography-mass spectrometry), GC-MS (gas chromatography-mass spectrometry) and NMR (nuclear magnetic resonance), which enhance quality control by verifying the authentication or adulteration and bioactive compounds stability during processing and storage (Emam et al. 2025). Post-harvest handling of fruits and berries is an important factor in the process of standardization, as it affects the nutritional content quality of many medicinal fruits like Rubus berries, blackberry, etc., which are a rich source of antioxidants. Application of standardized parameters like controlled environment conditions, non-thermal preservation techniques such as edible coatings, and heat treatment helps in maintaining nutritional quality (Shah et al. 2023). The *Prunus domestica* fruit has been also standardized by employing macroscopic and microscopic characterization, physicochemical assays (moisture, extractive values and ash content) and analytical methods including HPTLC (high-performance thin-layer chromatography) and GC-MS, which validates the absence of heavy metals and pesticide residues; however, phytochemical profiles have been mapped in order to guarantee the quality of the raw material and final product to be used in pharmaceutical and nutraceutical applications (Amir et al. 2022). The standardized procedures to test pesticides, heavy metals, microbial loads and aflatoxins in fruit berries and its products have become routine to meet regulatory limits of safety and to safeguard the health of consumers. Although regulatory challenges still exist regarding the claims about health benefits or therapeutic effect as well as in commercializing extracts of fruit berries, the demonstration of safety, efficacy, as well as compliance with food grade and pharmaceutical standards is necessary before making such claims (Carvalho et al. 2025). Stability, controlled release and bioaccessibility of fruit-derived bioactive compounds in different applications can be enhanced by further innovative studies, including the development of innovative delivery systems using polysaccharide-based carriers, nanoencapsulation and biopolymeric vehicles (de Oliveira et al. 2025).

1.9 SUSTAINABLE CULTIVATION AND PRODUCTION

The sustainable cultivation and production of medicinal fruit berries necessitate farming techniques that safeguard the environment, produce nutrient-rich, high-quality products and give financial stability to farmers. The cultivation practices for fruit berries would reduce the pressure on their natural population, but also satisfy the growing demand of consumers. The use of photovoltaic greenhouses to produce renewable energy while optimizing growing conditions can enhance phenolic content, anthocyanins and antioxidant capacity in berries without increasing sugar concentration (Blando et al. 2018). Introducing biofertilizers – live microorganisms that improve soil health by providing required nutrient availability by fixing nitrogen and solubilizing phosphorus and potassium into crops – promotes soil health and crop yields in a manner that minimizes the degradation of ecosystems caused by chemical fertilizers (Reddy et al. 2020).

Agronomic innovations like conservation agriculture reduce soil degradation and resource use. Although these concepts have been mostly researched in crops such as sugarcane, they can also be applied to medicinal fruit and berry, making them more resilient and productive in a sustainable manner (Tabriz et al. 2021). Soilless cultivation methods such as hydroponics, aeroponics and aquaponics also offer an efficient alternative by maximizing the efficiency of water and nutrient use while minimizing soil degradation, which enables high yields of medicinal fruits and berries with controlled environmental variables; however, more research is required to make these methods widely accessible (Kumar et al. 2024).

The use of smart agricultural technologies, such as automated fruit harvesting systems, can improve sustainability by reducing labour dependency and post-harvest losses and preserving fruit quality, which lowers costs and the environmental impacts in medicinal fruits and berry orchards (Rana et al. 2025). The usage of plant growth-promoting microorganisms and biofertilizers, such as nanomaterials, may further enhance nutrient uptake and productivity in a sustainable cultivation system for medicinal fruits and berries (Chandel et al. 2024).

1.10 CONCLUSION AND FUTURE REMARKS

Consumption of medicinal fruits and berries provides not only essential nutrients needed for life but also needed for health benefits and disease prevention. The presence of bioactive compounds and their effects in fruits and berries has great potential in plant and pharmacological sciences. In plant sciences, research on medicinal fruits and berries helps in understanding different aspects, e.g. phytochemistry, genetics and biotechnological enhancement of secondary metabolites. The nanoformulations with bioactive compounds that are derived from medicinal fruits and berries are a promising pharmacological advance to enhance therapeutic efficacy, increasing fruits' clinical utility, thus accelerating the translation of plant-based compounds into effective medicines.

Integrity and bioavailability of metabolites are further influenced by the method of processing and preservation, making the optimization of technologies like freeze-drying, high-pressure processing and green extraction approaches essential for retaining or enhancing the bioactive functionality. Human intervention studies often differ widely in the design, dosage, and duration, which indirectly contributes to variability and inconsistency in reported outcomes. Thus, there is a dire need for holistic research approaches that integrate the dietary patterns, long-term consumption, microbiome dynamics and personalized nutrition models, rather than targeted studies on isolated bioactive compounds. To consider fruit and berry products in clinical and commercial applications, standardization and quality control are highly necessary. In terms of further research, the preference needs to be placed on well-designed long-term human trials, standardized formulations and dose-response studies to support health benefits and explain therapeutic potential. Omics technologies, sustainable extraction strategies and new delivery systems would further improve the awareness and the significance of bioactives present in medicinal fruits and berries.

REFERENCES

Aaqil, M., Peng, C., Kamal, A., Nawaz, T., and Gong, J. 2024. Recent approaches to the formulation, uses, and impact of edible coatings on fresh peach fruit. *Foods.* 13: 267.

Aguilera, J. M. 2024. Berries as foods: Processing, products, and health implications. *Annu. Rev. Food Sci. Technol.* 15: 1–23.

Alasalvar, C., Chang, S. K., Kris-Etherton, P. M., Sullivan, V. K., Petersen, K. S., Guasch-Ferré, M., et al. 2023. Dried fruits: Bioactives, effects on gut microbiota, and possible health benefits—an update. *Nutrients.* 15: 1611.

Alvarez-Suarez, J. M., Giampieri, F., Mazzoni, L., Forbes-Hernandez, T. Y., Gasparrini, M., and Sabbadini, S. 2014. The effects of pre-harvest and post-harvest factors on the nutritional quality of strawberry fruits: A review. *J. Berry Res.* 4(1): 1–10. https://doi.org/10.3233/jbr-140068

Aly, A. A., Ali, H. G. M., and Eliwa, N. E. R. 2018. Phytochemical screening, anthocyanins and antimicrobial activities in some berries fruits. *J. Food Meas. Charact.* 13(2): 911–920. https://doi.org/10.1007/s11694-018-0005-0

Amir, M., Ameeduzzafar, Z., Ahmad, R., Ahmad, W., Sarafroz, M., Khalid, M., et al. 2022. Quality control standardization, contaminant detection and in vitro antioxidant activity of *Prunus domestica* Linn. fruit. *Plants.* 11: 706.

Asbaghi, O., Naeini, F., Moodi, V., Najafi, M., Shirinbakhshmasoleh, M., Kelishadi, M. R., et al. 2021. Effect of grape products on blood pressure: A systematic review and meta-analysis of randomized controlled trials. *Int. J. Food Prop.* 24: 627–645.

Azevedo, F. S., Pires, A. R., Lila, M. A., Valacchi, G., Hoskin, R. T., Sarandy, M. M., et al. 2025. Berry fruit extracts as topical cosmeceuticals for skin health applications: A systematic review. *Cosmetics.* 12: 87.

Baptista, S. D. L., Copetti, C. L. K., Di Pietro, P. F., and Cardoso, A. L. 2021. Biological activities of açaí (*Euterpe oleracea* Mart.) and juçara (*Euterpe edulis* Mart.) intake in humans: An integrative review of clinical trials. *Nutr. Rev.* 79(12): 1375–1391. https://doi.org/10.1093/nutrit/nuab002

Basu, A., Rhone, M., and Lyons, T. J. 2010. Berries: Emerging impact on cardiovascular health. *Nutr. Rev.* 68(3): 168–177. https://doi.org/10.1111/j.1753-4887.2010.00273.x

Battino, M., Laimer, M., Mezzetti, B., Beekwilder, J., McDougall, G. J., and Denoyes-Rothan, B. 2009. Bioactive compounds in berries relevant to human health. *Nutr. Rev.* 67(1): S145–S150. https://doi.org/10.1111/j.1753-4887.2009.00178.x

Bishayee, A., Haskell, Y., Do, C., Siveen, K. S., Mohandas, N., Sethi, G., and Stoner, G. D. 2015. Potential benefits of edible berries in the management of aerodigestive and gastrointestinal tract cancers: Preclinical and clinical evidence. *Crit. Rev. Food Sci. Nutr.* 56(10): 1753–1775. https://doi.org/10.1080/10408398.2014.982243

Blando, F., Gerardi, C., Castellano, S., Serio, F., and Renna, M. 2018. Characterisation of bioactive compounds in berries from plants grown under innovative photovoltaic greenhouses. *J. Berry Res.* 8(1): 55–69. https://doi.org/10.3233/jbr-170258

Błaszczyk, A., Sielicka, M., and Sady, S. 2018. The stilbene profile in edible berries. *Phytochem. Rev.* 18(1): 37–67. https://doi.org/10.1007/s11101-018-9580-2

Butt, E., Altemimi, A. B., Younas, A., Butt, M. S., Jalal, M., and Bhatty, M., et al. 2025. Guava (*Psidium guajava*): A brief overview of its therapeutic and health potential. *Food Chem.: X.* 31: 103027.

Carvalho, F., Lahlou, R. A., and Silva, L. R. 2025. Exploring bioactive compounds from fruit and vegetable by-products with potential for food and nutraceutical applications. *Foods.* 14: 3884.

Casati, C. B., Sánchez, V., and Baeza, R. 2019. Physicochemical properties and bioactive compounds content in encapsulated freeze-dried powders obtained from blueberry, elderberry, blackcurrant and maqui berry. *J. Berry Res.* 9(3): 431–447. https://doi.org/10.3233/jbr-190409

Celli, G. B., Ghanem, A., and Brooks, M. S. L. 2014. Haskap berries (*Lonicera caerulea* L.)—a critical review of antioxidant capacity and health-related studies for potential value-added products. *Food Bioprocess Technol.* 7(6): 1541–1554. https://doi.org/10.1007/s11947-014-1301-2

Chacón-Fuentes, M., Opazo-Navarrete, M., and Armijo-Godoy, G. 2025b. Bilberry bioactive compounds: Nutritional and medicinal benefits. *Food Sci. Nutr.* 3: 112–123.

Chacón-Fuentes, M., Opazo-Navarrete, M., Armijo-Godoy, G., Barra-Jiménez, M., and Mojica, L. 2025a. Bioactive compounds in elderberry and their nutritional significance. *Food Sci. Plant Nutr.* 3: 94–103.

Chandel, N., Kumar, A., and Kumar, R. 2024. Towards sustainable agriculture: Integrating agronomic practices, environmental physiology and plant nutrition. *Int. J. Plant Soil Sci.* 36(6): 492–503. https://doi.org/10.9734/ijpss/2024/v36i64651

De Amicis, R., Mambrini, S. P., Pellizzari, M., Foppiani, A., Bertoli, S., Battezzati, A., and Leone, A. 2022. Potential effect of berry intake on cognitive functions of healthy people: A systematic review. *Nutrients.* 14: 2977. https://doi.org/10.3390/nu14142977

De Oliveira, L. S., da Silva, D. V. T., da Trindade, L. R., Baião, D. D. S., de Almeida, C. C., Ferreira, V. F., and Paschoalin, V. M. F. 2025. Edible polysaccharides as stabilizers and carriers for the delivery of phenolic compounds and pigments in food formulations. *arXiv Prepr.* arXiv:2511.07264. https://doi.org/10.48550/arXiv.2511.07264

Del Bo, C., Martini, D., Porrini, M., Klimis-Zacas, D., and Riso, P. 2015. Berries and oxidative stress markers: An overview of human intervention studies. *Food Funct.* 6(9): 2890–2917. https://doi.org/10.1039/C5FO00657K

Di Lorenzo, C., Colombo, F., Biella, S., Stockley, C., and Restani, P. 2021. Polyphenols and human health: The role of bioavailability. *Nutrients.* 13: 273. https://doi.org/10.3390/nu13010273

Draelos, Z., Dahl, A., Yatskayer, M., Chen, N., Krol, Y., and Oresajo, C. 2013. Dyspigmentation, skin physiology, and a novel approach to skin lightening. *J. Cosmet. Dermatol.* 12: 247–253. https://doi.org/10.1111/jocd.12066

Emam, Y., Abdou, R. M., Younis, I. Y., and Farag, M. A. 2025. Metabolomics in guava: A comprehensive review of quality, agricultural traits, composition, and nutraceutical applications of *Psidium guajava* fruits and by-products. *Phytochem. Rev.* 1–23.

Folmer, F., Basavaraju, U., and Jaspars, M. 2014. Anticancer effects of bioactive berry compounds. *Phytochem. Rev.* 13: 295–322. https://doi.org/10.1007/s11101-013-9319-z

Giampieri, F., Brennan, R., Blyth, L., Agudo, P., Afrin, S., Santos, C. N., et al. 2018. Phytochemical composition and cytotoxic effects on liver hepatocellular carcinoma cells of different berries following a simulated in vitro gastrointestinal digestion. *Molecules.* 23(8): 1918. https://doi.org/10.3390/molecules23081918

Giampieri, F., Forbes-Hernández, T. Y., Gasparrini, M., Alvarez-Suarez, J. M., Afrin, S., and Bompadre, S., et al. 2015. Strawberry as a health promoter: An evidence-based review. *Food Funct.* 6: 1386–1398.

Golovinskaia, O., and Wang, C. K. 2021. Review of functional and pharmacological activities of berries. *Molecules.* 26(13): 3904. https://doi.org/10.3390/molecules26133904

González-Curbelo, M. Á., and Kabak, B. 2023. Occurrence of mycotoxins in dried fruits worldwide, with a focus on aflatoxins and ochratoxin A: A review. *Toxins.* 15(9): 576. https://doi.org/10.3390/toxins15090576

Grace, M. H., Mbeunkui, F., Lila, M. A., Massey, A. R., and Yousef, G. G. 2012. Comparison of health-relevant flavonoids in commonly consumed cranberry products. *J. Food Sci.* 77(8): H176–H183. https://doi.org/10.1111/j.1750-3841.2012.02788.x

Hazarika, T. K., Nautiyal, B. P., and Lalramchuana, L. 2012. Studies on wild edible fruits of Mizoram, India used as ethno-medicine. *Genet. Resour. Crop Evol.* 59(8): 1767–1776. https://doi.org/10.1007/s10722-012-9799-5

Hummer, K. E. 2010. Rubus pharmacology: Antiquity to the present. *HortScience.* 45(11): 1587–1591. https://doi.org/10.21273/hortsci.45.11.1587

Husen, A. 2023. *Antidiabetic medicinal plants and herbal treatments*. CRC Press, Boca Raton, FL. https://doi.org/10.1201/b23347

Husen, A. 2024a. *Antimalarial medicinal plants*. Taylor & Francis Group, Boca Raton, FL. https://doi.org/10.1201/9781003378396

Husen, A. 2024b. *Medicinal spices and condiment crops*. Taylor & Francis Group, Boca Raton, FL. https://doi.org/10.1201/9781003387046

Jacobo-Velázquez, D. A., and Cisneros-Zevallos, L. 2009. Correlations of antioxidant activity against phenolic content revisited: A new approach in data analysis for food and medicinal plants. *J. Food Sci.* 74(9): R107–R113. https://doi.org/10.1111/j.1750-3841.2009.01352.x

Jain, S., Thorakkattu, P., Sivapragasam, N., Maurya, A., Tiwari, S., Dwivedy, A. K., et al. 2025. Edible berries—an update on nutritional composition and biological activities (part I). *Curr. Nutr. Rep.* 14(1): 1–9. https://doi.org/10.1007/s13668-024-00607-y

Jamwal, K., Bhattacharya, S., and Puri, S. 2018. Plant growth regulator mediated consequences of secondary metabolites in medicinal plants. *J. Appl. Res. Med. Aromat. Plants.* 9: 26–38. https://doi.org/10.1016/j.jarmap.2017.12.003

Johnson, M. H., Gonzalez de Mejia, E., Lucius, A., and Meyer, T. 2011. Cultivar evaluation and effect of fermentation on antioxidant capacity and in vitro inhibition of α-amylase and α-glucosidase by highbush blueberry (*Vaccinium corymbosum*). *J. Agric. Food Chem.* 59(16): 8923–8930. https://doi.org/10.1021/jf201720z

Jurja, S., Pop, A., Pintea, A., Rugina, D., and Socaciu, C. 2025. Blueberries and honeysuckle berries: Anthocyanin-rich polyphenols for cardiovascular protection. *Nutrients.* 17: 3888. https://doi.org/10.3390/nu17243888

Kalt, W., Cassidy, A., Howard, L. R., Krikorian, R., Stull, A. J., Tremblay, F., and Zamora-Ros, R. 2020. Recent research on the health benefits of blueberries and their anthocyanins. *Adv. Nutr.* 11(2): 224–236. https://doi.org/10.1093/advances/nmz065

Kim, J., Lee, H., Park, Y., and Cho, S. 2025. Comprehensive phenolic profiling and antioxidant capacity of *Rubus* berries. *Antioxidants.* 14: 1012. https://doi.org/10.3390/antiox14081012

Knobloch, T. J., Uhrig, L. K., Pearl, D. K., Casto, B. C., Warner, B. M., Clinton, S. K., et al. 2016. Suppression of proinflammatory and prosurvival biomarkers in oral cancer patients consuming a black raspberry phytochemical-rich troche. *Cancer Prev. Res.* 9: 159–171. https://doi.org/10.1158/1940-6207.CAPR-15-0187

Kowalska, K., and Olejnik, A. 2016. Current evidence on the health-beneficial effects of berry fruits in the prevention and treatment of metabolic syndrome. *Curr. Opin. Clin. Nutr. Metab. Care.* 19(6): 446–452. https://doi.org/10.1097/MCO.0000000000000322

Kumar, S., Kumar, V., Singh, S., Gupta, A., and Tahseen, M. A. 2024. A comprehensive review of phytochemical and pharmacological properties of *Actinidia deliciosa. Int. J. Pharmacogn. Pharm. Res.* 6(1): 86–92.

Lamenza, F. F., Upadhaya, P., Roth, P., Shrestha, S., Jagadeesha, S., Horn, N., et al. 2024. Berries vs. disease: Revenge of the phytochemicals. *Pharmaceutics.* 17(1): 84. https://doi.org/10.3390/ph17010084

Lanjhiyana, R. 2024. A review on nutritional importance and health benefits of papaya (*Carica papaya* L.). *Int. J. Adv. Biochem. Res.* 8(7): 847–850

Laurindo, L. F., Rodrigues, V. D., Minniti, G., de Carvalho, A. C. A., Zutin, T. L. M., De Liberto, L. K., et al. 2024. Pomegranate phytochemicals target metabolic and inflammatory pathways. *J. Nutr. Biochem.* 131: 109670. https://doi.org/10.1016/j.jnutbio.2024.109670

Liu, C., Yan, J., Gu, M., Zhao, X., and Yuan, Z. 2019. Effects of salt stress on growth, photosynthesis, and mineral nutrients of 18 pomegranate (*Punica granatum*) cultivars. *Agronomy.* 10(1): 27. https://doi.org/10.3390/agronomy10010027

Liu, R. H. 2013. Health-promoting components of fruits and vegetables in the diet. *Adv. Nutr.* 4(3): 384S–392S. https://doi.org/10.3945/an.112.003517

Ma, Z. F., Zhang, H., Teh, S. S., Wang, C. W., Zhang, Y., Hayford, F., et al. 2019. Goji berries as a potential natural antioxidant medicine: An insight into their molecular mechanisms of action. *Oxid. Med. Cell. Longev.* 2019: 2437397. https://doi.org/10.1155/2019/2437397

Manchali, S., Chidambara Murthy, K. N., Vishnuvardana, and Patil, B. S. 2021. Nutritional composition and health benefits of various botanical types of melon (*Cucumis melo* L.). *Plants.* 10(9): 1755. https://doi.org/10.3390/plants10091755

Mao, T., Akshit, F. N. U., and Mohan, M. S. 2023. Effects of anthocyanin supplementation in diet on glycemic and related cardiovascular biomarkers in patients with type 2 diabetes: A systematic review and meta-analysis of randomized controlled trials. *Front. Nutr.* 10: 1199815. https://doi.org/10.3389/fnut.2023.1199815

Martini, D., Riso, P., Marino, M., Del Bo', C., Angelino, D., Porrini, M., et al. 2019. Role of berries in vascular function: A systematic review of human intervention studies. *Nutr. Rev.* 78(3): 189–206. https://doi.org/10.1093/nutrit/nuz053

Martins, M. S., Azevedo, R., Alves, G., Almeida, A., Guedes de Pinho, P., Garcia-Viguera, C., et al. 2025. Chemical, mineral and volatile profiling of blackberries and mulberries. *J. Food Compos. Anal.* 137: 106832. https://doi.org/10.1016/j.jfca.2024.106832

Maurer, T. R., Tonello, C. Z., Machado, B. O., Trentin, T. S., Bertol, C. D., Lângaro, N. C., and Chiomento, J. L. T. 2023. Phytochemical potential of berries: An overview. *LADEE.* 4(1): 11–28. https://doi.org/10.17981/ladee.04.01.2023.2

Mazzoni, L., Giampieri, F., Forbes-Hernandez, T. Y., Mezzetti, B., Alvarez-Suarez, J. M., Gasparrini, M., Quiles, J. L., Perez-Lopez, P., et al. 2015. The genetic aspects of berries: From field to health. *J. Sci. Food Agric.* 96(2): 365–371. https://doi.org/10.1002/jsfa.7216

Mezzetti, B., Balducci, F., Capocasa, F., Zhong, C.-F., Cappelletti, R., Di Vittori, L., et al. 2016. Breeding strawberry for higher phytochemicals content and claim it: Is it possible? *Int. J. Fruit Sci.* 16: 194–206. https://doi.org/10.1080/15538362.2016.1250695

Mujanović, I., Kovačević, D. B., Putnik, P., and Šubarić, D. 2024. Variability of polyphenols and anthocyanins in wild bilberries. *J. Food Compos. Anal.* 106732. https://doi.org/10.1016/j.jfca.2024.106732

Nasser, M. A., Darwish, D. B. E., Samaan, M. S. F., Hassan, K. M., Al-Harbi, N. A., Abd El-Gawad, H. G., El-Mogy, M. M., et al. 2022. Postharvest exogenous melatonin treatment of table grape berry enhances quality and maintains bioactive compounds during refrigerated storage. *Horticulturae.* 8(10): 860. https://doi.org/10.3390/horticulturae8100860

Nile, S. H., and Park, S. W. 2014. Edible berries: Bioactive components and their effect on human health. *Nutrition.* 30(2): 134–144. https://doi.org/10.1016/j.nut.2013.04.007

Noreen, S., Hashmi, B., Aja, P. M., and Atoki, A. V. 2025. Phytochemicals and pharmacology of pomegranate (*Punica granatum* L.): Nutraceutical benefits and industrial applications: A review. *Front. Nutr.* 12: 1528897. https://doi.org/10.3389/fnut.2025.1528897

Olennikov, D. N., Chirikova, N. K., Vladimirov, L. N., Koryakina, L. P., and Kashchenko, N. I. 2015. Bitter gentian teas: Nutritional and phytochemical profiles, polysaccharide characterisation and bioactivity. *Molecules.* 20(11): 20014–20030. https://doi.org/10.3390/molecules201119674

Pérez-Piñero, S., Muñoz-Carrillo, J. C., Echepare-Taberna, J., Herrera-Fernández, C., Muñoz-Cámara, M., Ávila-Gandía, V., and López-Román, F. J. 2024. Efficacy of a dietary supplement extracted from persimmon (*Diospyros kaki* L.) in overweight healthy adults: A randomized, double-blind, controlled clinical trial. *Foods.* 13(24): 4072. https://doi.org/10.3390/foods13244072

Perna, S., Gasparri, C., Infantino, V., and Rondanelli, M. 2025a. Berry-derived polyphenols and their role in inflammation, oxidative stress, and chronic diseases. *Nutrients.* 17(21): 3440. https://doi.org/10.3390/nu17213440

Perna, S., Gasparri, C., Infantino, V., and Rondanelli, M. 2025b. Functional components of goji berry: Nutritional and therapeutic relevance. *Nutrients.* 17(22): 3560. https://doi.org/10.3390/nu17223560

Rana, V., Thukral, S., Sharma, S., Thakur, K. K., and Sharma, K. 2025. An automatic fruit harvesting system for sustainable agricultural practices: Challenges and implementation. *N. Z. J. Crop Hortic. Sci.* 53(5): 2965–2989. https://doi.org/10.1080/01140671.2025.2522176

Rao, A. V., and Snyder, D. M. 2010. Raspberries and human health: A review. *J. Agric. Food Chem.* 58(7): 3871–3883. https://doi.org/10.1021/jf903484g

Reddy, G. C., Goyal, R. K., Puranik, S., Waghmar, V., Vikram, K. V., and Sruthy, K. S. 2020. Biofertilizers toward sustainable agricultural development. In *Plant microbe symbiosis*: 115–128. Springer. https://doi.org/10.1007/978-3-030-36248-5_7

Ren, Y., Lei, J., Frank, T., Grebenc, J. R., Kinghorn, A. D., Gao, Y. G., et al. 2022. Potential benefits of black chokeberry (*Aronia melanocarpa*) fruits and their constituents in improving human health. *Molecules.* 27(22): 7823. https://doi.org/10.3390/molecules27227823

Sayegh, M., Miglio, C., and Ray, S. 2014. Potential cardiovascular implications of sea buckthorn berry consumption in humans. *Int. J. Food Sci. Nutr.* 65(5): 521–528. https://doi.org/10.3109/09637486.2014.880672

Scalbert, A., Manach, C., Morand, C., Rémésy, C., and Jiménez, L. 2005. Dietary polyphenols and the prevention of diseases. *Crit. Rev. Food Sci. Nutr.* 45: 287–306.

Shah, H. M. S., Afrifa-Yamoah, E., Singh, Z., Kaur, J., Hasan, M. U., and Woodward, A. 2023. Trends in maintaining postharvest freshness and quality of *Rubus* berries. *Compr. Rev. Food Sci. Food Saf.* 22(6): 4600–4643. https://doi.org/10.1111/1541-4337.13235

Shan, M., Yu, S., Zhang, L., Chen, P., Yan, H., and Ding, A. 2016. A review of the botany, phytochemistry, pharmacology and toxicology of *Rubiae radix et rhizoma. Molecules.* 21(12): 1747. https://doi.org/10.3390/molecules21121747

Sharma, R., Patel, N., Gupta, A., and Munjal, K. 2024. Dragon fruit: The wonder fruit—a review on its phytochemistry and nutritional benefits. *Res. J. Pharm. Technol.* 17: 5642–5645.

Shi, X., Wang, X., Zheng, Y., and Fu, L. 2024. Advances in the study of bioactive compounds and nutraceutical properties of goji berry (*Lycium barbarum* L.). *Appl. Sci.* 15(1): 262. https://doi.org/10.3390/app15010262

Shinde, N., Pawar, V., Jiwtode, S., Joshi, M., and Rayyan, A. 2025. Banana: A comprehensive review of nutritional value, medicinal potential and antimicrobial properties. *Int. J. Agric. Food Sci.* 7: 94–101.

Singh, J., Nigam, R., and Katiyar, H. 2024. Nutritional and medicinal potential of dragon fruit (*Hylocereus polyrhizus*). *Int. J. Adv. Biochem. Res.* https://doi.org/10.1101/2025.07.24.666549

Sivapragasam, N., Thorakkattu, P., and Jain, S. 2025a. Edible berries—an update on nutritional composition and health benefits (part III). *Curr. Nutr. Rep.* 14: 24–38.

Sivapragasam, N., Thorakkattu, P., and Jain, S. 2025b. Mulberry (*Morus alba*) nutritional and medicinal potential: An update. *Curr. Nutr. Rep.* 14: 45–56.

Slavin, J. L., and Lloyd, B. 2012. Health benefits of fruits and vegetables. *Adv. Nutr.* 3: 506–516. https://doi.org/10.3945/an.112.002154

Svarcova, I., Valentova, K., and Heinrich, J. 2007. Berry fruits as a source of biologically active compounds: The case of *Lonicera caerulea. Biomed. Pap.* 151(2): 163–174. https://doi.org/10.5507/bp.2007.031

Swaroop, A., Bagchi, M., Moriyama, H., and Bagchi, D. 2018. Health benefits of mango (*Mangifera indica* L.) and mangiferin. *Jpn. J. Med.* 1: 149–154.

Tabriz, S. S., Rokonuzzaman, M., Kader, M. A., Hossen, M. S., and Awal, M. A. 2021. Prospects and challenges of conservation agriculture in Bangladesh for sustainable sugarcane cultivation. *Environ. Dev. Sustain.* 23(11): 15667–15694. https://doi.org/10.1007/s10668-021-01330-2

Thorakkattu, P., Jain, S., Sivapragasam, N., Maurya, A., Tiwari, S., Dwivedy, A. K., et al. 2025. Edible berries—an update on nutritional composition and health benefits (part II). *Curr. Nutr. Rep.* 14: 10–23. https://doi.org/10.1007/s13668-024-00608-x

Tripodi, P., Bianchi, G., Schiavi, M., Rotino, G. L., Cardi, T., Lo Scalzo, R., et al. 2018. Genetic and environmental factors underlying variation in yield performance and bioactive compound content of hot pepper varieties (*Capsicum annuum*) cultivated in two contrasting Italian locations. *Eur. Food Res. Technol.* 244(9): 1555–1567. https://doi.org/10.1007/s00217-018-3069-5

Wan, C., Langyan, S., Echeverría, J., Devkota, H. P., Tewari, D., Moosavi, M. A., et al. 2025. Edible fruits and berries as a source of functional polyphenols: Current scene and future perspectives. *Phytochem. Rev.* 24: 1197–1225. https://doi.org/10.1007/s11101-023-09892-x

Wei, H., Rui, J., Yan, X., Xu, R., Chen, S., Zhang, B., and Xiao, H. 2025. Plant polyphenols as natural bioactives for alleviating lipid metabolism disorder: Mechanisms and application challenges. *Food Res. Int.* 115682. https://doi.org/10.1016/j.foodres.2025.115682

Williamson, G., Kay, C. D., and Crozier, A. 2018. The bioavailability, transport, and bioactivity of dietary flavonoids: A historical perspective. *Compr. Rev. Food Sci. Food Saf.* 17: 1054–1112. https://doi.org/10.1111/1541-4337.12351

Woolf, E. K., Terwoord, J. D., Litwin, N. S., Vazquez, A. R., Lee, S. Y., Ghanem, N., and Johnson, S. A. 2023. Daily blueberry consumption for 12 weeks improves endothelial function in postmenopausal women with above-normal blood pressure through reductions in oxidative stress: A randomized controlled trial. *Food Funct.* 14: 2621–2641. https://doi.org/10.1039/D3FO00157A

Xu, C., Wang, Y., Hu, T., Wang, S., Zhang, H., Zhou, Z., and Cao, K. 2015. Identification and distribution of *Botryosphaeriaceae* species associated with blueberry stem blight in China. *Eur. J. Plant Pathol.* 143(4): 737–752. https://doi.org/10.1007/s10658-015-0724-6

Xue, H., Sang, Y., Gao, Y., Zeng, Y., Liao, J., and Tan, J. 2022. Research progress on absorption, metabolism, and biological activities of anthocyanins in berries: A review. *Antioxidants.* 12(1): 3. https://doi.org/10.3390/antiox12010003

Yan, Y., Pico, J., Sun, B., Pratap-Singh, A., Gerbrandt, E., and Castellarin, S. D. 2021. Phenolic profiles and their responses to pre- and post-harvest factors in small fruits: A review. *Crit. Rev. Food Sci. Nutr.* 63(19): 3574–3601. https://doi.org/10.1080/10408398.2021.1990849

Zhang, M., Song, Y., and Li, X. 2025a. Antioxidant components and health applications of common berries: A comprehensive review. *Food Res. Int.* 178: 110394.

Zhang, M., Song, Y., and Li, X. 2025b. Blackcurrant bioactive compounds and health applications. *Food Res. Int.* 179: 110500.

Zheng, J., Yang, B., Ou, S., Liu, P., Huang, C., and Kallio, H. 2018. Regulation of phytochemicals in fruits and berries by environmental variation—sugars and organic acids. *J. Food Biochem.* 43(6): e12642. https://doi.org/10.1111/jfbc.12642

Zurek, N., Pluta, S., Seliga, Ł., Lachowicz-Wiśniewska, S., and Kapusta, I. T. 2024. Comparative phytochemical composition of fruits of ten haskap berry cultivars. *Agriculture.* 14: 1734. https://doi.org/10.3390/agriculture14101734

2 Exploring Bioactive Compounds in Medicinal Fruits and Berries

Weria Weisany

2.1 INTRODUCTION

Medicinal fruits and berries have served as vital components of both nutrition and healthcare throughout human history. Their use in traditional medicine spans diverse cultures and continents, from Ayurvedic and Traditional Chinese Medicine to African folk healing systems and Native American ethnomedicine. These fruits have been valued not only for their nutritional content but also for their therapeutic properties, often being prescribed to manage inflammation, infections, digestive disorders, and fatigue. For example, the berries of *Sambucus nigra* (elderberry) have long been used as febrifuges and immune boosters (Mocanu and Amariei 2022), while *Punica granatum* (pomegranate) is revered for its anti-parasitic and anti-inflammatory potential in traditional Persian and Indian remedies (Maphetu et al. 2022).

With the advancement of modern phytochemical and pharmacological research, these ethnomedicinal uses have gained renewed scientific validation. Studies have elucidated the presence of numerous bioactive compounds in medicinal fruits such as flavonoids, anthocyanins, phenolic acids, tannins, alkaloids, and carotenoids that confer potent antioxidant, antimicrobial, antidiabetic, cardioprotective, and anticancer activities. As non-toxic, plant-derived agents, these compounds have become central to the development of functional foods, nutraceuticals, and pharmaceutical precursors (Husen 2021, 2022; Weisany et al. 2023a; Weisany et al. 2023b). The convergence of traditional knowledge and modern science thus positions medicinal fruits and berries as promising candidates in the global pursuit of sustainable, nature-based health solutions.

In recent decades, there has been a marked increase in the global interest and consumption of medicinal fruits and berries, driven by a growing awareness of their health-promoting potential and the rising demand for natural bioactives in functional diets. Berries such as acai (*Euterpe oleracea*), goji (*Lycium barbarum*), blueberry (*Vaccinium spp.*), and aronia (*Aronia melanocarpa*) have gained recognition as "superfruits" and are now widely incorporated into health supplements, beverages, and therapeutic formulations (Llorent-Martínez et al. 2013). The global superfruit market is projected to continue expanding, reflecting consumer preferences for ingredients with both nutritional and medicinal value.

Ethnobotanical records reveal that many of these species have deep cultural roots and ecological significance in indigenous communities, where they are not only food sources but also integral to traditional healing systems and seasonal rituals. For instance, Andean populations have long relied on *Myrciaria dubia* (camu camu) for its exceptionally high vitamin C content, while tribes in West Africa use *Irvingia gabonensis* (bush mango) to manage metabolic disorders (García-Chacón et al. 2023). Such indigenous knowledge provides a valuable foundation for bioprospecting and the discovery of novel therapeutic agents.

DOI: 10.1201/9781003688259-2

By exploring and documenting these bioactive-rich fruits within a cross-disciplinary framework that integrates ethnobotany, phytochemistry, and biomedical sciences, researchers can advance the development of novel plant-based therapeutics while promoting the conservation of both cultural knowledge and biodiversity.

2.2 PHYTOCHEMICAL PROFILE OF MEDICINAL FRUITS AND BERRIES

Medicinal fruits and berries possess a diverse and complex phytochemical composition that underpins their traditional and modern therapeutic applications. These bioactive constituents, including phenolics, terpenoids, alkaloids, pigments, vitamins, and minerals, contribute synergistically to a wide range of health-promoting effects such as antioxidant, anti-inflammatory, antimicrobial, cardioprotective, and anticancer activities (Kumar et al. 2023). Understanding the biochemical richness of these fruits is essential to unlocking their potential in functional food development and evidence-based phytotherapy.

Among the most studied phytochemicals in medicinal fruits and berries are phenolic compounds, particularly flavonoids, tannins, and phenolic acids. Flavonoids, such as quercetin, kaempferol, myricetin, rutin, and anthocyanins, are not only responsible for the vibrant colouration of berries but also exert strong antioxidant and anti-inflammatory actions (Dias et al. 2021). Anthocyanin-rich berries like blackcurrant (*Ribes nigrum*), elderberry (*Sambucus nigra*), and bilberry (*Vaccinium myrtillus*) have been associated with cardiovascular and cognitive benefits (Ockermann et al. 2021). Tannins, found in pomegranate (*Punica granatum*) and chokeberry (*Aronia melanocarpa*), are known for their astringent properties and their role in antimicrobial defence and gastrointestinal health (Saparbekova et al. 2023). Meanwhile, phenolic acids such as gallic acid, ferulic acid, and caffeic acid are abundant in fruits like blueberries and grapes and contribute to their capacity to scavenge free radicals and inhibit lipid peroxidation (Ashique et al. 2024).

Beyond phenolics, medicinal fruits and berries are a source of various non-phenolic secondary metabolites that further contribute to their medicinal value. Terpenoids, for instance, are prevalent in citrus fruits and sea buckthorn (*Hippophae rhamnoides*), and include bioactives such as limonene, ursolic acid, and β-caryophyllene, which exhibit anti-inflammatory, anticancer, and antiviral effects (Regolo et al. 2024). Although less widespread, alkaloids are present in some medicinal fruits and play unique pharmacological roles; examples include synephrine in bitter orange (*Citrus aurantium*) and harman alkaloids in passionfruit (*Passiflora spp.*), which affect neurotransmitter pathways and metabolism (Allison et al. 2005). Additionally, compounds such as saponins, glucosinolates, and coumarins, found in select fruit species, contribute to immune modulation, detoxification, and antimicrobial activity.

Complementing these secondary metabolites are essential micronutrients and pigments that contribute both nutritionally and functionally. Many medicinal fruits are rich in vitamin C, particularly acerola (*Malpighia emarginata*) and camu camu (*M. dubia*), while others provide fat-soluble vitamins such as vitamin A (as provitamin A carotenoids) and vitamin E (tocopherols) (Cunha-Santos et al. 2019). Minerals like potassium, calcium, iron, and magnesium are also prevalent, especially in berries such as mulberries (*Morus* spp.) and blackberries (*Rubus fruticosus*), and contribute to enzymatic processes, electrolyte balance, and overall metabolic support (Koyuncu 2004).

Pigments such as anthocyanins and carotenoids serve dual roles as visual markers of ripeness and health and as potent bioactives (Kapoor et al. 2022). Anthocyanins, water-soluble flavonoid pigments found in dark berries, exhibit strong antioxidant properties and are implicated in the prevention of diabetes, neurodegeneration, and cardiovascular disorders (Lakshmikanthan et al. 2024). Carotenoids, including β-carotene, lutein, and zeaxanthin, are lipophilic pigments abundant in fruits like goji (*Lycium barbarum*) and sea buckthorn. These compounds play a pivotal role in maintaining visual health, modulating immune response, and reducing oxidative stress (Visan et al. 2023).

2.3 ANALYTICAL TECHNIQUES FOR IDENTIFYING BIOACTIVE COMPOUNDS

The identification and quantification of bioactive compounds in medicinal fruits and berries require precise and reliable analytical techniques. Among these, chromatographic methods are the cornerstone for separating and analysing complex phytochemical mixtures (Bārzdiņa et al. 2025). High-performance liquid chromatography (HPLC) is widely used for quantifying phenolic compounds, flavonoids, and organic acids due to its high resolution and sensitivity (Mesquita and Monteiro 2018). Gas chromatography coupled with mass spectrometry (GC-MS) is particularly effective for volatile compounds, such as essential oils and terpenes, offering structural insights through mass spectral data (Quintanilla-Casas et al. 2025). Liquid chromatography-mass spectrometry (LC-MS), combining the separation power of HPLC with the structural elucidation capacity of MS, is invaluable for profiling complex secondary metabolites with high accuracy (Figure 2.1).

In addition to chromatographic techniques, spectroscopic methods provide complementary tools for compound identification. Ultraviolet-visible (UV-Vis) spectroscopy is commonly employed for the detection of pigments like anthocyanins and carotenoids. Fourier-transform infrared spectroscopy (FTIR) enables the characterization of functional groups, offering a fingerprint of bioactive constituents (Boggia et al. 2013). Nuclear magnetic resonance (NMR) spectroscopy provides detailed structural information, especially useful for novel compound elucidation and purity assessment (Khalil and Kashif 2023) (Figure 2.1).

Moreover, bioassay-guided fractionation serves as an integrative approach, combining chemical analysis with biological activity testing. This method involves sequential extraction and fractionation of fruit or berry extracts, followed by evaluation of bioactivity in vitro or in vivo, which helps isolate compounds with significant therapeutic potential. Together, these analytical platforms play a crucial role in advancing our understanding of the phytochemical composition and health-promoting effects of medicinal fruits and berries.

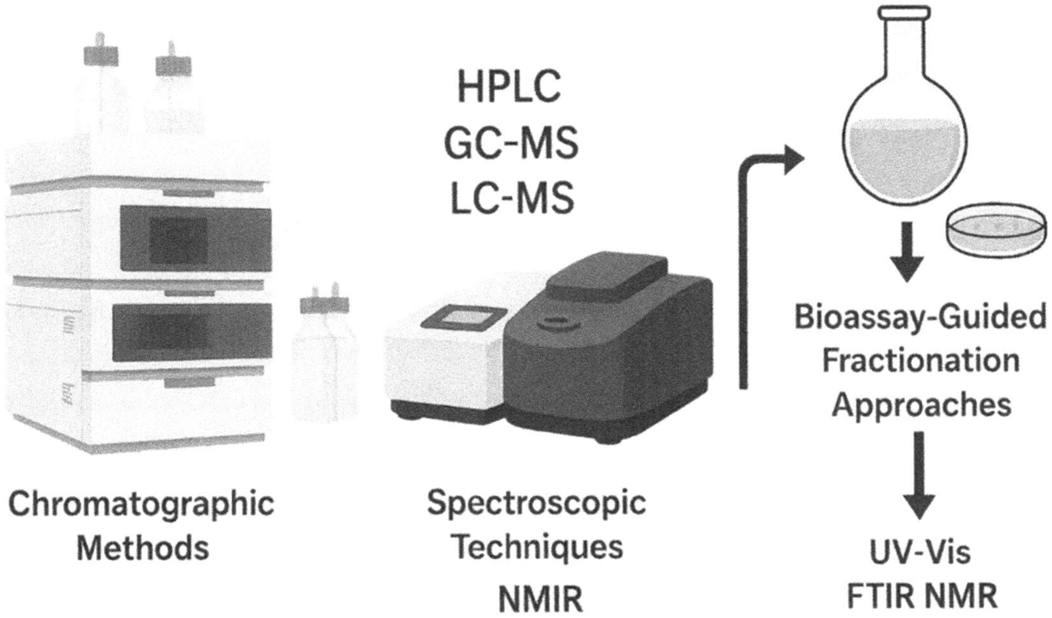

FIGURE 2.1 Key analytical techniques, including chromatography, spectroscopy, and bioassay-guided fractionation, used to identify and characterize bioactive compounds in medicinal fruits and berries.

2.4 BIOLOGICAL ACTIVITIES AND HEALTH BENEFITS

2.4.1 Antioxidant and Anti-inflammatory Properties

One of the most well-documented benefits of medicinal fruits and berries is their potent antioxidant activity, which stems primarily from their high content of flavonoids, phenolic acids, anthocyanins, and tannins. These compounds effectively neutralize reactive oxygen species (ROS) and reduce oxidative damage to DNA, proteins, and lipids (Zahra et al. 2024). This antioxidant protection is crucial for slowing down the progression of ageing and preventing chronic diseases such as cancer, cardiovascular disorders, and neurodegeneration (Rusu et al. 2022). Simultaneously, these fruits exhibit anti-inflammatory effects by downregulating inflammatory mediators such as TNF-α, IL-6, and COX-2. For instance, berries like blackcurrant, blueberry, and elderberry have been shown to suppress NF-κB signalling pathways, thereby reducing systemic inflammation and tissue damage in models of arthritis, colitis, and other inflammatory conditions (Felgus-Lavefve et al. 2022).

2.4.2 Antimicrobial and Antiviral Activities

A growing body of research supports the antimicrobial efficacy of medicinal fruit extracts against a broad spectrum of pathogens, including Gram-positive and Gram-negative bacteria, as well as fungi (Suriyaprom et al. 2022). Compounds such as ellagic acid, gallic acid, and various flavonoids disrupt bacterial cell walls, inhibit enzyme activity, and interfere with quorum-sensing mechanisms, reducing biofilm formation (Ratti et al. 2023). Moreover, these fruits possess significant antiviral properties. For example, elderberry (*Sambucus nigra*) has shown activity against influenza viruses by inhibiting haemagglutinin activity and blocking viral entry into host cells (Torabian et al. 2019). Polyphenols and polysaccharides from berries can also interfere with the replication cycles of viruses like HSV-1, rotavirus, and even SARS-CoV-2, highlighting their potential in developing natural antivirals (Chojnacka et al. 2021).

2.4.3 Anticancer and Cytoprotective Effects

Many medicinal fruits and berries exhibit anticancer properties by targeting multiple hallmarks of cancer. Bioactive compounds such as resveratrol, cyanidin-3-glucoside, and ursolic acid have been shown to induce apoptosis, inhibit cell proliferation, and suppress angiogenesis in cancer cells (Brockmueller et al. 2024). These compounds modulate signalling pathways, including PI3K/Akt, MAPK, and Wnt/β-catenin, which are commonly dysregulated in cancers (Tian et al. 2025). In addition, the cytoprotective effects of these fruits help safeguard normal cells against environmental stressors and cytotoxic agents. For instance, antioxidants in berries reduce DNA damage caused by UV radiation or toxins, while their anti-inflammatory components enhance cellular resilience, making them useful in both chemoprevention and adjuvant therapies (Stoner et al. 2008).

2.4.4 Cardioprotective, Neuroprotective, and Antidiabetic Effects

Regular consumption of medicinal fruits is linked to cardioprotective benefits, including the reduction of blood pressure, improvement of lipid profiles, and prevention of atherosclerosis (Zuraini et al. 2021). Anthocyanins from blueberries and chokeberries, for example, enhance endothelial function, increase nitric oxide production, and inhibit LDL oxidation, all contributing to heart health (Mozos et al. 2021). Their neuroprotective effects are equally significant. Flavonoids and polyphenols in fruits like strawberries and mulberries cross the blood-brain barrier and protect neurons from oxidative damage, inhibit neuroinflammation, and enhance memory and learning by modulating pathways such as BDNF and CREB (Minocha et al. 2022). In terms of antidiabetic activity, compounds such as quercetin and chlorogenic acid have been shown to regulate glucose metabolism by enhancing insulin secretion, improving insulin sensitivity, and inhibiting α-glucosidase and

FIGURE 2.2 Biological activities and health benefits associated with bioactive compounds in medicinal fruits and berries. The infographic highlights key mechanisms, including antioxidant defence, anti-inflammatory response, antimicrobial and antiviral actions, anticancer and cytoprotective effects, as well as protective roles in cardiovascular, neurological, and metabolic (antidiabetic) health.

α-amylase enzymes. This makes these fruits promising candidates for managing type 2 diabetes and preventing its complications (Figure 2.2).

2.5 CASE STUDIES OF NOTABLE MEDICINAL FRUITS AND BERRIES

2.5.1 Goji Berry (*Lycium barbarum*)

Goji berries, also known as wolfberries, are renowned in traditional Chinese medicine for promoting longevity and vitality. Rich in polysaccharides, flavonoids, carotenoids (notably zeaxanthin), and amino acids, they exhibit strong antioxidant, neuroprotective, and immunomodulatory activities (Vidović et al. 2022). Scientific studies have shown that goji berry supplementation can enhance retinal health, regulate glucose metabolism, and reduce fatigue-related oxidative stress (Vidović et al. 2022; Jurikova et al. 2025). Their adaptogenic properties also support resistance to various stressors, making them a popular functional food in modern wellness products.

2.5.2 Açaí (*Euterpe oleracea*)

Açaí berries, native to the Amazon rainforest, are celebrated for their exceptional antioxidant capacity due to high levels of anthocyanins, flavonoids, and essential fatty acids. These berries show promising anti-inflammatory and anticancer properties, attributed to their capacity to modulate oxidative pathways and suppress pro-inflammatory cytokines (Laurindo et al. 2023). Açaí extracts have also demonstrated potential in improving lipid profiles, protecting against neurological degeneration, and modulating immune responses, making them a focus in anti-ageing and cardiovascular research (Machado et al. 2019).

2.5.3 Sea Buckthorn (*Hippophae rhamnoides*)

Sea buckthorn is a nutrient-dense berry widely used in Ayurvedic and Tibetan medicine. It is rich in vitamins (especially vitamins C and E), carotenoids, flavonoids, sterols, and omega-7 fatty acids. These compounds contribute to its potent antioxidant, wound-healing, hepatoprotective, and cardioprotective properties (Veronique et al. 2021). Sea buckthorn oil is especially effective in skin repair and gastrointestinal health, with clinical evidence supporting its use in ulcer treatment and mucosal healing. Its antimicrobial and anti-inflammatory properties further broaden its therapeutic applications (Wang et al. 2022).

2.5.4 Elderberry (*Sambucus nigra*)

Elderberries have long been used for managing cold and flu symptoms due to their high anthocyanin and flavonoid content. They possess significant antiviral, immune-boosting, and anti-inflammatory effects (Veronique et al. 2021). Elderberry extracts have been shown to inhibit viral replication (particularly influenza and coronaviruses) and reduce the duration and severity of upper respiratory infections. Additionally, elderberry's polyphenols contribute to cardiovascular protection and free radical scavenging, reinforcing their role as a natural health supplement (Osman et al. 2023).

2.5.5 Pomegranate (*Punica granatum*)

Pomegranate is a fruit rich in punicalagins, ellagic acid, and anthocyanins, which contribute to its diverse pharmacological actions. It is known for its strong antioxidant, anticancer, and anti-atherogenic properties (Valero-Mendoza et al. 2023). Pomegranate juice has demonstrated efficacy in lowering blood pressure, reducing LDL oxidation, and inhibiting cancer cell proliferation, particularly in prostate and breast cancers. Its polyphenolic content also supports skin health and offers neuroprotection, making it a versatile medicinal fruit in both clinical and nutraceutical contexts (Caruso et al. 2020).

2.5.6 Others: Blueberry, Blackberry, Aronia, etc.

Numerous other berries, such as blueberry, blackberry, and aronia, are noteworthy for their high concentrations of anthocyanins, proanthocyanidins, and other phenolic compounds. Blueberries are particularly studied for their neuroprotective and cognitive-enhancing properties, while blackberries exhibit strong antibacterial and anticancer potential. Aronia (chokeberry), with its high ORAC (oxygen radical absorbance capacity) value, is being explored for antidiabetic and anti-inflammatory effects (Banach et al. 2020). Collectively, these berries underscore the diversity of health-promoting properties in medicinal fruits and their growing relevance in preventive nutrition and functional food development (Vignesh et al. 2024).

2.6 FACTORS INFLUENCING BIOACTIVE COMPOSITION

2.6.1 Genetic Variation and Species Differences

The genetic make-up of fruits and berries plays a foundational role in determining their bioactive composition. Different species and even cultivars within the same species can vary significantly in the type and concentration of phytochemicals they produce. For instance, wild blueberries tend to have higher anthocyanin content compared to cultivated varieties, while certain pomegranate cultivars are richer in punicalagins (Mena et al. 2011). These genetic differences influence not only the nutritional and medicinal quality but also the sensory attributes like colour, taste, and aroma (Diószegi et al. 2019). Selective breeding and molecular tools are increasingly being used to enhance the phytochemical profiles of medicinal fruits for targeted therapeutic benefits.

2.6.2 Environmental and Agronomic Conditions

Environmental factors such as soil composition, temperature, rainfall, altitude, and sunlight exposure have profound effects on the synthesis and accumulation of bioactive compounds. For example, berries grown under moderate abiotic stress (drought or UV exposure) often produce higher levels of secondary metabolites as a natural defence response (Ramakrishna and Ravishankar 2011). Agronomic practices, including irrigation regimes, organic versus conventional farming, and fertilizer application, also influence the metabolic pathways involved in phytochemical biosynthesis (Darakeh et al. 2022). Organic cultivation, in particular, has been associated with increased phenolic content in many fruits, likely due to the absence of synthetic inputs and greater reliance on plant defence mechanisms (Khalediyan et al. 2021) (Figure 2.3).

2.6.3 Postharvest Handling and Processing Methods

Postharvest handling, storage conditions, and processing techniques can significantly affect the stability and bioavailability of bioactive compounds in fruits and berries. Factors such as temperature, humidity, oxygen exposure, and light can lead to degradation of sensitive compounds like anthocyanins, vitamin C, and certain flavonoids (Xue et al. 2024). Minimal processing techniques such as freeze-drying, vacuum sealing, and cold-pressed juice extraction are preferred to preserve the phytochemical integrity. Conversely, heat-intensive methods such as pasteurization or prolonged drying can result in substantial nutrient losses. Moreover, value-added processing, such as fermentation or encapsulation, may enhance the bioavailability and functional efficacy of certain compounds, making postharvest management a crucial determinant of final medicinal value (Arabpoor et al. 2021; Yousefi et al. 2024; Weisany et al. 2019; Weisany et al. 2022b).

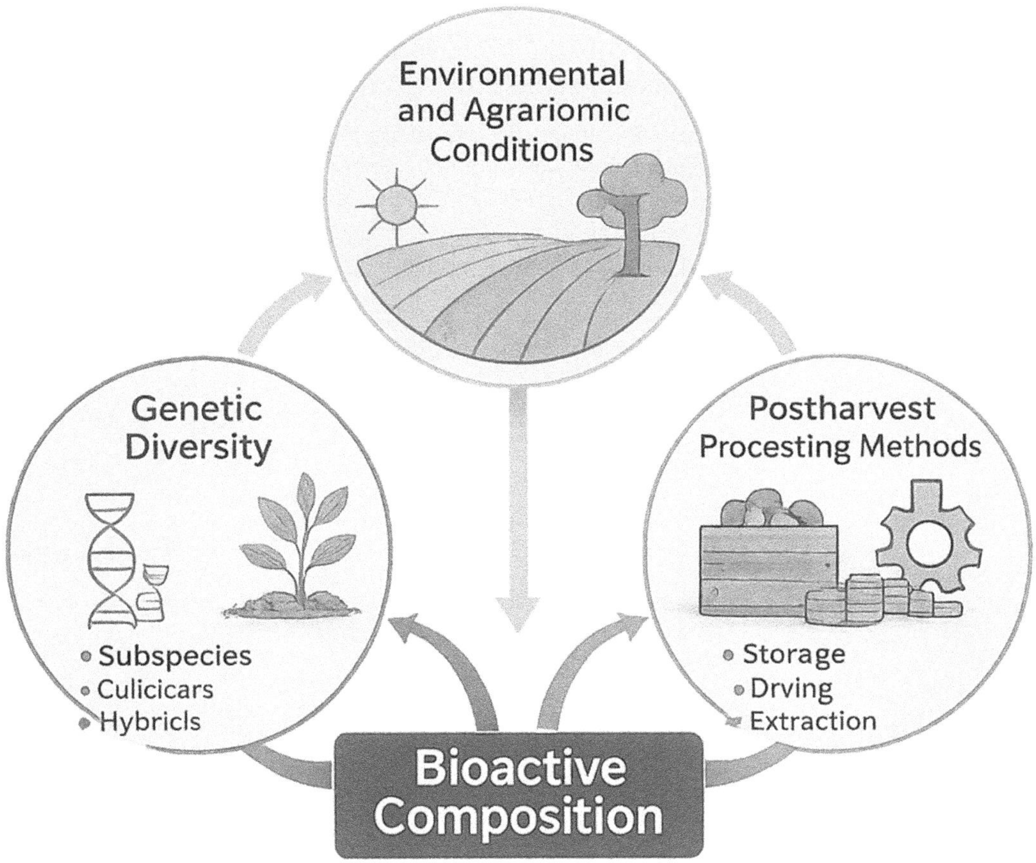

FIGURE 2.3 The key factors influencing the bioactive composition of medicinal fruits and berries, including genetic diversity, environmental conditions, agronomic practices, and postharvest processing methods.

2.7 ENCAPSULATION AND DELIVERY SYSTEMS FOR FUNCTIONAL USE

2.7.1 Nanoencapsulation of Fruit Extracts

Nanoencapsulation has emerged as a cutting-edge strategy to enhance the stability, bioavailability, and targeted delivery of bioactive compounds present in medicinal fruits and berries. This technique involves enclosing bioactive molecules such as polyphenols, flavonoids, and anthocyanins within nanocarriers such as liposomes, nanoemulsions, polymeric nanoparticles, or lipid-based systems. These carriers protect sensitive compounds from degradation due to environmental factors (light, heat, and oxygen), improve solubility in aqueous environments, and facilitate controlled release at target sites in the body (Soufiani et al. 2025; Yousefi et al. 2024; Weisany et al. 2024). Nanoencapsulation not only prolongs shelf life but also maximizes health benefits by improving intestinal absorption and cellular uptake of the bioactives (Figure 2.4).

2.7.2 Functional Foods and Nutraceutical Formulations

Bioactive-rich fruit and berry extracts are increasingly incorporated into functional foods and nutraceutical products aimed at preventing or managing chronic diseases. These include fortified beverages, dairy products, cereal bars, capsules, and gummies that deliver antioxidants, anti-inflammatory agents, or immune-boosting phytochemicals. Encapsulation technologies ensure that these compounds remain stable throughout food processing and storage while maintaining their bioefficacy upon consumption (Arabpoor et al. 2021; Weisany et al. 2022a). Formulations are often designed to deliver specific health outcomes such as cardioprotection, anti-ageing, or metabolic regulation.

2.7.3 Challenges in Stability and Bioavailability

Despite their potential, several challenges hinder the effective application of fruit-derived bioactives in functional products. These include low water solubility, poor permeability across biological membranes, chemical instability, and rapid degradation in the gastrointestinal tract (Rezagholizade-Shirvan et al. 2024). Additionally, selecting appropriate encapsulation materials, scaling up nanoformulation processes, ensuring safety and regulatory compliance, and preserving the sensory qualities

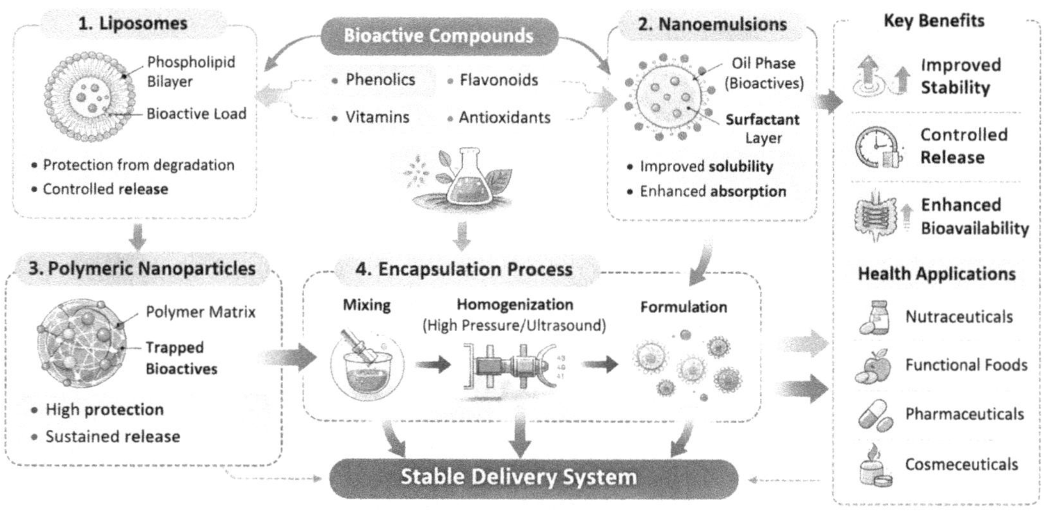

FIGURE 2.4 Encapsulation systems such as liposomes, nanoemulsions, and polymeric nanoparticles used to protect and deliver bioactive compounds from medicinal fruits and berries, improving their stability, bioavailability, and functionality in health-related applications.

of food products remain critical concerns. Ongoing research focuses on overcoming these limitations through novel encapsulation strategies, intelligent delivery systems, and advanced material engineering to optimize the functionality and therapeutic potential of medicinal fruit compounds (Yousefi et al. 2024; Weisany et al. 2022b, 2024).

2.8 FUTURE PERSPECTIVES AND RESEARCH GAPS

2.8.1 Emerging Technologies in Metabolomics and Biosensing

Advancements in metabolomics and biosensing are reshaping the way researchers explore bioactive compounds in medicinal fruits and berries. High-throughput metabolomic platforms, including LC-MS/MS, GC-MS, and NMR, enable comprehensive profiling of phytochemicals, offering new insights into their synergistic effects and health-promoting roles. Emerging biosensor technologies, particularly electrochemical and nanomaterial-based sensors, offer rapid, real-time, and on-site detection of specific antioxidants, flavonoids, or phenolics in fruit matrices. These technologies will enhance quality control, facilitate standardization, and enable more precise correlation of phytochemical content with biological activity.

2.8.2 Personalized Nutrition and Functional Ingredient Design

As the concept of personalized nutrition gains momentum, future research must focus on tailoring bioactive-rich products to individual genetic, microbiome, and metabolic profiles. This includes identifying which individuals respond most favourably to specific fruit-derived compounds, such as anthocyanins or ellagitannins, and developing delivery systems that cater to personalized absorption and metabolism. Artificial intelligence and machine learning are expected to play key roles in designing functional foods based on predictive health data, moving beyond a one-size-fits-all approach in nutraceutical development.

2.8.3 Integration into Sustainable Health Systems

The integration of medicinal fruits and berries into sustainable health strategies represents a major frontier. This includes promoting agroecological farming systems for medicinal fruit crops, supporting biodiversity, and utilizing by-products (skins, seeds) to reduce food waste and enhance circular bioeconomy models. From a healthcare perspective, clinical validation of fruit-based interventions will allow their incorporation into preventive medicine and dietary guidelines, especially in regions facing rising rates of non-communicable diseases. Bridging the gap between ethnobotany, evidence-based medicine, and sustainability requires transdisciplinary collaboration and public–private partnerships.

2.9 CONCLUSION

Medicinal fruits and berries are increasingly recognized for their dense concentrations of bioactive compounds, including phenolics, flavonoids, terpenoids, alkaloids, vitamins, and pigments, which collectively contribute to a wide range of health-promoting properties. This chapter has highlighted their phytochemical diversity, biological activities, including antioxidant, anti-inflammatory, antimicrobial, anticancer, cardioprotective, neuroprotective, and antidiabetic effects, and the analytical techniques used to characterize them. Additionally, emerging encapsulation and delivery systems were discussed as crucial technologies to enhance the stability, bioavailability, and targeted release of these bioactives in functional food and nutraceutical applications. The integration of advanced metabolomic tools, nanotechnology, and personalized nutrition frameworks presents exciting opportunities for future innovation. However, significant challenges remain, particularly regarding standardization, safety evaluations, regulatory approval, and consumer education. Genetic and environmental variability, postharvest handling, and processing also strongly influence the bioactive profiles and must be considered in future research and development. From a public health perspective, incorporating bioactive-rich fruits and berries into daily diets supported by clinical evidence can offer a sustainable and accessible strategy to prevent chronic diseases. Moreover, the natural product sector stands to benefit from innovations that align traditional knowledge with modern science, fostering safer, more effective, and eco-friendly therapeutic options.

REFERENCES

Allison DB, Cutter G, Poehlman ET, Moore DR, Barnes S (2005) Exactly which synephrine alkaloids does *Citrus aurantium* (bitter orange) contain? *International Journal of Obesity* 29 (4):443–446. https://doi.org/10.1038/sj.ijo.0802879

Arabpoor B, Yousefi S, Weisany W, Ghasemlou M (2021) Multifunctional coating composed of *Eryngium campestre* L. essential oil encapsulated in nano-chitosan to prolong the shelf-life of fresh cherry fruits. *Food Hydrocolloids* 111:106394. https://doi.org/10.1016/j.foodhyd.2020.106394

Ashique S, Mukherjee T, Mohanty S, Garg A, Mishra N, Kaushik M, Bhowmick M, Chattaraj B, Mohanto S, Srivastava S, Taghizadeh-Hesary F (2024) Blueberries in focus: exploring the phytochemical potentials and therapeutic applications. *Journal of Agriculture and Food Research* 18:101300. https://doi.org/10.1016/j.jafr.2024.101300

Banach M, Wiloch M, Zawada K, Cyplik W, Kujawski W (2020) Evaluation of antioxidant and anti-inflammatory activity of anthocyanin-rich water-soluble aronia dry extracts. *Molecules* 25 (18). https://doi.org/10.3390/molecules25184055

Bārzdiņa A, Prudņikova DP, Žogota M, Mauriņa B, Bandere D, Brangule A (2025) Combining chromatographic and spectroscopic fingerprinting with chemometrics and data fusion to characterize the phytochemical composition of anthocyanin-rich fruit extracts. *European Journal of Pharmaceutical Sciences* 212:107177. https://doi.org/10.1016/j.ejps.2025.107177

Boggia R, Casolino MC, Hysenaj V, Oliveri P, Zunin P (2013) A screening method based on UV–visible spectroscopy and multivariate analysis to assess addition of filler juices and water to pomegranate juices. *Food Chemistry* 140 (4):735–741.

Brockmueller A, Sajeev A, Koklesova L, Samuel SM, Kubatka P, Büsselberg D, Kunnumakkara AB, Shakibaei M (2024) Resveratrol as sensitizer in colorectal cancer plasticity. *Cancer Metastasis Reviews* 43 (1): 55–85. https://doi.org/10.1007/s10555-023-10126-x

Caruso A, Barbarossa A, Tassone A, Ceramella J, Carocci A, Catalano A, Basile G, Fazio A, Iacopetta D, Franchini C, Sinicropi MS (2020) Pomegranate: nutraceutical with promising benefits on human health. *Applied Sciences* 10 (19):6915.

Chojnacka K, Skrzypczak D, Izydorczyk G, Mikula K, Szopa D, Witek-Krowiak A (2021) Antiviral properties of polyphenols from plants. *Foods* 10 (10):2277.

Cunha-Santos ECE, Viganó J, Neves DA, Martínez J, Godoy HT (2019) Vitamin C in camu-camu (*Myrciaria dubia* (HBK) McVaugh): evaluation of extraction and analytical methods. *Food Research International* 115:160–166.

Darakeh SASS, Weisany W, Tahir NA-R, Schenk PM (2022) Physiological and biochemical responses of black cumin to vermicompost and plant biostimulants: arbuscular mycorrhizal and plant growth-promoting rhizobacteria. *Industrial Crops and Products* 188:115557. https://doi.org/10.1016/j.indcrop.2022.115557

Dias MC, Pinto D, Silva AMS (2021) Plant flavonoids: chemical characteristics and biological activity. *Molecules* 26 (17). https://doi.org/10.3390/molecules26175377

Diószegi J, Llanaj E, Ádány R (2019) Genetic background of taste perception, taste preferences, and its nutritional implications: a systematic review. *Frontiers in Genetics* 10:1272. https://doi.org/10.3389/fgene.2019.01272

Felgus-Lavefve L, Howard L, Adams SH, Baum JI (2022) The effects of blueberry phytochemicals on cell models of inflammation and oxidative stress. *Advances in Nutrition* 13 (4):1279–1309.

García-Chacón JM, Marín-Loaiza JC, Osorio C (2023) Camu camu (*Myrciaria dubia* (Kunth) McVaugh): an Amazonian fruit with biofunctional properties—A review. *ACS Omega* 8 (6):5169–5183. https://doi.org/10.1021/acsomega.2c07245

Husen A (2021) *Traditional herbal therapy for the human immune system*. CRC Press, Boca Raton, FL. https://doi.org/10.1201/9781003137955

Husen A (2022) *Herbs, shrubs and trees of potential medicinal benefits*. CRC Press, Boca Raton, FL. https://doi.org/10.1201/9781003205067

Jurikova T, Tinakova SM, Ziarovska J, Szekeres L, Mlcek J, Fatrcova-Sramkova K, Knazicka Z, Skrovankova S (2025) Polyphenolic spectrum of goji berries and their health-promoting activity. *Foods* 14 (8):1387.

Kapoor L, Simkin AJ, George Priya Doss C, Siva R (2022) Fruit ripening: dynamics and integrated analysis of carotenoids and anthocyanins. *BMC Plant Biology* 22 (1):27.

Khalediyan N, Weisany W, Schenk PM (2021) Arbuscular mycorrhizae and rhizobacteria improve growth, nutritional status and essential oil production in *Ocimum basilicum* and *Satureja hortensis*. *Industrial Crops and Products* 160:113163.

Khalil A, Kashif M (2023) Nuclear magnetic resonance spectroscopy for quantitative analysis: a review for its application in the chemical, pharmaceutical and medicinal domains. *Critical Reviews in Analytical Chemistry* 53 (5):997–1011.

Koyuncu F (2004) Organic acid composition of native black mulberry fruit. *Chemistry of Natural Compounds* 40 (4):367–369.

Kumar APN, Kumar M, Jose A, Tomer V, Oz E, Proestos C, Zeng M, Elobeid TKS, Oz F (2023) Major phytochemicals: recent advances in health benefits and extraction method. *Molecules* 28 (2). https://doi.org/10.3390/molecules28020887

Lakshmikanthan M, Muthu S, Krishnan K, Altemimi AB, Haider NN, Govindan L, Selvakumari J, Alkanan ZT, Cacciola F, Francis YM (2024) A comprehensive review on anthocyanin-rich foods: insights into extraction, medicinal potential, and sustainable applications. *Journal of Agriculture and Food Research* 17:101245. https://doi.org/10.1016/j.jafr.2024.101245

Laurindo LF, Barbalho SM, Araújo AC, Guiguer EL, Mondal A, Bachtel G, Bishayee A (2023) Açaí (*Euterpe oleracea* Mart.) in health and disease: a critical review. *Nutrients* 15 (4). https://doi.org/10.3390/nu15040989

Llorent-Martínez EJ, Fernández-de Córdova M, Ortega-Barrales P, Ruiz-Medina A (2013) Characterization and comparison of the chemical composition of exotic superfoods. *Microchemical Journal* 110:444–451.

Machado AK, Cadona FC, Assmann CE, Andreazza AC, Duarte MMMF, dos Santos Branco C, Zhou X, de Souza DV, Ribeiro EE, da Cruz IBM (2019) Açaí (*Euterpe oleracea* Mart.) has anti-inflammatory potential through NLRP3-inflammasome modulation. *Journal of Functional Foods* 56:364–371.

Maphetu N, Unuofin JO, Masuku NP, Olisah C, Lebelo SL (2022) Medicinal uses, pharmacological activities, phytochemistry, and the molecular mechanisms of *Punica granatum* L. (pomegranate) plant extracts: a review. *Biomedicine & Pharmacotherapy* 153:113256. https://doi.org/10.1016/j.biopha.2022.113256

Mena P, García-Viguera C, Navarro-Rico J, Moreno DA, Bartual J, Saura D, Martí N (2011) Phytochemical characterisation for industrial use of pomegranate (*Punica granatum* L.) cultivars grown in Spain. *Journal of the Science of Food and Agriculture* 91 (10):1893–1906.

Mesquita E, Monteiro M (2018) Simultaneous HPLC determination of flavonoids and phenolic acids profile in Pêra-Rio orange juice. *Food Research International* 106:54–63. https://doi.org/10.1016/j.foodres.2017.12.025

Minocha T, Birla H, Obaid AA, Rai V, Sushma P, Shivamallu C, Moustafa M, Al-Shehri M, Al-Emam A, Tikhonova MA, Yadav SK, Poeggeler B, Singh D, Singh SK (2022) Flavonoids as promising neuroprotectants and their therapeutic potential against Alzheimer's disease. *Oxidative Medicine and Cellular Longevity* 2022:6038996. https://doi.org/10.1155/2022/6038996

Mocanu ML, Amariei S (2022) Elderberries—a source of bioactive compounds with antiviral action. *Plants* 11 (6). https://doi.org/10.3390/plants11060740

Mozos I, Flangea C, Vlad DC, Gug C, Mozos C, Stoian D, Luca CT, Horbańczuk JO, Horbańczuk OK, Atanasov AG (2021) Effects of anthocyanins on vascular health. *Biomolecules* 11 (6). https://doi.org/10.3390/biom11060811

Ockermann P, Headley L, Lizio R, Hansmann J (2021) A review of the properties of anthocyanins and their influence on factors affecting cardiometabolic and cognitive health. *Nutrients* 13 (8). https://doi.org/10.3390/nu13082831

Osman AG, Avula B, Katragunta K, Ali Z, Chittiboyina AG, Khan IA (2023) Elderberry extracts: characterization of the polyphenolic chemical composition, quality consistency, safety, adulteration, and attenuation of oxidative stress- and inflammation-induced health disorders. *Molecules* 28 (7). https://doi.org/10.3390/molecules28073148

Quintanilla-Casas B, Torres-Cobos B, Bro R, Guardiola F, Vichi S, Tres A (2025) The volatile metabolome: gas chromatography–mass spectrometry approaches in the context of food fraud. *Current Opinion in Food Science* 61:101235. https://doi.org/10.1016/j.cofs.2024.101235

Ramakrishna A, Ravishankar GA (2011) Influence of abiotic stress signals on secondary metabolites in plants. *Plant Signaling & Behavior* 6 (11):1720–1731. https://doi.org/10.4161/psb.6.11.17613

Ratti A, Fassi EM, Forlani F, Mori M, Villa F, Cappitelli F, Sgrignani J, Roda G, Cavalli A, Villa S (2023) Mechanistic insights into the antibiofilm mode of action of ellagic acid. *Pharmaceutics* 15 (6):1757.

Regolo L, Giampieri F, Battino M, Armas Diaz Y, Mezzetti B, Elexpuru-Zabaleta M, Mazas C, Tutusaus K, Mazzoni L (2024) From by-products to new application opportunities: the enhancement of the leaves deriving from the fruit plants for new potential healthy products. *Frontiers in Nutrition* 11:1083759. https://doi.org/10.3389/fnut.2024.1083759

Rezagholizade-Shirvan A, Soltani M, Shokri S, Radfar R, Arab M, Shamloo E (2024) Bioactive compound encapsulation: characteristics, applications in food systems, and implications for human health. *Food Chemistry: X* 24:101953. https://doi.org/10.1016/j.fochx.2024.101953

Rusu ME, Fizeşan I, Vlase L, Popa DS (2022) Antioxidants in age-related diseases and anti-aging strategies. *Antioxidants* 11 (10). https://doi.org/10.3390/antiox11101868

Saparbekova AA, Kantureyeva GO, Kudasova DE, Konarbayeva ZK, Latif AS (2023) Potential of phenolic compounds from pomegranate (*Punica granatum* L.) by-product with significant antioxidant and therapeutic effects: a narrative review. *Saudi Journal of Biological Sciences* 30 (2):103553. https://doi.org/10.1016/j.sjbs.2022.103553

Soufiani SP, Yousefi S, Honarvar M, Weisany W, Asadi G (2025) Polyvinyl alcohol films incorporating cellulose nanocrystals, chitosan, and mesoporous silica nanoparticles for enhanced stability and controlled release of beet extracts. *Carbohydrate Polymer Technologies and Applications* 10:100740. https://doi.org/10.1016/j.carpta.2025.100740

Stoner GD, Wang LS, Casto BC (2008) Laboratory and clinical studies of cancer chemoprevention by antioxidants in berries. *Carcinogenesis* 29 (9):1665–1674. https://doi.org/10.1093/carcin/bgn142

Suriyaprom S, Mosoni P, Leroy S, Kaewkod T, Desvaux M, Tragoolpua Y (2022) Antioxidants of fruit extracts as antimicrobial agents against pathogenic bacteria. *Antioxidants* 11 (3):602.

Tian H, Wang P, Shao H (2025) Natural bioactive compounds targeting the Wnt/β-catenin pathway for the treatment of hepatocellular carcinoma. *Journal of Hepatocellular Carcinoma* 12:1613–1622. https://doi.org/10.2147/jhc.S523261

Torabian G, Valtchev P, Adil Q, Dehghani F (2019) Anti-influenza activity of elderberry (*Sambucus nigra*). *Journal of Functional Foods* 54:353–360.

Valero-Mendoza AG, Meléndez-Rentería NP, Chávez-González ML, Flores-Gallegos AC, Wong-Paz JE, Govea-Salas M, Zugasti-Cruz A, Ascacio-Valdés JA (2023) The whole pomegranate (*Punica granatum* L.), biological properties and important findings: a review. *Food Chemistry Advances* 2:100153. https://doi.org/10.1016/j.focha.2022.100153

Veronique T, Wing YM, Christian D (2021) Seabuckthorn polyphenols: characterization, bioactivities and associated health benefits. In: *Phenolic compounds—chemistry, synthesis, diversity, non-conventional industrial, pharmaceutical and therapeutic applications*, Edited by Farid Badria. Mansoura University, Egypt. London, IntechOpen.

Vidović BB, Milinčić DD, Marčetić MD, Djuriš JD, Ilić TD, Kostić A, Pešić MB (2022) Health benefits and applications of goji berries in functional food products development: a review. *Antioxidants* 11 (2). https://doi.org/10.3390/antiox11020248

Vignesh A, Amal TC, Sarvalingam A, Vasanth K (2024) A review on the influence of nutraceuticals and functional foods on health. *Food Chemistry Advances* 5:100749. https://doi.org/10.1016/j.focha.2024.100749

Visan S, Soritau O, Tatomir C, Baldasici O, Balacescu L, Balacescu O, Muntean P, Gherasim C, Pintea A (2023) The bioactive properties of carotenoids from lipophilic sea buckthorn extract (*Hippophae rhamnoides* L.) in breast cancer cell lines. *Molecules* 28 (11):4486.

Wang Z, Zhao F, Wei P, Chai X, Hou G, Meng Q (2022) Phytochemistry, health benefits, and food applications of sea buckthorn (*Hippophae rhamnoides* L.): a comprehensive review. *Frontiers in Nutrition* 9:1036295. https://doi.org/10.3389/fnut.2022.1036295

Weisany W, Samadi S, Amini J, Hossaini S, Yousefi S, Maggi F (2019) Enhancement of the antifungal activity of thyme and dill essential oils against *Colletotrichum nymphaeae* by nano-encapsulation with copper nanoparticles. *Industrial Crops and Products* 132:213–225. https://doi.org/10.1016/j.indcrop.2019.02.031

Weisany W, Samadi S, Tahir NA, Amini J, Hossaini S (2022a) Nano-encapsulation with mesoporous silica enhanced the antifungal activity of essential oil against *Botrytis cinerea* and *Colletotrichum nymphaeae*. *Physiological and Molecular Plant Pathology* 122:101902.

Weisany W, Samiakalantari K, Yousef S, Tahmasebi M (2023a) Antidiabetic features of *Allium cepa* L. (onion). In: *Antidiabetic medicinal plants and herbal treatments*, Edited By Azamal Husen, CRC Press, 47.

Weisany W, Yousefi S, Samiakalantari K, Tahmasebi M (2023b) Antidiabetic properties of *Cinnamomum verum* J. Presl (true cinnamon tree or Ceylon cinnamon tree): recent trends, progress, and challenges. In: *Antidiabetic medicinal plants and herbal treatments*, Edited By Azamal Husen, CRC Press, 411–427.

Weisany W, Yousefi S, Soufiani SP, Pashang D, McClements DJ, Ghasemlou M (2024) Mesoporous silica nanoparticles: a versatile platform for encapsulation and delivery of essential oils for food applications. *Advances in Colloid and Interface Science* 325:103116. https://doi.org/10.1016/j.cis.2024.103116

Weisany W, Yousefi S, Tahir NA, Golestanehzadeh N, McClements DJ, Adhikari B, Ghasemlou M (2022b) Targeted delivery and controlled release of essential oils using nanoencapsulation: a review. *Advances in Colloid and Interface Science* 303:102655.

Xue H, Zhao J, Wang Y, Shi Z, Xie K, Liao X, Tan J (2024) Factors affecting the stability of anthocyanins and strategies for improving their stability: a review. *Food Chemistry: X* 24:101883. https://doi.org/10.1016/j.fochx.2024.101883

Yousefi S, Weisany W, Hosseini SE, Ghasemlou M (2024) Mechanisms of nanoencapsulation to boost the antimicrobial efficacy of essential oils: a review. *Food Hydrocolloids* 150:109655. https://doi.org/10.1016/j.foodhyd.2023.109655

Zahra M, Abrahamse H, George BP (2024) Flavonoids: antioxidant powerhouses and their role in nanomedicine. *Antioxidants* 13 (8). https://doi.org/10.3390/antiox13080922

Zuraini NZA, Sekar M, Wu YS, Gan SH, Bonam SR, Mat Rani NNI, Begum MY, Lum PT, Subramaniyan V, Fuloria NK, Fuloria S (2021) Promising nutritional fruits against cardiovascular diseases: an overview of experimental evidence and understanding their mechanisms of action. *Vascular Health and Risk Management* 17:739–769. https://doi.org/10.2147/vhrm.S328096

3 Medicinal Fruits and Berries in Enhancing Immune System

Sharika Rajasekharan, Pandar Nartsawee, Balamurugan Krishnaswamy, and Siriporn Chuchawankul

3.1 INTRODUCTION

"An apple a day keeps the doctor away", a common proverb that stays true, as recent studies have shown the importance of a balanced diet, which includes fruits and berries, green vegetables, along with protein and carbohydrate-rich food. It is interesting to see how people are keen on looking into the benefits of eating fruits beyond being a snack. Post-COVID-19, people are embracing the idea of a nutritious and balanced diet to keep themselves healthy. Studies have shown that plant-based and fortified foods are capable of enhancing the immune system against viral infections like COVID-19 (Vishwakarma et al. 2022).

Regular intake of fruits can reduce the chances of chronic diseases as they are rich in antioxidants, flavonoids, tannins, carotenoids, anthocyanins, phenols, vitamins and minerals, and dietary fibres, which play a major role in immune system boosting (Husen 2021, 2022; Maheshwari et al. 2022; Maqsood et al. 2025). Recently, there has been a large increase in fruit-based beverages and supplements claiming their effectiveness against oxidative stress and pro-inflammatory markers. The bioactive compounds in cherries and blueberries have shown a protective effect against diabetes and inflammation through reducing the accumulation of antioxidant species and translocation of pro-inflammatory cytokines, reducing caspase-3 and -8 (pro-apoptotic genes), protecting DNA from damage, and controlling postprandial hyperglycaemia through inhibition of α-glucosidase, α-amylase, and dipeptidyl-peptidase IV. It is also known to protect the pancreatic cells and regulate insulin release (Gonçalves et al. 2022).

We are in a fast-moving world, where we depend more and more on ultra-processed food and drinks that disrupt the gut microbiota, which is a root cause of many diseases and disorders, such as cardiovascular disease, diabetes, and other inflammatory disorders. Regular consumption of fruits as a part of a balanced diet can alleviate the dysbiosis and associated disease conditions from worsening and can also prevent its occurrence, through immune system regulation and modulation (Maqsood et al. 2025). The present chapter focuses on the effect and mechanism of action of various fruits and berries on enhancing the immune system.

3.2 IMMUNITY AND NUTRITION

The immune system plays a great role in the daily function of the human body. Both innate and adaptive immunity are at work in protecting the body from various pathogens by being a barrier that stops the entry of the pathogen, sorting pathogenic and non-pathogenic microbes through molecular patterns or antigens, eliminating microbes identified as being harmful, and memorizing the molecular patterns to generate long-term maintenance of memory T-lymphocytes (T cells) and B-lymphocytes (B cells) for effective mounting of response in the future. Hence, a weakened immune system can lead to disastrous outcomes for the individual. Moreover, immune system plays the role of chemical mediators of inflammatory responses, which is a double-edged sword that can damage the pathogen,

DOI: 10.1201/9781003688259-3

as well as loss of regulation can lead to irreparable damage to the host and can cause or aggravate chronic diseases like asthma, allergy, cancer, autoimmune response, diabetes, neurodegeneration, and cardiovascular diseases (Miles and Calder 2021).

There is a complex relation between the immune system and nutrition, and improper nutrition often negatively influences the immune system due to the low essential nutrients absorbed, which may lead to the induction of unwanted immune responses, hence affecting the organism. A healthy diet includes the consumption of fruits, vegetables, berries, meat, and fish, while restricting the intake of processed food and refined sugar, to maintain a robust immune system (Nobs et al. 2020; Venter et al. 2020; Dunbar et al. 2023). Fruits are natural immunity boosters and contain various micronutrients like vitamins and minerals, and phytochemicals that can enhance and modulate the immune system. Vitamin C is important for both innate and adaptive immunity. The recommended intake is 150–200 mg per day. Vitamin C is enriched in several fruits and berries, such as oranges, limes, grapefruit, kiwi, and strawberries. Vitamin C enhances immune defence by enhancing collagen synthesis and stabilizing the skin barrier, protecting against ROS, enhancing phagocytic motility and phagocytosis, facilitating apoptosis, enhancing B- and T-cell differentiation and proliferation, and modulating cytokines. Neutrophils and monocytes actively accumulate vitamin C to protect them from oxidative damage after activation. It also plays a role in alternating oxidant generation and activation of nuclear factor κB (NF-κB) to regulate redox signalling pathway and maintain homeostasis (Carr and Maggini 2017). Vitamin A plays a major role in the functioning of major organs related to immunity and is abundantly present in fruits and vegetables. Through retinoic acid (RA), it strengthens host-barrier integrity, enhances innate immune response, and modulates adaptive immunity. It acts through the gut-immune system axis. RA imprints gut-homing specificity on T and B cells expressing $\alpha4\beta7$ and CCR9. Deficiency of vitamin A has been associated with many infectious diseases, dysbiosis, and reduced immune responses to vaccines. Impaired macrophage, neutrophil, and natural killer (NK) cell function, reduced mucosal barrier protein, increased inflammation, and enhanced production of cytokine, decreased secretion of immunoglobulin A, as well as dysregulation of adaptive immunity have been reported (Amimo et al. 2022; Ramanarayanan et al. 2023). Phytochemicals such as quercetin and resveratrol have also been identified in many fruits and are known to promote health and have several protective effects. Quercetin reportedly induces the expression of Th-1-derived interferon-γ (IFN-γ) and downregulates Th-2-derived interleukin 4 (IL-4) in peripheral blood mononuclear cells (PBMCs), which further results in immune-stimulation (Li et al. 2016). Resveratrol can regulate immune signalling pathways by inhibiting NF-κB. Moreover, it can also initiate cellular responses through CTL induction, cytokine formation, IL-2-inflicted LAH cells, and antigenically induced mitogenic effect to produce T cells (Zhang et al. 2021).

3.3 FRUITS AND BERRIES: MECHANISM OF ENHANCING IMMUNE SYSTEM

3.3.1 Citrus Fruits

Citrus fruits belong to the Rutaceae family, and are widely consumed fruits throughout the world. The trees are known to be cultivated in more than 140 countries and are mostly found in the subtropical regions, though they tend to thrive well in the northern hemisphere. There are about 150 genera and 1600 species, among which the lemon (*Citrus Limon*) (Figure 3.1A), lime (*C. Aurantifolia*), sweet orange *(C. Sinensis)* (Figure 3.1B), sour orange *(C. Aurantium)*, tangerine (*C. Reticulate*) (Figure 3.1C), grapefruit (*C. Paradise*) (Figure 3.1D), citron (*C. Medica*), and shaddock (*C. Maxima*) are economically important genus and popular varieties due to their taste and aroma. Citrus fruits are a staple in the daily diets of many due to its rich antioxidant content. They are eaten as fresh fruits, consumed as juices or in the form of candies, and are also used to add flavour to cooking and baking of other dishes as zest (Hussain et al. 2021; Lu et al. 2023).

Citrus fruits are valued for their sweet and tangy flavour, nutritional value, and its pharmaceutical importance. Some of the biological actives from the fruit often help in sustaining and enhancing the

(A) (B) (C) (D)

FIGURE 3.1 Photograph of lemon (*Citrus Limon*) (A), sweet orange (*C. Sinensis*) (B), tangerine (*C. Reticulate*) (C), and grapefruit (*C. Paradise*) (D).

Images kindly provided by Miss Pandar Nartsawee.

various functions of the human body. Sucrose, glucose, and fructose constitute about 80% of the total carbohydrate content, with trace amounts of mannose, rhamnose, heptulose, trehalose, and galactose. Although protein content is low, some phenolic and non-essential amino acids have been reported. Lipids like omega-6 (39%), oleic (37%), palmitic (25%), linoleic (18%), and palmitoleic (7%) have been reported. Fibre content comprises pectin, cellulose, hemicellulose, and lignin. Vitamin C is most abundant in citrus fruit, and traces of vitamin D and minerals are also present (Hussain et al. 2021).

Numerous reports have linked the regular intake of citrus fruits to a reduction in cancer risk (Cirmi et al. 2018). It is well established that citrus fruits are rich sources of vitamins such as ascorbic acid and folic acid, flavones (hesperidin, naringin, neohesperidin, and naringenin), polymethoxy flavones (nobiletin, tangeretin, sinesnsetin, and 5-hydroxy-nobiletin), essential oils (limonene, α-piene, α-terpinene, linalool, α-phellanderen, myrcene, γ-elemene, and α-cubebene), carotenoids (α-carotene, β-carotene, lutein, and β-cryptoxanthin), limonoids (limonin and nomilin), and synephrineus (p-synephrines, m-synephrine, and o-synephrine) (Lu et al. 2023).

Clinical studies have demonstrated that the consumption of orange juice can reduce oxidative stress, improve endothelial function, lower C-reactive protein levels, limit postprandial fat oxidation, and enhance lipid profiles (Kurowska et al. 2000; Morand et al. 2011; Stookey et al. 2012; Constans et al. 2015; Ribeiro et al. 2017). Notably, grapefruit and lemons were associated with fat loss and reduced insulin resistance in women, as well as beneficial effects in postmenopausal women (Wright et al. 2009; Dow et al. 2012, 2013; Habauzit et al. 2015; Kim et al. 2015).

Vitamin C is a well-known bioactive that plays a major role in both innate and adaptive immune system function by enhancing the epithelial barrier function, phagocytic cell function, B- and T-cell differentiation and proliferation, preventing systemic infection, promoting of wound healing, antioxidant activity, regulating gene expression and cell signalling, co-factor activity, and synthesis of catecholamine hormones (Carr and Maggini 2017). He et al. (2025) coined the term "vitcylation", referring to the direct post-translational modification of the lysine residue of STAT-1 by vitamin C, which ultimately leads to antitumour immunity in vitro and in vivo.

Hesperidin is well known for its antioxidant, anti-inflammatory, and immunoregulatory activity that provides a protective effect both in vitro and in vivo (Megahed et al. 2024). Hesperetin (from enzymatic hydrolysis of hesperidin) and hesperidin glucoside (from hesperidin through enzymatic transglycosylation) are also known for their protective effects (Choi et al. 2022). Sustained release of hesperidin using novel nano-hesperidin formula increased the expression of PPARγ mRNA, and the activation of the Nrf2/ARE/antioxidant pathways, as well as controlling inflammation by decreasing the levels of the pro-inflammatory cytokines TNF-α and IL-17, thereby showing a promising antidiabetic activity (El-Shahawy et al. 2021). Another bioactive naringin is found majorly in the peel of citrus fruits, as NG/nHA/COL scaffold was reported to promote effective bone growth and reconstruction in vivo (Zuo et al. 2022); additionally, it was also reported to have antioxidant, anti-inflammatory, hypoglycaemic, lipid-lowering, cardiovascular protection, neuroprotection, gastrointestinal protection, prevention of osteoporosis, and bone damage (He and Zhang 2023). The beneficial effects and mechanism of action related to the immune-enhancing effects of citrus fruits are given in Table 3.1.

3.3.2 POMEGRANATE (*PUNICA GRANATUM*)

Pomegranate (Figure 3.2) bioactive compounds share strong antioxidant activity, with additional anti-inflammatory, metabolic, immune, and antimicrobial effects. Punicalagin is a potent phenolic antioxidant that activates Nrf2-dependent enzymes (SOD, GSH), reduces ROS and lipid peroxidation, and simultaneously inhibits NF-κB/MAPK and NLRP3 inflammasome signalling (Lo et al., 2022), giving hepatoprotective (Ma et al. 2025), cardioprotective, and antibacterial benefits (Bi et al. 2026). Ellagic acid, released from ellagitannins, also acts as a high-capacity antioxidant while improving hepatic steatosis and NAFLD by blocking AKT/SREBP-1/FASN lipogenesis (Wojtunik-Kulesza et al. 2025), lowering insulin resistance and inflammatory markers such as hs-CRP and TNF-α (Wang et al., 2024).

Urolithin A, derived from ellagitannins by gut microbes, indirectly contributes to antioxidant defence by clearing damaged mitochondria via PINK1/Parkin-mediated mitophagy and enhancing mitochondrial biogenesis (Shreshtha Jolly and Marshall, 2024); this mitochondrial "clean-up" underlies its ability to reverse immune ageing (Denk et al. 2025) and improve muscle endurance in human trials (Wang and Yu, 2025). Anthocyanins from pomegranate peel (cyanidin, delphinidin, and pelargonidin glucosides) show very strong radical-scavenging and ferric-reducing power (Speer et al. 2020) but also exert antibacterial effects through membrane damage and anticancer activity by inhibiting VEGFR-2 and activating caspase-dependent apoptosis (Kowalczyk et al. 2024).

TABLE 3.1
Immune-Enhancing Mechanism of Fruits and Berries

Fruit/Berries	Bioactive/Extract	Mechanism of Action	Key References
Citrus fruits	Peel	Substantial increase in the levels of lysosomes and IgG and IgM in juvenile black rockfish	Lee et al. (2024)
	Blood orange juice by-product extract	Significant increase in levels of lysosomes and IgM was observed in juvenile black rockfish Showed increased survival rate when exposed to *V. harveyi*	Yun et al. (2025)
	Essential oils	Increased the growth of *C. macropomum* fry and their resistance to *A. hydrophila* was observed	Pereira Júnior et al. (2025)
	Essential oils (limonene, linalool, and citral)	Significant changes in the blood immune index, and an increase in H+K+-ATPase activity, IgA, IgG, and IL-2 activity	Wang et al. (2019)
	Hesperidin	• Prevented the increase of leucocytes in the blood • Higher T-cell proportion • Attenuated reduction of NK and NKT cells • Lower activated T cells in blood, lower $CD4^-CD8^+$ thymocyte ratio • Increased phagocytic monocytes • Increased IFN-γ, maintained IL-6	Ruiz-Iglesias et al. (2020)
		• Showed significantly higher levels of humoral antibody and significant enhancement of phagocytosis in immune suppressed rat • Increased the splenic lymphocytes proliferation • Reduced the MDA level and increased the level of enzymatic and non-enzymatic antioxidants • Attenuation of liver injury, re-stimulation, and protection of bone marrow	Berköz et al. (2021)
	Naringin	Suppressed interleukin-6 (IL-6) and enhanced arginase-1 (Arg-1) in RAW264.7	Xiong et al. (2023)
Jujube	Polysaccharides	• Improved the lymphocyte proliferation in the spleen • Decrease CD3+ and CD4 and ratio CD4+/CD8+ • Increased the levels of IL-2, IL-4, IL-10, IFN-γ, and TNF-α • Increased the levels of IgA and SIgA • Increased expressions of claudin-1 and occludin in cyclophosphamide-induced ICR mice	Han et al. (2020)
	Water-soluble lignins (JJC1 and JJC2)	Dose-dependent NO induced and enhanced the phagocytosis in RAW264.7 cells	Cai et al. (2017)
	HP1 or HP2	• Increased the spleen, thymus, and phagocytosis indices of mice • Improved serum haemolysin and enhanced cell-mediated immunity	Zou et al. (2018)
	Hydrolysed jujube extract	Reduced production of TNFα and IL-6 in THP-1 cells	Kim et al. (2020b)

(*Continued*)

TABLE 3.1 (Continued)

Fruit/Berries	Bioactive/Extract	Mechanism of Action	Key References
Mangosteen	Husk hot-water extracts	Immunostimulant activity of *Macrobrachium rosenbergii* promoted survival of the animal when infected with *L. garvieae*	Kitikiew et al. (2023)
	Peel extract, supernatant, and powder formulation	Immune-enhancing effects enhance neuroendocrine activity during hypo- or hyperthermal conditions	Kuo et al. (2023)
	Peel extract nanoemulsion	• Increase in total Ig content in Nile tilapia • Lysozyme activity and ACH50 significantly different from the control • Promoted survival of Nile tilapia against *A. veronii*	Yostawonkul et al. (2023)

	Isogarcinol	• Immunosuppressive effect:
		  • Suppression of Con A-induced lymphocyte proliferation and the MLR
		  • DNFB-induced delayed-type hypersensitivity
		  • Prolongation of allograft skin survival
			Cen et al. (2013)

	Pericarp ethanol extract	Improved the immune microenvironment in joints of CIA rats
		• Efficient suppression of TLR4 expression
		• Substantial decrease in the IL-17 level
		• Increased its expression of nicotinic acetylcholine receptor (α7nAChR) in the immune cells that had infiltrated the synovium
		• Alteration of T-cell differentiation in immune organs:
		  • Decrease in IFN-γ-positive cells
		  • Upregulated FOXP3 level
		• Increased the level of CD25+ Treg cells, and significantly induced the IL-10 level
			Zuo et al. (2018)

	α-Mangosteen	• Suppression of antiviral and cytokine/chemokine genes in vitro • Modulation of NF-κB and related cytokine production in DENV-infected cell line	Tarasuk et al. (2022)
	β-Mangosteen	• Promoted secretion of M1-related interleukin (IL)-1β and IL-6 • Improved αPD-1 effect against glycoma • In vitro and in vivo promoted STING phosphorylation in microglia	Yang et al. (2024)
	Arabinofuran (GMP90-1)	• Improved NO, TNF-α, and IL-6,1β in RAW264.7 cells • Stimulation of NO and ROS in the zebrafish embryo	Zhang et al. (2020)
Wax apple	Rhamnogalacturonan-I-rich pectic polysaccharides (alkaline extract)	• Rhamnogalacturonan-I-rich pectic polysaccharides (alkaline extract): immunostimulatory activity • NO, TNF-α, and IL-6 from RAW264.7 cells in a dose-dependent manner	Lu et al. (2025)
Water rose apple		Antitrypanosomal activity against the blood stage form of *T. brucei*, IC50 1.84 μg/mL	Jain et al. (2016)

(*Continued*)

TABLE 3.1 (Continued)

Fruit/Berries	Bioactive/Extract	Mechanism of Action	Key References
Pomegranate	Punicalagin	Inhibition of PI3K/Akt/mTOR activates SIRT1/FoxO3a–driven autophagy (LC3, Beclin-1), enhancing macrophage bacterial clearance of *E. coli* and *S. aureus*.	Bi et al. (2026)
	Urolithin A	• Reprogrammes immune metabolism towards FAO/AAO over glycolysis, improving mitochondrial function and expanding naïve-like $CD8^{+}$ T cells to counteract immune ageing. • Improves gut health by microbiota-derived activation of mitophagy, enhancing mitochondrial function, gut barrier integrity, and reducing inflammation	Denk et al. (2025); Manos et al. (2025)
	Anthocyanins	• Modulate immune responses by regulating NF-κB and MAPK pathways, balancing pro- and anti-inflammatory cytokine production, and influencing immune cell activation. • Improving gut health by modulating gut microbiota composition, increasing SCFA production, and strengthening intestinal barrier function via tight junction–related pathways.	Ijinu et al. (2023); Verediano et al. (2021)
	Punicic acid	• Exerts immunomodulatory effects by regulating immune cell activation and cytokine balance through NF-κB, MAPK, and STAT signalling. • Improving gut health by modulating gut microbiota composition, increasing SCFA production, and strengthening intestinal barrier integrity through regulation of tight junction proteins (e.g., ZO-1, occludin) and suppression of gut inflammation.	Almoraie et al. (2025)
	Gallic acid	Shows broad antimicrobial effects that indirectly reduce immune activation; also inhibits T-cell and macrophage NF-κB activation in some models, though pomegranate-specific immune data are limited.	Scaglione et al. (2024)
	Quercetin and kaempferol	Both flavonols modulate immune responses by regulating NF-κB, MAPK, and STAT signalling, thereby balancing pro- and anti-inflammatory cytokine production and influencing immune cell activation and function.	Septembre-Malaterre et al. (2022)
Noni	Ethanol extract	Decreased the production of TNF-α, IL-6/IL-10, and IL-1β. Expression of iNOS, COX-2, and NF-κB.	Ezzat et al. (2021)
		Immune stimulation enhances the NK population in vitro and in vivo. Activation of NF-κB and AP-1 in RAW264.7 and upregulation of their cytolytic activity.	Kim et al. (2022)
	Fermented Noni (oral administration)	Immune balancing, • Alleviated atopic dermatitis lesions, epidermal thickness, and infiltration of mast cells and eosinophils • Reduced levels of histamine, IgE, and IgG1/IgG2 modulation of cytokine expression Restoration of skin barrier–related proteins	Kim et al. (2020a)

FIGURE 3.2 Photograph of Pomegranate (*Punica granatum*).

Image kindly provided by Miss Pandar Nartsawee.

Punicic acid, the main fatty acid in pomegranate seed oil, limits membrane lipid peroxidation and simultaneously improves insulin sensitivity, lipid profiles, and inflammatory status via modulation of AKT/AMPK and NF-κB/COX-2 pathways (Almoraie et al. 2025; Ronasi et al. 2025).

Gallic acid in the peel adds further antioxidant capacity through Nrf2/ARE activation and metal chelation, while its COX-2 and NF-κB inhibition supports strong antimicrobial (including anti-ESKAPE) and possible antiviral activities in peel extracts (Hadidi et al. 2024; Scaglione et al. 2024). Flavonols such as quercetin and kaempferol enhance total antioxidant activity but also modulate immune and inflammatory signalling (NF-κB, JAK/STAT) and contribute to cytotoxic and anti-angiogenic actions in tumour models (Septembre-Malaterre et al. 2022). Key immune system-enhancing/immunomodulation studies have been listed in Table 3.1.

3.3.3 Mangosteen (*Garcinia mangostana* Linn.)

Garcinia mangostana Linn., "mangosteen" (Figure 3.3), is a tropical fruit tree and claims to be the best-tasting tropical fruit, hence referred to as the "queen of fruits", and mainly cultivated in Indonesia, Thailand, Myanmar, Malaysia, the Philippines, India, and Sri Lanka. The fruit is purplish-red in colour, with juicy white pulp, which is sweet and slightly acidic, and has a sweet aroma. It has been traditionally used to treat various ailments, including dysentery, wounds, skin infections, inflammation, ulcers, abdominal pain, and fever (Pedraza-Chaverri et al. 2008; Rizaldy et al. 2021).

Mangosteen has been widely explored for its phytochemical constituents, and several biologically active compounds have been identified, such as xanthones, benzophenones, flavonoids, and anthocyanins. It is also rich in minerals such as potassium, magnesium, and calcium, along with vitamins A, B2, B5, and C. The studies have shown that the pericarp of mangosteen and the fruit extract has shown a wide variety of pharmaceutical properties such as antibacterial, antioxidant, antitumour, antidiabetic, anti-obesity, anti-inflammatory, in inflammatory bowel disease, antimalarial, antiparasitic, wound healing, pulmonary fibrosis, chronic kidney disease, making it a promising drug candidate (Rizaldy et al. 2021; Matan et al. 2024). Some of the key immune system-enhancing/immunomodulation studies have been listed in Table 3.1.

FIGURE 3.3 Photograph of Mangosteen (*Garcinia mangostana* Linn).

Image kindly provided by Miss Pandar Nartsawee.

3.3.4 Wax Apple (*Syzygium samarangense*)

Syzygium samarangense (Blume) Merrill and Perry, commonly known as wax apple (Figure 3.4), is an important tropical fruit and widely cultivated in Asian countries such as Taiwan, India, Thailand, Indonesia, and Malaysia. The fruits are crispy and juicy, making it an interesting snack. It belongs to the genus Syzygium that comprises various similar and different species. Some of the similar species are *Syzygium aqueum* (water apple), *Syzygium jambosa* (rose apple), and *Syzygium malaccense* (Malay apple) (Banadka et al. 2022). The plant parts, including the leaves, roots, flowers, and fruits, have been traditionally used as medicines in different communities in Asian countries in treating a variety of ailments, including diarrhoea, fever, elevated oedema, and also used as an anthelminthic. Reportedly, a decoction containing root and bark has been used as an abortifacient, a menstrual flow stimulator, and in treating amenorrhoea (Banadka et al. 2022; Tarigan et al. 2022). A study in 2021 by Abdelfattah et al. revealed anti-trypanosomal activities both in vitro and in vivo by the leaf extract from *Syzygium samarangense*; in vivo, it was able to offset oxidative stress-related organ damage in a rat model (Abdelfattah et al. 2021). Apart from that, several phytochemicals have been isolated and characterized from the leaves of *Syzygium samarangense*.

The fruits are immensely nutritious and contain several phytochemicals with pharmacologically significant properties (Tarigan et al. 2022; Bao et al. 2023). Apart from the nutritional value the fruit also contains, phytochemicals such as, 2′,4′-dihydroxy-3′,5′-dimethyl-6′-methoxychalcone, 2′,4′-dihydroxy-6′-methoxchalcone (cardamonin), pinocembrin, quercetin, reynoutrin, hyperin, quercitrin, quercetin, myricetin, rutin, guaijaverin, epigallocatechin 3-O-gallate, epicatechin-3-O-gallate, samarangensis A, samarangensis B, anthocyanins cyanidin-3-glucoside, acids ellagic acid, gallic acid, jasmonic acid, and salicylic acid (Tarigan et al. 2022). In a recent study, an unusual 11-carbon δ-lactone derivative was isolated from the ethanolic extract from the ripened fruit called syzysamalactone, which was found to offer neuroprotective effects in the neuroblastoma cell line (Liu et al. 2022). Though several studies revealed the bioactive and protective activity of the plant as a whole, especially the leaves, there are only a few studies that have focused on the fruit, which is edible, and the phytochemicals isolated from the fruit and its protective effects are given in Table 3.1.

FIGURE 3.4 Photograph of wax apple (*Syzygium samarangense*).

Image kindly provided by Miss Pandar Nartsawee.

3.3.5 Water Rose Apple (*Syzygium aqueum*)

It is widely known as water rose apple, water apple, and Jambu in Malaysian and Indian languages, and belongs to the Syzygium genus. The trees are found to be native to the Pacific regions of the Philippines, Malaysia, Indonesia, and Java, and are also found to be grown in various parts of India and Thailand. It thrives well in a tropical climate. The fruiting season is typically from May to June. The fruit is colourful, glossy, fleshy, succulent, with a slight aroma, and sweet-acidic taste. They are eaten raw when ripe, or are pickled, or made into jam, or used to make sauces. The leaves are also found to be edible, raw or dried, especially the young ones. Reportedly, in Malaysia, the decoction from the bark has been used as a local application for thrush, and the dried powdered leaves are used for treating a cracked tongue. Interestingly, the leaves are used in reliving child birth pains, managing hyperglycaemia, and are also known to have antibiotic properties. Moreover, it has high dietary value too, as it can be consumed as a natural antioxidant, hydrating, high in terpenoids and γ-terpinene, and has phenolic content too (Sonawane 2018). The Seididamyeh et al. (2023) study reported an abundance of citric acid and quinic acid, which may be the reason for the immune-enhancing property of the fruit. The immune-enhancing mode of action has been discussed in Table 3.1.

3.3.6 Jujube (*Ziziphus jujuba* Mill.)

The fruit of *Ziziphus jujuba* Mill. is commonly known as Jujube, Chinese jujube, or red dates (Figure 3.5). It has been used as a folk medicine for 4000 years and is widely found in Europe and Asia. China is the largest producer of jujube, accounting for more than 90% of the total production in the world. It has high nutritional, as well as functional value, and is known to improve the quality of sleep, be beneficial to the skin, and eliminate toxins (Lu et al. 2021). A recent meta-analysis by Ahmadi et al. 2025 proposes its beneficial effect on metabolic disorders such as hyperlipidaemia and type 2 diabetes. Another meta-analysis also determined the effectiveness of jujube on antioxidant markers such as SOD, GSH-Px, and MDA levels, but pointed out the need for more methodological and long-term animal studies as well as clinical trials in the future (Zhu et al. 2024). Several studies have reported

FIGURE 3.5 Photograph of Jujube (*Ziziphus jujuba* Mill).

Image kindly provided by Miss Pandar Nartsawee.

the effectiveness against cancer, gastrointestinal-protective effects, hepatoprotective effects, and inflammation and anti-inflammatory effects. These effects are attributed to the presence of vitamin C, polysaccharides, hydroxycinnamic acids and benzoic acids, flavonols and flavan-3-ols, triterpenic acids, and nucleosides (Ji et al. 2017). The immune-enhancing function of jujube is given in Table 3.1.

3.3.7 Noni (*Morinda citrifolia* Linn.)

Noni (Figure 3.6) is also known as the "Indian mulberry", or "Yor" in Thailand, and is native to the Hawaiian and Polynesian countries. It is a tropical fruit and rich in functional metabolites (Nandhasri

FIGURE 3.6 Photograph of Noni (*Morinda citrifolia* Linn).

Image kindly provided by Miss Sirinapa Laobutsa.

et al. 2005; Ni et al. 2025). It is rich in phenolic compounds, polysaccharides, fatty acids, alkaloids, anthraquinones, esters, and triterpenoids, and has been known to have antibacterial, anti-inflammatory, immune enhancement, hypoglycaemic, and hepatoprotective activity. It is sought after its attractive list of functional compounds and rich nutrients; it has been listed as China's "new resource food". Though it is unsuitable for direct consumption, fermented noni fruit and juice have been marketed widely. It is rich in phenolic components like gallic acid, protocatechuic acid, and caffeic acid; flavonoids like rhinocerosin, naringin, eriocitrin, and kaempferol; lignins like vanillin moricitant; and coumarin like scopolamine. It is also rich in iridoids that are pharmacologically important. Noni is reported to regulate eNOS activity, and inflammatory cytokines regulate intestinal microbiota and lipid metabolism. It can enhance antioxidants like GSH and SOD, promote hepatic glycogen synthesis, and enhance insulin sensitivity, and also improve liver function (Ni et al. 2025). The immune-modulating function of Noni is given in Table 3.1.

3.4 DIETARY RECOMMENDATIONS AND PRECAUTIONS

Fruits are an essential part of a healthy, nutritious diet. Regular intake of fruits provides the body with sufficient vitamins, minerals, and fibres along with other phytopolyphenols, which are necessary for the function of the system. A meta-analysis study reported that approximately 5 servings of fruits and vegetables (not juice and potatoes) can lower the risk of mortality (Wang et al. 2021). Consumption of fruits and vegetables was reported to show positive effects on women's mental health (Guzek et al. 2022). Fruits and leafy vegetable intake also notably managed gestational diabetes in women (Jaworsky et al. 2023). chee

Allergy to raw fruits has been reported for many fruit varieties, due to adverse immune response, and is frequently associated with oral allergy syndrome (OAS). Some common fruits that cause allergy are apple, kiwi, peach, muskmelon, strawberry, sweet cherry, grapes, banana, mango, and pomegranate. Birch pollen-associated allergy is a known clinical phenomenon, which is caused by IgE cross-reaction with consumed fruits like apple, peach, cherry, or pear. Other known allergies are plant nsLTPs and TLPs, which are plant PR-14 and PR-5 defence proteins of plants. Class-1 chitinase, hevein, and prohevein are also reported. Most of these allergies are hard to avoid and elicit a reaction in susceptible individuals (Hassan and Venkatesh 2015). Hence, the suspected fruit variety must be avoided and should be evaluated for allergic response by testing under a physician.

3.5 CONCLUSION

In this fast-paced world, instant and ready-to-eat foods have become a normal essential in the daily life of common people. The nutritional value of these foods may not match the consumers' dietary requirements, which may lead to lifestyle diseases in the long term. Fruits and berries have phytocomponents that are pharmacologically important. Eating fruits and berries raw can help fight disease, starting from the oral cavity, through enhancing the immune system. Currently, there are studies that propose to increase the bioavailability of these bioactives to manage diseases as well as support the immune system.

REFERENCES

Abdelfattah, M. A., Ibrahim, M. A., Abdullahi, H. L., Aminu, R., Saad, S. B., Krstin, S., Wink, M., and Sobeh, M. (2021). Eugenia uniflora and Syzygium samarangense extracts exhibit anti-trypanosomal activity: Evidence from in silico molecular modelling, in vitro, and in vivo studies. *Biomedicine and Pharmacotherapy*, 138, p. 111508.

Ahmadi, M., Shirafkan, H., Mozaffarpur, S. A., and Rezghi, M. (2025). Impact of jujube fruit on serum lipid profile, glycemic index, and liver function: A systematic review and meta-analysis. *Nutrition and Diabetes*, 15(1), p. 22.

Almoraie, M., Spencer, J., and Wagstaff, C. (2025). Punicic acid: A potential nutraceutical compound in pomegranate seed oil and its cardiovascular benefits. *Foods*, 14(14), p. 2412.

Amimo, J. O., Michael, H., Chepngeno, J., Raev, S. A., Saif, L. J., and Vlasova, A. N. (2022). Immune impairment associated with vitamin A deficiency: insights from clinical studies and animal model research. *Nutrients*, 14(23), p. 5038.

Banadka, A., Wudali, N. S., Al-Khayri, J. M., and Nagella, P. (2022). The role of Syzygium samarangense in nutrition and economy: An overview. *South African Journal of Botany*, 145, pp. 481–492.

Bao, T., Karim, N., Ke, H., Tangpong, J., and Chen, W. (2023). Polysaccharide isolated from wax apple suppresses ethyl carbamate-induced oxidative damage in human hepatocytes. *Journal of Zhejiang University-Science B*, 24(7), pp. 574–586.

Berköz, M., Yalın, S., Özkan-Yılmaz, F., Özlüer-Hunt, A., Krośniak, M., Francik, R., Yunusoğlu, O., Adıyaman, A., Gezici, H., Yiğit, A., and Ünal, S. (2021). Protective effect of myricetin, apigenin, and hesperidin pretreatments on cyclophosphamide-induced immunosuppression. *Immunopharmacology and Immunotoxicology*, 43(3), pp. 353–369.

Bi, K., Xu, B., Cao, D., Xu, K., and Zhang, Y. (2026). Punicalagin enhances autophagy through Sirtuin 1/FoxO3a axis to inhibit intracellular Mycobacterium abscessus infection. *Advanced Science*, p. e11734.

Cai, Y., Zhou, X., Han, A., Chen, P., and Bai, H. (2017). In vitro immunological and anti-complementary activities of two water-soluble lignins from Zizyphus jujuba cv. Jinchangzao. *International Journal of Biological Macromolecules*, 105, pp. 204–212.

Carr, A. C., and Maggini, S. (2017). Vitamin C and immune function. *Nutrients*, 9(11), p. 1211.

Cen, J., Shi, M., Yang, Y., Fu, Y., Zhou, H., Wang, M., Su, Z., and Wei, Q. (2013). Isogarcinol is a new immunosuppressant. *PLoS One*, 8(6), p. e66503.

Choi, S. S., Lee, S. H., and Lee, K. A. (2022). A comparative study of hesperetin, hesperidin and hesperidin glucoside: Antioxidant, anti-inflammatory, and antibacterial activities in vitro. *Antioxidants*, 11(8), p. 1618.

Cirmi, S., Navarra, M., Woodside, J. V., and Cantwell, M. M. (2018). Citrus fruits intake and oral cancer risk: A systematic review and meta-analysis. *Pharmacological Research*, 133, pp. 187–194.

Constans, J., Bennetau-Pelissero, C., Martin, J. F., Rock, E., Mazur, A., Bedel, A., Morand, C., and Bérard, A. M. (2015). Marked antioxidant effect of orange juice intake and its phytomicronutrients in a preliminary randomized cross-over trial on mild hypercholesterolemic men. *Clinical Nutrition*, 34(6), pp. 1093–1100.

Denk, D., Singh, A., Kasler, H. G., D'Amico, D., Rey, J., Alcober-Boquet, L., Gorol, J. M., Steup, C., Tiwari, R., Kwok, R., and Argüello, R. J. (2025). Effect of the mitophagy inducer urolithin A on age-related immune decline: A randomized, placebo-controlled trial. *Nature Aging*, pp. 1–14.

Dow, C. A., Going, S. B., Chow, H. H. S., Patil, B. S., and Thomson, C. A. (2012). The effects of daily consumption of grapefruit on body weight, lipids, and blood pressure in healthy, overweight adults. *Metabolism*, 61(7), pp. 1026–1035.

Dow, C. A., Wertheim, B. C., Patil, B. S., and Thomson, C. A. (2013). Daily consumption of grapefruit for 6 weeks reduces urine F2-isoprostanes in overweight adults with high baseline values but has no effect on plasma high-sensitivity C-reactive protein or soluble vascular cellular adhesion molecule. *The Journal of Nutrition*, 143(10), pp. 1586–1592.

Dunbar, C.L., Aukema, H.M., Calder, P.C., Gibson, D.L., Henrickson, S.E., Khan, S., Mailhot, G., Panahi, S., Tabung, F.K., Tom, M. and Upton, J.E., 2023. Nutrition and immunity: perspectives on key issues and next steps. *Applied Physiology, Nutrition, and Metabolism*, 48(7), pp.484–497.

El-Shahawy, A. A., Abdel-Moneim, A., Ebeid, A. S., Eldin, Z. E., and Zanaty, M. I. (2021). A novel layered double hydroxide-hesperidin nanoparticles exert antidiabetic, antioxidant and anti-inflammatory effects in rats with diabetes. *Molecular Biology Reports*, 48(6), pp. 5217–5232.

Ezzat, M. I., Hassan, M., Abdelhalim, M. A., El-Desoky, A. M., Mohamed, S. O., and Ezzat, S. M. (2021). Immunomodulatory effect of noni fruit and its isolates: Insights into cell-mediated immune response and inhibition of LPS-induced THP-1 macrophage inflammation. *Food and Function*, 12(7), pp. 3170–3179.

Gonçalves, A. C., Nunes, A. R., Flores-Félix, J. D., Alves, G., and Silva, L. R. (2022). Cherries and blueberries-based beverages: Functional foods with antidiabetic and immune booster properties. *Molecules*, 27(10), p. 3294.

Guzek, D., Głąbska, D., Groele, B., and Gutkowska, K. (2022). Fruit and vegetable dietary patterns and mental health in women: A systematic review. *Nutrition Reviews*, 80(6), pp. 1357–1370.

Habauzit, V., Verny, M. A., Milenkovic, D., Barber-Chamoux, N., Mazur, A., Dubray, C., and Morand, C. (2015). Flavanones protect from arterial stiffness in postmenopausal women consuming grapefruit juice for 6 mo: A randomized, controlled, crossover trial. *The American Journal of Clinical Nutrition*, 102(1), pp. 66–74.

Hadidi, M., Liñán-Atero, R., Tarahi, M., Christodoulou, M. C., and Aghababaei, F. (2024). The potential health benefits of gallic acid: Therapeutic and food applications. *Antioxidants*, 13(8), p. 1001.

Han, X., Bai, B., Zhou, Q., Niu, J., Yuan, J., Zhang, H., Jia, J., Zhao, W., and Chen, H. (2020). Dietary supplementation with polysaccharides from Ziziphus jujuba cv. Pozao intervenes in immune response via regulating peripheral immunity and intestinal barrier function in cyclophosphamide-induced mice. *Food and Function*, 11(7), pp. 5992–6006.

Hassan, A. K., and Venkatesh, Y. P. (2015). An overview of fruit allergy and the causative allergens. *European Annals of Allergy and Clinical Immunology*, 47(6), pp. 180–187.

He, J., and Zhang, H. P. (2023). Research progress on the anti-tumor effect of naringin. *Frontiers in Pharmacology*, 14, p. 1217001.

He, X., Wang, Q., Cheng, X., Wang, W., Li, Y., Nan, Y., Wu, J., Xiu, B., Jiang, T., Bergholz, J. S., and Gu, H. (2025). Lysine vitcylation is a vitamin C-derived protein modification that enhances STAT1-mediated immune response. *Cell*, 188(7), pp. 1858–1877.

Husen, A. (2021). *Traditional herbal therapy for the human immune system*. CRC Press. https://doi.org/10.1201/9781003137955

Husen, A. (2022). *Herbs, shrubs and trees of potential medicinal benefits*. CRC Press. https://doi.org/10.1201/9781003205067

Hussain, S. Z., Naseer, B., Qadri, T., Fatima, T., and Bhat, T. A. (2021). Citrus fruits—morphology, taxonomy, composition and health benefits. In *Fruits grown in highland regions of the Himalayas: Nutritional and health benefits* (pp. 229–244). Springer International Publishing.

Ijinu, T. P., De Lellis, L. F., Shanmugarama, S., Pérez-Gregorio, R., Sasikumar, P., Ullah, H., Buccato, D. G., Di Minno, A., Baldi, A., and Daglia, M. (2023). Anthocyanins as immunomodulatory dietary supplements: A nutraceutical perspective and micro-/nano-strategies for enhanced bioavailability. *Nutrients*, 15(19), p. 4152.

Jain, S., Jacob, M., Walker, L., and Tekwani, B. (2016). Screening North American plant extracts in vitro against Trypanosoma brucei for discovery of new antitrypanosomal drug leads. *BMC Complementary and Alternative Medicine*, 16(1), p. 131.

Jaworsky, K., Devillez, P., Alexander, J. M., and Basu, A. (2023). Effects of an eating pattern including colorful fruits and vegetables on management of gestational diabetes: A randomized controlled trial. *Nutrients*, 15(16), p. 3624.

Ji, X., Peng, Q., Yuan, Y., Shen, J., Xie, X., and Wang, M. (2017). Isolation, structures and bioactivities of the polysaccharides from jujube fruit (Ziziphus jujuba Mill.): A review. *Food Chemistry*, 227, pp. 349–357.

Kim, H., Rahmawati, L., Hong, Y. H., Choi, S. Y., and Cho, J. Y. (2022). NK cell-mediated immunostimulatory effects of ethanol extract of Morinda citrifolia (noni) fruit. *BMC Complementary Medicine and Therapies*, 22(1), p. 222.

Kim, M. J., Hwang, J. H., Ko, H. J., Na, H. B., and Kim, J. H. (2015). Lemon detox diet reduced body fat, insulin resistance, and serum hs-CRP level without hematological changes in overweight Korean women. *Nutrition Research*, 35(5), pp. 409–420.

Kim, S. H., Seong, G. S., and Choung, S. Y. (2020a). Fermented Morinda citrifolia (noni) alleviates DNCB-induced atopic dermatitis in NC/Nga mice through modulating immune balance and skin barrier function. *Nutrients*, 12(1), p. 249.

Kim, Y., Oh, J., Jang, C. H., Lim, J. S., Lee, J. S., and Kim, J. S. (2020b). In vivo anti-inflammatory potential of viscozyme®-treated jujube fruit. *Foods*, 9(8), p. 1033.

Kitikiew, S., Chieng, Z. X., Li, C. Y., Kuo, H. W., and Cheng, W. (2023). Injection administration of mangosteen, Garcinia mangostana, husk extract enhances physiological and immune-neuroendocrinological regulation of Macrobrachium rosenbergii. *Fish and Shellfish Immunology*, 141, p. 109019.

Kowalczyk, T., Muskała, M., Merecz-Sadowska, A., Sikora, J., Picot, L., and Sitarek, P. (2024). Anti-inflammatory and anticancer effects of anthocyanins in in vitro and in vivo studies. *Antioxidants*, 13(9), p. 1143.

Kuo, H. W., Li, C. Y., Chieng, Z. X., and Cheng, W. (2023). Dietary administration of mangosteen, Garcinia mangostana, peel extract enhances the growth, and physiological and immunoendocrinological regulation of prawn, Macrobrachium rosenbergii. *Fish and Shellfish Immunology*, 140, p. 108982.

Kurowska, E. M., Spence, J. D., Jordan, J., Wetmore, S., Freeman, D. J., Piché, L. A., and Serratore, P. (2000). HDL-cholesterol-raising effect of orange juice in subjects with hypercholesterolemia. *The American Journal of Clinical Nutrition*, 72(5), pp. 1095–1100.

Lee, T. H., Kim, K. T., Oh, H. Y., Park, S. Y., Lee, G. J., Kim, H. S., and Kim, H. S. (2024). Effect of blood orange (Citrus sinensis L. Osbeck) peel waste as a feed additive on the growth performance, digestive enzyme activity, antioxidant capacity, and immune response in juvenile black rockfish (Sebastes schlegelii). *Antioxidants*, 13(12), p. 1452.

Li, Y., Yao, J., Han, C., Yang, J., Chaudhry, M.T., Wang, S., Liu, H., and Yin, Y. (2016). Quercetin, inflammation and immunity. *Nutrients*, 8(3), p. 167.

Liu, Y. P., Xie, Z., Guan, R. Q., Du, M. R., Qiao, Z. H., Suo, X. Y., Liu, Z. Y., Bian, Y., Qiang, L., and Fu, Y. H. (2022). Syzysamalactone, an unusual 11-carbon δ-lactone derivative from the fresh ripe fruits of Syzygium samarangense (wax apple). *Journal of Natural Products*, 85(8), pp. 2100–2103.

Lu, X., Zhao, C., Shi, H., Liao, Y., Xu, F., Du, H., Xiao, H., and Zheng, J. (2023). Nutrients and bioactives in citrus fruits: Different citrus varieties, fruit parts, and growth stages. *Critical Reviews in Food Science and Nutrition*, 63(14), pp. 2018–2041.

Lu, Y., Bao, T., Mo, J., Ni, J., and Chen, W. (2021). Research advances in bioactive components and health benefits of jujube (Ziziphus jujuba Mill.) fruit. *Journal of Zhejiang University Science B*, 22(6), pp. 431–449.

Lu, Y., He, S., Zhao, Z., Liu, C., Lei, Y., Liu, M., Zhang, Q., Lin, D., Liu, Y., Lin, S., and Lu, X. (2025). Structural characteristics, gelling properties, in vitro antioxidant activity and immunomodulatory effects of rhamnogalacturonan-I rich pectic polysaccharides alkaline-extracted from wax apple (Syzygium samarangense). *Foods*, 14(7), p. 1227.

Lo, J., Liu, C. C., Li, Y. S., Lee, P. Y., Liu, P. L., Wu, P. C., et al. (2022). Punicalagin attenuates LPS-induced inflammation and ROS production in microglia by inhibiting the MAPK/NF-kappaB signaling pathway and NLRP3 inflammasome activation. *Journal of Inflammation Research*, 15, 5347–5359.

Ma, L., Sun, H., Xiang, N., and Xu, Q. (2025). Punicalagin relieves hepatic injury by antioxidation and enhancement of autophagy in diet-induced nonalcoholic steatohepatitis. *Scientific Reports*, 15(1), p. 14516.

Maheshwari, S., Kumar, V., Bhadauria, G., and Mishra, A. (2022). Immunomodulatory potential of phytochemicals and other bioactive compounds of fruits: A review. *Food Frontiers*, 3(2), pp. 221–238.

Manos, A. (2025). Urolithin A: The microbiome-derived molecule that's redefining health, aging, and performance 2025 [updated 08 Dec 2025. https://healthpath.com/gut-health/urolithin-a/.

Matan, N., Kongchoosi, N., Sinthupachee, A., and Chaidech, P. (2024). Nutritional and bioactive compound analysis of mangosteen fruit in hill and flat land plantations, during both the season and off-season, in provinces along the Gulf of Thailand and the Andaman Sea. *NFS Journal*, 36, p. 100182.

Maqsood, S., Arshad, M. T., Ikram, A., and Gnedeka, K. T. (2025). Fruit-based diet and gut health: A review. *Food Science & Nutrition*, 13(5), p. e70159.

Megahed, A., Gadalla, H., Filimban, W. A., Albukhari, T. A., Sembawa, H., Bagadood, R. M., Sindi, G., Abdelhamid, F. M., El-Boshy, M. E., and Risha, E. F. (2024). Hesperidin ameliorates thioacetamide-induced liver fibrosis via antioxidative and anti-inflammatory mechanisms targeting TGF-β/α-SMA pathways in rats. *International Journal of Immunopathology and Pharmacology*, 38, p. 1–15.

Miles, E. A., and Calder, P. C. (2021). Effects of citrus fruit juices and their bioactive components on inflammation and immunity: A narrative review. *Frontiers in Immunology*, 12, p. 712608.

Morand, C., Dubray, C., Milenkovic, D., Lioger, D., Martin, J. F., Scalbert, A., and Mazur, A. (2011). Hesperidin contributes to the vascular protective effects of orange juice: A randomized crossover study in healthy volunteers. *The American Journal of Clinical Nutrition*, 93(1), pp. 73–80.

Nandhasri, P., Pawa, K.K., Kaewtubtim, J., Jeamchanya, C., Jansom, C., and Sattaponpun, C., 2005. Nutraceutical properties of Thai "Yor" Morinda citrifolia and "Noni" juice extract. *Songklanakarin Journal of Science and Technology*, 27(2), pp. 579–586.

Ni, Q., Zhang, Z., Niu, L., Yang, R., Xiong, L., Li, D., and Dai, Z., 2025. Research progress on nutritional properties of noni (Morinda citrifolia L.) fruit and its fermented foods. *Fermentation*, 11(7), p. 358.

Nobs, S.P., Zmora, N., and Elinav, E. (2020). Nutrition regulates innate immunity in health and disease. *Annual Review of Nutrition*, 40(1), pp. 189–219.

Pedraza-Chaverri, J., Cárdenas-Rodríguez, N., Orozco-Ibarra, M., and Pérez-Rojas, J. M. (2008). Medicinal properties of mangosteen (Garcinia mangostana). *Food and Chemical Toxicology*, 46(10), pp. 3227–3239.

Pereira Júnior, J. D. A., Costa, D. S., Silva, A. D. S. D., Santos, G. G. D., Santos, A. F. L. D., Silva, A. D. C. D., Couto, M. V. S. D., Cordeiro, C. A. M., Martins, M. L., and Sousa, N. D. C. (2025). Enriched diet with orange essential oil Citrus sinensis for tambaqui Colossoma macropomum promotes growth performance and resistance against Aeromonas hydrophila. *Journal of Fish Diseases*, 48(1), p. e14039.

Ramanarayanan, P., Heine, G., and Worm, M. (2023). Vitamin A and vitamin D induced nuclear hormone receptor activation and its impact on B cell differentiation and immunoglobulin production. *Immunology Letters*, 263, pp. 80–86.

Ribeiro, C., Dourado, G., and Cesar, T. (2017). Orange juice allied to a reduced-calorie diet results in weight loss and ameliorates obesity-related biomarkers: A randomized controlled trial. *Nutrition*, 38, pp. 13–19.

Rizaldy, D., Hartati, R., Nadhifa, T., and Fidrianny, I. (2021). Chemical compounds and pharmacological activities of mangosteen (Garcinia mangostana L.)—updated review. *Biointerface Research in Applied Chemistry*, 12(2), pp. 2503–2516.

Ronasi, S., Mahdavi, A. H., Varnosfaderani, S. R., Kowsar, R., Jafarpour, F., and Nasr-Esfahani, M. H. (2025). Punicic acid alleviates methylglyoxal-induced oocyte dysfunction during in vitro maturation in mouse species. *PLoS One*, 20(3), p. e0314602.

Ruiz-Iglesias, P., Estruel-Amades, S., Camps-Bossacoma, M., Massot-Cladera, M., Franch, À., Pérez-Cano, F. J., and Castell, M. (2020). Influence of hesperidin on systemic immunity of rats following an intensive training and exhausting exercise. *Nutrients*, 12(5), p. 1291.

Scaglione, E., Sateriale, D., Mantova, G., Di Rosario, M., Continisio, L., Vitiello, M., Pagliarulo, C., Colicchio, R., Pagliuca, C., and Salvatore, P. (2024). Antimicrobial efficacy of Punica granatum Lythraceae peel extract against pathogens belonging to the ESKAPE group. *Frontiers in Microbiology*, 15, p. 1383027.

Seididamyeh, M., Phan, A. D. T., Sivakumar, D., Netzel, M. E., Mereddy, R., and Sultanbawa, Y. (2023). Valorisation of three underutilised native Australian plants: Phenolic and organic acid profiles and in vitro antimicrobial activity. *Foods*, 12(3), p. 623.

Septembre-Malaterre, A., Boumendjel, A., Seteyen, A. L. S., Boina, C., Gasque, P., Guiraud, P., and Sélambarom, J. (2022). Focus on the high therapeutic potentials of quercetin and its derivatives. *Phytomedicine Plus*, 2(1), p. 100220.

Shreshtha Jolly, S. B., and Marshall, R. N. (2024). The science of mitophagy: Urolithin A's impact on mitochondrial function and aging. *Healthspan.*

Sonawane, M. S. (2018). Dietary benefits of watery rose apple (Syzygium aqueum (Burm. f.) Alston). *International Archives of Applied Science and Technology*, 9(4), pp. 126–129.

Speer, H., D'Cunha, N. M., Alexopoulos, N. I., McKune, A. J., and Naumovski, N. (2020). Anthocyanins and human health: A focus on oxidative stress, inflammation and disease. *Antioxidants*, 9(5).

Stookey, J. D., Hamer, J., Espinoza, G., Higa, A., Ng, V., Tinajero-Deck, L., Havel, P. J., and King, J. C. (2012). Orange juice limits postprandial fat oxidation after breakfast in normal-weight adolescents and adults. *Advances in Nutrition*, 3(4), pp. 629S–635S.

Tarasuk, M., Songprakhon, P., Chieochansin, T., Choomee, K., Na-Bangchang, K., and Yenchitsomanus, P. T. (2022). Alpha-mangostin inhibits viral replication and suppresses nuclear factor kappa B (NF-κB)-mediated inflammation in dengue virus infection. *Scientific Reports*, 12(1), p. 16088.

Tarigan, C., Pramastya, H., Insanu, M., and Fidrianny, I. (2022). Syzygium samarangense: Review of phytochemical compounds and pharmacological activities. *Biointerface Research in Applied Chemistry*, 12(2), pp. 2084–2107.

Venter, C., Eyerich, S., Sarin, T. and Klatt, K.C., 2020. Nutrition and the immune system: a complicated tango. *Nutrients*, 12(3), p. 818.

Verediano, T. A., Stampini Duarte Martino, H., Dias Paes, M. C., and Tako, E. (2021). Effects of anthocyanin on intestinal health: A systematic review. *Nutrients*, 13(4), p. 1331.

Vishwakarma, S., Panigrahi, C., Barua, S., Sahoo, M., and Mandliya, S. (2022). Food nutrients as inherent sources of immunomodulation during COVID-19 pandemic. *LWT*, 158, p. 113154.

Wang, D. D., Li, Y., Bhupathiraju, S. N., Rosner, B. A., Sun, Q., Giovannucci, E. L., Rimm, E. B., Manson, J. E., Willett, W. C., Stampfer, M. J., and Hu, F. B. (2021). Fruit and vegetable intake and mortality: Results from 2 prospective cohort studies of US men and women and a meta-analysis of 26 cohort studies. *Circulation*, 143(17), pp. 1642–1654.

Wang, L., Zhang, Y., Fan, G., Ren, J. N., Zhang, L. L., and Pan, S. Y. (2019). Effects of orange essential oil on intestinal microflora in mice. *Journal of the Science of Food and Agriculture*, 99(8), pp. 4019–4028.

Wang, X., Zhou, X., and Zhang X. (2024). Effects of ellagic acid on glucose and lipid metabolism: A systematic review and meta-analysis. *Journal of Nutrition and Metabolism*, 2024, p. 5558665.

Wang, M., and Yu, L. (2025). Emerging evidence of urolithin A in sports nutrition: Bridging preclinical findings to athletic applications. *Frontiers in Nutrition*, 12, p. 1585922.

Wojtunik-Kulesza, K., Niziński, P., Krajewska, A., Oniszczuk, T., Combrzyński, M., and Oniszczuk, A. (2025). Therapeutic potential of ellagic acid in liver diseases. *Molecules*, 30(12), p. 2596.

Wright, S. C., Maree, J. E., and Sibanyoni, M. (2009). Treatment of oral thrush in HIV/AIDS patients with lemon juice and lemon grass (Cymbopogon citratus) and gentian violet. *Phytomedicine*, 16(2–3), pp. 118–124.

Xiong, W., Yuan, L., Huang, J., Pan, B., Guo, L., Qian, G., Shuai, C., and Zeng, Z. (2023). Direct osteogenesis and immunomodulation dual function via sustained release of naringin from the polymer scaffold. *Journal of Materials Chemistry B*, 11(45), pp. 10896–10907.

Yang, Y., Luo, X., Wang, Y., Xu, A., Peng, L., Zhang, X., Wang, Z., Ying, Y., and Li, K. (2024). β-Mangostin targets and suppresses glioma via STING activation and tumor-associated microglia polarization. *Biomedicine and Pharmacotherapy*, 177, p. 117074.

Yostawonkul, J., Kamble, M. T., Sakuna, K., Madyod, S., Sukkarun, P., Medhe, S. V., Rodkhum, C., Pirarat, N., and Sewaka, M. (2023). Effects of mangosteen (Garcinia mangostana) peel extract loaded in nanoemulsion on growth performance, immune response, and disease resistance of Nile tilapia (Oreochromis niloticus) against Aeromonas veronii infection. *Animals*, 13(11), p. 1798.

Yun, A., Oh, H. Y., Lee, T. H., Kang, D. Y., Kim, K. T., Kim, H. S., and Kim, H. S. (2025). Blood orange (Citrus sinensis L. Osbeck) juice by-product extract as a functional feed additive: Effects on growth performance, digestive enzyme activity, antioxidant status, immune parameters, and disease resistance against Vibrio harveyi in juvenile black rockfish (Sebastes schlegelii). *Antioxidants*, 14(6), p. 745.

Zhang, L. X., Li, C. X., Kakar, M. U., Khan, M. S., Wu, P. F., Amir, R. M., Dai, D. F., Naveed, M., Li, Q. Y., Saeed, M., and Shen, J.Q. (2021). Resveratrol (RV): A pharmacological review and call for further research. *Biomedicine & Pharmacotherapy*, 143, p. 112164.

Zhang, S., Li, Z., Wang, X., An, L., Bao, J., Zhang, J., Cui, J., Li, Y., Jin, D. Q., Tuerhong, M., and Abudukeremu, M. (2020). Isolation, structural elucidation, and immunoregulation properties of an arabinofuranan from the rinds of Garcinia mangostana. *Carbohydrate Polymers*, 246, p. 116567.

Zhu, D., Zhu, Y., Tan, H., Ding, R., Dai, Q., Du, X., Liu, Y., and Yue, R. (2024). Effects of jujube (Ziziphus jujuba Mill.) fruit extracts on oxidative stress: A systematic review and meta-analysis of rodent studies. *Food Science and Nutrition*, 12(8), pp. 5312–5328.

Zou, M., Chen, Y., Sun-Waterhouse, D., Zhang, Y., and Li, F. (2018). Immunomodulatory acidic polysaccharides from Zizyphus jujuba cv. Huizao: Insights into their chemical characteristics and modes of action. *Food Chemistry*, 258, pp. 35–42.

Zuo, J., Yin, Q., Wang, L., Zhang, W., Fan, Y., Zhou, Y. Y., Li, Y., and Wang, G. D. (2018). Mangosteen ethanol extract alleviated the severity of collagen-induced arthritis in rats and produced synergistic effects with methotrexate. *Pharmaceutical Biology*, 56(1), pp. 455–464.

Zuo, Y., Li, Q., Xiong, Q., Li, J., Tang, C., Zhang, Y., and Wang, D. (2022). Naringin release from a nano-hydroxyapatite/collagen scaffold promotes osteogenesis and bone tissue reconstruction. *Polymers*, 14(16), p. 3260.

4 Traditional Medicinal Uses of Fruits and Berries in Indigenous Healing

Shah Alam, Nighat Anjum, Jamal Akhtar, and N. Zaheer Ahmed

4.1 INTRODUCTION

From the earliest stages of human history, primitive man relied on wild fruits to satisfy hunger. Over time, through repeated experiences, he discovered that certain fruits were not only edible but also possessed healing properties. Recognizing their dual role as food and medicine, early humans began conserving, propagating, and utilizing these fruits for therapeutic purposes as well as sustenance. With the advancement of biological sciences, the nutritive value and physiological effects of fruits became better understood, further enhancing their demand as both dietary staples and natural remedies (George and Benny 2010). A balanced diet that includes fruits, vegetables, cereals, millets, and pulses is essential for maintaining good health, promoting rejuvenation, and supporting longevity. The COVID-19 pandemic served as a wake-up call, heightening awareness of the strong link between nutrition and overall well-being, particularly in the context of today's sedentary and stressful lifestyle. Bioactive compounds present in fruits and berries are secondary metabolites that exert significant health-promoting effects by interacting with numerous physiological systems in the human body. These compounds include vitamins, minerals, phenolics, flavonoids, carotenoids, antioxidants, and dietary fiber, which play a crucial role in preventing and managing chronic conditions, including diabetes, cardiovascular disorders, and cancer (Singh 2023).

In recent years, fruits have attracted considerable scientific interest for their medicinal potential. Their widespread availability, biological activity, ease of absorption, and minimal side effects make them an exceptional component of the human diet. Extensive global research has consistently highlighted that fruits contain potent chemical constituents essential for human well-being and longevity. Nutritionally, fruits are indispensable to human health. They are rich in water, natural sugars, starches, pectin, and organic acids. Carbohydrates form the predominant macronutrient, while fats and proteins are present in negligible amounts. Fruits are also a primary source of vital vitamins. Moreover, wide varieties are abundant in insoluble fibers such as cellulose, which promote healthy digestion, improve bowel function by increasing stool bulk, and stimulate beneficial anaerobic bacterial fermentation. These attributes collectively underscore the nutritional and medicinal importance of fruits in maintaining a healthy life (George and Benny 2010).

Extensive research has demonstrated that fruits not only prevent malnutrition but also contribute to the maintenance of optimal health through a wide array of chemical compounds, particularly secondary metabolites. Regular consumption of fruits has been associated with a reduced incidence of chronic illnesses such as cardiovascular disease, stroke, diabetes, obesity, and hypertension. Fruits are abundant in polyphenols – potent antioxidants that enhance blood circulation and safeguard skin tissues. Citrus fruits, in particular, are rich in phenols, flavonoids, and carotenoids, which demonstrate anti-inflammatory, anticancer, and stress-mitigating effects by regulating

DOI: 10.1201/9781003688259-4

FIGURE 4.1 Common fruits and berries.

(Grapes, pomegranate, mulberry, gooseberry, and fig). Images A–E adapted from Namuwiki (2026), ExportersIndia (n.d.-b), WebMD (2026a); Shutterstock (2015a), Amazon (n.d.), TheTakeout (2025a); Pixabay (2024, n.d.), Amazon (n.d.); ExoticFlora (n.d.), Vecteezy (n.d.), Pixabay (2021); and Amazon (n.d.), OrganicSeeds (n.d.).

signaling pathways and interacting with key molecules such as P53 and P21 (Tongling, 2022). Moreover, certain phytochemicals present in fruits exhibit antioxidant properties by neutralizing free radicals – highly reactive molecules capable of damaging biomolecules, including lipids, proteins, and nucleic acids. This free radical scavenging activity plays a pivotal role in mitigating oxidative stress, thereby delaying the onset of degenerative processes and aging (George and Benny 2010).

Furthermore, across diverse cultures, fruits have historically occupied a central role in healing rituals and dietary practices. In ancient Egypt, figs and pomegranates were esteemed for both their symbolic and medicinal significance, while in Native American traditions, berries were incorporated into wellness practices and spiritual healing. Beyond their palatable flavors, fruits have long been recognized as integral components of traditional medicine, serving as natural remedies that embody a holistic approach to health and well-being. In the contemporary era, as society increasingly embraces natural therapies, the exploration of the healing properties of fruits and berries has become particularly relevant. This chapter examines the role of common fruits and berries, viz., grapes, pomegranates, mulberries, gooseberries and figs in traditional medicine, their distinctive bioactive constituents, and their medicinal applications in Indigenous healing, thus promoting human health (Figure 4.1).

4.2 MEDICINAL SIGNIFICANCE OF COMMON FRUITS

4.2.1 Grapes

Grapes are widely recognized as medicinal fruits in traditional systems of medicine such as Unani, Ayurveda, and Siddha. Valued for both their nutritional and therapeutic properties, they hold a special place in Ayurveda, where they are honored with the title "Phalottama," meaning the supreme fruit among all fruits. The grape belongs to the berry family as it is found attached to the stem. Many berries make up a cluster or bunch of grapes. The fruit ranges from green to dark brown or black in color, 1.5–2.5 cm long, and 0.5–1.5 cm wide. It occurs as soft clusters of pulpy berries that are oval or oblong, ellipsoidal to globose, with the skin adhering to the pulp. It is juicy, sweet, or sour. The external surface is fleshy when fresh, but becomes irregularly wrinkled with ridges and furrows when dry. Seeds are two to four in numbers, pyriform, or sometimes absent (Yeola et al. 2023). Fruits are consumed as ripe, unripe, and partly dried ones, i.e., raisins (Rizvi et al. 2019). It is the most widely produced fruit in the world. The cultivation of grapes is known as viticulture.There are about 10,000 varieties of grapes in the world. Worldwide, it is grown mainly for wine production (82% of production), followed by raisin production (10%), and the remainder for table purposes (8%) (Yeola et al. 2023).

4.2.1.1 Biological Source

Grapes are the fruits of *Vitis vinifera* Linn. belonging to the family Vitaceae.

4.2.1.2 Distributions

Vitis vinifera is a member of the Vitaceae (grape) family, native to Europe and Central Asia. It can be found in the wild in Afghanistan, Baluchistan, and the northwestern Himalaya. It is a deciduous climber, mostly cultivated in northwestern India in Himachal Pradesh, Maharashtra, Uttar Pradesh and Kashmir. It is also found in China, India, Iran, Egypt, Turkey, Brazil, Mexico, Central and Southern Europe, Western Asia, Africa, Northern Mediterranean Coast, South Africa, North Africa, California, Chile, Argentina, Uruguay, Peru. Most wine is produced in Germany, France, Italy, Canada, USA, New Zealand. Raisin production is popular in countries such as Iran, Turkey, India, USA (Yeola et al. 2023).

4.2.1.3 Phytochemical Composition

Different bioactive compounds are present in different parts of the grape fruit as follows: **Grape seeds**–procyanidins, gallic acid, quercetin, and flavan-3-ols (catechin and epicatechin);white grapes has flavonol glycosides; and black grapes has resveratrol, flavonol glycoside, anthocyanidins, phenolic compounds, coumaric acid, caffeic acid, coutaric acid, ferulic acid, fertaric acid, routine, quercitrin, quercetin-3-beta-D-glucoside, catechin, myricetin, epicatechin, primaric acid, caffeic acid, linoleic acid, p-hydroxy-phenylacetic acid, and gallic acid. **Grape skin**–flavonols, flavan-3-ols, anthocyanins, stilbenes, phenolic acid, quercetin, vanillic acid, proanthocyanidins, kaempferol, syringic acid, and gallic acid. **Grape juice**–coumaric acid, ferulic acid, caffeic acid, caftaric acid, fertaric acid, coutaric acid, epicatechin, catechin, resveratrol, procyanidin, flavonols, quercetin, rutin, kaempferol, quercitin-3-O-glucoronide, and quercitin-3-O-glucoside. **Grapevine canes**–gallic acid, ellagic acid, protocatechuic acid, vanillic acid, caffeic acid, caftaric acid, coutaric acid, syringic acid, ferulic acid, flavan-3-ol (procyanidin B1, procyanidin A1, procyanidin C1, procyanidin B2, catechin, and epicatechin), and flavonols,such as quercetin-3-O-galactoside, quercetin-3-O-arabinose, stilbenic compound, trans-resveratrol, trans-resveratrol-2-C-glucoside, and ampelopsin A and D (Parihar and Sharma 2021; Singh et al. 2023). Resveratrol is a well-known stilbenoid that is found in high concentrations in grape skins, particularly in *V. vinifera* grapes, which are used to make red wine (Singh et al. 2015).

4.2.1.4 Pharmacological Properties of Grape Fruits

Grape fruits demonstrated a wide range of pharmacological properties, viz., antioxidant, anti-inflammatory, antimicrobial, antidiabetic, and anticancer activities (Table 4.1).

TABLE 4.1
Pharmacological Properties of Common Fruits and Berries

Fruit – Source/Bioactive Components	Pharmacological Properties	Key References
Grapes–fruit skin and seeds/resveratrol	Anti-inflammatory	Fouad and Rizk (2019)
	Antioxidant	Xia et al. (2017); Tan and Ismail (2020)
	Anticontraction	Tabarrai et al. (2019)
	Skin aging	Kanigur et al. (2019); Petersen and Smith (2016)
	Anticancer	Tan and Ismail (2020)
	Antimicrobial	Kanigur et al. (2019); Petersen and Smith (2016)
	Antiobesity	Akaberi and Hosseinzadeh (2016);Mahdavi et al. (2020)
	Antifungal	Tabarrai et al. (2019)
Grapes – fruit skin and seeds/anthocyanins	Anti-inflammatory	Insanu et al. (2021)
Quercetin	Antioxidant	Di Lorenzo et al. (2019)
	Skinaging	Petersen andSmith (2016)
Grapes –seed extracts/procyanidin or proanthocyanidins	Skinaging	Petersen andSmith (2016)
	Anticancer	Grace et al. (2018)
	Antimicrobial	Nassiri and Hosseinzadeh (2016)
	Antiobesity	Santos et al. (2017)
	Antidiabetic	Akaberiand Hosseinzadeh (2016)
Grapes –fruit skin and seeds/gallic acid, catechin,epicatechin	Antioxidant	Yilmaz and Toledo (2004)
Grapes –skin and seed/ epigallocatechin-3-gallate	Antineoplastic	Borska et al. (2003)
Grape –juice/flavonoids	Antioxidant	O'Byrne et al. (2002)
Pomegranate –fruit exocarp extract	Antibacterial	Alanis et al. (2005)
Pomegranate –fruit shell extract	Antibacterial against *E. coli*	Voravuthikunchaiet al. (2004)
Pomegranate –peel extract	Antibacterial against *H. pylori*	Hajimahmoodiet al. (2011); Mayyaset al. (2021)
Pomegranate –fruit exocarp extract	Antiprotozoal	Calzada et al. (2006)
Pomegranate – juice	Antiproliferative	Banerjee et al. (2013)
Pomegranate –fruit extract	Anticancer	Sudha et al. (2021)
Pomegranate –peel extract	Anticancer	Cheshomiet al. (2022)
Pomegranate –peel extract	Antidiarrheal	Qnais et al. (2007)
Pomegranate –peel extract	Anti-inflammatory	Mastrogiovanni et al. (2019)
Pomegranate –fruit, juice, total extract	Antidiarrheal	Souli et al. (2015)
Pomegranate –mesocarp decoction	Chemopreventive	Tortora et al. (2018)
Pomegranate –mesocarp decoction, ellagitannins	Anti-inflammatory analgesic	Parisio et al. (2020)
Pomegranate – juice	Antihypertensive	Aviram and Dornfeld (2001)

(Continued)

TABLE 4.1 (Continued)

Fruit – Source/Bioactive Components	Pharmacological Properties	Key References
Pomegranate –peel extract	Hypoglycemic	Khalil (2004)
Pomegranate – juice	Antiobesity	Esmaillzadeh et al. (2006)
Mulberry – fruit	Antihyperlipidemic	Sirikanchanarod et al. (2016)
Mulberry – Juice	Antiobesity	Wu et al. (2013)
Mulberry fruit – aqueous extract	Antihyperlipidemic	Peng et al. (2011)
Mulberry fruits (*M. nigra*)/anthocyanins	Anti-inflammatory	Chen et al. (2016)
Mulberry fruit – extract	Antidiabetic	Wang et al. (2013); Chen et al. (2016);Sarikaphuti et al. (2013)
Mulberry fruit – extract	Antioxidant	Wang et al. (2013,2018); Raman et al. (2016)
Mulberry fruit – freeze-dried powder	Antioxidant	Yang et al. (2010)
	Antiaging	Pal et al. (2017)
	Anticancerous	Mahata et al. (2013); Wiart (2013)
	Antidiabetic	Kumar et al. (2012); Nain et al. (2012); Fatima et al. (2017); Sharma et al. (2020)
	Antimicrobial	Nath et al. (2012); Dinesh et al. (2017); Singh et al. (2019)
	Antimutagenic	Sumitra et al. (2009); Agrawal et al. (2012)
	Antipyretic, analgesic, anti-inflammatory	Muthuraman et al. (2011); Gupta et al. (2013);Asmilia et al. (2020)
	Antioxidant	Nain et al. (2012); Rose et al. (2018); Majeed et al. (2020)
Indian gooseberry fruits	Antiulcerous, wound healing	Mehrotra et al. (2011); Chatterjee et al.(2012)
	Cardioprotective	Ojha et al. (2012); Rajak et al. (2004)
	Chemoprotective	Sharma and Sharma (2011)
	Hepatoprotective	Sultana et al. (2005); Pramyothin et al. (2006); Baliga et al. (2019)
	Immunomodulatory	Sai et al. (2002); Srikumar et al. (2005); Patel et al. (2017)
	Memory enhancing	Vasudevan and Parle (2007); Ali et al. (2013)
	Neuroprotective	Reddy et al. (2011); Mathew and Subramanian (2014);Shalini and Sharma (2015); Justin et al. (2016)
Figs fruit – distilled water extract	Antiacne	Salma et al. (2020)
Ripe dried fruit – aqueous-ethanolic extract	Antispasmodic	Rahmani and Aldebasi (2017)
F. carica latex	Antiaging, wound healing	Ayuso et al. (2022)
F. carica aqueous acetonic extract	Antiepileptic and Anticonvulsant	Fazel et al. (2024)
F. carica paste	Laxative	Al-Snafi (2017)
Aqueous methanol extract of *F. carica* fruit	Antihypertensive	Alamgeer et al. (2017)
Fruit paste	Anti-inflammatory and analgesic	Naaz et al. (2024)
Fig latex	Anticancer	Naaz et al. (2024)
F. carica fruit	Alzheimer's disease	Naaz et al. (2024)

4.2.1.5 Traditional Uses

Unripe grapes: valued as a tonic and antiseptic; considered nutritive yet mildly constipating. Traditionally described as demulcent, refrigerant, hemostatic, restorative, emollient, intellect-enhancing, cardiotonic, hematinic, aphrodisiac, rejuvenating, immune-boosting, nervine tonic, antispasmodic, expectorant, digestive, stomachic, adipogenic, laxative, nutritive, concoctive, diuretic, and general tonic. **Fresh ripe grapes**: regarded as aphrodisiac and restorative; recommended for weakness and general debility. Juice of fresh grapes: strained juice given to children for constipation during teething and to prevent convulsions; formerly used in Europe for epilepsy; useful in thrush, severe colds, and fevers. Juice of sour grapes: applied for bruises and sprains. **Raisins (Dried Grapes)** are considered more laxative than fresh grapes. Used as ademulcent, expectorant, attenuant, nutritive, refrigerant, and blood purifier; Helpful in catarrh, thirst, and subacute enlargement of the liver and spleen. **Grape seeds**: traditionally described as refrigerant, astringent to the bowels, and aphrodisiac (Rizvi et al. 2019; Yeola et al. 2023; Saeed et al. 2024)

4.2.1.6 Therapeutic Applications

Grapes and their preparations have been mentioned in traditional medicine for a wide range of conditions, including cardiac debility, heart diseases, varicose veins, hemorrhoids, and capillary fragility; gout, dysuria, dropsy; bilious dyspepsia, constipation, colic, flatulence, dysentery, and cholera; cough, asthma, hoarseness, and pharyngopathy; cachexia, cancer, smallpox, putrid fever, and leprosy. Other conditions include thirst,burning sensations, emaciation, general debility, amentia, ophthalmopathy, and heavy menstrual bleeding. For cyst tumor care, *Terminalia chebula* powder is traditionally consumed with grape juice, which is regarded as an indispensable vehicle enhancing its healing potential (Rizvi et al. 2019; Yeola et al. 2023; Saeed et al. 2024).

4.2.2 Pomegranate

Pomegranate holds deep ethnocultural and religious significance across civilizations, symbolizing fertility, prosperity, abundance, and eternal life. In the Old Testament, it was revered as a sacred fruit of fertility and fortune; in the Quran, it is described as a paradise fruit; Egyptians linked it to prosperity and the afterlife while also using it medicinally; and Hindus associate it with Goddess Earth. In Greek mythology, it symbolized life, regeneration, and marriage. The fruit is large-sized, globose, brownish yellow to red in color, and round in shape, with a 5–12 cm diameter and a crown of tubular calyx. A fruit or berry consists of two parts: an outer hard pericarp and an inner mesocarp of a spongy nature. The mesocarp is divided into several chambers in a non-symmetrical manner, where triangular soft seeds, usually red, pink, or white in color, are embedded. The seeds are fully covered by arils, which are the edible, juicy portion (Kumari et al. 2021). In Ayurvedic medicine, the pomegranate is regarded as a complete "pharmacy," with the whole fruit, rind, peel, and sun-dried seeds with the arils traditionally used for both therapeutic purposes and overall well-being (Suthar. 2022).

4.2.2.1 Biological Source

Pomegranate is a mature ripe fruit of *Punica granatum* Linn., belonging to the family Lythraceae.

4.2.2.2 Distribution

Pomegranate is one of the first domesticated fruits that has been cultivated for centuries. It is indigenous to Iran and neighboring Afghanistan, the central Asian regions, the Himalaya, Anatolia, and the Middle East. It is also cultivated in the Mediterranean region, South Asia, and the Middle East countries; Kandahar in Afghanistan is famous for its high-quality pomegranate. Today, pomegranate is cultivated in most regions of the world, including Iran, Spain, Italy, Afghanistan, India, America, China, Russia, Uzbekistan, Morocco, and Greece. Iran is one of the biggest producers of pomegranates in the world (Shaygannia et al. 2016).

4.2.2.3 Phytochemical Composition

Bark and rind of the fruit contain tannin (22–25%), and the root bark contains punicotannic acid (20–25%), mannitol, gallic acid, sugar gum pectin, ash (15%), an active liquid alkaloid "Pelletierine," and oil liquid "Isopelletierine," as well as two inactive alkaloids methyl pelletierine and pseudo-pelletierine. Different parts of the pomegranate contain different phytochemicals, such as seed oil, which contains 95% punicic acid, ellagic acid, other fatty acids, and sterols. The pericarp (peel, rind) contains phenolic punicalagins, fatty acids, and gallic acid; rutin, quercetin, and other flavonols; flavones, flavanones; and anthocyanidins. Leaves contain tannins (punicalin and punicafolin) and flavone glycosides, including apigenin and luteolin. Flowers contain gallic acid, ursolic acid, and triterpenoids, including asiatic and maslinic acid. Roots and bark contain ellagitannins, including punicalin and punicalagin, and numerous piperidine alkaloids. Juice contains anthocyanins, ascorbic acid, ellagic acid, gallic acid, glucose, caffeic acid, quercetin, rutin, catechin, numerous minerals, particularly iron, and amino acids (Alam et al. 2018).

4.2.2.4 Pharmacological Properties of Pomegranate Fruit

Pomegranate fruits possess various pharmacological properties, including antibacterial, antidiarrheal, anti-inflammatory, antiproliferative, antiobesity, etc. (Table4.1).

4.2.2.5 Traditional Uses and Therapeutic Applications

Traditionally, *Punica granatum* has been esteemed as an astringent, cooling, and drying agent with diverse therapeutic applications across bodily systems. Its juice and peel are widely employed to strengthen the stomach, heal intestinal wounds, arrest diarrhea, and expel parasites. Cultural practices from India and China highlight its use in dysentery, anemia, and gastrointestinal disorders. Sweet varieties are prescribed for chronic cough, throat roughness, and chest pain, while preparations from buds and peels are applied for wound healing, scar removal, and dermatological conditions such as scabies and thrush. The peel is valued for regulating excessive menstrual bleeding, and the fruit supports oral health by treating ulcers, gum bleeding, and halitosis. Ancient traditions also associate pomegranate with vision protection, liver detoxification, jaundice treatment, and hematopoiesis, with sour types particularly noted for balancing humors, reducing palpitations, and acting as vasodilators. Overall, the fruit has been revered as a tonic for digestive, respiratory, reproductive, dermatological, oral, and systemic health (Mohammadi et al. 2023). Beyond these general properties, the whole fruit has been described as a cardiac and hepatic tonic and as an aphrodisiac in cases of sexual debility. The rind strengthens teeth and gums, while seeds exhibit antiemetic and diuretic actions (Alam et al. 2018). Rich in tannins, the fruit demonstrates pronounced astringency. Infusions or decoctions of flowers have historically been used to manage ordinary diarrhea and vaginal discharge and, when combined with peel, to alleviate pancreatic inflammation. Syrup prepared from the juice serves as a febrifuge in malaria and seasonal fevers and as a cooling agent in bilious complaints. Daily consumption of the fruit has also been recommended as an aphrodisiac (Alam et al. 2018). Fresh juice is traditionally prescribed for gallbladder disorders, while tannin-rich decoctions are employed for gastrointestinal ailments, including diarrhea, dysentery, and stomach disturbances. Seed preparations have been applied in the treatment of vaginal discharge and for wound healing. Root bark and extracts of pomegranate have been utilized for their anthelmintic activity, attributed to alkaloid compounds (Shaygannia et al. 2016).

In folk medicine, *P. granatum* has been widely used across diverse regions. In Jammu and Kashmir, seeds are consumed to treat anemia, jaundice, and general weakness, while juice is taken as an appetizer. In Maharashtra, fruit rind is used for diarrhea, and among the tribals of the Western Ghats, fruit powder is ingested for stomachache. In coastal Pakistan, fruit skin infusions are administered for dysentery, vomiting, and food poisoning. Among tribal communities of the Lesser Himalayas, the fruit is traditionally used for diabetes, diarrhea, piles, and dysentery (Kumari et al. 2021).

4.2.3 Mulberry

Mulberry is a versatile fruit widely utilized across major traditional medical systems, including Chinese, Ayurvedic, Unani, and Persian medicine. In India, it is revered as KalpaVruksha because every part of the plant serves a purpose, while its fruit is popularly known as toot or shahtoot, meaning "king" or "superior" mulberry. Belonging to the genus *Morus*, the plant encompasses more than 100 varieties, with white mulberry (*Morus alba* L.), black mulberry (*Morus nigra* L.), and red mulberry (*Morus rubra* L.) being the most prevalent (Ercisli and Orhan, 2007). Nearly all species of mulberry are acknowledged in Unani, Ayurvedic, and Chinese traditions for their diverse pharmacological properties. In Unani medicine, the fruits of *M. nigra* hold particular importance and are referred to as *Tut Siyaah*. Mulberry fruit is precisely a cluster of small fruits that are organized longitudinally around the central axis, similar to that of blackberries or loganberries. Its fruit, or syncarp, comprises numerous drupes that are enclosed in a fleshy perianth, ovoid or subglobose, up to 5 cm long, white to pinkish white, purple, or black when ripe. Mulberries are consumed in multiple forms, such as juice, wine, tea, jams, and jellies, owing to their rich carbohydrate content and associated health benefits (Ramappa et al. 2020).

4.2.3.1 Biological Source

Mulberry fruit belongs to the genus *Morus* of the family Moraceae.

4.2.3.2 Distributions

Mulberry is native to China and Iran and cultivated in Asia, Africa, America, Europe, India, and Northwestern Himalayas and Kashmir (Kumar and Singh 2020; Naaz et al. 2025).

4.2.3.3 Phytochemical Composition

Mulberry is a vital source of nutraceutical products such as amino acids (alanine, leucine, glutamate, tyrosine, etc.), carbohydrates (polysaccharide -inulin), fatty acids (stearic acid, linoleic acid, etc.), vitamins (A,C,E,K, etc.), and minerals (copper, iron, magnesium, zinc, etc.). Important phytochemicals –anthocyanins, i.e., cyanidin-3-glucoside and cyanidin-3-rutinoside – are present in mulberries. Flavonols isolated from mulberry fruits are as follows: kaempferolhexoside, kaempferolhexosyl-hexoside, and quercetin morin. Flavanones, such as gallocatechin, epigallocatechin, gallocatechingallate, and naringin, are isolated from mulberry fruits. Phenolic compounds reported from mulberry fruits arehydroxycinnamic acids, such as caffeic acid, chlorogenic acid, *p*-coumaric acid, o-coumaric acid, and ferulic acid, and the derivatives of benzoic acid, viz., gallic acid, syringic acid, vallinic acid, protocatechuic acid, and *p*-hydroxybenzoic acid. Pyrrole alkaloids were reported from mulberry fruits, viz., 2-formyl-5-(hydroxymethyl)-1H-pyrrole-1-butanoic acid, 5-(hydroxymethyl)-1H-pyrrole-2 carboxaldehyde, 2-formyl-1H-pyrrole-1-butanoic acid, and 2-formyl-5-(methoxymethyl)-1H-pyrrole-1-butanoic acid (Ramappa et al. 2020).

4.2.3.4 Pharmacological Properties of Mulberry Fruits

Mulberry fruits exhibit wide pharmacological activities, viz., antihyperlipidemic, antidiabetic, antioxidant, anti-inflammatory, etc. (Table 4.1).

4.2.3.5 Traditional Uses and Therapeutic Applications

Fruits of the mulberry, particularly black and red varieties, are advantageous to the human body. However, the Ministry of Health of China in1985 recognized *Morus alba* as the first medicinal and edible fruit; also, its leaves and fruits were considered not only food but also as drugs. Chinese medicine utilizes mulberry fruits to strengthen the joints, lower blood pressure, protect the liver from damage, and treat fever and urine incontinence. Azerbaijan people utilize *M. nigra* fruits in the treatment of ailments, such as gallbladder, liver, and heart diseases (Jan et al. 2021). In Chinese markets, mulberry is often sold in the form of a paste called as *Sangshengao*. The paste is mixed into hot water to make tea to enrich the efficiency of the liver and kidneys and sharpen the hearing and brighten the eyes. The fruit of *M. alba* has a cooling and laxative property and is used for throat, dyspepsia, and melancholia. Ripe fruits are an appetizer and carminative. Fruits are also used for loss of appetite, flatulence, constipation, and intestinal worms, such as tapeworm. Syrup of ripe fruits is useful in heart diseases, bleeding disorders, burning sensations, debility, and antiaging. Decoctions also prepared from the fruits are used against cerebral arteriosclerosis, chronic nephritis, kidney asthenia, alopecia areata, central retinitis, and nasopharyngeal cancer. The fruit is used in compound formulations meant for the treatment of tonsillitis.The fruit juice of the *Morus* species has refrigerant and laxative properties. It is also used as a febrifuge, in diarrhea, cold, endemic malaria, and amoebiasis (Kumar and Chauhan 2008). Fruits have been used for diabetes (Kumar and Singh 2020). Black mulberry is commonly consumed in most Western countries as fresh, dried, or processed as marmalade, yogurt, wine, syrups, fruit juice, jam, galette, and fresh fruit cakes, various beverages, and cosmetic products (Naaz et al. 2025).

Mulberry fruits support liver and kidney health, alleviate weakness, exhaustion, and anemia, and are traditionally used to address urinary incontinence, tinnitus, dizziness, constipation in the elderly, sore throat, depression, and fever. Mulberry fruit and juice nourish the skin and blood, calm the nerves, balance internal secretions, promote alcohol metabolism, and boost immunity. Additionally, mulberries can be brewed into wine, believed to cleanse the blood and aid intestinal health, with daily consumption thought to help remove fecal residue (Kumar and Chauhan 2008; Ramappa et al. 2020). In Unani medicine, mulberries are traditionally used to relieve throat ailments such as dryness, irritation, scratchiness, chronic ulcers, blisters of the throat and mouth, and diphtheria; they soothe and stabilize the blood humor, quench thirst, and reduce throat inflammation caused by excess heat, while also exerting complementary effects on the liver, spleen, and blood vessels by dilating them, dissolving thick or viscous humors, clearing obstructions, and acting as a deobstruent (Naaz et al. 2025).

4.2.4 Indian Gooseberry

Indian Gooseberry or *Amla* is one of the most extensively used fruits in the traditional system of medicine. It is known as "Divya" in Ayurveda, which literally means fruit of heaven or nectar fruit. It is also called the sustainer or the fruit where the Goddess of prosperity resides. According to ancient Indian mythology, it is the first tree to be created in the universe. The fruit is 1.3–1.6 cm in diameter, fleshy, globose with six obscure vertical furrows, green, and it tends to light yellow or brick-red color when mature. Both dry and fresh fruits are used in traditional medicine (Masihuddin et al. 2019).

4.2.4.1 Biological Source

Indian gooseberry, also known as *Emblica officinalis* Gaertn. (syn. *Phyllanthus emblica* L.), belongs to the family Euphorbiaceae.

4.2.4.2 Distributions

E. officinalis, originating from India, is also cultivated in several other tropical and subtropical countries such as Bangladesh, South China, Malaysia, Mascarene Islands, Myanmar, Pakistan, Sri Lanka, and Uzbekistan (Thilaga et al. 2013).

4.2.4.3 Phytochemical Composition

Indian gooseberry represents the richest natural source of vitamin C in the plant kingdom. When consumed as the whole fruit rather than as an isolated extract, its vitamin C is readily absorbed by the human body (Nisha et al. 2004). Moreover, the vitamin C in goose is complexed with tannins, which safeguard it against degradation from heat and light. Amla fruit contains –ascorbic acid, aspartic acid, alanine, arginine, 3–6-di-O-galloyl-glucose, β-carotene, boron, calcium, carbohydrates, chibulinic acid, chebulaginic acid, chebulagic acid, chebulic acid, chloride, copper, corilagin, corilagic acid, cystine, d-fructose, d-glucose, ellagic acid, emblicanin-A, B, emblicol, ethyl gallate, gallic acid, gallic acid ethyl ester, gibberellin A1, gibberellin A3, gibberellin A4, gibberellin A7, gibberellin A9, glucogallin, glucose, glutamic acid, glycine, histidine, iron, isoleucine, leucine, lysine, l-malic acid 2-*O*-gallate, magnesium, manganese, methionine, myristic acid, myo-inositol, nitrogen, niacin, pectin, phenylalanine, phosphorus, phyllemblic acid, phyllemblinic acid, polysaccharide, potassium, proline, protein, quercetin, rutin, riboflavin, selenium, serine, silica, sodium, starch, sucrose, sulfur, threonine, terchebin, tannin, thiamin, trigalloyl glucose, tryptophan, tyrosine zinc, zeatin, zeatin riboside, zeatin nucleotide, phyllantine, and phyllantidine (Jagetia et al. 2004; Srikumar et al. 2007; Singh et al. 2011; Srinivasan et al. 2018). Its seed contains β-sitosterol, linoleic acid, linolenic acid, myristic acid, oleic acid, palmitic acid, stearic acid, flavonoid, and tannins (Srikumar et al. 2007; Khan 2009; Sriwatcharakul 2020)

4.2.4.4 Pharmacological Properties of Gooseberry Fruits

Amla or Indian Gooseberries exhibits a wide range of pharmacological activities, viz., antidiabetic, antimicrobial, antioxidant, anti-inflammatory, hepatoprotective, cardioprotective, neuroprotective, memory-enhancing properties, etc. (Table 4.1)

4.2.4.5 Traditional and Therapeutic Uses

In Ayurveda and Unani medicine, regular intake of amla berry enhances digestion, nutrient absorption, and iron assimilation, thereby supporting healthy blood. It is effective in calming mild-to-moderate hyperacidity and purifies bodily nutrient fluids as well as blood, strengthening liver function and detoxification (Pramyothin et al. 2006; Tasduq et al. 2005; Sultana et al. 2005). Gooseberries are recognized as a medicinal fruit that promotes brain health by improving acquisition, retention, and recall. It sharpens intellect, supports the nervous system, and fortifies the senses (Reddy et al. 2011; Vasudevan and Parle 2007). It also acts as a tonic for the lungs, with seeds traditionally used in asthma, bronchitis, and biliousness (Singh et al. 1979). Fresh ripe fruits (1–2 pieces) are widely used in Ayurveda and Unani systems as a gentle laxative, often preserved in sugar known as *Murabba Amla*. The herb nurtures reproductive health, enhancing fertility, strengthening the uterus, and supporting both ovaries and sperm. It also benefits the urinary system, easing mild burning sensations and promoting natural diuretic action (Tsarong and Tsewang, 1994). Rich in vitamin C and minerals, amla berry improves complexion by moisturizing the skin, cleansing tissues of toxins, and boosting immunity against bacterial infection. It enhances glow and luster. Additionally, it supports calcium absorption, strengthening bones, teeth, nails, and hair. It helps maintain youthful hair color, delays premature graying, and fortifies hair follicles, reducing thinning with age (Stuart, 1911). Amla berry balances all doshas and is particularly effective in cooling *Pitta* during hot seasons. Tibetan medicine also describes it as sour in taste with cooling potency (Tsarong and Tsewang 1994). Gooseberry aids the liver in flushing out chemicals and additives, increases vitality by cleansing blood and microchannels, and supports cellular regeneration, thereby reducing fatigue (Brun and Schumacher 1987). As a tridoshic tonic, Amla berry strengthens eye health. Leaf infusions are applied to sore eyes, seed infusions are used as collyrium for conjunctival inflammation, and fruit exudates are applied externally for eye complaints (Jayaweera, 1980; Biswas et al., 2001). Amla berry also acts as a powerful antioxidant and is used in managing diabetes. Fruits, seed infusions, and leaf decoctions are employed in treating diabetes mellitus (Akhtar et al. 2011; Sabu and Kuttan 2002). Externally, the pericarp is used in decoctions with other ingredients and applied with cow ghee to boils, promoting suppuration (Jayaweera, 1980).

4.2.5 Figs

The fig is known to be one of the first plants cultivated and the oldest known fruit tree by humans. By recent investigations, it has been indicated that fig has been cultivated for over 11,000 years, and conceivably existed before cereal grains. *Ficus carica* L. has the greatest commercial importance among all the species of fig and is known by the name of common fig or *anjeer*. Figs have significant importance in Indian mythology throughout the tropics, both as objects of worship and for their many everyday uses. *Ficus carica* has been indicated in the traditional systems of medicine such as Unani, Siddha, Ayurveda, and Chinese medicine for its therapeutic efficacies (Salma et al. 2020). *Ficus carica* is a species of the genus *Ficus*, having over 800 tropical and subtropical plant species. The shape of the fig fruit is ovoid, turbinate, and pear- shaped. It is 1–4 inches long and varies in color from yellowish-green to coppery, bronze, or dark purple. The unique feature of figs is that they have a gap called "eye"or ostiole. Unripe fruit is green in color, whereas ripe fruit becomes reddish brown or purplish, and the taste is sweet and delicious. Dried fruits of *Ficus carica* are compressed into a circular shape with a central hole, 4–6 cm in diameter and 1cm thick. The surface of the fruit is wrinkled and light yellow to brown in color. The fruit contains many small seeds in a pulpy mesocarp (Ansari and Munshi, 2023).The fruits of fig can be eaten raw, dried, canned, or in other preserved forms, such as in sweets and jam, which can be eaten fresh or used in the preparation of delicacies and cakes (Salma et al. 2020).

4.2.5.1 Biological Source

Figs are the mature fruits of *Ficus carica* Linn., belonging to the family Moraceae.

4.2.5.2 Distributions

Fig is distributed in Southwest Asia and the Eastern Mediterranean region, from Turkey in the East to Spain and Portugal in the West; it is also grown commercially in parts of the USA and Chile. About 26% of the world's total figs are grown in Turkey, and around 70% are produced together by Egypt, Iran, Morocco, Algeria, and Greece. The most significant species of figs found in India include *F. bengalensis*, *F. carica*, *F. racemosa*, and *F. elastica*. It is cultivated commercially in India near Pune in Maharashtra, and in the Bellary and Anantapur districts in South India. In Punjab, Uttar Pradesh, and Mysore, it is mostly grown scattered in gardens or in home yards (Pal 2020).

4.2.5.3 Phytochemical Composition

Fig fruit has a nutritive index of 11, which is higher than that of apple (9), resin (8), and date (6). Furthermore, dried figs are a concentrated source of various nutrients and thus are often referred to as a poor man's diet in Mediterranean countries. It is a rich source of proteins, amino acids, carbohydrates, minerals, vitamins, and fibers. In addition, it is a highly nutritious fruit, it is rich in calories and carbohydrates, lipids, phenolics, and enzymes. Studies revealed the presence of numerous bioactive compounds, viz., fruits – cyanidin-3-O-glucoside, cyanidin-3-O-rhamnoglucoside, cholesterol, saturated fat, sodium, insoluble sugars, protein, vitamin A, vitamin C, calcium, and iron; ripe dried fruit – alkaloids, flavonoids, coumarins, saponins, and terpenes; Dried seeds – fixed oil including fatty acids (Naaz et al. 2024).

4.2.5.4 Pharmacological Properties of Fig Fruits

Fig fruits exhibit a wide range of pharmacological properties, viz., antispasmodic, antiaging, wound healing, antiepileptic and anticonvulsant, laxative, antihypertensive properties, etc. (Table 4.1)

4.2.5.5 Traditional and Therapeutic Uses

Figs have long been recognized in traditional Unani and Ayurvedic medicine for their diverse pharmacological properties, including detergent, diuretic, nutritive, laxative, purgative, diaphoretic, agglutinant, livertonic, demulcent, deobstruent, concoctive, braintonic, and anti-inflammatory actions (Salma et al. 2020). Various forms of the fruits, viz., dried, ripe, or fresh, are employed in native medicinal systems to manage gastrointestinal disorders such as colic, indigestion, loss of appetite, and diarrhea, as well as respiratory ailments, including sore throat, cough, and bronchial conditions. The fruits, consumed fresh, dried, or processed into jam, are conventionally valued for their laxative, cardiovascular, respiratory, antispasmodic, and anti-inflammatory benefits. Preparations such as fig juice mixed with honey are traditionally administered for hemorrhage, while in Indian medicine, figs are regarded as mild laxatives, expectorants, and diuretics, with additional applications in liver and spleen diseases. Dry figs serve as dietary supplements for diabetics and are widely commercialized owing to their high sugar content. Topical applications of fig paste are reported to alleviate swellings, tumors, and inflammatory pain, and fig-based remedies have historically been emphasized in dermatological care for conditions such as freckles, chickenpox, measles, moles, pityriasis, tinea, melasma, scabies, urticaria, and psoriasis. Moreover, figs are traditionally prescribed for neuropsychiatric disorders (epilepsy, mania, paralysis), respiratory diseases (chronic cough, throat dryness, chest pain, breathlessness, asthma), cardiovascular complaints (palpitation, splenic obstruction), and urogenital disorders (kidney dystrophy, dysuria, dribbling of urine, uterine hardening). Their efficacy is also noted in constipation, colitis, backache, joint pain, and gout, underscoring the broad therapeutic spectrum attributed to *F. carica* in ethnomedicine (Naaz et al. 2024; Salma et al. 2020).

4.3 CONCLUSION AND FUTURE PROSPECTS

Fruits and berries have occupied a fundamental role in human living and medicine since antiquity. They have been cherished not only for their nutritional richness but also for their therapeutic potential. Across diverse traditional and indigenous systems, including Ayurveda, Unani, Siddha, Chinese, and Persian medicine, fruits have been revered as natural remedies that embody holistic approaches to health and well-being. Their abundance of bioactive compounds, such as flavonoids, anthocyanins, tannins, phenolic acids, and vitamins, underscores their capacity to prevent malnutrition, strengthen immunity, and mitigate chronic diseases. Modern biomedical research has increasingly validated these traditional claims, highlighting antioxidant, anti-inflammatory, cardioprotective, and neuroprotective properties that position fruits and berries as vital contributors to functional foods and nutraceutical development. Despite centuries of empirical use, the therapeutic applications of fruits and berries remain incompletely explored. The complexity of their phytochemistry and the synergistic interactions among secondary metabolites demand deeper scientific inquiry. Bridging traditional wisdom with modern evidence requires standardized formulations, rigorous clinical validation, and mechanistic studies to establish reproducible outcomes.

Large-scale, multicentric trials are essential to confirm the efficacy of fruit-based traditional formulations in chronic diseases such as diabetes, obesity, cardiovascular disease, neurodegeneration, etc. Isolation and formulation of bioactive compounds into standardized nutraceuticals and functional foods can enhance therapeutic accessibility and global acceptance. Fruits and berries can be incorporated into dietary guidelines and indigenous-inspired wellness programs, particularly for aging populations vulnerable to oxidative stress and metabolic disorders. Rising consumer preference for natural therapies positions fruit-based products as promising candidates in nutraceutical, cosmeceutical, and functional food industries.

In conclusion, fruits and berries represent a profound juncture of tradition and modern science. Their enduring medicinal legacy, coupled with emerging biomedical evidence, positions them as invaluable natural resources for future healthcare innovations, reaffirming their role as both nourishment and medicine in the pursuit of human longevity and well-being.

REFERENCES

Agrawal, R.C., Sharma, R. & Maheshwari, S.K. 2012. Antimutagenic and wound healing activity of *Emblica officinalis* extract in Swiss albino mice. *Int. J. Sci. Eng. Res.* 3:1–12.

Akaberi, M.& Hosseinzadeh, H. 2016. Grapes (*Vitis vinifera*) as a potential candidate for the therapy of the metabolic syndrome. *Phytother. Res.* 30(4): 540–556.

Akhtar, M.S., Ramzan, A., Ali, A. & Ahmad, M. 2011. Effect of amla fruit (*Emblica officinalis* Gaertn.) on blood glucose and lipid profile of normal subjects and type 2 diabetic patients. *Int. J. Food Sci. Nutr.* 62(6): 609–616. https://doi.org/10.3109/09637486.2011.560565.

Alam, S., Anjum, N., Akhtar, J., Bashir, F., Khan, A.A. & Parveen, N. 2018. Phytochemical and pharmacological investigations on rumman (*Punica granatum* L.). *Hippocratic J. Unani Med.* 13(4): 13–23.

Alamgeer, Iman, S., Asif, H. & Saleem, M. 2017. Evaluation of antihypertensive potential of *Ficus carica* fruit. *Pharm. Biol.* 55(1): 1047–1053. https://doi.org/10.1080/13880209.2017.1278611

Alanis, A.D., Calzada, F., Cervantes, J.A., Torres, J. & Ceballos, G.M. 2005. Antibacterial properties of some plants used in Mexican traditional medicine for the treatment of gastrointestinal disorders. *J. Ethnopharmacol.* 100: 153–157.

Ali, S.K., Hamed, A.R., Soltan, M.M., Hegazy, U.M., Elgorashi, E.E., El-Garf, I.A. & Hussein, A.A. 2013. In vitro evaluation of selected Egyptian traditional herbal medicines for treatment of Alzheimer disease. *BMC Complement. Altern. Med.* 13: 121.

Al-Snafi, A.E. 2017. Nutritional and pharmacological importance of *Ficus carica*: a review. *IOSR J. Pharm.* 7(3): 33–48.

Anonymous (2015). Pomegranate [*image*] Retrieved Jan 23, 2026, from https://www.shutterstock.com/image-photo/ripe-pomegranate-fruit-on-tree-branch-291445157

Anonymous (2021). Indian gooseberry seeds [photograph] Retrieved Jan 23, 2026, from https://pixabay.com/photos/indian-gooseberry-fruit-food-amla-6229076/

Anonymous (2024). Mulberry fruit tree [*image*] Retrieved Jan 23, 2026, from https://pixabay.com/photos/mulberry-nature-fruit-tree-sweet-9219447/

Anonymous (2025). Pomegranate seeds [*image*] Retrieved Jan 23, 2026, from https://www.thetakeout.com/1831904/how-to-freeze-pomegranate-seeds/

Anonymous (2026). *Raisins* [*image*] Retrieved Jan 23, 2026, from https://www.webmd.com/diet/raisins-good-for-you

Anonymous (n.d.-a). *Grapes varieties* [photograph] Retrieved Jan 23, 2026, from https://en.namu.wiki/w/ under CC BY-NC-SA 2.0 KR

Anonymous (n.d.-b). *Grapes bunches* [photograph] Retrieved Jan 23, 2026, from https://www.exportersindia.com/product-detail/net-bag-plastic-box-fresh-grapes-4450564.htm

Anonymous (n.d.-c). Pomegranate [*image*] Retrieved Jan 23, 2026, from https://www.amazon.in/MVKR-Live-Red-Kandhari-Anar/dp/B08TC5WWM9

Anonymous (n.d.-d). Black Mulberry [*image*] Retrieved Jan 23, 2026, from https://www.amazon.in/GROWHEAVEN-Mulberry-Shahtoot-Must-Have-Gardeners/dp/B0F66YX2CQ

Anonymous (n.d.-e). White Mulberry [*image*] Retrieved Jan 23, 2026, from https://pixabay.com/photos/alba-morus-mulberry-white-fruit-88456/

Anonymous (n.d.-f). Indian gooseberry [*image*] Retrieved Jan 23, 2026, from https://exoticflora.in/products/amla-banaras-or-pada-or-rati-fruit-plantstree?srsltid=AfmBOopAfUI9aX6nW-T9qqqvCtCLHDItVAk7h87feVnsiJuZxFZDD6PV

Anonymous (n.d.-g). Indian gooseberry [*image*] Retrieved Jan 23, 2026, from https://www.vecteezy.com/png/75525281-amla-fruit-indian-gooseberry-fresh-and-healthy-isolated

Anonymous (n.d.-h). Fig plant [*image*] Retrieved Jan 23, 2026, from https://www.amazon.in/ABEFARIYA-Khurtmani-Anjeer-Plant-gardening/dp/B0FGW612G7

Anonymous (n.d.-i). Fig fruit T. Section [*image*] Retrieved Jan 23, 2026, from https://organicseeds.top/how_to_grow_figs

Anonymous (n.d.-j). Dried fig fruit [photograph] Retrieved Jan 23, 2026, from https://mydukaan.io/kashmirstore/products/kashmiri-anjeer-figs?sku_id=15366983

Ansari, M.M. & Munshi, Y.I. 2023. Anjeer (*Ficus carica* Linn.): a famous fruit with distinct pharmacological properties: a review. *Int. J. Unani Integr. Med.* 7(1): 38–42.

Asmilia, N., Sutriana, A., Aliza, D. & Sudril, N. 2020. Anti-inflammatory activity of ethanol extract from malacca leaves (*Phyllanthus emblica*) in carrageenan induced male mice. In: *E3S Web Conf.* 151: 01066.

Aviram, M. & Dornfeld, L. 2001. Pomegranate juice consumption inhibits serum angiotensin converting enzyme activity and reduces systolic blood pressure. *Atherosclerosis* 158: 195–198.

Ayuso, M., Carpena, M., Taofiq, O., Albuquerque, T.G. & Simal-Gandara, J. 2022. Fig (*Ficus carica* L.) and its by-products: a decade evidence of their health-promoting benefits towards the development of novel food formulations. *Trends Food Sci. Technol.* 127: 1–3. https://doi.org/10.1016/j.tifs.2022.06.010

Baliga, M.S., Ramachandrayya, A., Shivashankara, K.R., Palatty, M.P., Princy, G., Thomas, R. & Suresh, B. 2019. Hepatoprotective effects of the Indian gooseberry (*Emblica officinalis* Gaertn.). In: *Dietary Interventions in Liver Disease*. Academic Press, Cambridge, pp. 193–201. https://doi.org/10.1016/B978-0-12-814466-4.00016-1

Banerjee, N., Kim, H., Talcott, S. & Mertens-Talcott, S. 2013. Pomegranate polyphenolics suppressed azoxymethane-induced colorectal aberrant crypt foci and inflammation: possible role of miR-126/VCAM-1 and miR-126/PI3K/AKT/mTOR. *Carcinogenesis* 34: 2814–2822.

Biswas, N.R., Gupta, S.K. & Das, G.K. 2001. Evaluation of Ophthacare eye drops: a herbal formulation in the management of various ophthalmic disorders. *Phytother. Res.* 15(7): 618–620.

Borska, S., Gebarowska, E., Wysocka, T., Drag-Zalesinska, M. & Zabel, M. 2003. Induction of apoptosis by EGCG in selected tumour cell lines in vitro. *Folia Histochem. Cytobiol.* 41: 229–232.

Brun, V. & Schumacher, T. 1987. *Traditional herbal medicine in Northern Thailand.* University of California Press, Berkeley.

Calzada, F., Yépez-Mulia, L. & Aguilar, A. 2006. In vitro susceptibility of *Entamoeba histolytica* and *Giardia lamblia* to plants used in Mexican traditional medicine for the treatment of gastrointestinal disorders. *J. Ethnopharmacol.* 108: 367–370.

Chatterjee, A., Chatterjee, S., Biswas, A., Bhattacharya, S., Chattopadhyay, S. & Bandyopadhyay, S.K. 2012. Gallic acid enriched fraction of *Phyllanthus emblica* potentiates indomethacin-induced gastric ulcer healing via eNOS-dependent pathway. *Evid. Based Complement. Alternat. Med. eCAM* 2012: 487380. https://doi.org/10.1155/2012/487380

Chen, H., Pu, J., Liu, D., Yu, W., Shao, Y., Yang, G., Xiang, Z. & He, N. 2016. Antiinflammatory and antinociceptive properties of flavonoids from the fruits of black mulberry (*Morus nigra* L.). *PLoS One.* https://doi.org/10.1371/journal.pone.0153080

Cheshomi, H., Bahrami, A.R., Rafatpanah, H. & Matin, M.M. 2022. The effects of ellagic acid and other pomegranate (*Punica granatum* L.) derivatives on human gastric cancer AGS cells. *Hum. Exp. Toxicol.* 41: 9603271211064534.

Di Lorenzo, C., Colombo, F., Biella, S., Orgiu, F., Frigerio, G., Regazzoni, L., De Sousa, L.P., Bavaresco, L., Bosso, A., Aldini, G. & Restani, P. 2019. Phenolic profile and antioxidant activity of different grape (*Vitis vinifera* L.) varieties. In: *BIO Web Conf.* 12: 04005.

Dinesh, M., Roopan, S.M., Selvaraj, C.I. & Arunachalam, P. 2017. *Phyllanthus emblica* seed extract mediated synthesis of PdNPs against antibacterial, haemolytic and cytotoxic studies. *J. Photochem. Photobiol. B Biol.* 167: 64–71.

Ercisli, S. & Orhan, E. 2007. Chemical composition of white (*Morus alba*), red (*Morus rubra*) and black (*Morus nigra*) mulberry fruits. *Food Chem.* https://doi.org/10.1016/j.foodchem.2006.10.054

Esmaillzadeh, A., Tahbaz, F. & Gaieni, I. 2006. Cholesterol-lowering effect of concentrated pomegranate juice consumption in type II diabetic patients with hyperlipidemia. *Int. J. Vitam. Nutr. Res.* 76: 147–151.

Fatima, N., Hafizur, R.M., Hameed, A., Ahmed, S., Nisar, M. & Kabir, N. 2017. Ellagic acid in *Emblica officinalis* exerts antidiabetic activity through action on β-cells of pancreas. *Eur. J. Nutr.* 56: 591–601.

Fazel, M.F., Abu, I.F., Mohamad, M.H., Mat Daud, N.A., Hasan, A.N. & Aboobakkar, Z. 2024. Physicochemistry, nutritional and therapeutic potential of *Ficus carica*: a promising nutraceutical. *Drug Des. Dev. Ther.* 31: 1947–1968.

Fouad, G.I. & Rizk, M.Z. 2019. Possible neuromodulating role of different grape (*Vitis vinifera* L.) derived polyphenols against Alzheimer's dementia: treatment and mechanisms. *Bull. Natl. Res. Cent.* 43(108): 1–13.

George, S. & Benny, P.J. 2010. A review on the medicinal significance of common fruits. *Int. J. Biomed. Res. Analys.*1(2): 60–64.

Grace, N.J., Evangeline, C.S., Swaminathan, A., Narendhirakannan, R.T. & Chatterjee, S. 2018. Cytotoxicity and apoptotic cell death induced by *Vitis vinifera* peel and seed extracts in A431 skin cancer cells. *Cytotechnology* 70: 537–554.

Gupta, M., Banerjee, D. & Mukherjee, A. 2013. Evaluation of analgesic, antipyretic and anti-inflammatory effects of methanol extract of traditional herbal medicine using rodents. *J. Pharmacogn. Phytother.* 5: 106–113.

Hajimahmoodi, M., Shams-Ardakani, M., Saniee, P., Siavoshi, F., Mehrabani, M. & Hosseinzadeh, H. 2011. In vitro antibacterial activity of some Iranian medicinal plant extracts against *Helicobacter pylori. Nat. Prod. Res.* 25: 1059–1066. https://doi.org/10.1080/15538362.2020.1784075

Insanu, M., Karimah, H., Pramastya, H. & Fidrianny, I. 2021. Phytochemical compounds and pharmacological activities of *Vitis vinifera* L.: an updated review. *Biointerface Res. Appl. Chem.* 11: 13829–13863.

Jagetia, G.C., Malagi, K.J., Baliga, M.S., Venkatesh, P. & Veruva, R.R. 2004. Triphala, an ayurvedic rasayana drug, protects mice against radiation-induced lethality by free-radical scavenging. *J. Altern. Complement. Med.* 10: 971–978.

Jan, B., Parveen, R., Zahiruddin, S., Khan, M.U., Mohapatra, S. & Ahmad, S. 2021. Nutritional constituents of mulberry and their potential applications in food and pharmaceuticals: a review. *Saudi J. Biol. Sci.* 28: 3909–3921. https://doi.org/10.1016/j.sjbs.2021.03.056

Jayaweera, D.M.A. 1980. *Medicinal plants used in Ceylon.* National Science Council of Sri Lanka, Colombo.

Justin, T.A., Dhivyabharathi, M., William, R.T.R., Manivasagam, T. & Essa, M.M. 2016. Tannoid principles of *Emblica officinalis* renovate cognitive deficits and attenuate amyloid pathologies against aluminum chloride induced rat model of Alzheimer's disease. *Nutr. Neurosci.* 19: 269–278.

Kanigur, S.G., Soydas, T. & Yenmis, G. 2019. NF-κB as the mediator of metformin's effect on ageing and ageing-related diseases. *Clin. Exp. Pharmacol. Physiol.*46(5): 413–422.
Khalil, E.A. 2004. Antidiabetic effect of an aqueous extract of pomegranate (*Punica granatum* L.) peels in normal and alloxan diabetic rats. *Egypt. Hosp. Med.* 16: 92–99.
Khan, K.H. 2009. Roles of *Emblica officinalis* in medicine: a review. *Bot. Res. Int.* 2: 218–228.
Kumar, K.S., Bhowmik, D., Dutta, A., Yadav, A.P., Paswan, S., Srivastava, S. & Deb, L. 2012. Recent trends in potential traditional Indian herbs *Emblica officinalis* and its medicinal importance. *J. Pharmacogn. Phytochem.* 1: 18–28.
Kumar, S. & Singh, B. 2020. Medicinal and traditional uses of shahtoot (*Morus indica* Linn.): a review. *Int. J. Unani Integr. Med.* 4(2): 40–47.
Kumar, V.R. & Chauhan, S. 2008. Mulberry: life enhancer. *J. Med. Plant Res.* 2(10): 271–278.
Kumari, I., Kaurav, H. & Chaudhary, H. 2021. *Punica granatum* L. (dadim): therapeutic importance of world's most ancient fruit plant. *J. Drug Deliv. Ther.* 11(3): 113–121. https://doi.org/10.22270/jddt.v11i3.4832.
Mahata, S., Pandey, A., Shukla, S., Tyagi, A., Husain, S.A., Das, B.C. & Bharti, A.C. 2013. Anticancer activity of *Phyllanthus emblica* Linn. (Indian gooseberry): inhibition of transcription factor AP-1 and HPV gene expression in cervical cancer cells. *Nutr. Cancer* 65: 88–97.
Mahdavi, A., Bagherniya, M., Fakheran, O., Reiner, Z., Xu, S. & Sahebkar, A. 2020. Medicinal plants and bioactive natural compounds as inhibitors of HMG-CoA reductase: a literature review. *Biofactors* 46(6): 906–926.
Majeed, M., Majeed, S., Mundkur, L., Nagabhushanam, K., Arumugam, S., Beede, K. & Ali, F. 2020. Standardized *Emblica officinalis* fruit extract inhibited the activities of α-amylase, α-glucosidase and dipeptidyl peptidase-4 and displayed antioxidant potential. *J. Sci. Food Agric.* 100: 509–516.
Masihuddin, Jafri, M.A., Siddiqui, A. & Chaudhary, S. 2019. Traditional uses, phytochemistry and pharmacological activities of amla with special reference to Unani medicine: an updated review. *Asian J. Pharm. Clin. Res.* 12(2): 70–74.
Mastrogiovanni, F., Mukhopadhya, A., Lacetera, N., Ryan, M.T., Romani, A. & Bernini, R. 2019. Anti-inflammatory effects of pomegranate peel extracts on in vitro human intestinal Caco-2 cells and ex vivo porcine colonic tissue explants. *Nutrients* 11: 548–563.
Mathew, M. & Subramanian, S. 2014. In vitro screening for anticholinesterase and antioxidant activity of methanolic extracts of ayurvedic medicinal plants used for cognitive disorders. *PLoS One* 9: e86804.
Mayyas, A., Abu-Sini, M., Amr, R., Akasheh, R.T., Zalloum, W. & Khdair, A. 2021. Novel in vitro and in vivo anti-*Helicobacter pylori* effects of pomegranate peel ethanol extract. *Vet. World* 14: 120–128.
Mehrotra, S., Jamwal, R., Shyam, R., Meena, D.K., Mishra, K., Patra, R., De, R., Mukhopadhyay, A., Srivastava, A.K. & Nandi, S.P. 2011. Anti-*Helicobacter pylori* and antioxidant properties of *Emblica officinalis* pulp extract: a potential source for therapeutic use against gastric ulcer. *J. Med. Plant Res.* 5: 2577–2583.
Mohammadi, M., Boghrati, Z., Emami, S.A. & Akaberi, M. 2023. Pomegranate: a review of the heavenly healer's past, present and future. *Iran. J. Basic Med. Sci.* 26: 1245–1264. https://doi.org/10.22038/IJBMS.2023.72816.15844
Muthuraman, A., Shailja, S.S. & Singla, S.K. 2011. The anti-inflammatory potential of phenolic compounds from *Emblica officinalis* L. in rat. *Inflammopharmacology* 19: 327–334.
Naaz, A., Begum, W. & Fathima, S. 2024. Anjeer: nature's gift of balanced nutrition for holistic well-being: a review. *Indo Am. J. Pharm. Res.* 14(7): 677–687.
Naaz, Q., Jilani, S., Jabin, A. & Aqsa, S. 2025. Unveiling the therapeutic potential of toot siyah (*Morus nigra* Linn.): a comprehensive review. *Int. J. Unani Integr. Med.* 9(3): 182–186.
Nain, P., Saini, V., Sharma, S. & Nain, J. 2012. Antidiabetic and antioxidant potential of *Emblica officinalis* Gaertn. leaves extract in streptozotocin-induced type 2 diabetes mellitus rats. *J. Ethnopharmacol.* 142: 65–71.
Nassiri, A.M. & Hosseinzadeh, H. 2016. Review of the pharmacological effects of *Vitis vinifera* (grape) and its bioactive constituents: an update. *Phytother. Res.* 30(9): 1392–1403.
Nath, A., Raghunatha, P. & Joshi, S.R. 2012. Diversity and biological activities of endophytic fungi of *Emblica officinalis*, an ethnomedicinal plant of India. *Mycobiology* 40: 8–13.
Nisha, P., Singhal, R.S. & Pandit, A.B. 2004. A study on degradation kinetics of ascorbic acid in amla (*Phyllanthus emblica* L.) during cooking. *Int. J. Food Sci. Nutr.* 55(5): 415–422.
O'Byrne, D.J., Devaraj, S., Grundy, S.M. & Jialal, I. 2002. Comparison of the antioxidant effects of concord grape juice flavonoids and alpha-tocopherol on markers of oxidative stress in healthy adults. *Am. J. Clin. Nutr.* 76: 1367–1374.
Ojha, S., Golechha, M., Kumari, S. & Arya, D.S. 2012. Protective effect of *Emblica officinalis* (amla) on isoproterenol-induced cardiotoxicity in rats. *Toxicol. Ind. Health* 28: 399–411.

Pal, I. 2020. A comprehensive review on *Ficus carica* L.: an unexplored medicinal plant. *Int. J. Adv. Res.*8(12): 876–881. https://doi.org/10.21474/IJAR01/12227

Pal, S., Ramamurthy, A., Rath, S. & Mahajon, B. 2017. Healing role of guduchi (*Tinospora cordifolia* (Willd.) Miers) and amalaki (*Emblica officinalis* Gaertn.) capsules in premature aging due to stress: a comparative open clinical trial. *Eur. J. Med. Plant* 2: 1–3.

Parihar, S. & Sharma, D. 2021. A brief overview on *Vitis vinifera. Sch. Acad. J. Pharm.* 10(12): 231–239. https://doi.org/10.36347/sajp.2021.v10i12.005

Parisio, C., Lucarini, E., Micheli, L., Toti, A., Khatib, M. & Mulinacci, N. 2020. Pomegranate mesocarp against colitis-induced visceral pain in rats: effects of a decoction and its fractions. *Int. J. Mol. Sci.* 21: 1–16.

Patel, P., Singh, H.S., Mishra, A. & Ansari, S.P. 2017. Can *Emblica officinalis* and *Tinospora cordifolia* supplementation possess immunomodulatory and adaptogenic properties in Murrah buffalo calves? *IndianJ. Anim. Res.* 51: 506–509.

Peng, C.H., Liu, L.K., Chuang, C.M., Chyau, C.C., Huang, C.N. & Wang, C.J. 2011. Mulberry water extracts possess an anti-obesity effect and ability to inhibit hepatic lipogenesis and promote lipolysis. *J. Agric. Food Chem.* https://doi.org/10.1021/jf1043508

Petersen, K.S. & Smith, C. 2016. Ageing-associated oxidative stress and inflammation are alleviated by products from grapes. *Oxid. Med. Cell. Longev.* 6236309. https://doi.org/10.1155/2016/6236309

Pramyothin, P., Samosorn, P., Poungshompoo, S. & Chaichantipyuth, C. 2006. The protective effects of *Phyllanthus emblica* Linn. extract on ethanol induced rat hepatic injury. *J. Ethnopharmacol.* 107: 361–364.

Qnais, E.Y., Elokda, A.S., Abu Ghalyun, Y.Y. & Abdulla, F.A. 2007. Antidiarrheal activity of the aqueous extract of *Punica granatum* (pomegranate) peels. *Pharm. Biol.* 45: 715–720.

Rahmani, A.H. & Aldebasi, Y.H. 2017. *Ficus carica* and its constituents role in management of diseases. *Asian J. Pharm. Clin. Res.* 10(6): 49–53. https://doi.org/10.22159/ajpcr.2017.v10i6.17832

Rajak, S., Banerjee, S.K., Sood, S., Dinda, A.K., Gupta, Y.K., Gupta, S.K. & Maulik, S.K. 2004. *Emblica officinalis* causes myocardial adaptation and protects against oxidative stress in ischemic-reperfusion injury in rats. *Phytother. Res.* 18: 54–60.

Raman, S.T., Ganeshan, A.K., Chen, C., Jin, C., Li, S.H. & Chen, H.J. 2016. In vitro and in vivo antioxidant activity of flavonoid extracted from mulberry fruit (*Morus alba* L.). *Pharmacogn. Mag.* 12(46): 128–133. https://doi.org/10.4103/0973-1296.177910

Ramappa, V.K., Srivastava, D., Singh, P., Kumar, U., Kumar, D., Gosipatala, S.B., Saha, S., Kumar, D. & Raj, R. 2020. Mulberries: a promising fruit for phytochemicals, nutraceuticals and biological activities. *Int. J. Fruit Sci.* 20(Suppl. 3): S1254–S1279.

Reddy, V.D., Padmavathi, P., Kavitha, G., Gopi, S. & Varadacharyulu, N. 2011. *Emblica officinalis* ameliorates alcohol-induced brain mitochondrial dysfunction in rats. *J. Med. Food* 14: 62–68.

Rizvi, Z.H., Naime, M., Akhtar, J. & Khan, M.A. 2019. Medicinal uses of *Vitis vinifera* in Unani system of medicine: an overview. *World J. Pharm. Med. Res.* 5(3): 219–222.

Rose, K., Wan, C., Thomas, A., Seeram, N.P. & Ma, H. 2018. Phenolic compounds isolated and identified from amla (*Phyllanthus emblica*) juice powder and their antioxidant and neuroprotective activities. *Nat. Prod. Commun.* 13: 1934578X1801301019.

Sabu, M.C. & Kuttan, R. 2002. Antidiabetic activity of medicinal plants and its relationship with their antioxidant property. *J. Ethnopharmacol.* 81(2): 155–160.

Saeed, M., Ali, D. & Anees, A. 2024. Grapes as a medicine in the light of prophetic and Unani medicine: a comprehensive review. *Int. J. Unani Integr. Med.* 8(3): 35–38. https://doi.org/10.33545/2616454X.2024.v8.i3a.295

Sai, R.M., Neetu, D., Yogesh, B., Anju, B., Dipti, P., Pauline, T., Sharma, S.K., Sarada, S.K., Ilavazhagan, G., Kumar, D. & Selvamurthy, W. 2002. Cytoprotective and immunomodulating properties of amla (*Emblica officinalis*) on lymphocytes: an in vitro study. *J. Ethnopharmacol.* 81: 5–10.

Salma, Y., Ansari, S. & Sadia, N. 2020. *Ficus carica* L.: a panacea of nutritional and medicinal benefits. *TANG* 10(1): 1–6. https://doi.org/10.5667/tang.2020.0001

Santos, I.B., de Bem, G.F., Cordeiro, V.S., da Costa, C.A., de Carvalho, L.C., da Rocha, A.P., da Costa, G.F., Ognibene, D.T., de Moura, R.S. & Resende, A.C. 2017. Supplementation with *Vitis vinifera* L. skin extract improves insulin resistance and prevents hepatic lipid accumulation and steatosis in high-fat diet–fed mice. *Nutr. Res.* 43: 69–81.

Sarikaphuti, A., Nararatwanchai, T., Hashiguchi, T., Ito, T., Thaworanunta, S., Kikuchi, K. & Tancharoen, S. 2013. Preventive effects of *Morus alba* L. anthocyanins on diabetes in Zucker diabetic fatty rats. *Exp. Ther. Med.* 6: 689–695. https://doi.org/10.3892/etm.2013.1203

Shalini, B. & Sharma, J. 2015. Beneficial effects of *Emblica officinalis* on fluoride-induced toxicity on brain biochemical indices and learning-memory in rats. *Toxicol. Int.* 22: 35–39.

Sharma, A. & Sharma, K.K. 2011. Chemoprotective role of triphala against 1,2-dimethylhydrazine dihydrochloride induced carcinogenic damage to mouse liver. *Indian J. Clin. Biochem.* 26: 290–295.

Sharma, P., Joshi, T., Joshi, T., Chandra, S. & Tamta, S. 2020. In silico screening of potential antidiabetic phytochemicals from *Phyllanthus emblica* against therapeutic targets of type 2 diabetes. *J. Ethnopharmacol.* 248: 112268.

Shaygannia, E., Bahmani, M., Zamanzad, B. & Kopaei, M.R. 2016. A review study on *Punica granatum* L. *J. Evid. Based Complement. Altern. Med.* 21(3): 221–227. https://doi.org/10.1177/2156587215598039

Singh, C.K., Liu, X. & Ahmad, N. 2015. Resveratrol, in its natural combination in whole grape, for health promotion and disease management. *Ann. N. Y. Acad. Sci.* 1348(1): 150–160. https://doi.org/10.1111/nyas.12798

Singh, E., Sharma, S., Pareek, A., Dwivedi, J., Yadav, S. & Sharma, S. 2011. Phytochemistry, traditional uses and cancer chemopreventive activity of amla (*Phyllanthus emblica*): the sustainer. *J. Appl. Pharm. Sci.* 2: 176–183.

Singh, J., Kaur, H., Kaur, R., Garg, R., Prasad, R., Assouguem, A., Kara, M. & Bahhou, J. 2023. A review on the nutritional value and health benefits of different parts of grape (*Vitis vinifera* L.). *Trop. J. Nat. Prod. Res.* 7(9): 3874–3880. https://doi.org/10.26538/tjnpr/v7i9.3

Singh, J.B. 2023. Composition of bioactive carotenoid, flavonoid and terpene compounds in selected fruits: a mini review. *Int. J. Plant Soil Sci.*

Singh, M., Sharma, N., Paras, H.S., Hans, N.S., Singh, N.P. & Sarin, A. 2019. Antioxidative potential of *Phyllanthus emblica* for oxidation stability of biodiesels. *Environ. Prog. Sustain. Energy* 38: 721–726.

Singh, M.P., Malla, S.B., Rajbhandari, S.B. & Manandhar, A. 1979. Medicinal plants of Nepal: retrospects and prospects. *Econ. Bot.* 33: 185–198.

Sirikanchanarod, A., Bumrungpert, A., Kaewruang, W., Senawong, T. & Pavadhgul, P. 2016. The effect of mulberry fruits consumption on lipid profiles in hypercholesterolemic subjects: a randomized controlled trial. *J. Pharm. Nutr. Sci.*

Souli, A., Sebai, H., Rtibi, K., Chehimi, L., Sakly, M. & Amri, M. 2015. Inhibitory effects of two varieties of Tunisian pomegranate (*Punica granatum* L.) extracts on gastrointestinal transit in rat. *J. Med. Food* 18: 1007–1012.

Srikumar, R., Parthasarathy, N.J., Shankar, E.M., Manikandan, S., Vijayakumar, R., Thangaraj, R., Vijayananth, K., Sheela, D.R. & Rao, U.A. 2007. Evaluation of the growth inhibitory activities of triphala against common bacterial isolates from HIV infected patients. *Phytother. Res.*21: 476–480.

Srikumar, R., Parthasarathy, N.J. & Sheela, D.R. 2005. Immunomodulatory activity of triphala on neutrophil functions. *Biol. Pharm. Bull.* 28: 1398–1403.

Srinivasan, P., Vijayakumar, S., Kothandaraman, S. & Palani, M. 2018. Antidiabetic activity of quercetin extracted from *Phyllanthus emblica* L. fruit: in silico and in vivo approaches. *J. Pharm. Anal.* 8: 109–118.

Sriwatcharakul, S. 2020. Evaluation of bioactivities of *Phyllanthus emblica* seed. *Energy Rep.* 6: 442–447.

Stuart, G.A. 1911. *Chinese materia medica: vegetable kingdom.* American Presbyterian Mission Press, Shanghai.

Sudha, T., Mousa, D.S., El-Far, A.H. & Mousa, S.A. 2021. Pomegranate (*Punica granatum*) fruit extract suppresses cancer progression and tumor angiogenesis of pancreatic and colon cancer in chick chorioallantoic membrane model. *Nutr. Cancer* 73: 1350–1356.

Sultana, S., Ahmad, S., Khan, N. & Jahangir, T. 2005. Effect of *Emblica officinalis* (Gaertn.) on CCl_4 induced hepatic toxicity and DNA synthesis in Wistar rats. *Indian J. Exp. Biol.* 43: 430–436.

Sumitra, M., Manikandan, P., Gayathri, V.S., Mahendran, P. & Suguna, L. 2009. *Emblica officinalis* exerts wound healing action through upregulation of collagen and extracellular signal-regulated kinases (ERK1/2). *Wound Repair Regen.* 17: 99–107.

Suthar, V. 2022. Therapeutic effects of pomegranate (*Punica granatum*). *J. Pharm. Res. Innov.* 2(2): 23–28.

Tabarrai, M., Mehriardestani, M., Hekmat, S., Nejatbakhsh, F. & Moradi, F. 2019. The potential role of grape (*Vitis vinifera* L.) in prevention of threatened abortion via immunomodulatory and anti-inflammatory abilities: a hypothesis. *Tradit. Med. Res.* 4(3): 131–139. https://doi.org/10.53388/TMR20190403113

Tan, S.J. & Ismail, I.S. 2020. Potency of selected berries, grapes and citrus fruit as neuroprotective agents. *Evid. Based Complement. Altern. Med.* 3582947. https://doi.org/10.1155/2020/3582947

Tasduq, S.A., Kaisar, P. & Gupta, D.K. 2005. Protective effect of a 50% hydroalcoholic fruit extract of *Emblica officinalis* against antituberculosis drugs induced liver toxicity. *Phytother. Res.* 19(3): 193–197.

Thilaga, S., Largia, M.J.V., Parameswari, A., Nair, R.R. & Ganesh, D. 2013. High frequency somatic embryogenesis from leaf tissue of *Emblica officinalis* Gaertn., a high valued tree for non-timber forest products. *Aust. J. Crop Sci.* 7: 1480.

Tongling, S. 2022. Role of bioactive compounds in hormonal bioregulation. https://doi.org/10.1007/978-981-19-2366-1_19

Tortora, K., Femia, A.P., Romagnoli, A., Sineo, I., Khatib, M. & Mulinacci, N. 2018. Pomegranate by-products in colorectal cancer chemoprevention: effects in APC-mutated PIRC rats and mechanistic studies in vitro and ex vivo. *Mol. Nutr. Food Res.* 62: 28948694.

Tsarong & Tsewang, J. 1994. *Tibetan medicinal plants*. Tibetan Medical Publications, Kalimpong, India.

Vasudevan, M. & Parle, M. 2007. Effect of anwalachurna (*Emblica officinalis* Gaertn.): an ayurvedic preparation on memory deficit rats. *Yakugaku Zasshi* 127: 1701–1707.

Voravuthikunchai, S., Lortheeranuwat, A., Jeeju, W., Sririrak, T., Phongpaichit, S. & Supawita, T. 2004. Effective medicinal plants against enterohaemorrhagic *Escherichia coli* O157:H7. *J. Ethnopharmacol.* 94: 49–54.

Wang, W., Li, X., Bao, X., Gao, L. & Tao, Y. 2018. Extraction of polysaccharides from black mulberry fruit and their effect on enhancing antioxidant activity. *Int. J. Biol. Macromol.* https://doi.org/10.1016/j.ijbiomac.2018.09.132

Wang, Y., Xiang, L., Wang, C., Tang, C. & He, X. 2013. Antidiabetic and antioxidant effects and phytochemicals of mulberry fruit (*Morus alba* L.) polyphenol enhanced extract. *PLoS One.* https://doi.org/10.1371/journal.pone.0071144

Wiart, C. 2013. Note on the relevance of *Emblica officinalis* Gaertn. for the treatment and prevention of cancer. *Eur. J. Cancer Prev.* 22: 198.

Wu, T., Tang, Q., Gao, Z., Yu, Z., Song, H., Zheng, X. & Chen, W. 2013. Blueberry and mulberry juice prevent obesity development in C57BL/6 mice. *PLoS One* 8(10): e77585. https://doi.org/10.1371/journal.pone.0077585

Xia, N., Daiber, A., Förstermann, U. & Li, H. 2017. Antioxidant effects of resveratrol in the cardiovascular system. *Br. J. Pharmacol.* 174(12): 1633–1646.

Yang, X., Yang, L. & Zheng, H. 2010. Hypolipidemic and antioxidant effects of mulberry (*Morus alba* L.) fruit in hyperlipidemic mice. *J. Agric. Food Chem.* 58: 12355–12359.

Yeola, K.A., Bhambar, R.S., Pande, S. & Karanjkar, D. 2023. A brief overview of *Vitis vinifera* L. (draksha) and its health benefits. *Int. J. Bot. Stud.* 8(4): 4–10. www.botanyjournals.com

Yilmaz, Y. & Toledo, R.T. 2004. Major flavonoids in grape seeds and skins: antioxidant capacity of catechin, epicatechin and gallic acid. *J. Agric. Food Chem.* 52: 255–260.

5 Phytochemicals in Medicinal Fruits and Berries

Bendjazia Rania, Dris Djemaa, and Bouabida Hayatte

5.1 INTRODUCTION

Phytochemicals constitute a diverse class of plant-derived chemical compounds that possess diverse biologically active properties (Pawase et al., 2024). In plants, these metabolites have vital ecological roles, including defense against herbivores and pathogens, in addition to participation in pollination and other adaptive processes (Dzigbor et al., 2025). Beyond their ecological functions, phytochemicals are widely recognized for their relevance to human health, as they demonstrate antioxidant, anti-inflammatory, antimicrobial, anticancer, and metabolic regulatory properties (Viola et al., 2024; Husen, 2023, 2024a, 2024b; Elsadek and Al-Numair, 2024). Major classes of phytochemicals, such as terpenoids, alkaloids, polyphenols, esters, and organic acids, are broadly dispersed across various plant organs, including leaves, bark, roots, seeds, and fruits (Bekir and Mars, 2024; Buthelezi et al., 2024; Sundari and Kavitha, 2024; Kemayou et al., 2025).

Fruits were among the earliest foods consumed by humans and have long been recognized for their remarkable health benefits. Rich in vitamins, enzymes, and essential nutrients, fruits contribute to blood purification, support digestive processes, and are readily absorbed by the body. Consequently, their regular consumption plays an important role in the prevention and management of diseases associated with unhealthy dietary habits (Roy et al., 2025). Further than their basic nutritional role, fruits serve as rich reservoirs of pharmacologically active secondary metabolites. These intricate phytochemical profiles underpin the medicinal potential of fruits in addition to determining their distinctive colors, flavors, and scents. Bioactive elements largely contribute to the therapeutic efficacy of fruits and their various anatomical parts by conferring prominent health-promoting effects. (Fabela-Morón, 2024; Noreen et al., 2025). Moreover, fruits contain diverse classes of phytochemicals, including phenolic acids, carotenoids, and flavonoids (Table 5.1), which exhibit a wide spectrum of biological activities and play crucial roles in protecting against chronic diseases by inhibiting cancer cell proliferation and modulating inflammatory and immune responses (Figure 5.1) (Danyo and Ivantsova, 2025).

Within this context, berries are widely considered to be particularly important due to their remarkable abundance of phytochemicals and substantial nutritional value. Consumed globally for their low caloric content and high nutrient density, they are particularly abundant in anthocyanins, flavonoids, and other polyphenols (Figure 5.2), which provide strong antioxidant activity and a broad range of health-promoting effects. Numerous studies have associated these constituents with the prevention and mitigation of chronic noncommunicable diseases, thereby positioning berries as model systems in contemporary phytochemical and nutraceutical research (Pedisić et al., 2025; Sagandyk et al., 2024). Typically, berries are small, highly perishable fruits with blue, purple, or red pigmentation, such as blueberries, cranberries, currants, raspberries, and blackberries. Their distinctive coloration and sensory attributes arise from a diverse array of secondary metabolites, including anthocyanins, flavonoids, carotenoids, and stilbenes, such as resveratrol. These compounds have garnered considerable interest because of their reported roles in preventing and managing diabetes,

DOI: 10.1201/9781003688259-5

TABLE 5.1
Major Phytochemicals in Fruits and Berries and Their Representative Compounds

Phytochemicals			Representative Compounds	Fruits and Berries Sources
Polyphenols	**Flavonoids**	Flavonols	Quercetin, Kaempferol Myricetin	Apples, grapes, peaches, and berries (goji berries).
		Flavones	Apigenin	Yellow and orange fruits and goji berries
		Flavanones	Naringenin Hesperetin	Oranges, grapefruit, lemon or lime, and tangelo.
		Flavan-3-ols	Catechin Epicatechin, Epigallocatechin	Apples, red grapes, berries (raspberries), peaches, mangoes, pears, plums, and nectarines
		Isoflavones	Daidzin Genistin Glycitin	Small amounts in citrus fruits, apples, pineapples, apricots, and cherries
		Anthocyanins	Cyanidin Peonidin Malvidin Delphinidin	Cranberries, blackcurrants, red and merlot grapes, raspberries, strawberries, blueberries, blackberries, chokeberries, bilberries, and pomegranates.
	Non-flavonoids	Phenolic acids	Ellagic acid Caffeic acid Ferulic acid Chlorogenic acid	Red fruits and berry fruits (strawberries, raspberries, blueberries, and goji berries)
		Stilbenes	Resveratrol Pterostilbene	Grapes, blueberries, raspberries, mulberries, bilberries, cranberries, redcurrants, and strawberries
		Lignans	Secoisolariciresinol	Small amounts in red fruits, skin, and seeds of blackberries, raspberries, and strawberries
		Tannins	Condensed tannins Hydrolyzable tannins	Berries (chokeberries, strawberries, raspberries, and blackberries), grape seeds, and pomegranate skins and seeds
Terpenoids	**Tetraterpene**	Carotenoids	β-Carotene Zeaxanthin	Yellow, orange, and red fruits, dragon fruit, chokeberry, and goji berries,

obesity, and hypertension, as well as their beneficial effects on cardiovascular health and the reduction of heart disease risk (Li et al., 2024). Overall, berry fruits possess a rich repertoire of bioactive molecules, underscoring their importance in health-oriented dietary patterns and emphasizing their value as functional foods and promising sources for the development of health-promoting nutraceutical formulations (Chacón-Fuentes et al., 2025). Among the most abundant classes of phytochemicals in fruits are polyphenols, which are typically categorized into flavonoids, stilbenes, phenolic acids, and lignans, with flavonoids being the predominant subgroup (Jain et al., 2025).

Considering the growing interest in natural remedies and functional foods around the world, a comprehensive understanding of the phytochemical composition and biological activities of medicinal fruits and berries has become increasingly crucial. This chapter offers an in-depth overview of the major phytochemical classes present in fruits, their biological functions, and their relevance to human health.

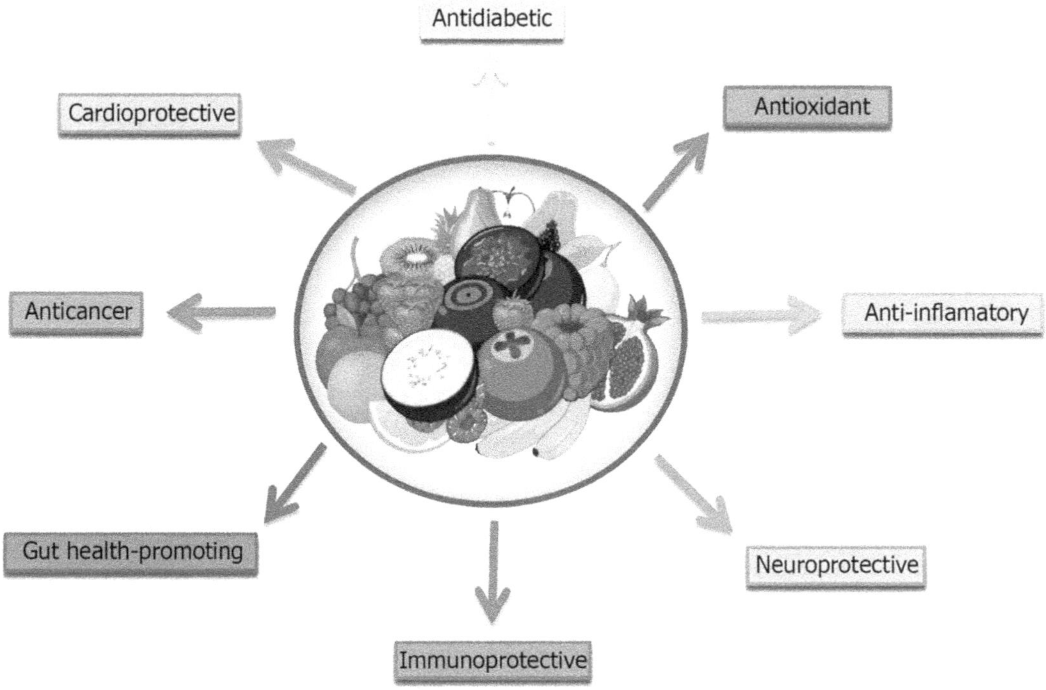

FIGURE 5.1 Health-promoting effects of medicinal fruits and berries.

5.2 MAIN PHYTOCHEMICALS IN MEDICINAL FRUITS AND BERRIES

5.2.1 Polyphenols

Polyphenols are the predominant and most widely distributed class of fruit phytochemicals (Roy et al., 2025). Typically, common fruits such as grapes, apples, cherries, pears, and various berries provide about 200–300 mg of polyphenols per 100 g fresh weight, significantly enhancing their capacity as antioxidants (Pawase et al., 2024). Within this group, berries are particularly important sources, notably black chokeberry (*Aronia melanocarpa*), black elderberry (*Sambucus nigra*), lowbush blueberries (*Vaccinium angustifolium*), highbush blueberries (*Vaccinium corymbosum*), and blackcurrant (*Ribes nigrum*) (Ruszkowska et al., 2024). Polyphenols are structurally characterized by one or more aromatic rings bearing multiple hydroxyl (–OH) groups, a feature that underpins much of their biological activity (Mandal & Domb, 2024).

Polyphenols are widely separated into flavonoids and non-flavonoids based on structural criteria. Flavonoids share a 15-carbon skeleton, consisting of two phenyl rings (A and B) linked by a heterocyclic, oxygen-containing ring (C), whereas non-flavonoids generally comprise simpler phenolic units or their polymers (Stankovic et al., 2025). Flavonoids include subclasses such as flavonols, flavones, flavanones, flavan-3-ols, isoflavones, and anthocyanidins, while non-flavonoids encompass phenolic acids, stilbenes, lignans, and tannins (El Oirdi, 2024).

The diversity of polyphenols contributes to their antimicrobial, antioxidant, antidiabetic, anticancer, and anti-aging activities, emphasizing their significance in disease prevention and human health (Bolat et al., 2024). Among polyphenol-rich fruits, pomegranate (*Punica granatum* L.) is a curative fruit that has health-promoting properties such as antibacterial, antifungal, antiobesity, antihypertensive, antidiabetic, anticancer, and anti-inflammatory (Noreen et al., 2025). Hawthorn (*Crataegus monogyna*) has traditionally been used for its antispasmodic, cardiotonic, hypotensive, and anti-atherosclerotic properties, while barberry (*Berberis vulgaris*) fruits are employed in the

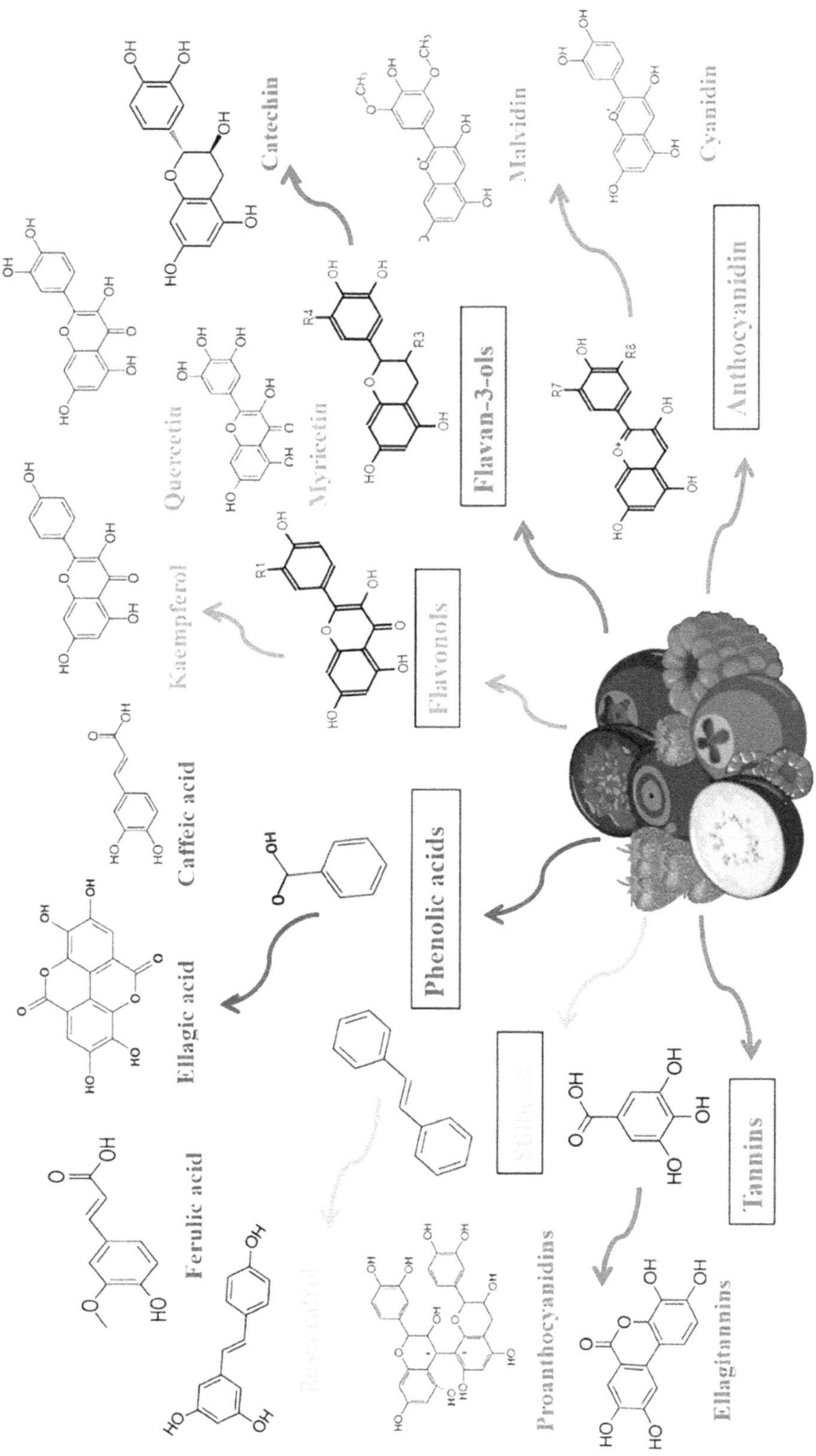

FIGURE 5.2 Key phenolic compounds in berries and their core structures.

management of high blood pressure. Black chokeberry is recognized for its antiviral activity, especially against various influenza virus strains. Elderberries are widely used in remedies for colds and influenza, and sloes, the fruits of blackthorn (*Prunus spinosa*), have long been applied as diuretic, astringent, and purgative agents (Szmagara et al., 2025). The existence of flavonoids and phenolic acids in blueberries strengthens their ability to reduce inflammation, prevent cancer, and protect the brain (Ashique et al., 2024).

The health benefits of these fruits and berries are primarily attributed to their high phenolic content, which collectively imparts significant medicinal and pharmacological properties (Dumitrașcu et al., 2024; Szmagara et al., 2025).

5.2.1.1 Flavonoids

Among polyphenols, flavonoids are the most abundant and widely investigated class (Eren et al., 2024). They are broadly distributed plant secondary metabolites, particularly abundant in fruits such as berries, cherries, plums, and apples (Abd El-Hack et al., 2026), where they contribute to color, flavor, and aroma. Structurally, flavonoids share a conserved C6–C3–C6 backbone composed of two phenolic rings linked by an oxygenated heterocycle, known as the flavan or phenylbenzopyran nucleus (El-Saadony et al., 2024; Jomova et al., 2025; Mandal & Domb, 2024). Structural modifications of this core bring about considerable molecular diversity and multiple subclasses, of which flavonols, flavones, flavanones, flavan-3-ols, isoflavones, and anthocyanidins are the most relevant in human nutrition (Stankovic et al., 2025). In plants, flavonoids are predominantly found as glycosides, with their glycosylation patterns affecting both stability and bioavailability (Rudrapal et al., 2024). Flavonoid-rich fruits and berries exhibit diverse biological activities in humans, including antioxidant, anti-inflammatory, cardioprotective, neuroprotective, anticancer, and potential anti-aging effects (Bolat et al., 2024; Pyo et al., 2024; Zheng et al., 2025).

5.2.1.1.1 Flavanones

Flavanones, a major flavonoid subclass, occur abundantly in citrus fruits as both aglycones and glycosides (Tang et al., 2025). Major representatives include naringenin (oranges, grapefruit, and lemon or lime peels), hesperetin (oranges, lime, lemon, and tangelo), and eriodictyol (lemons). These compounds display diverse biological activities, including antioxidant, anti-inflammatory, and anticancer effects, while naringenin additionally exhibits antimicrobial, antiviral, and anti-arthritic properties, as well as potential benefits in thyroid dysfunction (Ciupei et al., 2024; Abd El-Hack et al., 2026; Stankovic et al., 2025; Jomova et al., 2025; Otukwude et al., 2025).

5.2.1.1.2 Flavonols

Flavonols constitute the most prevalent subclass of flavonoids and are widely distributed in fruits, including apples, grapes, and berries (Otukwude et al., 2025; Abd El-Hack et al., 2026). Structurally, flavonols are phenolic compounds characterized by the presence of a 3-hydroxyl group, with quercetin, kaempferol, and myricetin being the principal representatives. These are the most abundant flavonols in goji berries (*Lycium barbarum* L.). These compounds exhibit a broad range of biological activities, including potent antioxidant, anticancer, antibacterial, and cardioprotective effects (Shi et al., 2024; Stankovic et al., 2025). In blueberries, quercetin 3-O-glucoside and quercetin 3-O-galactoside are reported as the most abundant flavonols (Ashique et al., 2024).

5.2.1.1.3 Anthocyanins

Anthocyanins are water-soluble flavonoid pigments responsible for the red, purple, and blue coloration of fruits, particularly berries, grapes, and pomegranates (Tang et al., 2025; Farheen et al., 2025). They are mainly localized in the outer cell layers of fruits such as cranberries, blackcurrants, red and Merlot grapes, raspberries, strawberries, blueberries, bilberries, and blackberries (Abd El-Hack et al., 2026; Otukwude et al., 2025). Structurally, anthocyanins are glycosylated derivatives of anthocyanidins, possessing a characteristic C6–C3–C6 backbone, and typically exist as flavylium ions under acidic conditions (El-Saadony et al., 2024). The six major anthocyanidins, cyanidin, delphinidin, pelargonidin, peonidin, petunidin, and malvidin, form glycosides whose structural variations impact pigment stability, color expression, and biological activity (Boyacıoğlu et al., 2025; Jomova et al., 2025).

The qualitative and quantitative profiles of anthocyanins differ widely among berry species. Chokeberries are particularly rich in cyanidin-3-galactoside, blueberries predominantly contain malvidin-derived anthocyanins such as malvidin-3-arabinoside, blackberries (*Rubus fruticosus*) mainly contain cyanidin-3-glucoside, strawberries (*Fragaria* × *ananassa*) mainly contain pelargonidin-3-glucoside, raspberries contain cyanidin-3-sophoroside, and cranberries contain a mixture of peonidin-3-glucoside and cyanidin-3-glucoside. Blackcurrants are rich in delphinidin-3-rutinoside, and bilberries (*Vaccinium myrtillus*) feature delphinidin and cyanidin derivatives, with malvidin-3-glucoside as the primary component (Chacón-Fuentes et al., 2025).

Beyond their role in pigmentation, anthocyanins exhibit multiple biological activities, including antioxidant, antimicrobial, cardioprotective, anti-inflammatory, anticancer, and neuroprotective effects (Saini et al., 2024; Stankovic et al., 2025).

5.2.1.1.4 Flavones

Flavones, light-sensitive polyphenols, commonly occur as glycosides in foods, with apigenin and luteolin as key representatives. Apigenin is abundant in some yellow and orange fruits and occurs at lower levels in goji berries (Shi et al., 2024; Stankovic et al., 2025). It has been demonstrated that flavones have a wide range of biological functions, including anti-allergic, anticancer, antibacterial, anti-inflammatory, and antispasmodic effects (Otukwude et al., 2025).

5.2.1.1.5 Flavan-3-ols

Flavan-3-ols, comprising catechin, epicatechin, epigallocatechin, catechin-3-gallate, and epigallocatechin-3-gallate, form an important flavonoid subclass widely found in fruits, particularly in apples, red grapes, berries (especially blueberries and raspberries), peaches, mangoes, pears, plums, and nectarines (Otukwude et al., 2025; Jomova et al., 2025; Ashique et al., 2024). Flavan-3-ols possess strong antioxidant properties and are associated with cardiovascular and metabolic health benefits, including improved blood pressure, lipid profiles, and glucose regulation (Stankovic et al., 2025).

5.2.1.1.6 Isoflavones

Isoflavones are plant-derived phenolic compounds belonging to the flavonoid family. Owing to their structural similarity to 17β-estradiol, they are classified as phytoestrogens. In nature, biologically active isoflavones are predominantly present in the aglycone form, including genistein, daidzein, biochanin A, and glycitein, as well as in their glycosylated derivatives, such as daidzin, genistin, and glycitin. These compounds occur in small quantities in various fruits, including apricots, cherries, apples, pineapples, and citrus fruits (Stankovic et al., 2025).

5.2.1.2 Non-Flavonoids

5.2.1.2.1 Stilbenes

Stilbenes, non-flavonoid polyphenols produced via the phenylpropanoid pathway, possess a C6–C2–C6 structure formed by two aromatic rings connected by an ethylene bridge (Bolat et al., 2024). They are naturally produced by plants and are commonly found in fruits, particularly grapes and berries (Stankovic et al., 2025). Resveratrol (3,4′,5-trihydroxystilbene) is the most extensively studied and is primarily present in the skins of grapes, blueberries, mulberries, bilberries, raspberries, cranberries (*Vaccinium oxycoccus*), redcurrants, and strawberries (Chacón-Fuentes et al., 2025; Li et al., 2024; Mandal & Domb, 2024; Ciupei et al., 2024). Pterostilbene, a naturally occurring dimethylated derivative of resveratrol, is located in blueberries and grapes (Eren et al., 2024). Other stilbenes include piceatannol, viniferin, pinosylvin, oxyresveratrol, polydatin, and rhapontigenin, all sharing structural resemblance with resveratrol (Rudrapal et al., 2024). Resveratrol demonstrates a broad spectrum of biological activities, including antioxidant, anti-inflammatory, antiproliferative, antimicrobial, cardioprotective, and neuroprotective effects (Roy et al., 2025).

5.2.1.2.2 *Lignans*

Lignans are a group of diphenolic compounds formed by the oxidative dimerization of two phenylpropanoid units (C6–C3) (Jain et al., 2025). In plants, these compounds are biosynthesized via the shikimate pathway and occur in both free and glycosylated forms, contributing to cell wall structure and enhancing resistance to environmental stresses (Rudrapal et al., 2024). The prime food sources of lignans are flaxseed, sesame seeds, whole grains, and a variety of vegetables and fruits, particularly berries (Burgberger et al., 2025). Lignans display a wide range of biological activities, including antioxidant, anticancer, antimicrobial, and phytoestrogenic effects, supporting their potential role in human health, such as cardiovascular protection and alleviation of menopausal symptoms (Bolat et al., 2024).

5.2.1.2.3 *Phenolic Acids*

Phenolic acids represent a major class of non-flavonoid polyphenols and are characterized by a phenolic aromatic ring linked to a carboxyl functional group. In plants, they are commonly present in conjugated forms, such as esters, amides, or glycosides. Structurally, phenolic acids are classified into two primary categories: hydroxybenzoic acids (C6–C1) and hydroxycinnamic acids (C6–C3), derived from benzoic and cinnamic acid skeletons, respectively (El-Saadony et al., 2024; Bolat et al., 2024). This family contains compounds such as p-hydroxybenzoic acid, salicylic acid, gallic acid, ellagic acid, caffeic acid, chlorogenic acid, and neochlorogenic acid, with ellagic acid representing the dominant phenolic acid in strawberries and raspberries (Chacón-Fuentes et al., 2025). In general, hydroxycinnamic acids, including caffeic, p-coumaric, ferulic, and sinapic acids, occur in higher concentrations than hydroxybenzoic acids (Mandal & Domb, 2024).

Fruits, specifically red-colored fruits, illustrate significant nutritional sources of phenolic acids, with caffeic and ferulic acids being among the most abundant (Stankovic et al., 2025; Rudrapal et al., 2024). Chlorogenic acid, a derivative of hydroxycinnamic acid, is stated as one of the predominant phenolic acids in goji berries (Shi et al., 2024). Blueberries are characterized by a diverse phenolic acid profile, including vanillic, ferulic, caffeic, chlorogenic, neochlorogenic, sinapic, ellagic, p-coumaric, and protocatechuic acids, whereas blackberries are distinguished by their high content of gallic acid and other hydroxybenzoic derivatives (Ashique et al., 2024).

Owing to their chemical structure, particularly the number and arrangement of hydroxyl groups, phenolic acids exhibit a wide range of biological activities, including antioxidant, antimicrobial, anticancer, antidiabetic, cardioprotective, and neuroprotective effects, highlighting their significance among fruit-derived phytochemicals (Roy et al., 2025).

5.2.1.2.4 *Tannins*

Tannins are water-soluble polyphenolic compounds widely distributed in fruits, leaves, bark, seeds, and roots (Stankovic et al., 2025). They are classified into hydrolyzable tannins, composed of phenolic acid esters and subdivided into gallotannins and ellagitannins, and non-hydrolyzable (condensed) tannins, also known as proanthocyanidins (Arzani et al., 2025). Berries, grape seeds, and pomegranate peels and seeds are particularly rich sources of tannins, which contribute to their astringent and bitter sensory characteristics (Rudrapal et al., 2024). Among berries, chokeberry contains particularly high levels of condensed tannins, whereas strawberries, raspberries (*Rubus idaeus*), and blackberries are notable for their high content of hydrolyzable tannins (Chacón-Fuentes et al., 2025). Ellagitannins are particularly abundant in blueberries, red raspberries, sea buckthorn, and strawberries, and are regarded as key bioactive constituents of these fruits. Additionally, blueberries, along with other berries such as cranberries, contain proanthocyanidins characterized by A-type interflavan bonds (Ashique et al., 2024). A broad spectrum of biological activities is exhibited by tannins and related proanthocyanidins in these fruits, encompassing antimicrobial, anti-inflammatory, antimutagenic, antioxidant, anticancer, and neuroprotective effects (Roy et al., 2025).

5.2.2 Carotenoids

Carotenoids constitute a broad and diverse family of naturally occurring pigments that are widely distributed in fruits and plant tissues, where they are responsible for the yellow, orange, and red colors characteristic of many fruits and vegetables (Danyo and Ivantsova, 2025). Several carotenoids, such as lycopene, β-carotene, ζ-carotene, β-cryptoxanthin, lutein, 5,6-epoxylutein, trans- and cis-violaxanthin, and neoxanthin, are generally present in low concentrations in berries, with chokeberries noted as among the richest sources (Chacón-Fuentes et al., 2025). The existence of β-carotene has also been documented in dragon fruit (*Hylocereus* spp.) (Katikala et al., 2025). In goji berries, zeaxanthin represents the predominant carotenoid, followed by β-carotene, whereas β-cryptoxanthin, neoxanthin, and lycophyll occur in smaller amounts (Shi et al., 2024). In addition, persimmons, sweet oranges (*Citrus sinensis* L.), and tangerines are recognized as important dietary sources of β-cryptoxanthin (Lalitha, 2024).

Beyond contributing to pigmentation, carotenoids play a crucial role in human health, primarily due to their antioxidant properties, their capacity to inhibit tumor development, and their ability to induce apoptosis in cancer cells (Li et al., 2024). Among these carotenoids, β-carotene, lutein, and zeaxanthin have been the most extensively studied and are especially recognized for their protective effects on eye health and their contribution to antioxidant defense mechanisms (Hossain et al., 2025).

5.3 CONCLUSION AND FUTURE REMARKS

In conclusion, there is a growing need to shift our perspective toward fruits and berries, considering them not only as essential components of a healthy diet but also as potent sources of natural medicinal compounds. The high phytochemical content of these fruits, particularly polyphenols such as flavonoids, phenolic acids, tannins, stilbenes, and lignans, as well as carotenoids, confers significant health-promoting effects, including antioxidant, anti-inflammatory, and disease-preventive activities. Consequently, increasing attention should be devoted to the exploitation of fruits and berries in both preventive nutrition and evidence-based phytotherapy. Although promising, current evidence is largely limited to in vitro and animal studies, necessitating human trials, standardized phytochemical analyses, and further investigation of bioavailability and synergistic interactions before clinical recommendations can be made.

REFERENCES

Abd El-Hack, M.E., A.K. Aldhalmi, M. Kamal, A.F. Khafaga, M. Moustafa and M. Al-Shehri et al. 2026. Flavonoids as a phytobiotic agent: sources, classifications, biological benefits, and useful impacts on broilers and layers. *Phytochem. Rev.* 25: 331–357.

Arzani, V., M. Soleimani, T. Fritsch, U.M. Jacob, V. Calabrese and A. Arzani. 2025. Plant polyphenols, terpenes, and terpenoids in oral health. *Open Med.* 20: 20251183.

Ashique, S., T. Mukherjee, S. Mohanty, A. Garg, N. Mishra and M. Kaushik et al. 2024. Blueberries in focus: exploring the phytochemical potentials and therapeutic applications. *J. Agric. Food Res.* 18: 101300.

Bekir, J. and M. Mars. 2024. Phytochemical composition and antioxidant, anti-alzheimer and anti-inflammatory activities of pomegranate fruit as function of variety and ripening. *J. Herb. Med.* 48: 100960.

Bolat, E., S. Sarıtaş, H. Duman, F. Eker, E. Akdaşçi and S. Karav et al. 2024. Polyphenols: secondary metabolites with a biological impression. *Nutrients.* 16: 2550.

Boyacıoğlu, O., B. Durmaz Kılınç and S.Ö. Boyacıoğlu. 2025. Cytotoxic and pro-apoptotic effects of malvidin anthocyanin in berry fruits on colon cancer. *Celal Bayar Univ. Sağlık Bilimleri Enstitüsü Derg.* 12(3): 434–440.

Burgberger, M., J. Mierziak, B. Augustyniak, W. Wojtasik and A. Kulma. 2025. The power of lignans: plant compounds with multifaceted health-promoting effects. *Metabolites.* 15(9): 589.

Buthelezi, L.G., S. Mavengahama, J. Sibiya, C.N. Mchunu and N.R. Ntuli. 2024. Phytochemical composition of Lagenaria siceraria fruits from KwaZulu-Natal and Limpopo, South Africa. *Food Chem. X.* 22: 101338.

Chacón-Fuentes, M., M. Opazo-Navarrete, G. Armijo-Godoy, M. Barra-Jiménez, L. Mojica and C. Burgos-Díaz. 2025. A current review on the impact of environmental and biological stresses on the production of bioactive compounds in berries. *Food Sci. Process.* 1: 2.

Ciupei, D., A. Colişar, L. Leopold, A. Stănilă and Z.M. Diaconeasa. 2024. Polyphenols: from classification to therapeutic potential and bioavailability. *Foods.* 13: 4131.

Danyo, E.K. and M.N. Ivantsova. 2025. Fruit phytochemicals: antioxidant activity and health-promoting properties. *Foods Raw Mater.* 13: 58–72.

Dumitraşcu, L., I. Banu, L. Patraşcu, I. Vasilean and I. Aprodu. 2024. The influence of processing on the bioactive compounds of small berries. *Appl. Sci.* 14: 8713.

Dzigbor, A., D. Neglo, C.S. Dzah and R. Sraha. 2025. Total phenolic content, phytochemical screening, antioxidant and antimicrobial activities of Borassus flabellifer and Borassus aethiopum fruits. *Food Chem. Adv.* 7: 100937.

El Oirdi, M. 2024. Harnessing the power of polyphenols: a new frontier in disease prevention and therapy. *Pharmaceuticals.* 17: 692.

El-Saadony, M.T., T. Yang, A.M. Saad, S.S. Alkafaas, S.S. Elkafas and G.S. Eldeeb et al. 2024. Polyphenols: chemistry, bioavailability, bioactivity, nutritional aspects and human health benefits: a review. *Int. J. Biol. Macromol.* 277: 134223.

Elsadek, M.F. and K.S. Al-Numair. 2024. Profiling of phytochemical constituents of Terminalia chebula fruit extract by different solvent effects and synchronized analysis of FTIR and GCMS. *J. King Saud Univ. Sci.* 36: 103414.

Eren, E., J. Das and T.O. Tollefsbol. 2024. Polyphenols as immunomodulators and epigenetic modulators. *Nutrients.* 16: 4143.

Fabela-Morón, M.F. 2024. Bioactive compounds, sensory attributes, and flavor perceptions involved in taste-active molecules in fruits and vegetables. *Front. Nutr.* 11: 1427857.

Farheen, H., A. Kumar, D. Kumar, S. Parmar, D. Pathak and P. Sharma et al. 2025. Anthocyanins: natural pigments with multifaceted health benefits for humans. *J. Sci. Innov. Nat. Earth.* 5: 1–3.

Hossain, M.S., M.A. Wazed, S. Asha, M.R. Amin and I.M. Shimul. 2025. Dietary phytochemicals in health and disease: mechanisms, clinical evidence, and applications a comprehensive review. *Food Sci. Nutr.* 13(3): e70101.

Husen, A. 2023. *Antidiabetic medicinal plants and herbal treatments.* CRC Press, Boca Raton, FL. https://doi.org/10.1201/b23347

Husen, A. 2024a. *Antimalarial medicinal plants.* Taylor & Francis Group. https://doi.org/10.1201/9781003378396

Husen, A. 2024b. *Medicinal spices and condiment crops.* Taylor & Francis Group. https://doi.org/10.1201/9781003387046

Jain, P.K., A.K. Parashar and V. Shrivastava. 2025. A review on exploring the health benefits and antioxidant properties of bioactive polyphenols. *Discover Food.* 5(1): 1–18.

Jomova, K., S.Y. Alomar, R. Valko, J. Liska, E. Nepovimova and K. Kuca et al. 2025. Flavonoids and their role in oxidative stress, inflammation, and human diseases. *Chem.-Biol. Interact.* 413: 111489.

Katikala, S., S. Nivedhitha, D. Rachel, C. Ch, W.A.A. Alawaad and B. Jayapriya et al. 2025. Dragon fruit (Hylocereus spp.): a comprehensive review. *Int. J. Pharm. Nat. Med.* 13: 70–76.

Kemayou, G.P.M., S.Q. Liu, S.C.N. Wouamba, B. Li, H.H. Yu and X. Tian et al. 2025. Schisandraceae family fruits: exploring chemophenetic significance, UPLC/MSn database, phytochemistry, and pharmacological potential. *Fitoterapia.*: 106415.

Lalitha, A. 2024. Phytochemicals and health benefits. *Ann. Geriatr. Educ. Med. Sci.* 11(1): 29–31.

Li, X., L. Zhao, B. Zhang and S. Wang. 2024. Berries and their active compounds in prevention of age-related macular degeneration. *Antioxidants.* 13: 1558.

Mandal, M.K. and A.J. Domb. 2024. Antimicrobial activities of natural bioactive polyphenols. *Pharmaceutics.* 16: 718.

Noreen, S., B. Hashmi, P.M. Aja and A.V. Atoki. 2025. Phytochemicals and pharmacology of pomegranate (Punica granatum L.). *Front. Nutr.* 12: 1528897.

Otukwude, F.O., A.A. Ahuchaogu and P.A. Otukwude. 2025. Chemistry of flavonoids and their therapeutic effects: a review. *Adv. J. Chem. Res.* 2(1): 1–10.
Pawase, P.A., C. Goswami, R. Shams, V.K. Pandey, A. Tripathi and S. Rustagi. 2024. A conceptual review on classification, extraction, bioactive potential and role of phytochemicals in human health. *Future Foods.* 9: 100313.
Pedisić, S., Z. Zorić, M. Repajić, B. Levaj, A. Dobrinčić and S. Balbino et al. 2025. Valorization of berry fruit by-products. *Foods.* 14: 1354.
Pyo, Y., K.H. Kwon and Y.J. Jung. 2024. Anticancer potential of flavonoids. *Foods.* 13: 2253.
Roy, S., S.C. Roy, M. Zehravi, S.H. Sweilam, R. Das and M. Palanisamy et al. 2025. Exploring the neuroprotective benefits of phytochemicals. *Anim. Models Exp. Med.* 8: 239–265.
Rudrapal, M., G. Rakshit, R.P. Singh, S. Garse, J. Khan and S. Chakraborty. 2024. Dietary polyphenols. *Antioxidants.* 13: 429.
Ruszkowska, J., W. Drygas and M. Kwaśniewska. 2024. The influence of berry-derived polyphenol supplementation. *Antioxidants.* 13: 1561.
Sagandyk, A.T., Â. Liberal, T.F. da Silveira, M.J. Alves, I.C. Ferreira and G.N. Zhakupova et al. 2024. Nutritional, phytochemical, and bioactive prospects of black chokeberry and saskatoon berry. *Appl. Food Res.* 4: 100564.
Saini, R.K., M.I. Khan, X. Shang, V. Kumar, V. Kumari and A. Kesarwani et al. 2024. Dietary sources, stabilization, health benefits, and industrial application of anthocyanins. *Foods.* 13: 1227.
Shi, X., X. Wang, Y. Zheng and L. Fu. 2024. Advances in the study of bioactive compounds of goji berry. *Appl. Sci.* 15: 262.
Stankovic, S., S. Mutavdzin Krneta, D. Djuric, V. Milosevic and D. Milenkovic. 2025. Plant polyphenols as heart's best friends. *Int. J. Mol. Sci.* 26: 915.
Sundari, T. and R. Kavitha. 2024. In vitro assessing of Cucumis pubescens Willd. *J. Indian Chem. Soc.* 101: 101176.
Szmagara, A., A. Krzyszczak-Turczyn and I. Sadok. 2025. Fruits of Polish medicinal plants as sources of antioxidants. *Appl. Sci.* 15: 6094.
Tang, S., B. Wang, X. Liu, W. Xi, Y. Yue and X. Tan et al. 2025. Structural insights and biological activities of flavonoids. *Food Front.* 6: 218–247.
Viola, E., G. Mannino, G. Serio, L. La Rosa, G. Garofalo and R. Schicchi et al. 2024. Phytochemical profiling of Sicilian sumac fruits. *Food Biosci.* 61: 104704.
Zheng, X., X. Zhang and F. Zeng. 2025. Biological functions and health benefits of flavonoids in fruits and vegetables. *Foods.* 14: 155.

6 Anti-Inflammatory Properties of Medicinal Fruits and Berries

Sharika Rajasekharan, Bynedi Seshadhri Chinna Mounish, Tewin Tencomnao, and Siriporn Chuchawankul

6.1 INTRODUCTION

Inflammation is triggered by the immune system in response to pathogenic invasion or exogenous material in the body. Pathogen recognition receptors (PRRs) recognize pathogen-associated molecular patterns (PAMPs) and damage-associated molecular recognition patterns (DAMPs), which are released from damaged cells, leukocyte degranulation substances, and heat shock proteins. PRRs, like TLR4 (toll-like receptors), recognize PAMPs like lipopolysaccharides (LPS) to initiate an immune response. Overstimulation of PRRs can cause systemic inflammatory response syndrome. Studies have shown that chronic inflammation is involved in cancer, ischemic heart diseases, and cerebrovascular disease, adding to the previous understanding of a weakened immune system due to aging (Suzuki 2019). Later stages of type 1 and type 2 diabetes are marked by diabetic retinopathy associated with inflammatory cytokine production (Dharmarajan et al. 2023). Most chronic non-communicable diseases are related to inflammation and can be reduced by a proper diet.

Nutrition and diet play a major role in the well-being of the body. Daily intake of a healthy diet consisting of sufficient carbohydrates, proteins, fats, dietary fibers, and other micronutrients, such as vitamins and minerals, has been associated with reduced risk of systemic chronic inflammation. Fruits (including berries) are known to be an excellent source of nutrients like vitamin C, carotene, potassium, dietary fibers, and other bio-actives like phenolic acids, flavonoids, stilbenes, coumarin, tannins, alkaloids, lectins, and isothiocyanates. Polyphenols and carotenoids are widely studied and are known to reduce the risk of cardiovascular diseases and to modulate glucose metabolism, immune system, cholesterol, inflammation, and the scavenging of oxidant species (Wallace et al. 2020; Majdan and Bobrowska-Korczak 2022). Several studies have highlighted the significance of anti-inflammatory diets in the regulation of systemic inflammatory responses, primarily through their modulatory effects on metabolic processes. Sustained inflammation can result in weakening of barrier function and dysregulation of the immune system, resulting in increased production of cytokines, chemokines, and C-reactive protein, which can be successfully reversed by consuming a diet balanced with fruits, legumes, vegetables, nuts, and other produce rich in n-3 fatty acids, antioxidants, and polyphenols (Yu et al. 2024).

6.2 KEY MOLECULAR PATHWAYS IN INFLAMMATION AND ITS REGULATION

Acute inflammation occurs mainly in response to impending injury or infection and is mediated by the local immune system, with symptoms such as pain, fever, swelling, redness, or loss of tissue function, after which the tissue returns to normal condition. But continuous or uncontrollable stimuli can lead to chronic inflammation, leading to chronic inflammatory diseases like diabetes, arthritis, allergy, asthma, cancer, bowel diseases, cardiovascular diseases, and neurological diseases

DOI: 10.1201/9781003688259-6

(Chen et al., 2017). There are several mechanisms involving the initiation and modulation of inflammation, which could be used to target for bioactives for protective anti-inflammatory activity.

The markers of inflammation are mainly inflammatory cytokines (IL-1, IL-6), IFNs, TNFs, TGFs, chemokines, and colony-stimulating factors primarily secreted by leukocytes. Other markers are mainly inflammatory proteins, and antioxidant defense system, like C-reactive protein (CRP), haptoglobin, serum amyloid A, fibrinogen, and alpha 1-acid glycoprotein, superoxide dismutase (SOD), glutathione peroxidase (GPx), NADPH oxidase (NOX), inducible nitric oxide synthase (iNOS), and cyclooxygenase (COX)-2 (Chen et al. 2017).

6.2.1 NF-κB Pathway

Nuclear factor kappa B (NF-κB) is activated through two different pathways: the canonical pathway via external stimulus and the non-canonical pathway activated by TNF superfamily receptors. The pathway plays a crucial role in immune and inflammatory response, cell differentiation and proliferation, apoptosis, and survival. IκB kinase (IKK), which activates NF-κB, has two homologous catalytic subunits, IKKα and IKKβ, and an NF-κB essential modulator (NEMO). Both IKKβ and NEMO are involved in the canonical pathway and are responsible for pro-inflammatory cytokine production, whereas IKKα is involved in non-canonical pathway activation. Activation of IKKβ results in the release of NF-κB dimers from cytoplasmic inhibition and their translocation into the nucleus, resulting in the expression of target genes in both the innate and adaptive immune systems. A well-regulated inflammatory response is advantageous to cells; however, chronic inflammation due to infection may also lead to the development of cancer and malignancy. For example, infection by *Helicobacter pylori* is often linked to lymphoma and gastric cancer. Other factors, such as immune dysregulation and diabetes, can also lead to chronic inflammation (Yu et al. 2020).

6.2.2 MAPK Pathway

External stimuli, such as stress or cytokines, are the main triggers for MAPK pathway–mediated inflammatory cytokine production. The mammalian MAPK family includes ERK1/2, JNK, and p38, and responds to various stimuli like osmotic stress, heat shock, mitogens, and inflammatory cytokines IL-1, IL-6, and TNF-α. ERK1/2 is activated by mitogen signals, whereas JNK and p38 pathways are stimulated by inflammatory stimuli or stress. Activation of ERK1/2, JN, and p38 is mediated by MKK1/2, MKK 4/7, and MKK3/6, respectively. ERK1/2 and JNK can also activate the p38 transcription factor, which further initiates the inflammatory response. p38 MAPK is associated with cell growth and differentiation, inflammation, and cell death. The external stimuli of p38 MAPK are mainly cytokines, pathogens, and LPS, viral particles, and toxins. Activation of p38 MAPK is important for inflammatory cytokine production and signaling for protective action through the initiation and recruitment of leukocytes. p38 MAPK has been implicated in various diseases, such as rheumatoid arthritis, systemic inflammation, chronic bowel disease, inflammation of the brain, and stroke. Targeting p38 MAPK can regulate both translation and post-transcriptional expression of pro-inflammatory cytokines (Kaminska 2005; Chen et al. 2017).

6.2.3 JAK-STAT Pathway

The Janus kinase–signal transducer and activator of transcription (JAK-STAT) pathway is a family of non-receptor tyrosine kinases (nTKs) comprising four members, namely JAK1, JAK2, JAK3, and TYK2. It facilitates signal transmission from the receptor to the nucleus and is involved in both the innate and adaptive immune systems. Dysregulation of the JAK-STAT pathway leads to the development of immunodeficiency, progression of infectious diseases, autoimmune disorders, and oncogenesis. Extracellular stimuli like IL-6 can bind to the JAK-STAT pathway, serving the translocation of STAT into the nucleus for the regulation of target gene expression. It is interesting to note that this pathway plays a regulatory role in brain development and functioning of neurons, astrocytes, and oligodendrocytes; however, studies have also revealed JAK-STAT pathway is also responsible for neuroinflammatory pathological conditions of the nervous system, including Alzheimer's disease, ischemic disease, epilepsy, and brain cancer (Chen et al. 2017; Sarapultsev et al. 2023).

6.3 FRUITS IN ACTION AGAINST INFLAMMATION

6.3.1 Apple (*Malus* spp.)

Apples (Figure 6.1) are a rich source of structurally diverse bioactive compounds whose health-promoting effects are primarily driven by strong antioxidant capacity, complemented by anti-inflammatory, immunomodulatory, gut-protective, metabolic, and anticancer activities. Among flavonoids, quercetin is a potent antioxidant that scavenges reactive oxygen species (ROS) and activates the Nrf2 signaling pathway, upregulating endogenous enzymes such as SOD, CAT, and GPx and thereby protecting cells from oxidative stress; it also exerts anti-inflammatory effects via inhibition of NF-κB and MAPK and displays anticancer activity by suppressing PI3K/Akt signaling and inducing mitochondria-mediated apoptosis and growth arrest (Gorjanović et al. 2020; Wang et al., 2024a). Catechin, cyanidin-3-O-galactoside, and procyanidins further strengthen antioxidant defenses by directly neutralizing ROS and stimulating Nrf2–ARE signaling, while simultaneously modulating gut microbiota composition, increasing short-chain fatty acid (SCFA) production, reinforcing epithelial tight junctions, and suppressing intestinal inflammation through NF-κB inhibition and mitochondrial-protective pathways such as PGC-1α activation and ERK1/2 suppression (Kalinowska et al. 2023; Di Pede et al. 2023; Liang et al. 2021a; Ma et al. 2025; Masuda et al. 2018; Liang et al. 2021b; Masumoto et al. 2016; Andre et al. 2012).

Phenolic acids in apples, including chlorogenic, caffeic, gallic, ferulic, p-coumaric, and protocatechuic acids, provide complementary antioxidant mechanisms by donating hydrogen or electrons, chelating pro-oxidant metals, and enhancing cellular antioxidant systems through Nrf2–ARE activation (Commisso et al. 2021; Shaw et al. 2021; Liaudanskas et al. 2014; Rana et al. 2025).

FIGURE 6.1 Photograph of Apple (*Malus* spp).

Image kindly provided by Miss Pandar Nartsawee.

These compounds also modulate immune responses and reduce inflammatory cytokines via inhibition of NF-κB and MAPK signaling, support gut barrier integrity by attenuating TLR-mediated inflammation, and, in the case of gallic acid, trigger ROS-dependent mitochondrial apoptosis and PI3K/Akt suppression in cancer cells (Commisso et al. 2021; Shaw et al. 2021; Liaudanskas et al. 2014; Qanash et al. 2022; Ongay et al. 2023). Ferulic acid additionally promotes immune tolerance and intestinal homeostasis by strengthening the epithelial barrier and dampening TLR4/NF-κB–driven Th2 responses, whereas p-coumaric acid primarily acts through direct ROS scavenging, and protocatechuic acid couples a dihydroxy structure with Nrf2 activation to amplify intracellular antioxidant defense (Liaudanskas et al. 2014; Rana et al. 2025).

Dihydrochalcones, notably phloretin and phloridzin, have partly unique functions. Phloretin provides moderate antioxidant effects by lowering oxidative stress and allowing endogenous enzymes (SOD, CAT, and GSH-Px) to work more efficiently, while exerting strong anti-inflammatory and gut-protective effects through NF-κB and MAPK suppression, downregulation of COX-2, PGE_2, and IL-8, preservation of tight-junction integrity, and reduction of mucosal cytokine and chemokine production, thereby improving intestinal homeostasis (Zielinska et al. 2019). Phloridzin, although a weaker direct antioxidant because its glucose moiety reduces redox efficiency, still scavenges radicals and mildly inhibits NF-κB. After enzymatic or microbial conversion to phloretin in the gut, it helps mitigate oxidative and glycation stress in intestinal tissues and contributes indirectly to gut health (Zielinska et al. 2019).

Apple peel is also enriched in triterpenes such as ursolic and oleanolic acids, which participate in antioxidant and anti-inflammatory defense by scavenging ROS, activating Nrf2-mediated antioxidant pathways, and inhibiting NF-κB and MAPK signaling, leading to reduced expression of pro-inflammatory mediators and improved metabolic and tissue homeostasis (Odun-Ayo et al. 2022; Woźniak et al., 2023). Beyond polyphenols and triterpenes, apple pectin—an abundant fermentable dietary fiber—plays a crucial supportive role in gut health by serving as a substrate for beneficial microbiota, promoting SCFA production, beneficially reshaping microbial composition, enhancing gut barrier function, and indirectly lowering oxidative stress and systemic inflammation through microbiota-driven metabolic pathways (Yao et al. 2024). Collectively, these data show that the antioxidant activity of apple bioactive compounds via direct ROS scavenging, metal chelation, and Nrf2–ARE activation—forms the central mechanism underpinning their multifunctional effects on inflammation control, immune balance, gut integrity, metabolic regulation, and cancer prevention, reinforcing apples as a valuable functional food with synergistic health benefits. Key bioactive compounds and their mechanism of action are listed in Table 6.1.

6.3.2 Blueberries (*Vaccinium corymbosum*)

Blueberries (Figure 6.2) are exceptionally rich in structurally diverse polyphenolic compounds whose primary and unifying biological function is antioxidant defense, operating through both direct redox activity and regulation of endogenous cytoprotective signaling networks, while secondarily exerting anti-inflammatory, metabolic, neuroprotective, gut modulatory, and anticancer effects (Zhu et al. 2023; Wang et al. 2024b; Rojas-Ocampo et al. 2021). Anthocyanins—particularly malvidin-3-galactoside, malvidin-3-glucoside, delphinidin-3-galactoside, and cyanidin-3-O-glucoside (C3G)—constitute the dominant antioxidant fraction (Zhu et al. 2023), directly neutralizing reactive oxygen species (ROS) via hydrogen atom donation and electron transfer while simultaneously activating the Keap1–Nrf2–ARE axis, resulting in transcriptional upregulation of phase II detoxifying and antioxidant enzymes including SOD, CAT, GPx, HO-1, and NQO1. These redox-sensitive actions are tightly coupled with MAPK signaling modulation (ERK1/2, p38, and JNK), enabling fine control over oxidative stress–induced inflammation, apoptosis, and cellular survival (Lin et al. 2020; Huang et al. 2014). Suppression of oxidative stress further attenuates NF-κB pathway activation, preventing IκBα degradation and p65 nuclear translocation, thereby reducing downstream pro-inflammatory mediators (COX-2, iNOS, TNF-α, IL-6, IL-1β, and

TABLE 6.1
Bioactive Compounds from Apple and their Antioxidant and Anti-Inflammatory Mechanisms

Bioactive/ Extract	Antioxidant Activity	Anti-Inflammatory	Key References
Quercetin	Quercetin acts as a potent antioxidant by scavenging ROS and activating the Nrf2 pathway, which upregulates endogenous antioxidant enzymes and protects cells from oxidative stress.	Reducing inflammation by inhibiting the NF-κB and MAPK pathways, leading to decreased expression of TNF-α, IL-6, COX-2, and iNOS.	Gorjanović et al. (2020); Wang et al. (2024a)
Catechin	Exhibit an antioxidant by scavenging reactive oxygen species (ROS) and activating the Nrf2–ARE pathway, which upregulates endogenous antioxidant enzymes and protects cells from oxidative stress.		Kalinowska et al. (2023)
Cyanidin-3-O-Galactoside	Acts as a potent antioxidant by directly neutralizing ROS and activating the Nrf2–ARE pathway to boost internal antioxidant enzyme defenses.		Liang et al. (2021b)
Procyanidin	Exhibit antioxidants by neutralizing ROS and activating pathways such as PGC-1α (enhancing mitochondrial defenses) and by inhibiting ERK1/2 signaling, thereby reducing oxidative stress.	Reducing inflammation by inhibiting NF-κB signaling, thereby lowering pro-inflammatory mediator expression.	Masuda et al. (2018); Andre et al. (2012)
Chlorogenic Acid	Exhibits antioxidant activity by scavenging reactive oxygen species (ROS) and activating the Nrf2–ARE pathway, which upregulates endogenous antioxidant enzymes and protects cells from oxidative stress.	Reducing inflammation by inhibiting the NF-κB and MAPK (ERK, JNK, and p38) pathways, leading to decreased production of pro-inflammatory cytokines such as TNF-α, IL-1β, and IL-6.	Commisso et al. (2021); Shaw et al. (2021)
Caffeic Acid	Neutralizes free radicals and chelates pro-oxidant metals; activates Nrf2–ARE, increasing expression of enzymes like SOD, CAT, and GPx.		Liaudanskas et al. (2014)
Gallic Acid	Donates H atoms to neutralize ROS and chelates metals; also engages Nrf2–ARE to upregulate endogenous antioxidant defenses.	Reducing inflammation by inhibiting NF-κB and MAPK pathways, decreasing pro-inflammatory cytokines, and inhibiting inflammatory enzyme expression.	Liaudanskas et al. (2014); Ongay et al. (2023)
Ferulic Acid	Donates electrons to interrupt free radical chain reactions; also influences Nrf2–ARE signaling to boost antioxidant enzymes (SOD, CAT, and GPx).		Liaudanskas et al. (2014)
p-Coumaric Acid	Scavenges free radicals through H-donation; its antioxidant signaling effects are less potent than caffeic/ferulic and primarily rely on neutralizing ROS directly.		Liaudanskas et al. (2014)
Protocatechuic Acid	Strong radical scavenger due to its dihydroxy structure; also linked to Nrf2 activation, leading to enhanced cellular antioxidant responses.		Liaudanskas et al. (2014)

(Continued)

TABLE 6.1 (Continued)

Bioactive/ Extract	Antioxidant Activity	Anti-Inflammatory	Key References
Phloretin	Lowering oxidative stress, cellular defense systems like SOD, CAT, and GSH-Px become more effective (though phloretin's effect may be weaker than classic antioxidants).	Exhibits strong anti-inflammatory effects by inhibiting the NF-κB and MAPK pathways, which leads to downregulation of COX-2 expression and reduced production of inflammatory mediators such as PGE_2 and IL-8.	Zielinska et al. (2019)
Phloridzin	Exhibits moderate antioxidant activity through direct free-radical scavenging. The attached glucose moiety reduces its redox efficiency compared with phloretin, resulting in weaker antioxidant potency.	Show mild anti-inflammatory effects, likely through weak modulation of NF-κB-related pathways, leading to limited reductions in inflammatory mediator production.	Zielinska et al. (2019)
Ursolic Acid	Scavenging reactive oxygen species (ROS) through hydrogen/electron donation and by supporting antioxidant defense pathways contributes to reduced oxidative stress.	Exhibit anti-inflammatory activity mainly by inhibiting NF-κB and MAPK signaling pathways, leading to reduced expression of pro-inflammatory cytokines and inflammation-related mediators.	Odun-Ayo et al. (2022); Woźniak et al. (2023)
Oleanolic Acid	Exhibit antioxidant activity by scavenging reactive oxygen species (ROS) and activating Nrf2-mediated antioxidant pathways, helping reduce oxidative stress.		Odun-Ayo et al. (2022)

FIGURE 6.2 Photograph of Blueberries (*Vaccinium corymbosum*).

Image kindly provided by Dr. Siriporn Chuchawankul.

VCAM-1), which links antioxidant capacity directly to anti-inflammatory efficacy (Huang et al. 2014; Goszcz et al. 2017; Sari et al. 2019).

Beyond cytosolic redox regulation, blueberry anthocyanins—most notably C3G—preserve mitochondrial homeostasis by reducing mitochondrial ROS generation, stabilizing membrane potential, and maintaining mitochondrial dynamics through modulation of fusion proteins (Mfn2, OPA1), while activating PI3K/Akt/GSK3β and AMPK/SIRT1 pathways to support mitochondrial biogenesis (PGC-1α/TFAM), energy metabolism, and neuronal survival (Yang et al. 2025a). These effects underpin the strong neuroprotective profile of blueberries, including suppression of NLRP3 inflammasome activation, inhibition of caspase-3–dependent apoptosis, promotion of autophagic flux (LC3-II, TFEB, and LAMP-1), and reduction of tau hyperphosphorylation via GSK-3β inhibition, collectively mitigating neurodegeneration and cognitive decline. Concurrently, anthocyanins and proanthocyanidins exert gut-centered antioxidant and anti-inflammatory actions by reshaping microbiota composition toward beneficial taxa (e.g., Akkermansia and Bifidobacterium), enhancing short-chain fatty acid production, strengthening tight junction integrity (ZO-1 and claudin-1), and suppressing intestinal NF-κB signaling, thereby reducing systemic oxidative burden through the gut–liver–brain axis (Wang et al. 2022c; Tian et al. 2023; Satheesh Babu et al. 2024; Morissette et al. 2020).

Flavonols such as quercetin, kaempferol, kaempferol-3-glucoside, and myricetin reinforce these antioxidant networks through potent Nrf2/HO-1 activation, supported by their multiple hydroxyl substitutions that confer high radical-scavenging efficiency, while also modulating MAPK, STAT, and NF-κB signaling to suppress chronic inflammation and oxidative damage in vascular and neural tissues (Bayazid and Lim 2022; Li et al., 2021). Stilbenes, particularly pterostilbene, provide complementary antioxidant protection through LOX-1/NF-κB inhibition and enhancement of enzymatic antioxidant activity, while exerting anti-inflammatory, neuroprotective, and anticancer effects via PI3K/Akt, MAPK, and p53/p21 signaling modulation (McCormack and McFadden 2013; Liu et al. 2023). Phenolic acids—including chlorogenic, caffeic, gallic, ferulic, and p-coumaric acids—contribute to the antioxidant pool by scavenging ROS, regulating glutathione homeostasis, and activating Nrf2 signaling (Wang et al. 2022b; de Armas-Ricard et al. 2019; Liu et al. 2020), while also influencing glucose metabolism (AMPK activation), autophagy-mediated inflammasome suppression (NLRP3), antiviral defense (Sekizawa et al. 2019), and DNA damage response pathways (Aiyer et al. 2008). Importantly, gut microbiota–derived metabolites of blueberry polyphenols further extend antioxidant and anti-inflammatory effects systemically by inhibiting endothelial NF-κB activation, reducing oxidative enzyme expression (NOX-4), and modulating GPCR-dependent metabolic and skeletal signaling (Satheesh Babu et al. 2024; Morissette et al. 2020).

Collectively, the health-promoting effects of blueberries arise from concerted activation of antioxidant defense systems centered on Nrf2 signaling, coupled with suppression of oxidative stress–driven inflammation via NF-κB inhibition, integration with mitochondrial quality control, metabolic regulation, and gut microbiota signaling, and downstream modulation of cell survival and apoptosis pathways (Zhu et al. 2023; Wang et al. 2024b). This multilayered, redox-centered mechanism explains the broad protective efficacy of blueberry bioactive compounds against chronic diseases characterized by oxidative stress, inflammation, and metabolic dysfunction. Key bioactive compounds and their mechanisms of action are listed in Table 6.2.

6.3.3 Kiwi (*Actinidia deliciosa*)

Kiwifruit (Figure 6.3) represents a complex functional food matrix enriched with diverse bioactive compounds whose biological effects are predominantly driven by robust antioxidant mechanisms that integrate with anti-inflammatory, metabolic, anticancer, and gut-regulatory pathways. The antioxidant capacity of kiwifruit is mediated through both direct and indirect mechanisms. Direct antioxidant action is primarily attributed to vitamin C and polyphenols, which scavenge reactive oxygen and nitrogen species (ROS/RNS), inhibit lipid peroxidation, protect DNA from oxidative damage,

TABLE 6.2
Bioactives from Blueberries and their Antioxidant and Anti-Inflammatory Mechanisms

Bioactive/Extract	Antioxidant Activity	Anti-inflammatory	Key References
Malvidin-3-Galactoside (M3G)	Exhibit antioxidant activity by scavenging ROS and activating PI3K/Akt–Nrf2 signaling, which increases endogenous antioxidant enzymes (e.g., SOD, HO-1) and reduces oxidative stress.	Showing anti-inflammatory activity by inhibiting NF-κB signaling, reducing expression of pro-inflammatory molecules, and inflammatory responses.	Zhu et al. (2023); Huang et al. (2014)
Malvidin-3-Glucoside	Exhibit antioxidant activity by scavenging reactive oxygen species and activating the Nrf2/ARE pathway, enhancing endogenous antioxidant defenses, and reducing oxidative stress.	Inhibition of NF-κB signaling, leading to reduced expression of pro-inflammatory mediators.	Zhu et al. (2023); Wang et al. (2024b)
Delphinidin-3-Galactoside	Showing antioxidant activity by scavenging reactive oxygen species and enhancing endogenous antioxidant defenses, thereby reducing oxidative stress.		Goszcz et al. (2017)
Cyanidin-3-O-Glucoside (C3G)	Reducing oxidative stress by direct ROS scavenging and ERK-dependent activation of the Nrf2/ARE pathway, increasing endogenous antioxidant enzyme expression.	inhibiting NF-κB signaling and reducing pro-inflammatory cytokine expression.	Yang et al. (2025a); Wang et al. (2022b)
Peonidin-3-Glucoside		Peonidin-3-glucoside may inhibit TNF-α signaling by binding to the TNF-α receptor and disrupting TNF-α–receptor interaction, thereby potentially reducing downstream pro-inflammatory signaling linked to NF-κB activation.	Sari et al. (2019)
Pterostilbene	Exhibits antioxidant activity by scavenging reactive oxygen species and activating Nrf2-mediated antioxidant pathways, leading to increased endogenous antioxidant enzyme expression and reduced oxidative stress.	Exhibit anti-inflammatory activity by suppressing NF-κB signaling and reducing pro-inflammatory cytokine expression.	McCormack and McFadden (2013); Liu et al. (2024)
Quercetin	Showing antioxidant activity by scavenging ROS and activating the Nrf2/HO-1 pathway, enhancing endogenous antioxidant defenses.	Exerts anti-inflammatory effects mainly by blocking NF-κB activation through inhibition of IκBα degradation, thereby suppressing transcription of pro-inflammatory cytokines and inflammatory enzymes.	Bayazid and Lim (2022)
Myricetin	Exhibit antioxidant activity by scavenging reactive oxygen species and activating the Nrf2/HO-1 pathway, enhancing endogenous antioxidant defenses, and reducing oxidative stress.		Li et al. (2021)

(*Continued*)

TABLE 6.2 (Continued)

Bioactive/Extract	Antioxidant Activity	Anti-inflammatory	Key References
Catechins	Exhibit antioxidant activity by scavenging ROS and activating the Nrf2/ARE pathway, enhancing endogenous antioxidant defenses.		Tian et al. (2023)
Chlorogenic Acid	Reducing oxidative stress by directly scavenging reactive oxygen species and activating the Keap1-Nrf2/ARE antioxidant signaling pathway, leading to increased expression of endogenous antioxidant enzymes.		Wang et al. (2022b)
Gallic Acid	Showing antioxidant activity by scavenging reactive oxygen species and activating Nrf2/ARE-mediated antioxidant defenses, thereby reducing oxidative stress.		Rojas-Ocampo et al. (2021)
Caffeic Acid	Showing antioxidant activity by scavenging reactive oxygen species and activating the Nrf2/ARE pathway, enhancing endogenous antioxidant defenses.		de Armas-Ricard et al. (2019)
Ellagic Acid	Scavenges reactive oxygen species and chelates metal ions, reducing oxidative stress and protecting cellular components.	Suppresses inflammation mainly by inhibiting NF-κB signaling, leading to reduced expression of pro-inflammatory cytokines and enzymes.	Aiyer et al. (2008)
p-Coumaric Acid	Directly scavenging reactive oxygen species, inhibiting lipid peroxidation, chelating metal ions, and reducing oxidants, it also enhances endogenous antioxidant defenses via activation of Nrf2-mediated signaling that upregulates enzymes to protect cells from oxidative stress.		Liu et al. (2020)

and maintain intracellular redox balance by preserving glutathione levels. Vitamin C, one of the most abundant antioxidants in kiwifruit, acts as a rapid electron donor, suppresses membrane lipid oxidation, and protects erythrocytes and epithelial cells from oxidative injury, contributing to systemic antioxidant defense (Vlasiuk et al. 2024).

Phenolic acids constitute a major class of kiwifruit antioxidants, including chlorogenic, neochlorogenic, gallic, caffeic, ferulic, syringic, p-coumaric, quinic, and ellagic acids, which exhibit strong free radical scavenging capacity, metal ion chelation, and ferric reducing activity. These compounds also exert indirect antioxidant effects through activation of the Nrf2/ARE signaling pathway, leading to upregulation of endogenous antioxidant enzymes such as superoxide dismutase (SOD), catalase (CAT), glutathione peroxidase (GPx), heme oxygenase-1 (HO-1), and NAD(P)H quinone oxidoreductase-1 (NQO-1), thereby reinforcing cellular resistance to oxidative stress (Zhang et al. 2021; Wu et al. 2023; Yang et al. 2024; Li et al. 2018; El Azab and Mostafa 2021). Neochlorogenic and chlorogenic acids further contribute to metabolic health by inhibiting α-glucosidase activity, linking antioxidant protection with improved glycemic control.

Flavonoids in kiwifruit, including quercetin, kaempferol, catechin, epicatechin, rutin, quercetin-3-rhamnoside, and anthocyanins (notably cyanidin-3-glucoside), exhibit pleiotropic bioactivities by

FIGURE 6.3 Photograph of Kiwi (*Actinidia deliciosa*).

Image kindly provided by Miss Pandar Nartsawee.

coupling antioxidant signaling with suppression of pro-oxidant and pro-inflammatory pathways. These flavonoids inhibit NF-κB nuclear translocation, suppress MAPK cascades (ERK, p38, and JNK), and modulate PI3K/Akt and Wnt/β-catenin signaling, resulting in reduced inflammatory cytokine production, attenuation of oxidative stress-induced apoptosis, and inhibition of cancer cell proliferation (Lee et al. 2010; Peng et al. 2020; Tsouh Fokou et al. 2025; Sip et al. 2022). Anthocyanins additionally downregulate COX-1 and COX-2 expression and reduce endothelial inflammation, contributing to vascular protection and anti-atherogenic effects. Procyanidin B1 and B2 (proanthocyanidins) strongly activate the Nrf2/ARE pathway while decreasing malondialdehyde (MDA) accumulation and ROS levels, protecting intestinal epithelial cells from oxidative damage and demonstrating anti-proliferative and anti-tyrosinase activities (Li et al. 2018; Chai et al. 2014; Lv et al. 2022).

Beyond polyphenols, lipid-soluble antioxidants such as δ-tocomonoenol, a unique vitamin E isoform abundant in kiwifruit peel, protect membrane phospholipids from oxidative chain reactions, inhibit TBARS (thiobarbituric acid reactive substances) formation, and stabilize cellular membranes under oxidative stress (Fiorentino et al. 2009). Additional bioactive components, including cysteine protease inhibitor (CPKE), exhibit potent anti-inflammatory activity through suppression of Akt phosphorylation, MAPK signaling, and NF-κB activation, demonstrating therapeutic relevance in inflammatory skin disorders (Yang et al. 2025b). Dietary fibers and pectic polysaccharides further contribute to antioxidant and anti-inflammatory homeostasis by modulating gut microbiota composition, increasing short-chain fatty acid production, restoring intestinal barrier integrity, and indirectly reducing systemic oxidative stress and inflammation (Yuan et al. 2021; Wang et al. 2022a; Parkar et al. 2010; Jiang et al. 2025).

Collectively, the bioactive compounds in kiwifruit act synergistically to establish a comprehensive antioxidant-centered protective network that extends beyond redox regulation to encompass inflammation control, metabolic regulation, cardiovascular protection, gut health, neuroprotection, and cancer prevention. This multifactorial mechanism positions kiwifruit as a scientifically substantiated functional food with broad therapeutic potential, emphasizing the importance of whole-fruit consumption to maximize synergistic bioactivity. Key bioactive compounds and their mechanisms of action are listed out in Table 6.3.

TABLE 6.3
Bioactives from Kiwi and their Antioxidant and Anti-inflammatory Mechanisms

Bioactive/Extract	Antioxidant Activity	Anti-Inflammatory	Key References
Cysteine Protease Inhibitor (CPKE)		Exerts anti-inflammatory activity by inhibiting TNF-α–induced MAPK (ERK1/2, p38) and Akt signaling and suppressing NF-κB activation, thereby reducing pro-inflammatory responses.	Yang et al. (2025)
Vitamin C (Ascorbic Acid)	Exhibits strong antioxidant activity by directly scavenging ROS and activating the Nrf2/ARE pathway, which enhances endogenous antioxidant enzymes and protects respiratory tissues by reducing oxidative stress and inflammation in the airway epithelium.		Vlasiuk et al. (2024)
Quercetin	Exerts antioxidant activity by scavenging ROS and inhibiting ERK1/2–Cx43 signaling, thereby preventing H_2O_2-induced disruption of gap-junction intercellular communication.		Lee et al. (2010)
Polyphenol Extract		Reducing inflammation by modulating gut microbiota and inhibiting TLR2/4-mediated inflammatory signaling, leading to decreased pro-inflammatory cytokine production.	Yuan et al. (2021)
Chlorogenic Acid	Exhibits antioxidant activity by scavenging ROS and activating the Nrf2/ARE pathway, enhancing endogenous antioxidant defenses.	Suppressing inflammation in Caco-2 cells by inhibiting NF-κB signaling and activating Nrf2-mediated antioxidant pathways, reducing pro-inflammatory cytokine production.	Zhang et al. (2021); Wu et al. (2023)
Neochlorogenic Acid	Exhibiting antioxidant activity by scavenging ROS and activating the AMPK–Nrf2/ARE pathway, enhancing endogenous antioxidant defenses.	Exerts anti-inflammatory activity by inhibiting NF-κB/MAPK signaling and activating the AMPK–Nrf2 pathway, reducing pro-inflammatory mediators.	Yang et al. (2024)
Procyanidin B1 and B2 (Proanthocyanidins)	Exhibit antioxidant activity by scavenging ROS and activating the Nrf2/ARE pathway, enhancing endogenous antioxidant defenses.		Li et al. (2018)
Anthocyanins (Cyanidin-3-Glucoside)		Exerts anti-inflammatory and respiratory benefits by suppressing NF-κB–mediated inflammatory signaling and reducing chemokine release involved in allergic airway inflammation.	Peng et al. (2020)

(Continued)

TABLE 6.3 (Continued)

Bioactive/Extract	Antioxidant Activity	Anti-Inflammatory	Key References
Vitamin E (δ-Tocomonoenol)	Acts as a membrane-bound antioxidant that scavenges free radicals and interrupts lipid peroxidation pathways, thereby protecting cells from oxidative damage.		Fiorentino et al. (2009)
--Pectin (Pectic Polysaccharides—HG and RG-I Domains)		Suppressing gut inflammation via reduced NF-κB–dependent cytokine signaling, thereby maintaining colonic homeostasis.	Parkar et al. (2010); Jiang et al. (2025)
p-Coumaric Acid	Showing antioxidant activity by directly scavenging ROS and activating the Nrf2/ARE pathway, leading to increased expression of antioxidant enzymes (e.g., SOD, GPx, HO-1) and reduced oxidative stress.		El Azab and Mostafa (2021)
Ferulic Acid	Acts as a potent antioxidant by scavenging ROS and activating the Nrf2/ARE pathway, which upregulates antioxidant enzymes (SOD, CAT, HO-1) and protects cells from oxidative damage.		El Azab and Mostafa (2021)
Syringic Acid	Exhibit antioxidant by scavenging ROS and activating the Nrf2/ARE pathway, enhancing cellular antioxidant defenses and reducing oxidative stress.		El Azab and Mostafa (2021)
Catechin	Showing antioxidant activity by scavenging ROS and activating the Nrf2/ARE pathway, leading to upregulation of endogenous antioxidant enzymes such as SOD and CAT.		Tsouh Fokou et al. (2025)
Gallic Acid	Exhibits antioxidant activity by scavenging ROS and activating the Nrf2/ARE pathway, leading to upregulation of endogenous antioxidant enzymes.	Exerts anti-inflammatory effects by inhibiting the NF-κB pathway, leading to reduced expression of pro-inflammatory cytokines and enzymes.	Yang et al. (2024)
Caffeic Acid	Showing antioxidant activity by scavenging ROS and activating the Nrf2/ARE pathway, enhancing cellular antioxidant defenses.	Exerts anti-inflammatory effects by suppressing NF-κB and MAPK signaling, reducing pro-inflammatory cytokine production.	Yang et al. (2024)

6.3.4 Pineapple (*Ananas comosus*)

Pineapple (Figure 6.4), *A. comosus*, is an antioxidant-rich fruit with a significant amount of vitamin C, a water-soluble vitamin. Bromelain, a protease enzyme extracted from *A. comosus*, is reported to have antioxidant properties and anti-inflammatory activity. Nady et al. (2025) studied the anti-inflammatory activity of bromelain against 3-nitropropionic acid (3-NP)-induced Huntington's disease-like symptoms in rats, which is regulated via the SIRT1/PGC-1α and FOXO3a pathway. This isolated compound also shows anti-inflammatory activity against atherosclerosis and shows protective benefits against non-alcoholic fatty liver (Chen et al. 2025b). It also activates the AMPK

FIGURE 6.4 Photograph of Pineapple (*Ananas comosus*).

Image kindly provided by Mr. Nattapong Kawjit.

signaling pathway to increase nitric oxide availability, along with the activation of autophagy pathways. The effect of Bromelain against the excessive oxidative stress and inflammatory responses caused by ischemia–reperfusion injury in male Wistar rats is studied by Sezen et al. (2025). Ortega-Hernández et al. (2023) discussed the extraction of phenolic compounds from the pineapple peels through solid-state fermentation with the use of certain bacteria, which increased the release of phenolic compounds, and, in turn, increased the cellular antioxidant and anti-inflammatory potential of the extract when tested on human colon cells and murine macrophage cells. Chhabile et al. (2025) compared the anti-inflammatory and antioxidant properties of *A. comosus, Malus pumila, Manilkara zapota, M. indica*, and *M. pumila*. The study concluded that *A. comosus* has higher antioxidant activity.

Concluding that pineapple and its derivatives have anti-oxidant and anti-inflammatory properties. Apart from this, they also act as natural anti-parasitic agent also helps lower the complications of irritable bowel syndrome (Mehraj et al. 2024). Key anti-inflammatory mechanisms are listed in Table 6.4.

6.3.5 Cherries (*Prunus* spp.)

Tart cherry (Figure 6.5), *Prunus cerasus*, is a polyphenol-rich fruit that has been studied by Dufour et al. 2025 for its intracellular ROS-scavenging activity. The study used human hepatocyte cells (HepG2), concluding that the extract of *P. cerasus* has antioxidant properties and plays both direct and indirect roles in pathways that reduce ROS levels via antioxidant response element (ARE).

TABLE 6.4
Bioactives from Pineapple and Anti-Inflammatory Mechanism

Bioactive	Anti-Inflammatory	Key References
Bromelain	Regulates SIRT1/PGC-1α and FOXO3a pathway	Nady et al. (2025)
Bromelain	Activates AMPK signaling pathway	Chen et al. (2025b)

FIGURE 6.5 Photograph of Cherries (*Prunus* spp).

Image kindly provided by Miss Waraporn Sonthi.

Perillyl alcohol, a plant-derived monoterpene present in cherry, cranberry, and mint, has been reported to have strong antioxidant, anti-inflammatory, and anti-proliferative properties (Puppala et al. 2022a). It is being reported to downregulate the pro-inflammatory cytokines such as IL-1β, IL-6, and TNF-α in serum, and to reverse pathological changes in the colon caused by ulcerative colitis, a most prevalent chronic inflammatory immune bowel disease. It also inhibits neuroinflammation and improves depressive behavior in mice by modulating NF-κB and STAT3 in both isolated brain and colon tissues. Apart from this, the perillyl alcohol also has a role in lipid metabolism and reduces obesity in mice (Chen et al. 2025a). Puppala et al. (2022b) reported the anti-inflammatory activity of perillyl alcohol against rheumatoid arthritis, a chronic and idiopathic autoimmune disorder, by downregulating NF-κB and Nrf2 via the Ras/Raf/MAPK pathways. The hydroethanolic extract of sweet cherry, *P. avium*, shows strong antioxidant activity in RAW 264.7 macrophage cells. The extract shows no cytotoxicity and exhibits anti-inflammatory activity by reducing ROS and NO levels through inhibition of NF-kB nuclear translocation and MAPK signaling. Mahmoud et al. (2013) reported that sour cherry (*P. cerasus*) seed extracts modulate CD3+ T lymphocyte activity and downregulate the inflammatory cytokines, including TNF-α and IL-8. Key anti-inflammatory mechanisms are listed in Table 6.5.

TABLE 6.5
Bioactives from Cherry and their Anti-Inflammatory Mechanisms

Bioactive	Anti-Inflammatory	Key References
Perillyl Alcohol	Downregulate the pro-inflammatory cytokines such as IL-1β, IL-6, and TNF-α	Puppala et al. (2022a)
Perillyl Alcohol	Downregulates NF-κB and Nrf2 via Ras/Raf/MAPK pathways	Puppala et al. (2022b)
Prunus cerasus Seed Extracts	Modulates CD3+ T lymphocyte activity and downregulates the inflammatory cytokines, including TNF-α and IL-8	Mahmoud et al. (2013)

6.3.6 Watermelon (*Citrullus lanatus*)

Watermelon (Figure 6.6A and B) rind extract has been reported to show antioxidant effects. Michael et al. (2021) studied lead acetate toxicity, which causes hepato-renal injury in male Wistar rats, and reported that it was reversed to normal upon treatment with watermelon rind ethanol extract via a uric acid/nitric oxide-dependent pathway. Lycopene, a non–pro-vitamin A carotenoid, is a potent antioxidant. It is extracted from plants like tomatoes, watermelons, red carrots, and papayas, and has remarkable health benefits. Khalaf and Awad (2023) discussed the anti-cancer properties of lycopene via modulation of the insulin-like growth factor system, PI3K/Akt pathway, and the regulation of mitochondrial behaviors. Watermelon juice is reported to inhibit α-glucosidase and α-amylase and also possesses antioxidant properties along with its anti-diabetic activity (Ajiboye et al. 2020). Shanely et al. (2020) conducted a study that concluded that consumption of 100% watermelon puree daily for 6 weeks improved markers linked to cardiovascular disease, without changing insulin resistance in overweight and obese post-menopausal women, a group at higher risk of insulin resistance and cardiovascular disease. Watermelon supplementation improved vascular health markers (sVCAM-1) and had a protective effect against cardiovascular disease.

6.3.7 Grapefruit (*Citrus paradisi*)

Zargar et al. (2018) discussed the synergistic effect of grapefruit (Figure 6.7) juice in combination with the antipsychotic drug aripiprazole (ARI). This study shows that the grapefruit juice significantly increases the anti-inflammatory activity of the drug even at very minimal doses. Thus, this method of using synergistic effects of natural compounds along with drugs can reduce the side effects associated with high drug doses. Rhoifolin, derived from various plants such as *Rhus succedanea*, bitter orange, lemon, grapefruit, tomatoes, lupinus, bananas, and grapes, exhibits antioxidant, anti-inflammatory, and neuroprotective properties. Yang et al. (2025b) studied its therapeutic impacts against herpes simplex encephalitis (HSE), caused by herpes simplex virus type 1 (HSV-1) in a mouse model. The results show that the derived compound has a positive result in reducing the effects of HSC by enhancing the expression of SESN2 and regulating downstream Keap1/Nrf2 and

(A) (B)

FIGURE 6.6 (A and B) Photograph of Watermelon (*Citrullus lanatus*).

Image kindly provided by Dr. Siriporn Chuchawankul.

FIGURE 6.7 Photograph of Grapefruit (*Citrus paradisi*).

Image kindly provided by Miss Pandar Nartsawee.

AMPK/Nox pathways. Naringenin and (S)-naringenin were purified from *C. paradisi*. From these purified compounds, 7,4-O-disenecioic ester naringenin, (S)-7,4-O-disenecioic ester naringenin, and 7-O-senecioic ester naringenin were synthesized. These extracted compounds showed anti-inflammatory and antioxidant activity by increasing glutathione and reducing GFAP and nitric oxide production (Soares et al., 2025). Also, Zhang et al. (2024) reported that naringenin reduces cigarette smoke extract-induced structural disruption of airway cilia by modulating the IL-17 signaling pathway. *C. sinensis* and *C. paradisi* showed a highly significant fall in cholesterol, low-density lipoprotein, and triglycerides in the murine model (Mallick and Khan 2016). Thus, the *C. paradisi* and its derivatives have effects on hyperlipidemia. Key anti-inflammatory mechanisms are listed in Table 6.6.

6.3.8 Mango (*Mangifera indica*)

Chhabile et al. (2025) compared the anti-inflammatory and antioxidant properties of *A. comosus*, *M. pumila*, *M. zapota*, *M. indica*, and *M. pumila* and concluded that *M. indica* and *M. pumila* have antioxidant activity. Mangiferin, a polyphenol present in *M. indica* (Figure 6.8), has been reported to have neuroprotective activity (Mustafa et al. 2025). It has high antioxidant, anti-inflammatory, and anti-apoptotic activities, which pave the way to study them against cognitive impairments and neuronal damage in disorders such as Alzheimer's disease, Parkinson's disease, Huntington's disease, and others. To enhance their bioavailability, researchers are developing nanoparticle-mediated

TABLE 6.6
Bioactive from Grapefruit and Anti-Inflammatory Mechanism

Bioactive	Anti-Inflammatory	Key References
Rhoifolin	Regulates Keap1/Nrf2 and AMPK/Nox pathways	Yang et al. (2025b)
Naringenin	Modulates IL-17 signaling pathway	Zhang et al. (2024)

Figure 6.8 Photograph of Mango (*Mangifera indica*).

Image kindly provided by Miss Pandar Nartsawee.

drug delivery systems. The anti-neuroinflammatory effects of the isolated benzophenones from the bark of *M. indica* were studied by Liao et al. (2025). The benzophenones were evaluated in BV-2 microglial cells. This study shows that the benzophenones exhibit significant inhibitory effects on the production of nitric oxide and inhibit NF-κB nuclear translocation, thus reducing the levels of iNOS, COX-2, IL-1β, IL-6, and CCL2. The fruit of *M. indica* is reported to have antioxidant, anti-inflammatory, and anti-senescent functions; thus, it is also used in anti-aging studies (Menezes et al. 2025). El-Sayed et al. (2025) studied the hepatoprotective effect of *M. indica* against methotrexate-induced liver injury in rats. Key anti-inflammatory mechanisms are listed in Table 6.7.

6.3.9 Java Plum (*Syzygium cumini*)

Syzygium cumini contains significant amounts of quercetin, rutin, and gallic acid, which contribute to its hypoglycemic, antioxidant, and anti-inflammatory properties. Kaur and Chaudhary (2025) reported that the leaves of *S. cumini* (Figure 6.9) can be used to treat diabetes, instead of synthetic drugs. The role of *S. cumini* (L.) Skeels leaves in the treatment of Parkinson's disease was studied by Phan et al. (2025). It reveals that phenolic compounds are major components, and molecular docking analysis suggests that dopaminergic neuron protection, dopamine degradation prevention, and dopaminergic receptor activation are the possible mechanisms for its therapeutic effect. Nijum et al. (2025) conducted a study to develop stable, quality-retaining probiotic-enriched fruit powders, especially using jamun (*S. cumini*) juice powder.

TABLE 6.7
Bioactive from Mango and Anti-Inflammatory Mechanism

Bioactive	Anti-Inflammatory	Key Reference
Benzophenones	Inhibits NF-κB nuclear translocation and reduces levels of inflammatory cytokines.	Liao et al. (2025)

FIGURE 6.9 Photograph of Java plum (*Syzygium cumini*).

Image kindly provided by Miss Sirinapa Laobutsa.

6.4 FUTURE PERSPECTIVE AND CONCLUSION

The future research on fruit-derived bioactive compounds appears highly promising, supported by rapid advancements in analytical chemistry, biotechnology, and personalized nutrition (Gómez-Zavaglia and Cassani 2025). The application of omics technologies, coupled with artificial intelligence, is facilitating the discovery of novel phytochemicals and enabling more sustainable production through plant cell cultures and microbial biosynthesis. Current investigations are also elucidating the molecular pathways through which fruit-based phytochemicals exert their health benefits, particularly in relation to inflammation, metabolic regulation, and gut microbiota signaling (Pérez-Jiménez and Saura-Calixto 2018). Novel extraction methods and nano-delivery approaches, such as nano-encapsulation, emulsification, and polymer-based carrier systems, are being developed to enhance the stability and bioavailability of key compounds, including polyphenols, carotenoids, and flavonoids (Pan et al. 2025; Williamson 2025). There is also an increasing focus on the interactions between fruit-derived bioactives and the gut microbiome, given their emerging roles in immune modulation, metabolic homeostasis, and brain–gut communication, which together support the development of microbiome-informed functional foods and nutraceuticals (Jia et al. 2025). Collectively, these advances make fruit bioactive compounds important contributors to future nutraceutical innovations, functional food design, and preventive health strategies that can be used for the prevention of chronic inflammation.

REFERENCES

Aiyer, H.S., Vadhanam, M.V., Stoyanova, R., Caprio, G.D., Clapper, M.L., and Gupta, R.C. (2008). Dietary berries and ellagic acid prevent oxidative DNA damage and modulate expression of DNA repair genes. *International Journal of Molecular Sciences*, 9(3), pp. 327–341.

Ajiboye, B.O., Shonibare, M.T., and Oyinloye, B.E. (2020). Antidiabetic activity of watermelon (*Citrullus lanatus*) juice in alloxan-induced diabetic rats. *Journal of Diabetes and Metabolic Disorders*, 19(1), pp. 343–352.

Andre, C.M., Greenwood, J.M., Walker, E.G., Rassam, M., Sullivan, M., Evers, D., et al. (2012). Anti-inflammatory procyanidins and triterpenes in 109 apple varieties. *Journal of Agricultural and Food Chemistry*, 60(42), pp. 10546–10554.

Bayazid, A.B., and Lim, B.O. (2022). Quercetin is an active agent in berries against neurodegenerative diseases progression through modulation of Nrf2/HO1. *Nutrients*, 14(23), p. 5132.

Chai, W.M., Shi, Y., Feng, H.L., Xu, L., Xiang, Z.H., Gao, Y.S., et al. (2014). Structure characterization and anti-tyrosinase mechanism of polymeric proanthocyanidins fractionated from kiwifruit pericarp. *Journal of Agricultural and Food Chemistry*, 62(27), pp. 6382–6389.

Chen, L., Deng, H., Cui, H., Fang, J., Zuo, Z., Deng, J., et al. (2017). Inflammatory responses and inflammation-associated diseases in organs. *Oncotarget*, 9(6), p. 7204.

Chen, S., Chen, S., Li, W., Wang, M., Pan, R., Duan, Y., et al. (2025a). Perillyl alcohol ameliorates high-fat diet-induced obese mice by modulating gut microbiota and activating IRF4-mediated brown fat thermogenesis. *Phytomedicine*, 148, p. 157400.

Chen, W.H., Wu, Y.Y., Hsu, M.C., Chen, C.H., Hsu, J.C.N., et al. (2025b). Bromelain enhances nitric oxide bioavailability: bradykinin's link to TRPV1/Ca2+/AMPK/autophagy signaling. *Biomedicine and Pharmacotherapy*, 190, p. 118376.

Chhabile, S., Dodamani, A., Vishwakarma, P., Jadav, H., Gadekar, A., and Chintale, S. (2025). Anti-inflammatory and antioxidant properties of six common fruit extracts: an in vitro study. *Bioinformation*, 21(7), p. 1986.

Commisso, M., Bianconi, M., Poletti, S., Negri, S., Munari, F., Ceoldo, S., et al. (2021). Metabolomic profiling and antioxidant activity of fruits representing diverse apple and pear cultivars. *Biology*, 10(5), p. 380.

de Armas-Ricard, M., Ruiz-Reyes, E., and Ramírez-Rodríguez, O. (2019). Caffeates and caffeamides: synthetic methodologies and their antioxidant properties. *International Journal of Medicinal Chemistry*, 2019(1), p. 2592609.

Dharmarajan, S., Carrillo, C., Qi, Z., Wilson, J.M., Baucum, A.J., Sorenson, C.M., et al. (2023). Retinal inflammation in murine models of type 1 and type 2 diabetes with diabetic retinopathy. *Diabetologia*, 66(11), pp. 2170–2185.

Di Pede, G., Mena, P., Bresciani, L., Achour, M., Lamuela-Raventós, R.M., Estruch, R., et al. (2023). Revisiting the bioavailability of flavan-3-ols in humans: a systematic review and comprehensive data analysis. *Molecular Aspects of Medicine*, 89, p. 101146.

Dufour, C., Rigal, M., Gironde, C., Plattner, S., and Furger, C. (2025). Tart cherry (*Prunus cerasus*) extract exerts high intracellular ROS scavenging activity and repression of ARE pathway in human hepatocytes. *International Journal of Molecular Sciences*, 26(22), p. 10827.

El Azab, E.F., and Mostafa, H.S. (2021). Phytochemical analysis and antioxidant defense of kiwifruit (*Actinidia deliciosa*) against pancreatic cancer and AAPH-induced RBCs hemolysis. *Food Science and Technology*, 42, p. e06021.

El-Sayed, N.M., Menze, E.T., Tadros, M.G., and Hanna, D.M. (2025). Mangiferin mitigates methotrexate-induced liver injury and suppresses hepatic stellate cells activation in rats: imperative role of Nrf2/NF-κB/NLRP3 signaling axis. *Journal of Ethnopharmacology*, 340, p. 119296.

Fiorentino, A., Mastellone, C., D'Abrosca, B., Pacifico, S., Scognamiglio, M., Cefarelli, G., et al. (2009). δ-Tocomonoenol: a new vitamin E from kiwi (*Actinidia chinensis*) fruits. *Food Chemistry*, 115(1), pp. 187–192.

Gómez-Zavaglia, A., and Cassani, L. (2025). Advances in fruit and vegetable quality, bioactive compounds, and nutritional value. *Foods*, 14(15), p. 2665.

Gorjanović, S., Micić, D., Pastor, F., Tosti, T., Kalušević, A., Ristić, S., et al. (2020). Evaluation of apple pomace flour obtained industrially by dehydration as a source of biomolecules with antioxidant, antidiabetic and antiobesity effects. *Antioxidants*, 9(5), p. 413.

Goszcz, K., Deakin, S.J., Duthie, G.G., Stewart, D., and Megson, I.L. (2017). Bioavailable concentrations of delphinidin and its metabolite, gallic acid, induce antioxidant protection associated with increased intracellular glutathione in cultured endothelial cells. *Oxidative Medicine and Cellular Longevity*, 2017(1), p. 9260701.

Huang, W.Y., Liu, Y.M., Wang, J., Wang, X.N., and Li, C.Y. (2014). Anti-inflammatory effect of blueberry anthocyanins malvidin-3-glucoside and malvidin-3-galactoside in endothelial cells. *Molecules*, 19(8), pp. 12827–12841.

Jia, Y., Liu, Y., Wu, Y., Ren, F., and Liu, H. (2025). From diet to microbiota: how fruits and vegetables influence gut microbiota. *International Journal of Food Science and Technology*, 60(1), p. vvae008.

Jiang, K.X., Geng, J.L., Yuan, Q., Li, J., Liu, W., Liu, H.Y., et al. (2025). Pectic polysaccharides from unripe and mature kiwifruit alleviate ulcerative colitis by modulating gut microbiota and restoring the intestinal barrier. *Food Bioscience*, 71, p. 107315.

Kalinowska, M., Gołebiewska, E., Zawadzka, M., Choińska, R., Koronkiewicz, K., Piasecka-Jóźwiak, K., et al. (2023). Sustainable extraction of bioactive compound from apple pomace through lactic acid bacteria fermentation. *Scientific Reports*, 13(1), p. 19310.

Kaminska, B. (2005). MAPK signalling pathways as molecular targets for anti-inflammatory therapy—from molecular mechanisms to therapeutic benefits. *Biochimica et Biophysica Acta (BBA) - Proteins and Proteomics*, 1754(1–2), pp. 253–262.

Kaur, S., and Chaudhary, R. (2025). A comprehensive review of the effects of *Psidium guajava* and *Syzygium cumini* leaves on diabetes mellitus. *Preventive Nutrition and Food Science*, 30(3), p. 209.

Khalaf, R.A., and Awad, M. (2023). Lycopene as a potential bioactive compound: chemistry, extraction, and anticancer prospective. *Current Cancer Drug Targets*, 23(8), pp. 634–642.

Lee, D.E., Shin, B.J., Hur, H.J., Kim, J.H., Kim, J., Kang, N.J., et al. (2010). Quercetin, the active phenolic component in kiwifruit, prevents hydrogen peroxide-induced inhibition of gap-junction intercellular communication. *British Journal of Nutrition*, 104(2), pp. 164–170.

Li, H.Y., Yuan, Q., Yang, Y.L., Han, Q.H., He, J.L., Zhao, L., et al. (2018). Phenolic profiles, antioxidant capacities, and inhibitory effects on digestive enzymes of different kiwifruits. *Molecules*, 23(11), p. 2957.

Li, J., Xiang, H., Huang, C., and Lu, J. (2021). Pharmacological actions of myricetin in the nervous system: a comprehensive review of preclinical studies in animals and cell models. *Frontiers in Pharmacology*, 12, p. 797298.

Liang, Y., Chen, G., Zhang, F., Yang, X., Chen, Y., Duan, Y., et al. (2021a). Procyanidin B2 reduces vascular calcification through inactivation of ERK1/2-RUNX2 pathway. *Antioxidants*, 10(6), p. 916.

Liang, Z., Liang, H., Guo, Y., and Yang, D. (2021b). Cyanidin 3-O-galactoside: a natural compound with multiple health benefits. *International Journal of Molecular Sciences*, 22(5), p. 2261.

Liao, R., Cai, J., Zhang, W., Wang, Y., Xu, J., and He, X. (2025). Anti-inflammatory benzophenones from the barks of mango (*Mangifera indica* L.). *Fitoterapia*, 183, p. 106575.

Liaudanskas, M., Viškelis, P., Raudonis, R., Kviklys, D., Uselis, N., and Janulis, V. (2014). Phenolic composition and antioxidant activity of *Malus domestica* leaves. *The Scientific World Journal*, 2014(1), p. 306217.

Lin, J., Tian, J., Shu, C., Cheng, Z., Liu, Y., Wang, W., et al. (2020). Malvidin-3-galactoside from blueberry suppresses the growth and metastasis potential of hepatocellular carcinoma cell Huh-7 by regulating apoptosis and metastases pathways. *Food Science and Human Wellness*, 9(2), pp. 136–145.

Liu, J., Zhuang, Y., Hu, Y., Xue, S., Li, H., Chen, L., et al. (2020). Improving the color stability and antioxidation activity of blueberry anthocyanins by enzymatic acylation with p-coumaric acid and caffeic acid. *LWT*, 130, p. 109673.

Liu, P., Tang, W., Xiang, K., and Li, G. (2023). Pterostilbene in the treatment of inflammatory and oncological diseases. *Front Pharmacol.*;14:1323377

Liu, P., Tang, W., Xiang, K., and Li, G. (2024). Pterostilbene in the treatment of inflammatory and oncological diseases. *Frontiers in Pharmacology*, 14, p. 1323377.

Lv, J.M., Gouda, M., Bekhit, A.E.D., He, Y.K., Ye, X.Q., and Chen, J.C. (2022). Identification of novel bioactive proanthocyanidins with potent antioxidant and anti-proliferative activities from kiwifruit leaves. *Food Bioscience*, 46, p. 101554.

Ma, J., Zhang, L., Ren, X., Luo, Z., Zhao, M., Tong, P., et al. (2025). High hydrostatic pressure pretreated fermented apple juice attenuated anaphylaxis by improving gut microbiota and metabolic regulation. *Food Bioscience*, 64, p. 105844.

Mahmoud, F.F., Al-Awadhi, R., Haines, D.D., Dashti, A., Dashti, H., Al-Ozairi, E., et al. (2013). Sour cherry seed kernel extract increases heme oxygenase-1 expression and decreases representation of CD3+ TNF-α+ and CD3+ IL-8+ subpopulations in peripheral blood leukocyte cultures from type 2 diabetes patients. *Phytotherapy Research*, 27(5), pp. 767–774.

Majdan, M., and Bobrowska-Korczak, B. (2022). Active compounds in fruits and inflammation in the body. *Nutrients*, 14(12), p. 2496.

Mallick, N., and Khan, R.A. (2016). Antihyperlipidemic effects of *Citrus sinensis*, *Citrus paradisi*, and their combinations. *Journal of Pharmacy and Bioallied Sciences*, 8(2), pp. 112–118.

Masuda, I., Koike, M., Nakashima, S., Mizutani, Y., Ozawa, Y., Watanabe, K., et al. (2018). Apple procyanidins promote mitochondrial biogenesis and proteoglycan biosynthesis in chondrocytes. *Scientific Reports*, 8(1), p. 7229.

Masumoto, S., Terao, A., Yamamoto, Y., Mukai, T., Miura, T., and Shoji, T. (2016). Non-absorbable apple procyanidins prevent obesity associated with gut microbial and metabolomic changes. *Scientific Reports*, 6(1), p. 31208.

McCormack, D., and McFadden, D. (2013). A review of pterostilbene antioxidant activity and disease modification. *Oxidative Medicine and Cellular Longevity*, 2013(1), p. 575482.

Mehraj, M., Das, S., Feroz, F., Waheed Wani, A., Dar, S.Q., Kumar, S., et al. (2024). Nutritional composition and therapeutic potential of pineapple peel—a comprehensive review. *Chemistry and Biodiversity*, 21(5), p. e202400315.

Menezes, G.M.S.D., Maia, M.D.S., Sousa, N.F.D., Santos, A.C.D., Cahino, L.M., Scotti, M.T., et al. (2025). Multiple targets directed screening of compounds from northeastern Brazilian fruits with activity in oxidative stress pathways associated with vascular damage throughout ageing. *Natural Product Research*, 8, pp. 1–5.

Michael, O.S., Bamidele, O., Ogheneovo, P., Ariyo, T.A., Adedayo, L.D., Oluranti, O.I., et al. (2021). Watermelon rind ethanol extract exhibits hepato-renal protection against lead induced-impaired antioxidant defenses in male Wistar rats. *Current Research in Physiology*, 4, pp. 252–259.

Morissette, A., Kropp, C., Songpadith, J.P., Junges Moreira, R., Costa, J., Mariné-Casadó, R., et al. (2020). Blueberry proanthocyanidins and anthocyanins improve metabolic health through a gut microbiota-dependent mechanism in diet-induced obese mice. *American Journal of Physiology-Endocrinology and Metabolism*, 318(6), pp. E965–E980.

Mustafa, A.M., Bastawesy, G.A., Hatem, S., Moussa, R.A.G., Hal, D.M., Fawzy, M.H., et al. (2025). Polyphenolic protection: the role of mangiferin in mitigating neurodegeneration and neuroinflammation. *Inflammopharmacology*, 33(8), pp. 4535–4552.

Nady, O.M., Said, S.E., Ibrahim, S.M., and Safwat, M.H. (2025). Bromelain alleviates 3-nitropropionic acid-induced Huntington's disease-like symptoms in rats through the interplay of SIRT1/PGC-1α and FOXO3a pathways. *BioFactors*, 51(6), p. e70054.

Nijum, M.R., Islam, M.Z., Chowdhury, A.M., Monalisa, K., and Jahan, M.I. (2025). Impact of foam mat drying methods on physicochemical properties, anthocyanin stability, and probiotic viability in jamun (*Syzygium cumini*) juice powders. *Journal of Food Science*, 90(12), p. e70730.

Odun-Ayo, F., Chetty, K., and Reddy, L. (2022). Determination of the ursolic and oleanolic acids content with the antioxidant capacity in apple peel extract of various cultivars. *Brazilian Journal of Biology*, 82, p. e258442.

Ongay, K.K., Granato, D., and Barreto, G.E. (2023). Comparison of antioxidant capacity and network pharmacology of phloretin and phlorizin against neuroinflammation in traumatic brain injury. *Molecules*, 28(3), p. 919.

Ortega-Hernández, E., Martinez-Alvarado, L., Acosta-Estrada, B.A., and Antunes-Ricardo, M. (2023). Solid-state fermented pineapple peel: a novel food ingredient with antioxidant and anti-inflammatory properties. *Foods*, 12(22), p. 4162.

Pan, C., Ni, H., Li, Z., Yang, K., Li, X., and Kakeya, H. (2025). Recent advances in bioactive natural products from plant fruits: structures, activities, and future prospects. *Plant Foods for Human Nutrition*, 80(4), pp. 1–17.

Parkar, S.G., Redgate, E.L., Wibisono, R., Luo, X., Koh, E.T.H., and Schröder, R. (2010). Gut health benefits of kiwifruit pectins: comparison with commercial functional polysaccharides. *Journal of Functional Foods*, 2(3), pp. 210–218.

Peng, Y., Cordiner, S.B., Sawyer, G.M., McGhie, T.K., Espley, R.V., Allan, A.C., et al. (2020). Kiwifruit with high anthocyanin content modulates NF-κB activation and reduces CCL11 secretion in human alveolar epithelial cells. *Journal of Functional Foods*, 65, p. 103734.

Pérez-Jiménez, J., and Saura-Calixto, F. (2018). Fruit peels as sources of non-extractable polyphenols or macromolecular antioxidants: analysis and nutritional implications. *Food Research International*, 111, pp. 148–152.

Phan, D.T.A., Le, H.P., Nguyen, H.T.T., Tran, T.H., Van Le, U., Van Le, M., et al. (2025). Therapeutic potential of phenolic-rich crude extract from *Syzygium cumini* (L.) Skeels leaves for Parkinson's disease. *Journal of Ethnopharmacology*, 1:358, p. 121012.

Puppala, E.R., Aochenlar, S.L., Shantanu, P.A., Ahmed, S., Jannu, A.K., Jala, A., et al. (2022a). Perillyl alcohol attenuates chronic restraint stress aggravated dextran sulfate sodium-induced ulcerative colitis by modulating TLR4/NF-κB and JAK2/STAT3 signaling pathways. *Phytomedicine*, 106, p. 154415.

Puppala, E.R., Jain, S., Saha, P., Rachamalla, M., Np, S., Yalamarthi, S.S., et al. (2022b). Perillyl alcohol attenuates rheumatoid arthritis via regulating TLR4/NF-κB and Keap1/Nrf2 signaling pathways. *Phytomedicine*, 97, p. 153926.

Qanash, H., Yahya, R., Bakri, M.M., Bazaid, A.S., Qanash, S., Shater, A.F., and Abdelghany, T.M. (2022). Anticancer, antioxidant, antiviral and antimicrobial activities of kei apple (*Dovyalis caffra*) fruit. *Scientific Reports*, 12(1), p. 5914.

Rana, T.S., Bansode, R.R., Rana, J.P., and Williams, L.L. (2025). A systematic review: polyphenol's effect on food allergy via microbiome modulation. *Frontiers in Microbiology*, 16, p. 1673472.

Rojas-Ocampo, E., Torrejón-Valqui, L., Muñóz-Astecker, L.D., Medina-Mendoza, M., Mori-Mestanza, D., and Castro-Alayo, E.M. (2021). Antioxidant capacity, total phenolic content and phenolic compounds of pulp and bagasse of four Peruvian berries. *Heliyon*, 7(8), p. e07703.

Sarapultsev, A., Gusev, E., Komelkova, M., Utepova, I., Luo, S., and Hu, D. (2023). JAK-STAT signaling in inflammation and stress-related diseases: implications for therapeutic interventions. *Molecular Biomedicine*, 4(1), p. 40.

Sari, D.R.T., Cairns, J.R.K., Safitri, A., and Fatchiyah, F. (2019). Virtual prediction of delphinidin-3-O-glucoside and peonidin-3-O-glucoside as anti-inflammatory agents of TNF-α signaling. *Acta Informatica Medica*, 27(3), p. 152.

Satheesh Babu, A.K., Petersen, C., Iglesias-Carres, L., Paz, H.A., Wankhade, U.D., Neilson, A.P., et al. (2024). Blueberry intervention mitigates trimethylamine N-oxide by modulating gut microbes. *BioFactors*, 50(2), pp. 392–404.

Sekizawa, H., Ikuta, K., Ohnishi-Kameyama, M., Nishiyama, K., and Suzutani, T., 2019. Identification of the components in a *Vaccinium oldhamii* extract showing inhibitory activity against influenza virus adsorption. *Foods*, 8(5), p. 172.

Sezen, S.C., Demirtas, H., Yıldırım, A.K., Özer, A., Dursun, A.D., Kucuk, A., et al., 2025. Protective role of bromelain's antioxidant and anti-inflammatory effects in experimental lower limb ischemia–reperfusion injury. *Scientific Reports* 15, pp. 45355.

Shanely, R.A., Zwetsloot, J.J., Jurrissen, T.J., Hannan, L.C., Zwetsloot, K.A., Needle, A.R., et al., 2020. Daily watermelon consumption decreases plasma sVCAM-1 levels in overweight and obese postmenopausal women. *Nutrition Research*, 76, pp. 9–19.

Shaw, O.M., Hurst, R.D., Cooney, J., Sawyer, G.M., Dinnan, H., and Martell, S., 2021. Boysenberry and apple juice concentrate reduced acute lung inflammation and increased M2 macrophage-associated cytokines in an acute mouse model of allergic airways disease. *Food Science & Nutrition*, 9(3), pp. 1491–1503.

Sip, S., Gościniak, A., Szulc, P., Walkowiak, J., and Cielecka-Piontek, J., 2022. Assisted extraction with cyclodextrins as a way of improving the antidiabetic activity of *Actinidia* leaves. *Pharmaceutics*, 14(11), p. 2473.

Soares, J.R.P., Dos Santos, C.C., de Oliveira, L.M.G., Rocha Neto, H., Victor, M.M., França, E.L., et al., 2025. Synthesis of naringenin and senecioic acid ester derivatives and biological evaluation of the astrocyte antioxidant mechanism and reactivity after inflammatory stimulus. *International Journal of Molecular Sciences*, 26(5), p. 2215.

Suzuki, K., 2019. Chronic inflammation as an immunological abnormality and effectiveness of exercise. *Biomolecules*, 9(6), p. 223.

Tian, Z.X., Li, Y.F., Long, M.X., Liang, Q., Chen, X., Huang, D.M., et al., 2023. Effects of six different microbial strains on polyphenol profiles, antioxidant activity, and bioaccessibility of blueberry pomace with solid-state fermentation. *Frontiers in Nutrition*, 10, p. 1282438.

Tsouh Fokou, P.V., Kamdem Pone, B., Appiah-Oppong, R., Ngouana, V., Bakarnga-Via, I., Ntieche Woutouoba, D., et al., 2025. An update on antitumor efficacy of catechins: from molecular mechanisms to clinical applications. *Food Science & Nutrition*, 13(4), p. e70169.

Vlasiuk, E., Zawari, M., Storer, M., Maze, M.J., Williman, J., Chambers, S.T., et al., 2024. SunGold kiwifruit consumption restores adequate to optimal vitamin C status in people with a history of severe respiratory infections. *Antioxidants*, 13(3), p. 272.

Wallace, T.C., Bailey, R.L., Blumberg, J.B., Burton-Freeman, B., Chen, C.O., Crowe-White, K.M., et al., 2020. Fruits, vegetables, and health: a comprehensive narrative umbrella review of the science and recommendations for enhanced public policy to improve intake. *Critical Reviews in Food Science and Nutrition*, 60(13), pp. 2174–2211.

Wang, D., Wang, G., Lu, X., Liu, Z., Sun, S., Guo, H., et al., 2024a. Identification and mining of functional components of polyphenols in fruits of *Malus* germplasm resources based on multivariate analysis. *Foods*, 13(21), p. 3465.

Wang, K., Wang, Y., Chen, S., Gu, J., and Ni, Y., 2022a. Insoluble and soluble dietary fibers from kiwifruit (*Actinidia deliciosa*) modify gut microbiota to alleviate high-fat diet and streptozotocin-induced type 2 diabetes in rats. *Nutrients*, 14(16), p. 3369.

Wang, L., Lan, W., and Chen, D., 2024b. Blueberry (*Vaccinium* spp.) anthocyanins and their functions, stability, bioavailability, and applications. *Foods*, 13(17), p. 2851.

Wang, L., Pan, X., Jiang, L., Chu, Y., Gao, S., Jiang, X., et al., 2022b. The biological activity mechanism of chlorogenic acid and its applications in the food industry: a review. *Frontiers in Nutrition*, 9, p. 943911.

Wang, Y., Chen, J., Wang, Y., Zheng, F., Qu, M., Huang, Z., et al., 2022c. Cyanidin-3-O-glucoside extracted from the Chinese bayberry (*Myrica rubra* Sieb. et Zucc.) alleviates antibiotic-associated diarrhea by regulating gut microbiota and down-regulating inflammatory factors in the NF-κB pathway. *Frontiers in Nutrition*, 9, p. 970530.

Williamson, G., 2025. Bioavailability of food polyphenols: current state of knowledge. *Annual Review of Food Science and Technology*, 16. 16(1), p. 315–332.

Woźniak, Ł., Szakiel, A., Głowacka, A., Rozpara, E., Marszałek, K., and Skąpska, S., 2023. Triterpenoids of three apple cultivars—biosynthesis, antioxidative and anti-inflammatory properties, and fate during processing. *Molecules*, 28(6), p. 2584.

Wu, Q., Kong, Y., Liang, Y., Niu, M., Feng, N., Zhang, C., et al., 2023. Protective mechanism of fruit vinegar polyphenols against AGEs-induced Caco-2 cell damage. *Food Chemistry: X*, 19, p. 100736.

Yang, H.R., Zahan, M.N., Hwang, D.H., Prakash, R.L.M., Ravi, D.A., Hong, I.H., et al., 2025. The therapeutic potential of kiwi extract as a source of cysteine protease inhibitors on DNCB-induced atopic dermatitis in mice and human keratinocyte HaCaT cells. *International Journal of Molecular Sciences*, 26(4), p. 1534.

Yang, L., Zhang, L., Dong, L., Ma, Y., Zhao, L., Xu, R., et al., 2025a. Protective effects of cyanidin-3-O-glucoside against neurotoxin acrylamide through alleviating mitochondrial dysfunction. *Foods*, 14(22), p. 3826.

Yang, Q.N., Deng, W., Wu, D.T., Li, J., Liu, H.Y., Yan, H.L., et al., 2024. Characterization, antioxidant capacity, and anti-inflammatory activity of polyphenol-enriched extracts obtained from unripe, mature, and overripe fruits of red-fleshed kiwifruit cultivars. *Foods*, 13(18), p. 2860.

Yang, W., Chen, X., Wang, D., Liu, L., Ma, H., Wang, T., et al., 2025b. Rhoifolin prevents herpes simplex virus encephalitis by modulating SESN2 to regulate the Keap1/Nrf2 and AMPK/Nox pathways to inhibit neuro-oxidative damage in mice. *Phytomedicine*, p. 157621.

Yao, H., Flanagan, B.M., Williams, B.A., Wu, X., Mikkelsen, D., and Gidley, M.J., 2024. Differential effects of pectin-based dietary fibre type and gut microbiota composition on in v-itro fermentation outcomes. *Carbohydrate Polymers*, 339, p. 122284.

Yu, H., Lin, L., Zhang, Z., Zhang, H., and Hu, H., 2020. Targeting the NF-κB pathway for the therapy of diseases: mechanism and clinical study. *Signal Transduction and Targeted Therapy*, 5(1), p. 209.

Yu, X., Pu, H., and Voss, M., 2024. Overview of anti-inflammatory diets and their promising effects on noncommunicable diseases. *British Journal of Nutrition*, 132(7), pp. 898–918.

Yuan, M., Chen, X., Su, T., Zhou, Y., and Sun, X., 2021. Supplementation of kiwifruit polyphenol extract attenuates high-fat diet-induced intestinal barrier damage and inflammation via reshaping gut microbiome. *Frontiers in Nutrition*, 8, p. 702157.

Zargar, S., Al-Majed, A.R.A., and Wani, T.A., 2018. Potentiating and synergistic effect of grapefruit juice on the antioxidant and anti-inflammatory activity of aripiprazole against hydrogen peroxide-induced oxidative stress in mice. *BMC Complementary and Alternative Medicine*, 18(1), p. 106.

Zhang, J., Fan, W., Wu, H., Yao, Y., Jin, L., Chen, R., et al., 2024. Naringenin attenuated airway cilia structural and functional injury induced by cigarette smoke extract via IL-17 and cAMP pathways. *Phytomedicine*, 126, p. 155053.

Zhang, J., Gao, N., Shu, C., Cheng, S., Sun, X., Liu, C., et al., 2021. Phenolics profile and antioxidant activity analysis of kiwi berry (*Actinidia arguta*) flesh and peel extracts from four regions in China. *Frontiers in Plant Science*, 12, p. 689038.

Zhu, C.W., Lü, H., Du, L.L., Li, J., Chen, H., Zhao, H.F., et al., 2023. Five blueberry anthocyanins and their antioxidant, hypoglycemic, and hypolipidemic effects in vitro. *Frontiers in Nutrition*, 10, p. 1172982.

Zielinska, D., Laparra-Llopis, J.M., Zielinski, H., Szawara-Nowak, D., and Giménez-Bastida, J.A., 2019. Role of apple phytochemicals, phloretin and phloridzin, in modulating processes related to intestinal inflammation. *Nutrients*, 11(5), p. 1173.

7 Medicinal Fruits and Berries in Cardiovascular Health

Arpan Modi, Malik Nasibullah, Chaachouay Noureddine, Pankaj Sharma, and Azamal Husen

7.1 INTRODUCTION

Cardiovascular diseases (CVDs) are currently the most commonly acknowledged cause of death and are still a great concern for public health, despite the progress of medical science and technology. According to the World Health Organization (2023), CVDs are responsible for one-third of all deaths, with ischemic heart disease and stroke being the leading causes. The increasing incidence of CVDs is inextricably linked to modifiable lifestyle risk factors, including unhealthy dietary habits, a sedentary lifestyle, obesity, and metabolic disorders. As a result, dietary modification has become the cornerstone of CVD prevention and treatment. Among the nutritional factors, fruits and berries have been extensively studied for their regular and crucial role in lowering cardiovascular risk. Epidemiological studies and meta-analyses have clearly shown that increased fruit consumption is inversely associated with the incidence of coronary heart disease, stroke, and cardiovascular deaths (Dauchet et al., 2006; He et al., 2007; David et al., 2025). Fruits are a key component of heart-healthy diets like the Mediterranean diet and the Dietary Approaches to Stop Hypertension (DASH) diet, which have been found to lower lipid levels, blood pressure, and inflammation (Diab et al., 2023). Fruits are also an excellent source of essential nutrients like dietary fiber, potassium, vitamins, and a host of phytochemicals, which work synergistically on cardiovascular physiology.

In the nutritional and functional food context, medicinal fruits can be described as fruits that have health-promoting properties in addition to their nutritional value because of their high content of biological compounds. Berries, a type of fruit, are known to be rich in polyphenols and anthocyanins and have a high cardioprotective property. Medicinal berries such as blueberries, strawberries, cranberries, and blackberries have high flavonoid content, making them different from other fruits. The high flavonoid content in these fruits makes them distinct from other fruits. The evidence from prospective cohort studies supports the use of berries in the prevention of cardiovascular risk (Cassidy, 2018). The cardioprotective properties of berries are associated with their multifaceted mechanisms, which include the suppression of oxidative stress, chronic inflammation, augmentation of nitric oxide levels, and prevention of LDL cholesterol oxidation. From a public health standpoint, the promotion of fruit and berry consumption is regarded as a cost-effective dietary approach. Inadequate fruit consumption is identified as one of the primary risk factors for non-communicable diseases by the World Health Organization (2023). The application of medicinal fruits and berries in people's diets is an effective approach to counteract the risk factors of CVDs. The aim of this book chapter is to emphasize the importance of medicinal fruits and berries as natural and alternative remedies in the prevention and treatment of CVDs, among others.

7.2 BIOACTIVE COMPONENTS IN FRUITS AND BERRIES

Fruits and berries are known to be a rich source of various bioactive compounds, which have an important role in maintaining cardiovascular health. Bioactive compounds have various

 DOI: 10.1201/9781003688259-7

mechanisms of action, which include the regulation of oxidative stress, inflammation, lipid metabolism, endothelial function, and vascular homeostasis. The health benefits of fruits and berries are primarily attributed to their high content of phenolic compounds, vitamins, dietary fibers, minerals, and phytonutrients (**Liu**, 2013).

7.2.1 Phenolic Compounds, Flavonoids, Anthocyanins, and Tannins

Phenolic compounds are one of the most significant bioactive compounds found in different fruits and berries. Phenolic compounds are classified into phenolic acids, flavonoids, anthocyanins, and tannins. Polyphenols have the ability to scavenge reactive oxygen species (ROS), chelate metal ions, and inhibit or activate enzyme activity associated with oxidative stress (Rao and Zheng, 2025). Flavonoids, including quercetin, catechins, and kaempferol, are found in apples, grapes, citrus fruits, and berries. Gao et al. (2021) have reported that the increased consumption of flavonoid-rich foods is associated with a lower risk of myocardial infarction and hypertension. These compounds have increased the bioavailability of endothelial nitric oxide, thus improving vasodilation and vascular function. Anthocyanins, responsible for the red, blue, and purple colors of berries, namely blueberries, strawberries, blackberries, and cranberries, have been well studied for their cardiovascular effects. Various investigations have exhibited that the consumption of anthocyanins improves endothelial function, reduces arterial stiffness, and reduces blood pressure (Jennings et al., 2012; Fairlie-Jones et al., 2017; Mozos et al., 2021). Tannins, which are noticed in grapes, pomegranates, and berries, have lipid-lowering and anti-atherogenic properties because of the inhibition of low-density lipoprotein (LDL) oxidation (Chung et al., 1998; Asgary et al., 2018; Islam et al., 2021).

7.2.2 Vitamins, Dietary Fiber, Minerals, and Phytonutrients

Fruits and berries are also rich in terms of essential vitamins such as C and E, which play important roles in maintaining cardiovascular health. Vitamin C, found in citrus fruits, strawberries, and guava, is a powerful antioxidant that helps to regenerate endothelial nitric oxide, thereby improving vascular integrity (Carr & Maggini, 2017; Doseděl et al., 2021). Vitamin E, found in fruits such as avocado and kiwi, protects cell membranes from oxidative damage and inhibits lipid peroxidation (Ameer, 2016; Abdelqader et al., 2023). Another important component of fruits and berries is dietary fiber, which has been proven to provide important benefits to cardiovascular health. Soluble fibers such as pectin and beta-glucans reduce blood cholesterol levels by binding to bile acids and inhibiting cholesterol absorption in the intestine (Anderson et al., 2009). The consumption of fruits that are rich in dietary fiber has been demonstrated to lower lipid abnormalities, blood pressure, and the prevalence of coronary heart disease (Threapleton 2014; Nweze et al., 2021). The minerals potassium, magnesium, and calcium in bananas, berries, and citrus fruits play a role in regulating blood pressure and heart rate. Potassium, in particular, has been demonstrated to have an inverse relationship with the risk of stroke and hypertension (Aburto et al., 2013). Moreover, they contain a wide range of phytochemicals such as carotenoids and phytosterols, which play a role in anti-inflammatory and cholesterol-lowering mechanisms (Bohn et al., 2017).

7.2.3 Mechanisms of Action: Antioxidant and Anti-Inflammatory Effects

The anti-cardiac assets of fruits and berries are primarily due to their antioxidant and anti-inflammatory features. Oxidative stress is one of the major contributing factors in the development of atherosclerosis, endothelial dysfunction, and myocardial injury. Antioxidants in fruits work by counteracting the effects of free radicals, preventing lipid peroxidation, and protecting vascular cells from oxidative damage (Basu and Lyons, 2012; Boeing et al., 2012). Inflammation is another major contributing factor in the progression of CVDs. Polyphenols and flavonoids have been shown to modulate the inflammatory pathways by inhibiting the activation of nuclear factor-kappa B (NF-κB) and reducing the expression of pro-inflammatory cytokines like interleukin-6 (IL-6) and tumor necrosis factor-alpha (TNF-α) (Zamora-Ros et al., 2016, 2018).

7.2.4 Bioavailability and Metabolic Aspects

Bioavailability of bioactive compounds in fruits and berries is a concern, although their health benefits are well recognized (Figure 7.1). The high metabolism of polyphenols in the gastrointestinal tract and liver results in the formation of metabolites with different biological activities (Manach et al., 2004). The contribution of gut microbiota to the transformation of complex polyphenols into more bioavailable metabolites should not be underestimated, as it influences individual differences in the response to fruit consumption. The food matrix, processing, and co-consumption of fruits with lipids or proteins may affect the absorption and biological activity of bioactive compounds. Although fresh fruits have been found to have optimal bioavailability, processed fruits such as juices and extracts may influence the stability of polyphenols (Crozier et al., 2009).

7.3 MECHANISMS OF CARDIOVASCULAR PROTECTION

The cardiovascular protective features of medicinal fruits and berries are attributed to a number of biological mechanisms. These mechanisms are centered on the major pathological processes associated with CVDs, which include oxidative stress, inflammation, endothelial dysfunction, dyslipidemia, hypertension, and thrombosis. The combined effect of bioactive compounds, for

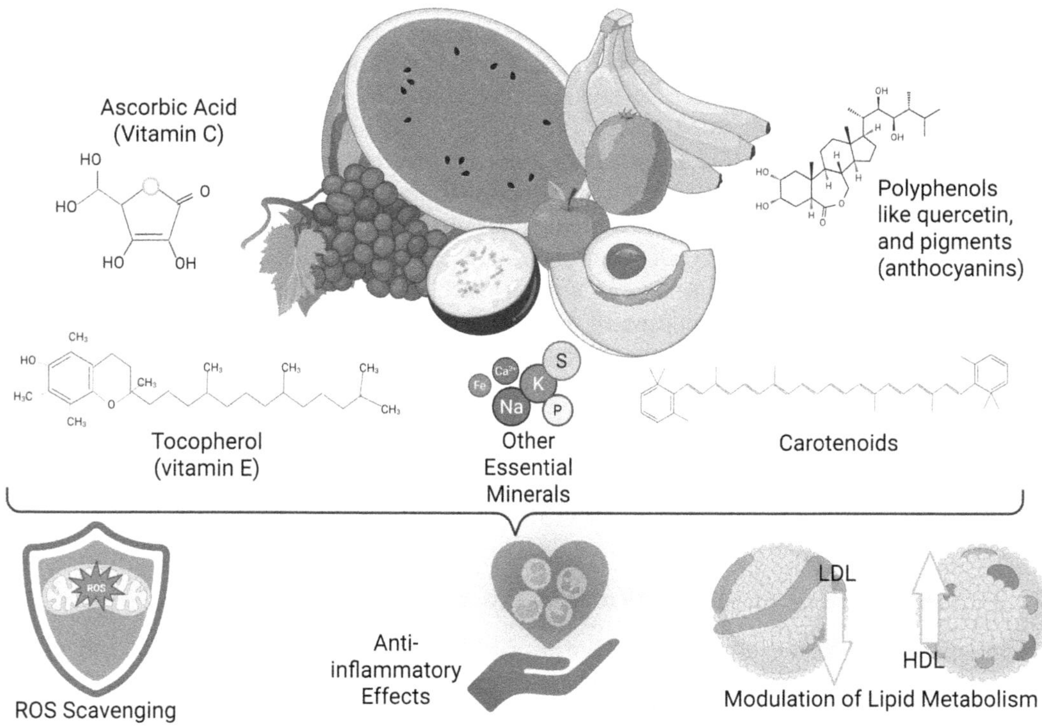

FIGURE 7.1 Bioactive compounds reported in fruits and berries and their cardioprotective mechanisms.

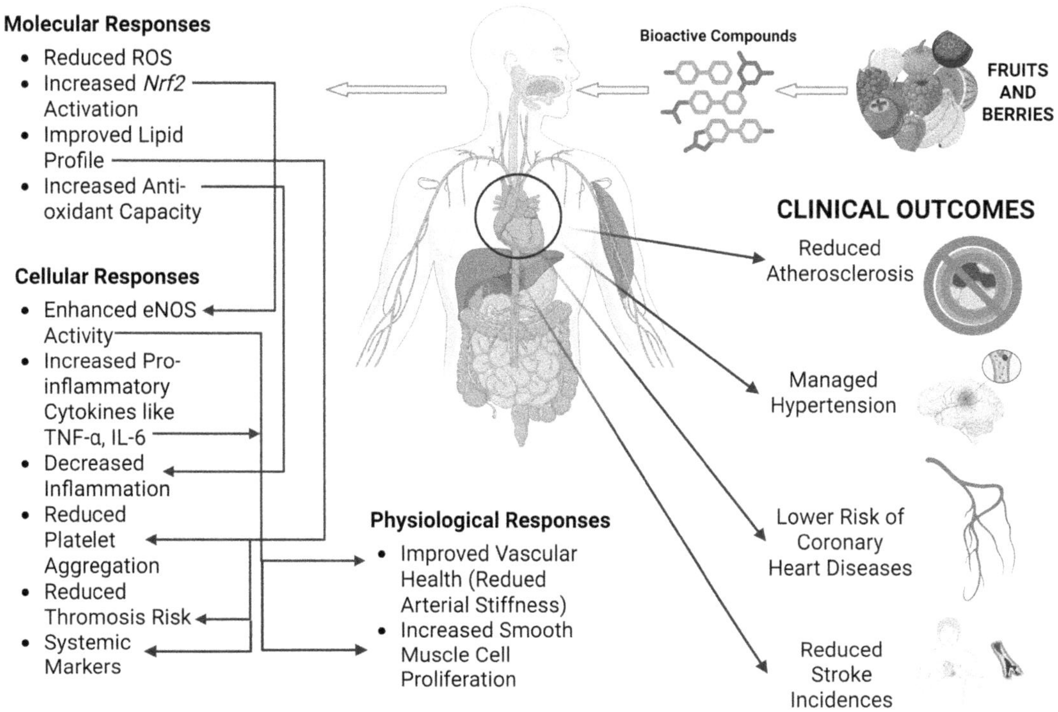

FIGURE 7.2 Mechanistic pathways linking fruits and berries to improved cardiovascular outcomes.

example, polyphenols, flavonoids, anthocyanins, vitamins, and dietary fibers, is attributed to the cardioprotective properties of fruit- and berry-rich diets (Liu, 2013; Rodríguez-Mateos et al., 2014a) (Figure 7.2).

7.3.1 Reduction of Oxidative Stress and Free Radical Scavenging

Oxidative stress is an important contributing factor in the development and progression of atherosclerosis and other CVDs. The overproduction of ROS results in lipid peroxidation, endothelial dysfunction, and the oxidation of LDL, which plays an important role in the formation of plaques. Fruits and berries have been shown to be rich in antioxidants that neutralize free radicals and enhance the body's antioxidant defenses (Lobo et al., 2010). Polyphenols and flavonoids work by suppressing ROS by donating electrons to stabilize reactive molecules. Based on Manach et al. (2004), these molecules also increase the expression of antioxidant enzymes such as superoxide dismutase, catalase, and glutathione peroxidase. Berries rich in anthocyanins, such as blueberries and blackcurrants, have been shown to have a potent antioxidant effect, which reduces the biomarkers of oxidative stress and LDL oxidation. By reducing oxidative stress, fruits and berries reduce cardiovascular risk by maintaining the integrity of blood vessels.

7.3.2 Anti-Inflammatory Pathways

Low-grade chronic inflammation is one of the key elements of CVD, leading to endothelial dysfunction, plaque instability, and thrombosis. The bioactive compounds in fruits and berries have anti-inflammatory features at the molecular and or cellular levels. Polyphenols suppress key inflammatory transcription factors, including NF-κB, thereby reducing the expression of pro-inflammatory cytokines such as TNF-α, IL-6, and C-reactive protein (Zamora-Ros et al., 2016). Higher dietary intake of anthocyanins was reported to be associated with decreased concentrations of inflammatory biomarkers and improved arterial stiffness, as reported by Cassidy et al. (2016). It has also been noticed that the pomegranate polyphenols have been proven to suppress macrophage activation and foam cell formation, which is a crucial event in the pathogenesis of atherosclerosis (Aviram & Rosenblat, 2013). Therefore, the anti-inflammatory properties of medicinal fruits and berries suppress plaque instability and inflammation of the vasculature.

7.3.3 Improvement of Endothelial Function and Nitric Oxide Signaling

Endothelial dysfunction is the first sign of cardiovascular disease, which is defined by decreased bioavailability of nitric oxide (NO). NO is a major vasodilator that regulates vascular tone, inhibits platelet aggregation, and reduces smooth muscle cell proliferation. However, oxidative stress and inflammation decrease NO synthesis and increase its degradation, resulting in vascular stiffness and hypertension. Fruits and berries are rich in bioactive compounds that have a positive regulatory effect on endothelial function due to their ability to increase NO production and bioavailability. Rodríguez-Mateos et al. (2014b) demonstrated that the consumption of anthocyanin-rich berries exerts a significant effect on enhancing flow-mediated dilation, which is a major indicator of endothelial function. Flavonoids, such as quercetin and catechins, enhance the activation of endothelial nitric oxide synthase (eNOS), which increases NO bioavailability and stimulates vasodilation (Heiss et al., 2010).

7.3.4 Modulation of Lipid Profile

Dyslipidemia, which is high LDL cholesterol and low levels of high-density lipoprotein (HDL) cholesterol, is a major risk factor for the progression of CVDs. Fruits and berries have been demonstrated to affect lipid metabolism in various ways, including the inhibition of cholesterol absorption, the stimulation of the secretion of bile acids, and the protection of LDL from oxidation. The dietary fiber present in fruits stimulates the secretion of bile acids in the intestines, thus increasing the secretion of cholesterol and decreasing the serum levels of LDL cholesterol (Anderson et al., 2009). Polyphenols have also been demonstrated to inhibit the synthesis of cholesterol in the liver and to stimulate the function of HDL. Efficacy trials have demonstrated that the regular consumption of berries and pomegranate juice can significantly decrease the levels of total and LDL cholesterol while increasing the levels of HDL (Aviram et al., 2000; Basu et al., 2010). The lipid-lowering effect is of major importance in the management of atherosclerosis.

7.3.5 Regulation of Blood Pressure

Hypertension is a major contributor to cardiovascular morbidity and mortality globally. Fruits and berries are known to affect blood pressure due to their potassium content, antioxidant properties, and vascular effects. Potassium helps in the excretion of sodium in the urine, lowers vascular resistance, and helps maintain normal cardiac function (Aburto et al., 2013). In addition, flavonoids enhance the distensibility of arteries and decrease oxidative stress in the arterial wall, causing a subsequent decrease in systolic and diastolic blood pressure. Meta-analyses conducted by Cassidy et al. (2011) and Rodríguez-Mateos et al. (2014a) showed that the consumption of berry fruits and flavonoid-rich fruits is linked to small but significant reductions in blood pressure, especially in people with prehypertension or metabolic syndrome.

7.3.6 Effects on Platelet Aggregation and Thrombosis

Platelet aggregation and thrombosis are major pathophysiologies in acute cardiovascular events such as myocardial infarction and stroke. The antithrombotic properties of bioactive compounds in fruits and berries inhibit platelet activation and aggregation. Flavonoids inhibit platelet signaling pathways, decreasing thromboxane A2 production and platelet adhesion to injured endothelium (Hubbard et al., 2004). Grape and berry polyphenols have been demonstrated to increase clotting time and decrease platelet hyperactivity, thus lowering the possibility of thrombotic events (Keevil et al., 2000). The antiplatelet property, together with improved endothelial function and lipid properties, further adds to the cardioprotective properties of medicinal fruits and berries.

7.4 EVIDENCE FROM HUMAN STUDIES AND CLINICAL TRIALS

Human studies and clinical trials represent a significant source of evidence regarding the cardiovascular effects of medicinal fruits and berries. Although animal studies and observational research have revealed a high cardioprotective potential, randomized clinical trials and meta-analyses provide more precise information on their impact on blood pressure, lipid levels, glycemic indices, endothelial function, and arterial stiffness. In recent years, there has been a growing tendency toward berry-based therapies because of their high polyphenol and anthocyanin content (Rodríguez-Mateos et al., 2014a; Basu et al., 2010).

7.4.1 Evidence from Randomized Controlled Trials

7.4.1.1 Effects on Blood Pressure

Among the effects of fruit and berry consumption that have been most extensively studied is the blood pressure-lowering effect. Several randomized control trials (RCTs) have shown that the use of cranberry, cherry, and berry supplements results in a modest but significant decrease in systolic blood pressure (SBP) and diastolic blood pressure (DBP). In a randomized crossover study, Dohadwala et al. (2011) showed that the daily consumption of blueberry drinks significantly improved endothelial function and decreased SBP in patients with risk factors for CVDs. In a similar study, Basu et al. (2010) showed that the use of blueberry supplements for eight weeks decreased SBP and DBP in obese patients with metabolic syndrome. The antihypertensive properties of tart cherry supplements have also been validated, and this can be attributed to the high anthocyanin and potassium levels (Keane et al., 2016). Trials conducted on cranberry supplements have also produced similar outcomes. A clinical trial conducted by Ruel et al. (2008) demonstrated that the consumption of cranberry juice reduced blood pressure and enhanced vascular function in men with coronary artery disease. The results of the above studies cumulatively indicate the importance of polyphenols derived from berries in the regulation of blood pressure by enhancing nitric oxide bioavailability and suppressing oxidative stress.

7.4.1.2 Lipid Profile and Glycemic Outcomes

Numerous clinical trials have examined the impact of medicinal fruits and berries on lipid profiles and glycemic indices, which are two major risk factors for CVDs. Pomegranate juice has been widely investigated for its lipid-lowering abilities. Aviram et al. (2000) demonstrated significant decreases in LDL oxidation and total cholesterol in patients with carotid artery stenosis who were administered pomegranate juice. Berry supplements have also been found to be effective in enhancing lipid profiles. In a randomized clinical trial, Erlund et al. (2008) demonstrated that the administration of mixed berries significantly increased HDL cholesterol and decreased LDL cholesterol in healthy individuals. Similarly, cranberry and blueberry supplements have been linked to decreased triglyceride concentrations and improved LDL particle size (Basu et al., 2010). In terms of glycemic control, berries have been found to be highly effective owing to their low glycemic index and high polyphenol content. Törrönen et al. (2010) have reported that berry supplements decreased postprandial glucose and insulin concentrations in patients with impaired glucose tolerance. This is particularly important, given the close link between diabetes, metabolic syndrome, and cardiovascular disease.

7.4.1.3 Endothelial Function and Arterial Stiffness

Numerous clinical trials have investigated the impact of medicinal fruits and berries on lipid profiles and glycemic indices, which are two major risk factors for CVDs. Pomegranate juice has been widely investigated for its lipid-lowering abilities. Aviram et al. (2000) demonstrated significant decreases in LDL oxidation and total cholesterol in patients with carotid artery stenosis who were administered pomegranate juice. Berry supplements have also been found to be effective in enhancing lipid profiles. In a randomized clinical trial, Erlund et al. (2008) demonstrated that the administration of mixed berries significantly increased HDL cholesterol and decreased LDL cholesterol in healthy individuals. Similarly, cranberry and blueberry supplements have been linked to decreased triglyceride concentrations and improved LDL particle size (Basu et al., 2010). In terms of glycemic control, berries have been found to be highly effective owing to their low glycemic index and high polyphenol content. Törrönen et al. (2012) demonstrated that berry supplements decreased postprandial glucose and insulin concentrations in patients with impaired glucose tolerance. This is particularly important, given the close link between diabetes, metabolic syndrome, and cardiovascular disease.

7.4.2 Evidence from Meta-Analyses and Systematic Reviews

Meta-analyses suggest a wider insight into the cardiovascular characteristic features of fruits and berries. Cassidy et al. (2011) conducted a meta-analysis and found that a higher dietary intake of flavonoids was significantly associated with a decrease in SBP and DBP, especially in people with high blood pressure. Berries were significantly effective in lowering LDL cholesterol, total cholesterol, and oxidative stress. In a similar meta-analysis, Huang et al. (2016) noticed improvements in fasting glucose and insulin sensitivity with the consumption of berries, but the effect was small on BMI. The reduction in LDL cholesterol and SBP seems to be the most consistent result, while the effect on BMI and glycemic parameters seems to be more inconsistent. These results highlight the importance of intervention duration, dose, and population characteristics in shaping the results (Zamora-Ros et al., 2016).

7.4.3 Strengths and Limitations of the Evidence

The developing human literature provides strong support for the cardioprotective effects of medicinal fruits and berries. The strengths of the current literature include the increasing number of high-quality RCTs, the use of standardized biomarkers, and the assessment of clinically relevant outcomes such as blood pressure, lipids, and endothelial function. However, there are still some limitations of the present literature. The vast majority of the present trials are of a short-term nature and small size, and hence it is difficult to generalize the findings. There is also some variability in the fruits used, the dose, the processing, and the bioactive compounds, and hence it is difficult to compare the findings of the different studies (Manach et al., 2005). In addition, there could be some individual differences in the metabolic breakdown of the bioactive compounds present in fruits. Despite these constraints, the level of consistency in the findings, particularly with respect to blood pressure reduction, LDL cholesterol lowering, and improvement of endothelial function, attests to the use of medicinal fruits and berries in the prevention and treatment of CVD. A brief summary of key clinical trials on berries and CVD is provided in Table 7.1. However, future large-scale clinical trials are required to establish the appropriate levels of consumption.

TABLE 7.1
Summary of Key Clinical Trials on Berries and CVD Markers.

Conditions	Berry Type	Primary Outcomes
Patients with carotid artery stenosis	Pomegranate	Decreased LDL oxidation, total cholesterol, and systolic blood pressure (SBP)
Obese adults with metabolic syndrome	Blueberry	Decreased SBP and diastolic blood pressure (DBP), oxidative stress, and improved insulin sensitivity
Patients with coronary artery disease	Cranberry	Increased Endothelial function and decreased SBP
Healthy adults	Mixed berries	Increased HDL cholesterol, decreased platelet aggregation, and blood pressure
Middle-aged adults (cohort-based intervention)	Anthocyanin-rich berries	Decreased risk of myocardial infarction, improved vascular health
Healthy and at-risk adults	Tart cherry	Decreased arterial stiffness and SBP
Men with coronary artery disease	Cranberry	Decreased cardiometabolic risk markers and increased vascular function
Healthy adults and individuals at CVD risk	Blueberry	Increased nitric oxide bioavailability and Flow-mediated dilation (FMD)
Individuals with impaired glucose tolerance	Mixed berries	Decreased postprandial glucose and insulin response
Meta-analysis of RCTs	Various berries	Decreased LDL cholesterol, SBP, and oxidative stress

7.5 SPECIFIC MEDICINAL FRUITS AND BERRIES AND THEIR CARDIOVASCULAR PROPERTIES

There is some variation in the phytochemical content of medicinal fruits and berries, but there are similarities in their cardioprotective effects, which are ascribed to polyphenols, flavonoids, vitamins, and dietary fibers. Human and animal research has demonstrated that the regular consumption of specific fruits and berries can have a beneficial effect on lipid metabolism, vascular function, inflammation, and oxidative stress, which are significant factors in cardiovascular health. This chapter will highlight the cardiovascular properties of specific medicinal berries and fruits that have been proven by clinical and scientific evidence.

7.5.1 Berries

7.5.1.1 Blueberry: Anthocyanins and Lipid Protection

Blueberries (*Vaccinium* spp.) are among the most extensively studied berries in cardiovascular research due to their exceptionally high anthocyanin content. Anthocyanins are very potent flavonoids, which are responsible for the deep blue color and high antioxidant capacity of blueberries. Cassidy et al. (2013) reported that a higher regular dietary intake of anthocyanin-rich foods, particularly blueberries, was significantly related to a reduced risk of myocardial infarction. Clinical trials have also established their lipid-protective effects. Basu et al. (2010) have shown that freeze-dried blueberry supplementation in obese patients with metabolic syndrome significantly reduced systolic and DBP and improved insulin sensitivity, in addition to reducing markers of oxidative stress. Moreover, blueberry polyphenols have been shown to inhibit the oxidation of LDL cholesterol, a key event in atherogenesis, and to enhance HDL-associated antioxidant activity (Rodríguez-Mateos et al., 2014a).

7.5.1.2 Strawberry: Antioxidant and Anti-Inflammatory Effects

Strawberries (*Fragaria × ananassa*) have been reported to have a high antioxidant activity due to their richness in vitamin C, ellagic acid, anthocyanins, and flavonols. Epidemiological investigations have suggested that the regular consumption of strawberries is associated with reduced levels of inflammatory biomarkers, including C-reactive protein (CRP) and interleukin-6 (IL-6) (Giampieri et al., 2014). In human intervention investigations, the consumption of strawberries has a positive effect on lipid profiles and oxidative stress. Alvarez-Suarez et al. (2014) noticed that the total cholesterol and LDL-cholesterol levels were reduced in hyperlipidemic patients who received strawberry supplements. The anti-inflammatory effects of strawberries are ascribed to the inhibition of NF-κB signaling and the suppression of pro-inflammatory cytokines, which result in the reduction of vascular inflammation and endothelial dysfunction.

7.5.1.3 Cranberry: Blood Pressure and Endothelial Function

Cranberries (*Vaccinium macrocarpon*) are recognized for their high content of proanthocyanidins and flavonols, which have strong protective properties against the vascular system. The use of cranberries has been shown to improve endothelial function, which is an indicator of the progression of cardiovascular disease. Dohadwala et al. (2011) reported that the regular consumption of cranberry juice significantly improved flow-mediated dilation in patients with coronary artery disease. Moreover, Ruel et al. (2008) reported that the supplementation of low-calorie cranberry juice significantly reduced SBP and oxidized LDL levels in men with cardiovascular risk factors. The polyphenols in cranberries increase nitric oxide availability, leading to improved vasodilation and reduced arterial stiffness. These reports suggest the potential use of cranberries as a functional food ingredient for blood pressure regulation.

7.5.1.4 Blackberry and Raspberry: Polyphenols and Cholesterol Regulation

Blackberries (*Rubus fruticosus*) and raspberries (*Rubus idaeus*) are recognized for their ellagitannins, anthocyanins, and dietary fiber content, which makes them an important source for cholesterol regulation and metabolism. As stated by Erlund et al. (2008), the intake of a combination of berries, including blackberries and raspberries, exhibited the ability to increase HDL cholesterol levels and prevent platelet aggregation in healthy individuals. The high fiber content in berries also makes them a potential source for cholesterol-lowering by promoting the secretion of bile acids and inhibiting the absorption of cholesterol in the intestines. It was noticed that ellagitannins are metabolized to urolithins by gut microbiota, which possess anti-inflammatory and anti-atherogenic properties (Selma et al., 2009). Although there is no study available on the individual properties of blackberries and raspberries, there is sufficient evidence available to support their effectiveness in improving lipid profiles and lowering the risk of CVDs.

7.5.2 Other Medicinal Fruits

7.5.2.1 Pomegranate: Polyphenols and Nitric Oxide Preservation

Pomegranate (*Punica granatum*) has a high content of ellagitannins, punicalagins, and anthocyanins, which have a strong antioxidant activity. Aviram et al. (2000) were the first to demonstrate the clinical potency of pomegranate juice, which significantly decreased LDL oxidation, total cholesterol, and SBP in patients with carotid artery stenosis. The polyphenols of pomegranate have been shown to preserve the bioactivity of nitric oxide by inhibiting its oxidative degradation. This improves endothelial function and vascular tone. Chronic consumption of pomegranate juice has also been associated with reduced carotid intima-media thickness, suggesting a slower rate of atherosclerosis progression (Aviram et al., 2004).

7.5.2.2 Grapes: Resveratrol and LDL Oxidation

Grapes (*Vitis vinifera*), mainly red and purple grapes, have been recognized for their richness in resveratrol, flavanols, and anthocyanins. Resveratrol has been extensively investigated for its cardio-protective properties, which include the suppression of LDL oxidation, improvement of endothelial function, and attenuation of platelet aggregation (Baur & Sinclair, 2006). Clinical investigations have shown the efficacy of grape and grape-based supplements, such as red grape juice and wine polyphenols, in improving vascular function and reducing oxidative stress markers (Zern & Fernandez, 2005). Resveratrol has also been verified to activate SIRT1 signaling, which plays a role in vascular aging and metabolism.

7.5.2.3 Goji Berries: HDL Cholesterol Increase and Triglyceride Lowering

Goji berries (*Lycium barbarum*) have recently gained popularity due to their cardiometabolic effects. Goji berries are rich in polysaccharides, carotenoids (zeaxanthin), flavonoids, and vitamins. Clinical trials have shown that the supplementation of goji berries can improve lipid metabolism by increasing HDL cholesterol levels and reducing triglycerides (Amagase & Farnsworth, 2011). Besides lipid management, goji berries possess antioxidant and anti-inflammatory activities, which can potentially prevent endothelial dysfunction. The polysaccharides in goji berries have been shown to improve the activity of antioxidant enzymes and reduce oxidative stress, which is directly linked to the development of cardiovascular disease (Potterat, 2010). Although human clinical trials are limited in number, available evidence supports the use of goji berries as a functional fruit for the enhancement of cardiovascular health. Medicinal fruits and berries possess cardioprotective effects through synergistic mechanisms that include lipid management, antioxidant effects, anti-inflammatory effects, and endothelial function modulation.

7.6 DIETARY PATTERNS AND RECOMMENDATIONS

Dietary change is one of the most effective and sustainable approaches for the prevention and management of CVDs. In this regard, fruits and berries play a crucial role owing to their high bioactive compound, fiber, vitamins, and mineral content, which work synergistically to provide cardiovascular protection. Epidemiological investigations have unequivocally exhibited that dietary habits rich in fruits and berries can significantly lower the risk of developing hypertension, dyslipidemia, coronary heart disease, and stroke. Heart-healthy eating habits, including the Mediterranean diet, the DASH diet, and plant-based diets, highlight the importance of regular fruit and berry consumption. These diets recommend the consumption of whole fruits as snacks, berries in breakfast meals (yogurt and oats), and fruit-based ingredients in salads and desserts as healthy substitutes for processed foods. Berries like blueberries, strawberries, and cranberries are specifically recommended due to their high polyphenol density per unit of energy. Moreover, from the point of view of cardiovascular health, whole fruits are always preferred over fruit juices and extracts. The presence of fiber in whole fruits plays a crucial role in reducing cholesterol levels, regulating glucose, and promoting satiety, which cannot be achieved through fruit juices. Moreover, fruit juices are rich in free sugars, which can negate the potential benefits of fruits in protecting cardiovascular health due to excessive intake. Fruit extracts and supplements may be rich in concentrated bioactive compounds; however, their effectiveness depends on their formulation, bioavailability, and dose, and they should be used as supplements, not as substitutes for whole fruits (Manach et al., 2005). It has also been observed that the cardiovascular effects of fruits and berries can be enhanced when they are used in an organized, holistic dietary program. Fruits and berries, when used in combination with vegetables, whole grains, nuts, seeds, and healthy fats such as olive oil, can help in the enhancement of bioavailability and anti-inflammatory and antioxidant effects (Jacobs Jr. and Tapsell, 2013). For example, berries and nuts can have a synergistic effect on lipid metabolism, and berries and vegetables that are rich in nitrates can further enhance the production of endothelial nitric oxide (Liu, 2013). These effects emphasize the importance of considering the overall dietary patterns and not just the individual nutrients. However, for the maximum prevention of CVD, it is recommended that one emphasizes diversity, seasonality, and regularity of fruit and berry consumption. There should be a focus on fresh or minimally processed fruits, with a special emphasis on berries and polyphenol-rich fruits such as pomegranate and grapes. In patients with existing CVD risk factors, the use of fruits in organized dietary programs may have additional benefits.

7.7 FUTURE PERSPECTIVES AND CONCLUSION

Despite the overwhelming evidence available on the cardiovascular efficacy of medicinal fruits and berries, there are still some gaps that make it difficult to apply them in clinical practice. Most of the human studies conducted on fruits and berries are of a short-term nature and conducted on small populations, making it difficult to form a conclusive opinion about their long-term cardiovascular efficacy. There is also some confusion regarding the dosage and route of administration. The studies conducted vary greatly in the level of intake, processing, and polyphenol concentration. Moreover, the bioavailability and metabolism of polyphenols remain an issue since polyphenols are largely metabolized by the gut microbiota, resulting in the production of compounds with diverse biological activities. In conclusion, medicinal fruits and berries are a promising dietary approach for the prevention of cardiovascular disease. The biological compounds in medicinal fruits and berries have been shown to alleviate oxidative stress and inflammation, improve endothelial dysfunction, modulate lipid metabolism, and prevent platelet aggregation. The consumption of a variety of fruits and berries as part of a healthy diet is a safe, practical, and cost-effective approach to promote long-term cardiovascular health.

REFERENCES

Abdelqader, A., Obeidat, M. D., Al-Rawashdeh, M. S. and Alrazak, A. A. (2023). The role of vitamin E as an antioxidant and preventing damage caused by free radicals. *Journal of Life Science and Applied Research*, *4*(2), pp. 88–95.

Aburto, N. J., Hanson, S., Gutierrez, H., Hooper, L., Elliott, P. and Cappuccio, F. P. (2013). Effect of increased potassium intake on cardiovascular risk factors and disease: systematic review and meta-analyses. *BMJ*, *346*, p. f1378.

Alvarez-Suarez, J. M., Giampieri, F., Tulipani, S., Casoli, T., Di Stefano, G., González-Paramás, A. M., Santos-Buelga, C., Busco, F., Quiles, J. L., Cordero, M. D. and Bompadre, S. (2014). One-month strawberry-rich anthocyanin supplementation ameliorates cardiovascular risk, oxidative stress markers and platelet activation in humans. *The Journal of Nutritional Biochemistry*, *25*(3), pp. 289–294.

Amagase, H. and Farnsworth, N. R. (2011). A review of botanical characteristics, phytochemistry, clinical relevance in efficacy and safety of *Lycium barbarum* fruit (goji). *Food Research International*, *44*(7), pp. 1702–1717.

Ameer, K. (2016). Avocado as a major dietary source of antioxidants and its preventive role in neurodegenerative diseases. In: Essa, M., Guillemin, G. (eds) *The Benefits of Natural Products for Neurodegenerative Diseases*, pp. 337–354. Springer, Cham.

Anderson, J. W., Baird, P., Davis Jr., R. H., Ferreri, S., Knudtson, M., Koraym, A., Waters, V. and Williams, C. L. (2009). Health benefits of dietary fiber. *Nutrition Reviews*, *67*(4), pp. 188–205.

Asgary, S., Rastqar, A. and Keshvari, M. (2018). Functional food and cardiovascular disease prevention and treatment: a review. *Journal of the American College of Nutrition*, *37*(5), pp. 429–455.

Aviram, M., Dornfeld, L., Rosenblat, M., Volkova, N., Kaplan, M., Coleman, R., Hayek, T., Presser, D. and Fuhrman, B. (2000). Pomegranate juice consumption reduces oxidative stress, atherogenic modifications to LDL, and platelet aggregation: studies in humans and in atherosclerotic apolipoprotein E–deficient mice. *The American Journal of Clinical Nutrition*, *71*(5), pp. 1062–1076.

Aviram, M. and Rosenblat, M. (2013). Pomegranate for your cardiovascular health. *Rambam Maimonides Medical Journal*, *4*(2), p. e0013.

Aviram, M., Rosenblat, M., Gaitini, D., Nitecki, S., Hoffman, A., Dornfeld, L., Volkova, N., Presser, D., Attias, J., Liker, H. and Hayek, T. (2004). Pomegranate juice consumption for 3 years by patients with carotid artery stenosis reduces common carotid intima–media thickness, blood pressure and LDL oxidation. *Clinical Nutrition*, *23*(3), pp. 423–433.

Basu, A., Du, M., Leyva, M. J., Sanchez, K., Betts, N. M., Wu, M., Aston, C. E. and Lyons, T. J. (2010). Blueberries decrease cardiovascular risk factors in obese men and women with metabolic syndrome. *The Journal of Nutrition*, *140*(9), pp. 1582–1587.

Basu, A. and Lyons, T. J. (2012). Strawberries, blueberries, and cranberries in the metabolic syndrome: clinical perspectives. *Journal of Agricultural and Food Chemistry*, *60*(23), pp. 5687–5692.

Baur, J. A. and Sinclair, D. A. (2006). Therapeutic potential of resveratrol: the in vivo evidence. *Nature Reviews Drug Discovery*, *5*(6), pp. 493–506.

Boeing, H., Bechthold, A., Bub, A., Ellinger, S., Haller, D., Kroke, A., Leschik-Bonnet, E., Müller, M. J., Oberritter, H., Schulze, M. and Stehle, P. (2012). Critical review: vegetables and fruit in the prevention of chronic diseases. *European Journal of Nutrition*, *51*(6), pp. 637–663.

Bohn, T., Desmarchelier, C., Dragsted, L. O., Nielsen, C. S., Stahl, W., Rühl, R., Keijer, J. and Borel, P. (2017). Host-related factors explaining interindividual variability of carotenoid bioavailability and tissue concentrations in humans. *Molecular Nutrition & Food Research*, *61*(6), p. 1600685.

Carr, A. C. and Maggini, S. (2017). Vitamin C and immune function. *Nutrients*, *9*(11), p. 1211.

Cassidy, A., 2018. Berry anthocyanin intake and cardiovascular health. *Molecular Aspects of Medicine*, 61, pp. 76–82.

Cassidy, A., Bertoia, M., Chiuve, S., Flint, A., Forman, J. and Rimm, E. B. (2016). Habitual intake of anthocyanins and flavanones and risk of cardiovascular disease in men. *The American Journal of Clinical Nutrition*, *104*(3), pp. 587–594.

Cassidy, A., Mukamal, K.J., Liu, L., Franz, M., Eliassen, A.H. and Rimm, E.B., 2013. High anthocyanin intake is associated with a reduced risk of myocardial infarction in young and middle-aged women. *Circulation*, *127*(2), pp. 188–196.

Cassidy, A., O'Reilly, É. J., Kay, C., Sampson, L., Franz, M., Forman, J. P., Curhan, G. and Rimm, E. B. (2011). Habitual intake of flavonoid subclasses and incident hypertension in adults. *The American Journal of Clinical Nutrition*, *93*(2), pp. 338–347.

Chung, K. T., Wong, T. Y., Wei, C. I., Huang, Y. W. and Lin, Y. (1998). Tannins and human health: a review. *Critical Reviews in Food Science and Nutrition*, *38*(6), pp. 421–464.

Crozier, A., Jaganath, I. B. and Clifford, M. N. (2009). Dietary phenolics: chemistry, bioavailability and effects on health. *Natural Product Reports*, *26*(8), pp. 1001–1043.

Dauchet, L., Amouyel, P., Hercberg, S. and Dallongeville, J. (2006). Fruit and vegetable consumption and risk of coronary heart disease: a meta-analysis of cohort studies. *The Journal of Nutrition*, *136*(10), pp. 2588–2593.

David, F. T., Abah, M. A., Oladosu, M. A., Agida, O. D., Clive, N. O., Adebukola, A. M., Nnabuko, O. M. and Maryanne, O. U. (2025). The impact of plant-based diets on cardiovascular health: a comprehensive review. *Plant Science Archives*, *4*. https://doi.org/10.51470/PSA

Diab, A., Dastmalchi, L. N., Gulati, M. and Michos, E. D. (2023). A heart-healthy diet for cardiovascular disease prevention: where are we now? *Vascular Health and Risk Management*, *19*, pp. 237–253.

Dohadwala, M. M., Holbrook, M., Hamburg, N. M., Shenouda, S. M., Chung, W. B., Titas, M., Kluge, M. A., Wang, N., Palmisano, J., Milbury, P. E. and Blumberg, J. B. (2011). Effects of cranberry juice consumption on vascular function in patients with coronary artery disease. *The American Journal of Clinical Nutrition*, *93*(5), pp. 934–940.

Doseděl, M., Jirkovský, E., Macáková, K., Krčmová, L. K., Javorská, L., Pourová, J., Mercolini, L., Remião, F., Nováková, L., and Mladěnka, P. (2021). Vitamin C—sources, physiological role, kinetics, deficiency, use, toxicity, and determination. *Nutrients*, *13*(2), p. 615.

Erlund, I., Koli, R., Alfthan, G., Marniemi, J., Puukka, P., Mustonen, P., Mattila, P. and Jula, A. (2008). Favorable effects of berry consumption on platelet function, blood pressure, and HDL cholesterol. *The American Journal of Clinical Nutrition*, *87*(2), pp. 323–331.

Fairlie-Jones, L., Davison, K., Fromentin, E. and Hill, A. M. (2017). The effect of anthocyanin-rich foods or extracts on vascular function in adults: a systematic review and meta-analysis of randomised controlled trials. *Nutrients*, *9*(8), p. 908.

Gao, Q., Dong, J. Y., Cui, R., Muraki, I., Yamagishi, K., Sawada, N., Iso, H., Tsugane, S. and Japan Public Health Center-Based Prospective Study Group. (2021). Consumption of flavonoid-rich fruits, flavonoids from fruits and stroke risk: a prospective cohort study. *British Journal of Nutrition*, *126*(11), pp. 1717–1724.

Giampieri, F., Alvarez-Suarez, J. M. and Battino, M. (2014). Strawberry and human health: effects beyond antioxidant activity. *Journal of Agricultural and Food Chemistry*, *62*(18), pp. 3867–3876.

He, F. J., Nowson, C. A., Lucas, M. and MacGregor, G. A. (2007). Increased consumption of fruit and vegetables is related to a reduced risk of coronary heart disease: meta-analysis of cohort studies. *Journal of Human Hypertension*, *21*(9), pp. 717–728.

Heiss, C., Keen, C. L. and Kelm, M. (2010). Flavanols and cardiovascular disease prevention. *European Heart Journal*, *31*(21), pp. 2583–2592.

Huang, H., Chen, G., Liao, D., Zhu, Y. and Xue, X. (2016). Effects of berries consumption on cardiovascular risk factors: a meta-analysis with trial sequential analysis of randomized controlled trials. *Scientific Reports*, *6*(1), p. 23625.

Hubbard, G. P., Wolffram, S., Lovegrove, J. A. and Gibbins, J. M. (2004). Ingestion of quercetin inhibits platelet aggregation and essential components of the collagen-stimulated platelet activation pathway in humans. *Journal of Thrombosis and Haemostasis*, *2*(12), pp. 2138–2145.

Islam, S. U., Ahmed, M. B., Ahsan, H. and Lee, Y. S. (2021). Recent molecular mechanisms and beneficial effects of phytochemicals and plant-based whole foods in reducing LDL-C and preventing cardiovascular disease. *Antioxidants*, *10*(5), p. 784.

Jacobs Jr., D. R. and Tapsell, L. C. (2013). Food synergy: the key to a healthy diet. *Proceedings of the Nutrition Society*, *72*(2), pp. 200–206.

Jennings, A., Welch, A. A., Fairweather-Tait, S. J., Kay, C., Minihane, A. M., Chowienczyk, P., Jiang, B., Cecelja, M., Spector, T., MacGregor, A. and Cassidy, A. (2012). Higher anthocyanin intake is associated with lower arterial stiffness and central blood pressure in women. *The American Journal of Clinical Nutrition*, *96*(4), pp. 781–788.

Keane, K. M., George, T. W., Constantinou, C. L., Brown, M. A., Clifford, T. and Howatson, G. (2016). Effects of Montmorency tart cherry (*Prunus cerasus* L.) consumption on vascular function in men with early hypertension. *The American Journal of Clinical Nutrition*, *103*(6), pp. 1531–1539.

Keevil, J. G., Osman, H. E., Reed, J. D. and Folts, J. D. (2000). Grape juice, but not orange juice or grapefruit juice, inhibits human platelet aggregation. *The Journal of Nutrition*, *130*(1), pp. 53–56.

Liu, R. H. (2013). Health-promoting components of fruits and vegetables in the diet. *Advances in Nutrition*, *4*(3), pp. 384S–392S.

Lobo, V., Patil, A., Phatak, A. and Chandra, N. (2010). Free radicals, antioxidants and functional foods: impact on human health. *Pharmacognosy Reviews*, *4*(8), p. 118.

Manach, C., Scalbert, A., Morand, C., Rémésy, C. and Jiménez, L. (2004). Polyphenols: food sources and bioavailability. *The American Journal of Clinical Nutrition*, *79*(5), pp. 727–747.

Manach, C., Williamson, G., Morand, C., Scalbert, A. and Rémésy, C. (2005). Bioavailability and bioefficacy of polyphenols in humans. I. Review of 97 bioavailability studies. *The American Journal of Clinical Nutrition*, *81*(1), pp. 230S–242S.

Mozos, I., Flangea, C., Vlad, D. C., Gug, C., Mozos, C., Stoian, D., Luca, C. T., Horbańczuk, J. O., Horbańczuk, O. K. and Atanasov, A. G. (2021). Effects of anthocyanins on vascular health. *Biomolecules*, *11*(6), p. 811.

Nweze, C. C., Nebechukwu, E. W. and Bawa, M. Y. (2021). Dietary fiber and risk of coronary heart diseases. *GSC Advanced Research and Reviews*, *9*(3), pp. 1–9.

Potterat, O. (2010). Goji (*Lycium barbarum* and *L. chinense*): phytochemistry, pharmacology and safety in the perspective of traditional uses and recent popularity. *Planta Medica*, *76*(01), pp. 7–19.

Rao, M. J. and Zheng, B. (2025). The role of polyphenols in abiotic stress tolerance and their antioxidant properties to scavenge reactive oxygen species and free radicals. *Antioxidants*, *14*(1), p. 74.

Rodriguez-Mateos, A., Heiss, C., Borges, G. and Crozier, A. (2014a). Berry (poly)phenols and cardiovascular health. *Journal of Agricultural and Food Chemistry*, *62*(18), pp. 3842–3851.

Rodriguez-Mateos, A., Pino-García, R. D., George, T. W., Vidal-Diez, A., Heiss, C. and Spencer, J. P. (2014b). Impact of processing on the bioavailability and vascular effects of blueberry (poly)phenols. *Molecular Nutrition & Food Research*, *58*(10), pp. 1952–1961.

Ruel, G., Pomerleau, S., Couture, P., Lemieux, S., Lamarche, B. and Couillard, C. (2008). Low-calorie cranberry juice supplementation reduces plasma oxidized LDL and cell adhesion molecule concentrations in men. *British Journal of Nutrition*, *99*(2), pp. 352–359.

Selma, M. V., Espín, J. C. and Tomas-Barberan, F. A. (2009). Interaction between phenolics and gut microbiota: role in human health. *Journal of Agricultural and Food Chemistry*, *57*(15), pp. 6485–6501.

Threapleton, D. E. (2014). *Cardiovascular disease risk and dietary fibre intake in the United Kingdom Women's Cohort Study*. University of Leeds.

Törrönen, R., Sarkkinen, E., Tapola, N., Hautaniemi, E., Kilpi, K. and Niskanen, L. (2010). Berries modify the postprandial plasma glucose response to sucrose in healthy subjects. *British Journal of Nutrition*, *103*(8), pp. 1094–1097.

Törrönen, R., Sarkkinen, E., Niskanen, T., Tapola, N., Kilpi, K. and Niskanen, L., 2012. Postprandial glucose, insulin and glucagon-like peptide 1 responses to sucrose ingested with berries in healthy subjects. *British Journal of Nutrition, 107*(10), pp. 1445–1451.

World Health Organization. (2023). Cardiovascular diseases (CVDs). https://www.who.int/news-room/fact-sheets/detail/cardiovascular-diseases-(cvds)

Zamora-Ros, R., Cayssials, V., Jenab, M., Rothwell, J. A., Fedirko, V., Aleksandrova, K., Tjønneland, A., Kyrø, C., Overvad, K., Boutron-Ruault, M. C. and Carbonnel, F. (2018). Dietary intake of total polyphenol and polyphenol classes and the risk of colorectal cancer in the European Prospective Investigation into Cancer and Nutrition (EPIC) cohort. *European Journal of Epidemiology, 33*(11), pp. 1063–1075.

Zamora-Ros, R., Knaze, V., Rothwell, J. A., Hémon, B., Moskal, A., Overvad, K., Tjønneland, A., Kyrø, C., Fagherazzi, G., Boutron-Ruault, M. C. and Touillaud, M. (2016). Dietary polyphenol intake in Europe: the European Prospective Investigation into Cancer and Nutrition (EPIC) study. *European Journal of Nutrition, 55*(4), pp. 1359–1375.

Zern, T. L. and Fernandez, M. L. (2005). Cardioprotective effects of dietary polyphenols. *The Journal of Nutrition, 135*(10), pp. 2291–2294.

8 Fruits and Berries as Natural Detoxifiers for the Body

Muhammad Zahoor, Sumaira Naz, and Sajad Khan

8.1 INTRODUCTION

Human nutrition is the process by which food and fruit substances can be transferred into body tissues. These substances are essential for providing energy to the body, significantly supporting physical and mental activities (Hazim et al. 2024). Detoxification, being one of the basic natural life processes, removes various types of toxins present in the body (Aralbaeva et al. 2024). The human body possesses several important detoxification organs and systems, such as the kidneys, liver, skin, and digestive system (Ramalho Mercé 2023). Toxins arise both exogenously and endogenously. Exogenous sources include air pollutants, pesticides, insecticides, etc., while metabolic pathways produce them endogenously. Constant exposure to these sources increases the risk of high levels of toxins (Shuai et al. 2022). The detoxification system removes toxins through various mechanisms. Proper diet and regular exercise are key factors in the efficiency of the detoxification system. Fruits and berries are well-known natural detoxifiers due to the presence of larger amounts of antioxidants, dietary fiber, and polyphenols. It also encompasses vital phytochemicals that can help to keep us healthy. Similarly, it also protects against various diseases and provides a healthy, well-balanced, and active lifestyle. Daily intake of a higher amount of fruit and vegetables is effective in reducing obesity while maintaining a perfect weight, as well as reducing cholesterol by lowering blood pressure (Rudrapal et al. 2024). Among foods, particularly fruits and berries, are a natural and easily available way to support the detox organs (Urugo and Tringo 2023). A high level of water content is found in these that promote hydration. It has enriched dietary fiber, which improves bowel movements, and antioxidants for protecting cells from oxidative stress. Several foods consist of organic acids, enzymes, and other vital natural compounds that promote digestion, enhance liver function, and help in removing harmful substances from the body (Lakra and Gahlawat 2022). The major detox organs, including the liver, kidneys, digestive system, and skin, are found in humans to eliminate toxins. Among these, the liver plays a key role in metabolism, preserving and regulating glucose and lipid levels, and supporting energy metabolism. The liver is the center of key processes like glycolysis, glycogenosis, and gluconeogenesis, which are the main mechanisms in maintaining the blood sugar level under a wide range of environmental conditions (Alamri 2018). Kidneys carry out blood purification, water and solute balance, thereby playing a key role in homeostasis. It is the most pivotal and highly specialized organ that helps in the proper regulation of electrolytes, water, and acid-base balance. It effectively controls the ionic composition and volume of body fluids, their pH level, and osmotic concentration, as well as significantly produces hormones, reabsorbs ions, nutrients, and water from the plasma ultrafiltrate (Gerich 2010).

8.2 UNDERSTANDING THE CONCEPT OF THE DETOXIFICATION SYSTEM

8.2.1 Detoxification

Detoxification is the process of transforming harmful waste products and toxins into useful forms for safe elimination from the body (Black 2020). Detoxification or biotransformation of xenobiotics

DOI: 10.1201/9781003688259-8

and drugs is generally associated with liver metabolism, but colonocytes of the gastrointestinal tract are also equally important in mediating these processes (Beyerle et al. 2015). Detoxification involves two major metabolic pathways, such as phase I and phase II enzyme systems.

8.2.2 Phase I Cytochrome P450 Enzymes

Primary phase I process of detoxification, oxygen is added to toxic compounds to make a reactive site on it, in which the cytochrome P450 super-family of enzymes (CYP450) is normally the first defense utilized by the body to bio-transform xenobiotics, steroid-hormones, and other drugs. The microsomal membrane-bound, heme thiolate proteins are found primarily in the liver, as well as inside the lung, kidney, enterocytes, and even in the brain. They perform reduction, oxidation, and peroxidation of numerous exogenous and endogenous substrates (Danielson 2005; Standish 2006). Among these, CYP450 enzymes have a pivotal role in addition of reactive groups like carboxyl, hydroxyl, or an amino-containing group through reduction, oxidation, by/or hydrolysis reactions (Hodges and Minich 2015). The initial reaction forms a reactive electrophilic species, which causes oxidative damage to the cellular system. The genetic expression of such enzymes depends on the 90% metabolization of presently utilized drugs (Chen et al. 2011). Another important CYP1 enzyme, which plays a role in metabolizing hormones, procarcinogens, and pharmaceuticals. The significant function of these enzymes is the bioactivation of carcinogenic polycyclic aromatic hydrocarbons (PAHs), polychlorinated biphenyls (PCBs), heterocyclic aromatic amines/amides, and many harmful environmental toxins (James et al. 2008). The low activity of this enzyme enhances the risk of testicular cancer (Vistisen et al. 2004). Another important CYP2A-E enzyme has been involved in the metabolism of hormones, pharmaceuticals, xenobiotics, as well as other endogenous compounds such as fatty acids, glycerol, and ketones (Nissar et al. 2013). It also has a role in several morbidities. 2-E1 metabolizes neural agents like ethanol, chlorzoxazone, and isoflurane, halothane, and bioactivates procarcinogenic aflatoxin B1 and nitrosamines (Sheweita 2005). CYP2E1-induced oxidative stress reduces the action of insulin by suppressing GLUT-4 expression (Armoni et al. 2014). Similarly, the CYP3A4 enzyme is expressed mostly in the liver but less in the kidney, and has an important function in the catabolism of more than 50% of all drugs (Leibelt et al. 2003). CYP4 enzymes perform their function in drug metabolism (Figure 8.1). It significantly metabolizes MCTs (medium-chain triglycerides) and bioactivates carcinogenic and pneumotoxic compounds (Ioannides 1999; Baer and Rettie 2006).

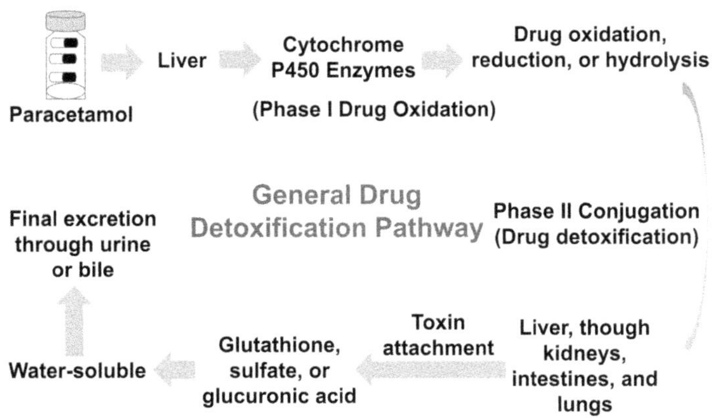

FIGURE 8.1 The general drug detoxification pathway involves phase I and phase II enzymes.

8.2.3 Phase II Conjugation Enzymes

When CYP450 enzymes metabolize xenobiotics by making them hydrophilic through reactions, then their reactive sites bind with endogenous water-loving substances, and the reaction is known as phase-II detoxification. The conjugation has been involved in the transformation of some hydrophilic compounds (through their concerned enzymes), such as acetyl-group (N acetyl transferases), sulfate (sulfotransferases), a methyl-group (N and O methyl transferases, glutathione (glutathione-transferases, glucuronic acid (glucuronyl-transferases), and an amino-acid (amino acid-transferases). The overall activity of these enzymes causes higher hydrophilicity of the metabolite and improved excretion in the bile and/or urine (Xu et al. 2005). UDP-Glucuronosyltransferases consist of multiple proteins and even subfamilies, which effectively eliminate the bio-transformed toxins in feces and urine, along with metabolizing bilirubin and steroid hormones (Jenkinson et al. 2013). The glucuronidation process mostly takes place inside the liver and also in tissues like the small intestine, which primarily metabolize the covalent link of glucuronic acid from UDP glucuronic acid to various functional groups on molecules (Strassburg et al. 2000). UGT1A1 primarily conjugates bilirubin in hepatocytes after excreting it alongside bile into the intestinal lining (Wells et al. 2004). Among the enzymes, sulfo-transferases (SULTs) are a superfamily containing enzymes that transfer the sulfuryl group given by 3′-phosphoadenosine-5′-phosphosulfate (PAPs) to hydroxyl or amine groups, known as sulfonation or sulfurylation (James and Ambadapadi 2013). It is a silly occurrence in the adrenal gland, liver, intestine, brain, and skin tissues. The reduction in productivity of these enzymes is associated with interference in androgen, estrogen, and thyroid hormone levels (Wang and James 2005; Ung and Nagar 2007). The SULT-enzyme activity depends on a decreasing reserve of inorganic sulfates (McFadden 1996). Glutathione S-transferases involve the process of attaching a glutathione group to a biotransformed metabolite, which induces reactive oxygen species productivity and gene transcription containing the antioxidant-responsive element (ARE) and the xenobiotic-responsive element (XRE) (Hayes and Pulford 1995). The diverse types of amino acids, including glycine and taurine, whether endogenous or of exogenous origin from the dietary sources, can be employed in attaching to molecules to facilitate excretion. N-acetyltransferases (NAT) are a type of enzyme that transfers an acetyl group to alter aromatic amines or hydrazines into aromatic amides as well as hydrazides. Methyltransferases are phase II enzymes that perform methylation to reduce disease risk (Figure 8.1) (Jancova et al. 2010). S-adenosyl-L-methionine (SAMe) donates a conjugating donor compound like the methionine group in methyltransferase reactions (Kohalmy and Vrzal 2011). Catechol O-methyltransferase (COMT) has received considerable attention due to its important function in the detoxification of estrogen (Yager 2015).

8.2.4 Major Detox Organs

8.2.4.1 Liver

The liver is the primary organ responsible for detoxification in the body. It is effectively converting drugs, alcohols, and other toxins into less or non-harmful forms (Granados et al. 2021). The presence of a unique enzyme system in the liver, particularly cytochrome P450, allows it to combat a vast variety of toxins (Lee et al. 2025). The liver also performs other important functions, such as neutralization of ammonia into urea for safe excretion via the urethral system (Lynch and Cohen 1995). Similarly, it also plays a significant role in storing vital nutrients as well as filtering heavy metals. The production of bile is responsible for removing waste through the digestive tract (Maldonado-Valderrama et al. 2011). The liver produces antioxidants to diminish oxidative stress, boosts immune defense to clear pathogens, and helps to regulate lipid metabolism and blood glucose during detoxification. The liver is considered the most vital metabolic center as it controls several imperative metabolic functions, such as detoxification and nutrient metabolism. The function of these fundamental hepatic processes is well balanced to "a rhythmically modifying systemic environment" (Reinke and Asher 2016). The liver exploits it to anticipate persistent, wide-ranging ecological modifications (Zhang and Kay 2010). After a meal, digestion and glucose absorption from the intestine into the circulation result in a higher level of blood glucose (10mM) in the portal vein, and movement of glucose into the hepatocytes through liver glucose transporter 2 (GLUT2). This enables the excessive transformation of glucose into glycogen by the mechanism known as glycogenesis. Insulin stimulates the glycogenesis process, which dephosphorylates and triggers glycogen synthase (Mitra and Metcalf 2012). Blood glucose levels drop to a normal level two to three hours after meal consumption (the postprandial period), and initiation of glycogenolysis due to a low ratio of glucagon and insulin, while the hormone associated with glucagon boosts the action of adenylyl cyclase to elevate the level of cyclic adenosine monophosphate, which boosts protein kinase A. The liver regulates glucose homeostasis, as well as the skeletal muscle, brain, and pancreas. The current regulation involves important protocols, such as glucagon and insulin secretion, as well as glucose uptake and synthesis. Liver cells store glucose, ketone bodies, and fatty acids, which carry them out to the other tissues to produce energy (Chiang 2014). After dietary protein digestion, free amino acids become the final product, and the cellular breakdown of protein further moves into the stage of cell degradation. In this process, the alpha-amino group is removed and transformed into ammonia, as well as integrated into urea for excretion. The urea cycle enzymes play a role in the urea cycle. In this cycle, one molecule of CO_2 and two molecules of NH_3 are utilized to remove the toxic ammonia. Approximately 80% of urea is formed in the liver, and its excretion occurs through urination by the kidneys. However, all wastes that are produced in the liver are excreted through urination, but some of it is transferred through colonic bacteria into ammonia. It is a pivotal phenomenon in the de novo protein synthesis, particularly in hunger, while 10% of the urea is excreted through feces from the body (Rostom 2018). When dietary lipids are digested and absorbed, the free fatty acids and other compounds that are obtained from the adipose tissue are finally entered into the liver. Later on, it is used to synthesize membranes, energy, or deposited in the form of triglycerides. The diffusion of free fatty acids (FFAs) occurs within the hepatocytes. Long-chain fatty acids are enabled by the addition of fatty acid translocase (FAT) and fatty acid transport protein (FATP). The binding of FFA to fatty acid binding protein (FABP)-1 inside the hepatocytes. This causes FFAs to move cytoplasm to diverse compartments within the cell either for metabolism or for regulation of gene expression through interaction with transcription factors, like peroxisome proliferator-activated receptor (PPAR)α, thereby defending the liver from lipotoxicity (C. H. Chen et al. 2006).

8.2.4.2 Kidney

Kidneys are highly specialized organs playing a central role in maintaining electrolytes and water balance, together with acid-base level (Imenez Silva and Mohebbi 2022). Kidneys have a unique system that filters blood and removes metabolic wastes, thus aiding in the detoxification system of the body. Its functions are secreting various hormones, reabsorbing ions, water, and minerals from blood plasma in ultrafiltration (Ahmed et al. 2025). Excess water, minerals, salts, and toxins are removed in the urine. Kidneys effectively maintain the physiological range of minerals and keep the PH balanced. Its primary role is to remove various nitrogenous wastes, particularly urea and creatinine. The kidney secretes vital hormones to regulate blood pressure, and also performs its function as the main detoxification system by clearing water-soluble toxins. The primary function of the kidneys is to excrete the waste solutes through the filtration (glomeruli) and active secretion (tubules) processes (Natochin Iu 1973). The wastes are composed of endogenous and exogenous wastes product like drugs, toxins, and environmental pollutants. The liver processes the drug's metabolites, which are later properly excreted by the kidneys through the urinary system. Toxic accumulation is almost negligible in healthy kidneys. It also regulates the productivity of red blood cells through erythropoietin. Nephron is the basic and functional unit of the kidney, that are composed of several parts; each part performs a specific function. The glomerulus is the first part through which blood passes. Water and solutes with a size of ≈60 kDa travel in capillaries due to the high pressure of blood in the capillaries. After the glomerulus, the filtrate (size, ≈120 mL min^{-1} to ≈170 L d^{-1}) passes through the proximal tubule for the reabsorption of water and important components. Furthermore, the proximal tubule actively secretes solute/toxin, produces hormones, and activates metabolic processes. Specialized transporters are present at proximal tubule epithelial cells (PTEC) that direct the basolateral uptake, luminal release, and protein-bound solutes with high selectivity and capacity. These exceptional properties enable PTEC to be especially sensitive to ischemia and xenobiotic-induced toxicity and subsequent AKI. Lastly, blood passes from the proximal tubule to the loop of Henle, where the reabsorption of excessive water and solutes (electrolytes) occurs, with this process making urine more concentrated and blood more purified, thereby maintaining fluid and electrolyte homeostasis. The urine is then removed through the urinary system (Nakhoul and Batuman 2011; Masereeuw et al. 2014).

8.2.4.3 Skin

Skin is the largest and most important detoxifier organ of the body, performing its function as a protective barrier between the environment and our inner organs. Skin is also an excretory organ that is composed of 3 to 4 million eccrine sweat glands with a similar weight as one kidney (i.e., 100 g). These are distributed over the whole surface of the human body. A single person perspires around 10 liters per day (Sato 1977). Sweating is an important phenomenon for the elimination of water-soluble endogenous and exogenous, both toxic and bioactive substances, such as drugs, metals, steroids, and cytokines (Zhou et al. 2012). Xenobiotic metabolism results in the production of ROS and toxic intermediate products (Zhou et al. 2009). The effective removal of xenobiotics through sweating might cause low-toxic intermediate products and hence prevent oxidative stress. Skin contains two key layers, namely the epidermis and the dermis. The topmost layer of the skin, termed the epidermis, is composed of the stratum corneum, which is rich in lipids, allowing the skin to function as a barrier (Ismayilzada et al. 2024). As an excretory organ, skin possesses sweat glands which, during sweating, remove excess water, salts, and some urea. Sweating also lowers the temperature of the body, thus helping in thermoregulation during hot conditions. Its vital function is to prevent harmful substances from entering the body (Baker 2019). The skin secretes sebum, which acts as a main trapping agent and eradicates impurities. Dead cells are constantly removed from the skin, resulting in the formation of new cells. Drinking an adequate amount of water regulates hydration, which effectively removes toxins through sweat. Skin health is responsible for maintaining the efficacy of the detox system. Proper blood circulation to the skin enhances the supply of nutrients and oxygen. In a study, it was reported that nicotinamide loading significantly improves toxic intermediate metabolite N^1-methylnicotinamide in the sweat (Zhou et al. 2009). Another study stated that sauna causes a higher skin temperature and promotes sweating to protect us from oxidative stress, as well as ease symptoms of intoxication (Ross and Sternquist 2012). Sebum production occurs in the sebaceous gland, which is found all over the human body except on the palms and feet. Sebum consists of squalene, cholesterol, triglycerides, fatty acids, and wax esters. Human sebum contains main components such as fatty acids and triglycerides (57%). These are significantly higher than those of many other species, such as rabbits and rodents (triglycerides and free fatty acids <10%) (Smith and Thiboutot 2008). The higher fat diet increases sebum production, while caloric deprivation reduces its productivity (Downing et al. 1972). The higher energy intake mostly enhances the triglycerides, esters, and cholesterol excretion in sebum, but not of squalene (Pochi et al. 1970). The primary function of sebum secretion is the removal of excess cholesterol and lipids from the body, and hence it properly maintains lipid and cholesterol homeostasis. Overall, perspiration is considered a crucial antioxidant mechanism, particularly for those individuals with a genetic enzyme deficiency in dealing with ROS and xenobiotics (Zech et al. 1983).

8.2.5 Digestive System

The digestive system plays a major role in eliminating solid waste from the body by breaking down food into the smallest particles and helps in absorbing crucial nutrients (Moini and Ferdowsi 2024).

The presence of the gut microbiome supports neutralization of pathogens and toxins (Pickard et al. 2017). The daily intake of fiber-enriched food promotes toxin removal through feces. The intestines are the central organ of the digestive system that stops the reabsorption of harmful substances. Digestive enzymes provide support in the detoxification of food components, in which both the gallbladder and liver work together closely with digestion (Hyder et al. 2023). A detox balance is required for regular bowel movement. The presence of gut barriers is essential in protecting against the entry of toxins. Healthy digestion encompasses gastric emptying, intestinal transit, and bile secretion, which show a potential effect on detoxification. The conjugated toxin in the intestinal tract and during first-pass metabolism in the liver needs healthy fecal production through bile. Healthy excretion requires dietary fiber to effectively eliminate bio-transformed toxins. The binding of some toxins directly occurs with fibers and thus promotes their elimination before the occurrence of central absorption (Harris et al. 1998). Nutrients (food-based) have an important role in the modulation of metabolic pathways, which are involved in detoxification processes. Numerous studies have been conducted on the detoxification and removal of toxins from the body using nutrients in cell and animal models (Baer-Dubowska and Szaefer 2013). The specific foods might be responsible for upregulation or a favorable balance of metabolic pathways to contribute to toxin biotransformation and further eradication (Moon et al. 2006). Many whole foods, such as garlic, berries, turmeric, etc., are generally recommended as functional medicines and naturopathic agents (James et al. 2012).

8.2.6 Lymphatic System

The lymphatic system is the key detoxification system that transports waste products away from tissues. It affects immunity, detoxification, and overall wellness, particularly as we get older (Rahm 2024). The lymphatic system plays an important role in maintaining health by boosting immunity, absorbing fats, maintaining body fluid balance, and removing toxins and waste from the body. It is also involved in the collection of toxins, pathogens, and cellular debris (Singh 2016). The activation of the immune response and its filtration are performed by lymph nodes (Sainte-Marie 2010). The lymphatic system acts as an excellent detox that transfers waste to excretory organs. Proper lymph flow prevents the accumulation of toxins and fluid, as well as eliminates excess fats and proteins. Physical movement is required for proper lymphatic circulation, and it supports tissue repair and inflammation control (Bookwatch 2023).

8.2.7 The Impact of Diet and Dietary Fibers on Detoxification

Among other things, nutrients are very crucial for promoting enzyme activity, waste disposal, and antioxidant protection. The most advantageous are fruits and berries, particularly due to their significance in bringing down antioxidant stress, taking care of irritable bowel syndrome, kidney function, and enzymatic function, and contributing to DNA and tissue protection. Plants are a rich source of dietary fiber, including non-digestible carbohydrates and the complex fiber lignin, which affects human physiology in many ways. Dietary fibers are divided into two categories, soluble and insoluble, based on chemical and physical properties (Anderson et al. 2009; Chutkan et al. 2012). Soluble dietary fibers have additional forms, namely, viscous and non-viscous. Further, it is subdivided into carbohydrates (both long and short chain), as well as fermentable and non-fermentable types (Muir et al. 2009; El-Salhy et al. 2017).

8.2.8 Antioxidants to Reduce Cellular Stress

Fruit and berry bioactive components play a central role against oxidative stress, performing their function as scavengers of free radicals. Oxidative stress emerges due to the imbalance between the cellular production of oxidant molecules and the antioxidants, which can effectively combat these stresses (Del Bo et al. 2015). Oxidative stress is mainly developed by the contribution of reactive oxygen species (ROS) and reactive nitrogen species (RNS). It involves hydrogen peroxide (H_2O_2), nitric oxide (NO), superoxide anion (O_2•–), peroxynitrite (ONOO–), hydroxyl radical (OH•), and hypochlorous acid (HOCl). ROS production is involved in apoptosis, necrosis, and cell damage, due to the oxidation and nitration of cellular DNA, proteins, and lipids, which leads to loss of cell function (Wiwanitkit 2014).

8.2.9 Dietary Fiber Promotes Irritable Bowel Movement

Dietary fiber (DF) is the food component that human digestive enzymes fail to break down properly (Ito et al. 2023). Every food consists of dietary fibers, but some contain large amounts, such as grains, beans, vegetables, and dried fruits (Slavin 2013). Its daily intake is highly important due to its potential role in reducing the risk of diseases and conditions such as cardiovascular disease, hypertension, stroke, diabetes, and obesity (Anderson et al. 2009). Irritable bowel syndrome (IBS) is one of the complex disorders of the brain–gut interaction (DGBI) due to its wide range of symptoms, such as bloating, abdominal pain, and changing bowel patterns. Although the disorder is widespread and has negative impacts on the quality of life, its specific pathogenesis remains unexplored, and the treatment options usually do not provide full but partial relief to the patients (Saha 2014; Dumitrascu et al. 2021). In the current time, dietary changes have become a major focus in the management of IBS symptoms, as they directly influence the composition of the microbiota, visceral hypersensitivity, and gut motility (Chiarioni et al. 2023). A huge group of secondary metabolites is polyphenols that are present in many fruits and vegetables of different colors. Polyphenols are composed of subclasses characterized by different chemical structures and biological activities, including flavonoids (flavanones, flavones, isoflavones, anthocyanins, and flavanols), phenolic acids (hydroxybenzoic acids and hydroxycinnamic acids), stilbenes, tannins, and ligands (Croft 2016). Berries (raspberries, cranberries, blueberries, blackberries, and strawberries), plums, apples, grapes, cherries, citrus fruits, kiwifruit, pomegranates, black tea, green tea, dark chocolate, kale, onions, extra virgin olive oils, coffee, spinach, broccoli, artichokes, and red cabbage are the highest sources of polyphenols (Tomás-Barberán and Andrés-Lacueva 2012). Polyphenols have a prebiotic effect, stimulating the growth of good bacteria while simultaneously preventing the proliferation of the bad ones. Such modulation of the gut microbiota has a positive effect on gut health, with subsequent benefits for immunity and overall health (Roca-Saavedra et al. 2018). Fruits and berries are rich dietary sources of polyphenolic compounds that provide health benefits, help with bowel movement, and possess antioxidant properties (Tsao 2010).

8.2.10 Vitamins Support Enzyme Function

The fruits and berries that we eat not just give us antioxidants and natural sugars, but they also supply us with the essential vitamins and minerals needed for the human body in varying roles of activators, cofactors, or stabilizers of different enzymes. This group of micronutrients is essential for life, as they are required for detoxification, metabolism, energy production, and cellular repair. Moreover, the intake of vitamins and minerals is insufficient; thereby, enzyme activity is low, and, on the other hand, the presence of enzymes in large amounts is not enough for normal functioning. Whole fruits, whether fresh, canned, frozen, or dried, have a large fiber content and are regarded as the main sources of essential nutrients like potassium and vitamin C, and phytochemicals such as polyphenols and carotenoids. These nutrients work together to provide health benefits (Slavin and Lloyd 2012; Padayachee et al. 2017). The amount of daily fruits intake which is considered normal (e.g., 1–1 ½ cups for children 2–8 years old, 1 ½ cups for adolescent girls and boys, 1 ½–2 cups for women, and 2 cups for men) is approximately half the recommended level, with juice consumption accounting for one-third of this level for adults and one-half for children (Burton-Freeman et al. 2018). Fruits and berries provide vitamins and minerals, which are an indispensable and balanced supply necessary for the proper functioning of enzymes. The presence of micronutrients ensures that functions such as detoxification, tissue maintenance, metabolism, and antioxidant defense, which require enzymes, are carried out. Thus, it is advisable to regularly consume a variety of fruit and berries to not only support good enzymatic activity but also to promote the body's physiological health (Dreher 2018).

8.2.11 Phytochemicals Protecting DNA and Tissues

Usually, free radical neutralizers and antioxidants such as vitamins, polyphenols, and carotenoids, along with some trace metals like Zn and Mo, are the main contributors to ROS elimination and help prevent the formation of the extensive oxidation processes (Singh et al. 2019). *C. carandas* leaves methanol extract is linked to the suppression of DNA damage. The water extract of *Ganoderma lucidum* exhibits remarkable antioxidant activity and the capability to safeguard DNA from radiation- or chemical-induced damage. The medicinal plants mentioned above harbor vital secondary metabolites such as flavonoids, flavones, polyphenols, terpenes, and alkaloids, which could serve as adjuncts to cancer therapy. Arecoline, an alkaloid constituent found in areca nut, has strong potential in the treatment of oral and pharyngeal cancers. A crucial constituent of plant chelates binds to heavy metals and protects DNA from damage caused by them. Furthermore, several vitamins, including vitamins C and E, are effective quenchers of free radicals and are involved in inhibiting DNA-damaging properties in living cells (Gupta et al. 2015). In a study, the potential of a few natural products, such as indole-3-carbinol, resveratrol, curcumin, and ellagic acid, to alter the DNA-damaging ability of the alkylating carcinogen N-methyl-N′-nitro-N-nitrosoguanidine (MNNG) was explored. The cell culture report showed that the preincubation of cells treated with any of these compounds was considerably effective in the treatment of MNNG-mediated DNA damage, which shows their accurate chemopreventive function (James and Ambadapadi 2013).

8.3 DETOXIFYING FRUITS: DETAILED OVERVIEW

8.3.1 Citrus Fruits (Lemon, Orange, Grapefruit, and Lime)

The genus *Citrus* belongs to the Rutaceae family, commonly known as citrus fruits. It is found in different forms and sizes, such as round or oblong, including limes, mandarins, lemons, citrons, and grapefruits. The sensory characteristics of fruits (sweetness, color, astringency, and bitterness) constitute pivotal organoleptic and commercial properties (García-Salas et al. 2013). The availability of Citrus species is in several forms, such as fresh or raw juices or canned. Furthermore, the uses of citrus fruits are wide but are generally used for the food, cosmetic, and pharmacological industries as spices, cosmetic ingredients, additives, and chemopreventive drugs, respectively (Lv et al. 2015). Citrus fruits are a highly nutritious source of fruits with an abundant amount of vitamin C. In addition, the fruits are sufficient in other macronutrients, such as potassium, dietary fiber, sugars, calcium, thiamin, folate, niacin, vitamin B6, riboflavin, phosphorus, pantothenic acid, copper, and magnesium (Martí et al. 2009). Citrus fruit contains phytochemicals and secondary metabolites such as limonoids, alkaloids, flavonoids, phenolic acids, carotenoids, essential oils, and cumarins. These secondary metabolites perform several biological activities that are important for human health, such as cardiovascular protective effects, anti-cancer, anti-inflammatory, antioxidant, and neuroprotective effects. Additionally, citrus fruits are a famous traditional medicinal herb used in numerous Asian countries, for example, Japan, China, and Korea (Lu et al. 2006). The peels or complete fruits, either mature or immature, of *C. medica* L. *var. sarcodactylis Swingle, C. reticulata Blanco, Citrus aurantium* L., *C. sinensis Osbeck, C. wilsonii Tanaka*, and *C. medica* L. are used to treat skin inflammation, cough, muscle pain, indigestion, ringworm infections, and lower blood pressure (Benedetto et al. 2023). Several research studies confirmed that citrus genotypes contain many bioactive compounds, among which limonoids are active and inhibit various types of cancer. Additionally, the structures of citrus limonoids are responsible for inducing the activity of phase II detoxifying enzymes (Perez et al. 2010).

8.3.2 Apples

Apples are among the most significant dietary sources of flavonoids, which are strongly associated with lower mortality (Boyer and Liu 2004). Apples contain several important antioxidant compounds, such as quercetin-3-galactoside, cyanidin-3-galactoside, quercetin-3-glucoside, quercetin-3-rhamnoside, chlorogenic acid, gallic acid, catechin, coumaric acid, epicatechin, phloridzin, and procyanidin. The highly enriched compounds of apple peels are phloridzin, epicatechin, procyanidins, chlorogenic acid, and catechin. Its peel involves phloridzin, epicatechin, catechin, and procyanidin, nevertheless considerably less in amount compared to the peels. Chlorogenic acid quercetin conjugates are present in the flesh and peel of apples, respectively (Escarpa and González 1998). Due to the rich source of flavonoids, people in the United States and Europe include them in their diet. Twenty-two percent of the U.S. population consumes phenolics derived from fruits, particularly apples, making them the major source of phenolics (Vinson et al. 2001). In a study, the intake of flavonoids among 10,000 people has shown significate reduction in overall mortality (Knekt et al. 2002). Apples also possess strong antioxidant properties and are ranked as the second most concentrated source of phenolic compounds (Sun et al. 2002). Several studies on the consumption of apples have suggested a lower cancer risk, particularly lungcancer. The consumption of one apple and one pear daily is associated with a considerably lower risk of lung cancer (Feskanich et al. 2000). Apple peels contain antioxidants that have the potential to inhibit the growth of liver and colon cancer cells (Wolfe et al. 2003). The peel of apples encompasses 83 μmol of vitamin C, suggesting that 100 g of apples contains around 1500 mg, with about 5.7 mg of vitamin C (Eberhardt et al. 2000). The crude extracts of immature apples have an important role in the inhibition of the enzymatic activities of cholera toxin in a dose-dependent manner (Saito et al. 2002).

8.3.3 Grapes

Grapes are extensively grown fruits, with a total worldwide productivity of around 60 million tons (Matthäus 2008). Grape seeds are a rich source of phenolic compounds, which show promising effects on human health, for example, by providing protection against peptic ulcers (Kim et al. 2013). Further, it consists of flavan-3-ol, such as catechins and proanthocyanidins, exhibiting scavenging of superoxide radicals (El-Beshbishy et al. 2009). Proanthocyanidins are found in considerable amounts in grape seeds and have strongly attracted consumers due to their potential health effects (Varzakas et al. 2016). Grapes are well-known natural detoxifiers due to the presence of a rich composition of bioactive compounds, minerals, vitamins, and dietary fiber, which assist the human body in performing well in detoxification systems. Grapes are well-known natural detoxifiers due to the presence of a rich composition of bioactive compounds, minerals, vitamins, and dietary fiber, which assist the human body in performing well in detoxification systems. Grapes are involved in enhancing the function of toxin-removing organs and enzymes, particularly the liver, kidneys, and digestive system. The person who consumes a daily red and purple grape significantly improves the metabolic balance and cellular protection. Compound in grapes, including quercetin, resveratrol, catechins, and anthocyanins, play a key role in neutralization of reactive oxygen species and reduction of oxidative stress. Polyphenols are found in large amounts in grapes, which highly support phase I and phase II detoxification enzymes in the liver, such as cytochrome P450 and conjugation enzymes, including glutathione S-transferase. These pathways are activated by grapes, enabling the liver to transform lipid-soluble substances into water-soluble compounds for efficient removal from the body. Grapes contain a large amount of nutrients, including fiber, which binds toxic chemicals, metabolic waste, and bile acids, thereby regulating their absorption into the blood and promoting their removal through feces. Therefore, in this way, the toxicity burden is reduced on the liver, regulating bowel movement, and providing a normal environment to the intestine. The key components of human detoxification system are summarized in Table 8.1.

TABLE 8.1
Key Components of the Human Detoxification System, Associated Organs, and Dietary Support

Detoxification Component	Major Organs/ Sites	Key Enzymes or Processes	Role in Detoxification	Dietary Support
Phase I Detoxification	Liver (primary), intestine	Cytochrome P450 enzymes	Converts lipophilic toxins into reactive intermediates	Vitamins B2, B3, B6; flavonoids; cruciferous vegetables
Phase II Detoxification	Liver, kidney	Conjugation enzymes (GST, UGT, SULT)	Neutralizes and enhances the solubility of toxins for excretion	Sulfur-containing compounds, polyphenols, and amino acids
Liver Function	Liver	Biotransformation, bile production	Central organ for toxin metabolism and elimination	Antioxidants, citrus fruits, apples
Renal Excretion	Kidney	Filtration and urinary excretion	Removes water-soluble toxins and metabolites	Adequate hydration, berries
Digestive Elimination	Gastrointestinal tract	Fiber-mediated binding and elimination	Prevents reabsorption of toxins	Dietary fibers, fruits, and whole grains
Skin Detoxification	Skin	Sweat secretion	Minor elimination route for toxins	Hydration, antioxidants
Lymphatic Support	Lymphatic system	Immune-mediated clearance	Transports waste products and supports immunity	Phytochemicals, anti-inflammatory nutrients
Antioxidant Defense	Cellular level	ROS scavenging systems	Reduces oxidative stress during detoxification	Berries, grapes, vitamins C and E
Dietary Fibers	Intestine	Toxin binding and motility regulation	Enhances fecal excretion of toxins	Apples, citrus fruits, and whole grains
Phytochemicals	Systemic	DNA protection and enzyme modulation	Protects tissues and regulates detox enzymes	Grapes, berries, citrus fruits

8.4 DETOXIFYING BERRIES: DETAILED OVERVIEW

Detoxifying berries encompass a valuable and rich concentration of antioxidants, vitamins, minerals, and dietary fiber, playing a pivotal role in naturally supporting the body's detoxification processes. Several common detoxifying berries include blueberries, strawberries, blackberries, cranberries, and raspberries. These berries possess unique bioactive compounds that support metabolic balance and overall health.

8.4.1 Blueberries

Blueberries contain a high amount of minerals, vitamins, polyphenols, and dietary fiber (Golovinskaia and Wang 2021). The gut microbiota is favorably modified by the dietary variations, such as chronic consumption of blueberries. Anthocyanins and other polyphenols found in blueberries perform wide-ranging metabolism through intestinal microbiota, improving the possibility that microbiota is induced by polyphenol metabolites, which are associated with mediating the health-promoting effects of blueberries (Petersen et al. 2022). In a study, it was reported that the consumption of blueberries is considerably associated with improved gut health by boosting intestinal morphology, inhibiting oxidative stress, ameliorating gut inflammation, lowering gut permeability, and modulating the composition and function of gut microbes (Della Lucia et al. 2023). Several studies suggested that consumption of anthocyanins (a class of flavonoids) through dietary intake provides broad-ranging protective effects against the detrimental impact of foodborne contaminants (Jiang et al. 2024). Blueberry extracts exhibit strong antioxidant potential and enhance phase II enzyme activities. A study showed that 1% blueberry flavonoid consumption reduced the oxidative DNA damage in the liver (Dulebohn et al. 2008). The TLR4-MAPK signaling pathway is efficient in protecting against such damage through inhibiting TLR4 (Nair et al. 2014).

8.4.2 Strawberries

Strawberries are the most widely consumed fruit, possessing a good source of bioactive compounds such as β-carotene, vitamins (vitamin C), and phenolic compounds (flavonoids, phenolic acids, and anthocyanins) (Oszmiański and Wojdyło 2009). Large amounts of anthocyanins and ellagitannins are used to reduce oxidative stress and allergic reactions and to lower the risk of heart-related diseases; in addition, they help control weight gain and high levels of glucose and inhibit cancer (Ariza et al. 2016). Strawberries contain a large amount of antioxidant like polyphenolic compounds (anthocyanins, flavonoids, and ellagic acid) and ascorbic acid. Free radicals generated during the metabolism of toxins are neutralized by antioxidants present in strawberries. Strawberries greatly support the preservation of the detoxification enzyme activity in the liver by reducing oxidative stress, particularly phase I and phase II metabolic pathways (Irshad et al. 2025). Ellagic acid improves the liver's overall health by improving antioxidant defense and diminishing the inflammatory properties. Strawberries possess vitamin C, which supports detoxification by enhancing the synthesis and recycling of glutathione. It is an important antioxidant that detoxifies pesticides, heavy metals, and other environmental pollutants (Wojtunik-Kulesza et al. 2025). Strawberries are a natural detoxifier due to their high antioxidant, vitamin C, and polyphenolic content, including flavonoids, anthocyanins, and ellagic acid. Strawberry antioxidants neutralize free radicals generated during toxin metabolism. Strawberries provide significant support for detoxification.

8.4.3 Blackberries

Blackberries contain polyphenolic compounds, as well as carbohydrates and various essential vitamins. The primary organic acid in blackberries is malic acid, while other acids, such as ascorbic acid are found in the fruit (Kafkas et al. 2006). Blackberries are a fruit of interest due to their high content of ellagitannins (ETs), anthocyanins, and phenolic compounds, which contribute to their high antioxidant activity (Kaume et al. 2012). Blackberries possess a high number of bioactive compounds and nutrients, vitamins A, C, and E, and minerals including calcium, sodium, phosphorus, potassium, iron, manganese, magnesium, and copper, as well as phenolic and dietary fiber (Paredes-López et al. 2010). Blackberries offer health-promoting effects, including excellent chemopreventive and chemotherapeutic properties, as well as protection against both chronic and non-chronic diseases (Surh 2003). The presence of compounds in blackberries activates PXR, which is considerably expressed in the liver. The regulation of gene transcription depends on ligand-activated PXR, which encodes detoxification-induced proteins and transporters involved in metabolite elimination (Xiao et al. 2022). Blackberries are low in calories and fat, with enriched contents of dietary fiber to effectively remove toxins. The fiber is essential for intestinal motility, as it binds to heavy metals, bile acids, and metabolic waste products, and removes them through feces. Blackberries also contain various important vitamins, including vitamin C, vitamin K, vitamin A in the form of beta-carotene, and folate. Vitamin C is considered one of the most important natural antioxidants that is used to neutralize dangerous compounds by forming glutathione (Martins et al. 2023). Folate are very important for normalization and destroying the chemical substances by initiating the methylation process, and also for the functioning of enzymes and metabolism in the liver. The crucial component is vitamin K. Additionally, minerals like potassium, magnesium, manganese, and iron are found in blackberries, which function as an activating factor for cleansing enzymes and antioxidants (Yang et al. 2024). For a better detoxification process, some useful polyphenolic compounds, such as ellagitannins. anthocyanins, ellagic acid, and flavonols. Anthocyanins are promising antioxidants found in the dark purple color of blackberries that neutralize reactive oxygen species (Bader Ul Ain et al. 2022). ROS production usually occurs when exposed to xenobiotics, heavy metals, and environmental pollutants. The free radicals are the substances that prevent DNA, protein, and lipids from oxidative cellular damage. Blackberry polyphenols are important for the regulation of cytochrome P450 activity by preventing the production of toxic compounds. Glutathione, sulfate, or glucuronic acid binds with these toxic compounds by making them water-soluble and preparing them for excretion in phase II detoxification. Blackberries boost the phase II enzymes activity, primarily glutathione-S-transferase, thus speeding up the process of toxin elimination (Manful et al. 2025).

8.4.4 Cranberries

The cranberry fruit is used for many purposes, but is extensively used for food products, such as fresh or dried fruits, sauces, juices, and powder forms in tablets and capsules. The extract of cranberries contains a high number of flavonoids, particularly myricetin, flavonols, and quercetin, as well as encompassing biological properties that provide promising human health benefits. The ripe fruits are used by native for the treatment of bladder and kidney diseases (Bodet et al. 2008). Cranberry juice is ability and attracted researchers because of its valuable role in the prevention of urinary tract infections (UTIs) (Raz et al. 2004). For the prevention of uropathogenic fimbriated *E. coli* interacting with uroepithelial cells in the urinary system, the condensed tannins called proanthocyanidins are present in cranberry beverages (Howell et al. 2005). Cranberries (*Vaccinium macrocarpon*) are used to be famous for their beneficial properties, especially for the health of the urinary system. According to modern research cranberry considered to be a natural agent of detoxification. Detoxification includes several organs, such as the kidneys, antioxidant defense system, and gastrointestinal tract, used to be considered a very complex process in humans (Chettaoui et al. 2021). Some of the essential elements are used to be found in cranberries include flavonoids, proanthocyanins, anthocyanins, and polyphenols. The combined work of phytochemicals is used to improve their function at both the cellular and molecular levels for better removal and neutralization of toxins. Cranberries contain high amounts of flavonols like quercetin and myricetin, anthocyanins, and specially structured A-type proanthocyanidins (Nemzer et al. 2022). ROS cause damage to DNA, proteins, and lipids produced in the metabolic process when exposed to external toxins. For the protection of the detoxification organ, scavenging of free radicals, and reducing oxidative stress, the polyphenols of cranberries are considered to be very important. Also, they have a key role in urinary excretion and kidney function to remove metabolic waste and toxic chemicals, which is why they are called renal detoxifiers (Pizzino et al. 2017). The stickiness of bacteria to the surface of the urinary tract is limited by proanthocyanidins, which decrease infection caused by inflammation and also allow better filtration. The lowering of inflammatory signals and better urine flow can be attained by cranberries, which facilitate the metabolic waste excretion as well as the removal of water-soluble chemicals from the body that cause toxicity. Toxin absorption is inhibited by cranberries, which also allow the renal tubules to function normally and maintain their integrity (Howell 2002).

8.4.5 Raspberries

Flavonoids and anthocyanins, which are called major antioxidants, are found in high amounts in raspberries. The fruits are considered to be a rich source of dietary antioxidants because they contain high levels of phenolics such as anthocyanins and ellagitannins (Zhang et al. 2018). Polyphenols of raspberry not only possess antioxidant properties but also show anti-inflammatory and antibacterial activity against intestinal bacteria, improve lipid composition, and also inhibit cancer-causing cells in humans (Fotschki et al. 2015). The immune metabolic signals which are responsible for causing fats are used to be controlled by polyphenols of raspberries (Fotschki et al. 2018). Liver is used to be known for the detoxification process involving phase I and phase II. The pathways of phase II conjugation, which are enhanced by bioactive compounds present in raspberries, have a key role in the detoxification of the liver (Ab et al. 2024). The enhancement of the sulfation and glucuronidation process to remove toxins whose solubility increases by flavonoids and ellagitannins found in raspberries. The phase I enzymes are used to be prevented by antioxidant of raspberry lower the formation of some reactive intermediates that affect liver cells. The normal functioning of the liver and the removal of toxins are maintained by a balanced mechanism. A rich amount of fiber is found in raspberries, which are essential for normal detoxification of the GIT. The fibers have a chance to lose their reabsorption ability of metabolic waste and toxins when they bind to them to initiate the fecal removal. The liver's toxic burden is used to be reduced by this mechanism, which is why this mechanism is very important (Mullen et al. 2002). In addition to that, the removal of toxic chemicals and maintaining the normal function of the intestine on time, and also decreasing the exposure to the mucosa of the intestine, is maintained by fiber. For beneficial renal detoxification and normal kidney function, raspberries are found to be the most beneficial. High amounts of strong antioxidant and anti-inflammatory compounds are found in raspberries, which protect the tissues of the kidney from inflammation and oxidation, which can cause damage to filtration. The removal of metabolic wastes, medications, and other external factors through urination is related to normal kidney functions (Lopez-Corona et al. 2022).

8.4.6 Conclusion

Detoxification is the most effective natural process of the human body for removing toxins, metabolic waste, drugs, and other harmful byproducts. The detoxification system involves several important organs, including the digestive system, liver, kidneys, lymphatic system, and skin. Normally, the human body absorbs harmful toxins from the environment through varied sources such as exposure to pesticides, insecticides, alcohol usage, food /drug intake, and other polluted compounds. The constant exposure to these harmful toxins adversely affects the human body by severely disturbing the detoxifying organs, and thus, the normal detoxification mechanism loses its function in removing waste from the body. Therefore, it is essential to make the detoxifying organs healthy so that the body can effectively perform its function. The healthy detoxification system requires proper diets, daily nutritious intake, and regular exercise. Among foods, the fruits and berries play a central role in making detoxifying organs healthy and enable them to perform the detoxification process. Fruits and berries are natural sources of significant phytochemicals, antioxidants, bioactive compounds, and essential vitamins, minerals, and fibers, helping the detoxifying organs to remove toxins and waste from the body in the most convenient pathway. The major advantages of fruits and berries for the human body necessitate researchers to focus on their mechanism of action and their role in removing metabolic waste. The regular intake of daily fruits such as apples, grapes, citrus fruits, and berries will make the kidneys, liver, skin, lymphatic system, and digestive system healthy.

REFERENCES

Ab, H., Abbas, A. A., Umr, A., Tanko, H. O., Adgeboye, Y., and Kazeem, F. A. 2024. Importance of liver in the detoxification of xenobiotics. *IRASS Journal of Multidisciplinary Studies*, 1: 1–29.

Ahmed, K., Dubey, M. K., Kajal, Dubey, S., and Pandey, D. K. 2025. Chronic kidney disease: Causes, treatment, management, and future scope. In: *Computational Intelligence for Genomics Data*. Academic Press: pp. 99–111.

Alamri, Z. Z. 2018. The role of liver in metabolism: An updated review with physiological emphasis. *International Journal of Basic & Clinical Pharmacology*, 7: 2271.

Anderson, J. W., Baird, P., Davis, R. H., Ferreri, S., Knudtson, M., et al. 2009. Health benefits of dietary fiber. *Nutrition Reviews*, 67: 188–205.

Aralbaeva, A. N., Yeszhanova, G. A., Aralbayev, A. N., Zhamanbayeva, G. T., Zhaparkulova, N. I., et al. 2024. Overview on the heavy metal toxicity mechanisms and the role of alimentary factors in detoxification. *International Journal of Biology and Chemistry*, 17: 74–95.

Ariza, M. T., Reboredo-Rodríguez, P., Mazzoni, L., Forbes-Hernández, T. Y., Giampieri, F., et al. 2016. Strawberry achenes are an important source of bioactive compounds for human health. *International Journal of Molecular Sciences*, 17(7): 1–14.

Armoni, M., Harel, C., Ramdas, M., and Karnieli, E. 2014. CYP2E1 impairs GLUT4 gene expression and function: NRF2 as a possible mediator. *Hormone and Metabolic Research*, 46: 477–483.

Bader Ul Ain, H., Tufail, T., Javed, M., Tufail, T., Arshad, M. U., et al. 2022, December 31. Phytochemical profile and pro-healthy properties of berries. *International Journal of Food Properties*, 25: 1717–1735.

Baer, B., and Rettie, A. 2006, June 1. CYP4B1: An enigmatic P450 at the interface between xenobiotic and endobiotic metabolism. *Drug Metabolism Reviews*, 38: 451–476.

Baer-Dubowska, W., and Szaefer, H. 2013, August. Modulation of carcinogen-metabolizing cytochromes P450 by phytochemicals in humans. *Expert Opinion on Drug Metabolism and Toxicology*, 9(8): 927–941.

Baker, L. B. 2019, July 3. Physiology of sweat gland function: The roles of sweating and sweat composition in human health. *Temperature*, 6(3): 211–259.

Benedetto, N., Carlucci, V., Faraone, I., Lela, L., Ponticelli, M., et al. 2023 June 10. An insight into *Citrus medica* Linn.: A systematic review on phytochemical profile and biological activities. *Plants*, 12(12): 2267.

Beyerle, J., Frei, E., Stiborova, M., Habermann, N., and Ulrich, C. M. 2015, May 1. Biotransformation of xenobiotics in the human colon and rectum and its association with colorectal cancer. *Drug Metabolism Reviews*, , 47(2): 199–221.

Black, J. L. 2020. Detoxification and biotransformation. In: Noland, D., Drisko, J. A., and Wagner, L. (eds.). *Integrative and Functional Medical Nutrition Therapy*. Humana: pp. 205–212.

Bodet, C., Grenier, D., Chandad, F., Ofek, I., Steinberg, D., et al. 2008, August. Potential oral health benefits of cranberry. *Critical Reviews in Food Science and Nutrition*, 48(7): 672–680.

Bookwatch, S. B. 2023. Lymph health: The key to a strong immune system. Healing Arts Press, USA. Available at: https://books.google.com/books?id=8kiUEAAAQBAJ.

Boyer, J., and Liu, R. H. 2004. Apple phytochemicals and their health benefits. *Nutrition Journal*, 3: 5.

Burton-Freeman, B. M., Guenther, P. M., Oh, M., Stuart, D., and Jensen, H. H. 2018. Assessing the consumption of berries and associated factors in the United States using the National Health and Nutrition Examination Survey (NHANES), 2007–2012. *Food & Function*, 9: 1009–1016.

Chen, C. H., Huang, M. H., Yang, J. C., Nien, C. K., Yang, C. C., et al. 2006. Prevalence and risk factors of nonalcoholic fatty liver disease in an adult population of Taiwan: Metabolic significance of nonalcoholic fatty liver disease in nonobese adults. *Journal of Clinical Gastroenterology*, 40: 745–752.

Chen, Q., Zhang, T., Wang, J.-F., and Wei, D.-Q. 2011. Advances in human cytochrome P450 and personalized medicine. *Current Drug Metabolism*, 12: 436–444.

Chettaoui, R., Mayot, G., De Almeida, L., and Di Martino, P. 2021. Cranberry (*Vaccinium macrocarpon*) dietary supplementation and fecal microbiota of Wistar rats. *AIMS Microbiology*, 7: 257–270.

Chiang, J. 2014. Liver physiology: Metabolism and detoxification. In: *Pathobiology of Human Disease: A dynamic encyclopedia of disease mechanisms*. pp. 1770–1782.

Chiarioni, G., Popa, S. L., Ismaiel, A., Pop, C., Dumitrascu, D. I., et al. 2023, September 20. The effect of polyphenols, minerals, fibers, and fruits on irritable bowel syndrome: A systematic review. *Nutrients*, 18(18): 4070.

Chutkan, R., Fahey, G., Wright, W. L., and McRorie, J. 2012. Viscous versus nonviscous soluble fiber supplements: Mechanisms and evidence for fiber-specific health benefits. *Journal of the American Academy of Nurse Practitioners*, 24(8): 476–487.

Croft, K. D. 2016. Dietary polyphenols: Antioxidants or not? *Archives of Biochemistry and Biophysics*, 595: 120–124.

Danielson, P. B. 2005. The cytochrome P450 superfamily: Biochemistry, evolution and drug metabolism in humans. *Current Drug Metabolism*, 3: 561–597.

Del Bo, C., Martini, D., Porrini, M., Klimis-Zacas, D., and Riso, P. 2015, September 3 Berries and oxidative stress markers: An overview of human intervention studies. *Food & Function*, 6(9): 2890–2917.

Della Lucia, C. M., Oliveira, L. A., Dias, K. A., Pereira, S. M. S., da Conceição, A. R., and Anandh Babu, P. V. 2023, August 1. Scientific evidence for the beneficial effects of dietary blueberries on gut health: A systematic review. *Molecular Nutrition & Food Research*, 67(15): 2300096.

Downing, D. T., Strauss, J. S., and Pochi, P. E. 1972. Changes in skin surface lipid composition induced by severe caloric restriction in man. *The American Journal of Clinical Nutrition*, 25: 365–367.

Dreher, M. L. 2018, November 28. Whole fruits and fruit fiber: Emerging health effects. *Nutrients*, 10(12): 1–54.
Dulebohn, R. V., Yi, W., Srivastava, A., Akoh, C. C., Krewer, G., and Fischer, J. G. 2008. Effects of blueberry (*Vaccinium ashei*) on DNA damage, lipid peroxidation, and phase II enzyme activities in rats. *Journal of Agricultural and Food Chemistry*, 56: 11700–11706.
Dumitrascu, D. L., Baban, A., Bancila, I., Barboi, O., Bataga, S., et al. 2021. Romanian guidelines for nonpharmacological therapy of IBS. *Journal of Gastrointestinal and Liver Diseases*, 30(2): 291–306.
Eberhardt, M. V., Lee, C. Y., and Liu, R. H. 2000. Antioxidant activity of fresh apples. *Nature*, 405: 903–904.
El-Beshbishy, H. A., Mohamadin, A. M., and Abdel-Naim, A. B. 2009. In vitro evaluation of the antioxidant activities of grape seed (*Vitis vinifera*) extract, blackseed (*Nigella sativa*) extract, and curcumin. *Journal of Taibah University Medical Sciences*, 4: 23–35.
El-Salhy, M., Ystad, S. O., Mazzawi, T., and Gundersen, D. 2017, September 1. Dietary fiber in irritable bowel syndrome (review). *International Journal of Molecular Medicine*. Spandidos Publications.
Escarpa, A., and González, M. C. 1998. High-performance liquid chromatography with diode-array detection for the determination of phenolic compounds in peel and pulp from different apple varieties. *Journal of Chromatography A*, 23(1-12): 331–337.
Feskanich, D., Ziegler, R. G., Michaud, D. S., Giovannucci, E. L., Speizer, F. E., et al. 2000. Prospective study of fruit and vegetable consumption and risk of lung cancer among men and women. *Journal of the National Cancer Institute*, 92: 1812–1823.
Fotschki, B., Juśkiewicz, J., Sójka, M., Jurgoński, A., Zduńczyk, Z., et al. 2015. Ellagitannins and flavan-3-ols from raspberry pomace modulate caecal fermentation processes and plasma lipid parameters in rats. *Molecules*, 20: 22848–22862.
Fotschki, B., Laparra, J. M., and Sójka, M. 2018. Raspberry polyphenolic extract regulates obesogenic signals in hepatocytes. *Molecules*, 23(9): 1–14.
García-Salas, P., Gómez-Caravaca, A. M., Arráez-Román, D., Segura-Carretero, A., Guerra-Hernández, E., et al. 2013. Influence of technological processes on phenolic compounds, organic acids, furanic derivatives, and antioxidant activity of whole-lemon powder. *Food Chemistry*, 141: 869–878.
Gerich, J. E. 2010, February 1. Role of the kidney in normal glucose homeostasis and in the hyperglycaemia of diabetes mellitus: Therapeutic implications. *Diabetic Medicine*, 27(2): 136–142.
Golovinskaia, O., and Wang, C. K. 2021. Review of functional and pharmacological activities of berries. *Molecules*, 26(13): 1–25.
Granados, J. C., Richelle, A., Gutierrez, J. M., Zhang, P., Zhang, X., et al. 2021. Coordinate regulation of systemic and kidney tryptophan metabolism by the drug transporters OAT1 and OAT3. *Journal of Biological Chemistry*, 296: 100575.
Gupta, V. K., Singh, S., Agrawal, A., Siddiqi, N. J., and Sharma, B. 2015. Phytochemicals mediated remediation of neurotoxicity induced by heavy metals. *Biochemistry Research International*, 1(1–9): 534769.
Harris, P. J., Sasidharan, V. K., Roberton, A. M., Triggs, C. M., Blakeney, A. B., and Ferguson, L. R. 1998. Adsorption of a hydrophobic mutagen to cereal brans and cereal bran dietary fibres. *Mutation Research - Genetic Toxicology and Environmental Mutagenesis*, 412: 323–331.
Hayes, J. D., and Pulford, D. J. 1995. The glutathione S-transferase supergene family: Regulation of GST and the contribution of the isoenzymes to cancer chemoprotection and drug resistance part I. *Critical Reviews in Biochemistry and Molecular Biology*, 30: 445–520.
Hazim, H., Zalmai, M. R., and Bahir, H. 2024. Exploring the impact of nutrition and physical activity on human metabolism. *Sprin Journal of Arts, Humanities and Social Sciences*, 3: 39–42.
Hodges, R. E., and Minich, D. M. 2015. Modulation of metabolic detoxification pathways using foods and food-derived components: A scientific review with clinical application. *Journal of Nutrition and Metabolism*, 1: 760689.
Howell, A. B. 2002. Cranberry proanthocyanidins and the maintenance of urinary tract health. *Critical Reviews in Food Science and Nutrition*, 42(53): 273–278
Howell, A. B., Reed, J. D., Krueger, C. G., Winterbottom, R., Cunningham, D. G., and Leahy, M. 2005. A-type cranberry proanthocyanidins and uropathogenic bacterial anti-adhesion activity. *Phytochemistry*, 66(18): 2281–2291.
Hyder, I., Reddy, P. R. K., and Mukherjee, J. 2023. Physiology of Digestion. In: Das, P. K., Mukherjee, J., Banerjee, D., and Sejian, V. (eds.). *Textbook of Veterinary Physiology*. Springer: pp. 315–351.
Imenez Silva, P. H., and Mohebbi, N. 2022, May 5. Kidney metabolism and acid–base control: Back to the basics. *Pflugers Archiv European Journal of Physiology*, 478(8): 919–934.
Ioannides, C. 1999. Effect of diet and nutrition on the expression of cytochromes P450. *Xenobiotica*, 29(2): 104–154.

Irshad, R., Amir, R. M., Azam, M. T., Abbasi, F. A., Fatima, E., et al. 2025. Nutraceutical Potential of Strawberries, Its Phytochemicals and Health Benefits: A Comprehensive Review. *Indus Journal of Bioscience Research*, 3: 36–45.

Ismayilzada, N., Tarar, C., Dabbagh, S. R., Tokyay, B. K., Dilmani, S. A., et al. 2024, July 16. Skin-on-a-chip technologies towards clinical translation and commercialization. *Biofabrication*, 16(4): 1–37.

Ito, M., Yoshimoto, J., Maeda, T., Ishii, S., Wada, Y., et al. 2023. Effects of high-fiber food product consumption and personal health record use on body mass index and bowel movement. *Journal of Functional Foods*, 102: 105443.

James, D., Devaraj, S., Bellur, P., Lakkanna, S., Vicini, J., and Boddupalli, S. 2012. Novel concepts of broccoli sulforaphanes and disease: Induction of phase II antioxidant and detoxification enzymes by enhanced-glucoraphanin broccoli. *Nutrition Reviews*, 70: 654–665.

James, M. O., and Ambadapadi, S. 2013, November. Interactions of cytosolic sulfotransferases with xenobiotics. *Drug Metabolism Reviews*, 45(4): 401–414.

James, M. O., Sacco, J. C., and Faux, L. R. 2008. Effects of food natural products on the biotransformation of PCBs. *Environmental Toxicology and Pharmacology*, 25: 211–217.

Jancova, P., Anzenbacher, P., and Anzenbacherova, E. 2010. Phase II drug metabolizing enzymes. *Biomedical Papers*, 154: 103–116.

Jenkinson, C., Petroczi, A., and Naughton, D. P. 2013. Effects of dietary components on testosterone metabolism via UDP-glucuronosyltransferase. *Frontiers in Endocrinology*, 8(4): 1–4.

Jiang, Y., Li, X., Zhang, Y., Wu, B., Li, Y., et al. 2024, January 1. Mechanism of action of anthocyanin on the detoxification of foodborne contaminants—A review of recent literature. *Comprehensive Reviews in Food Science and Food Safety*, 23(1): e13259.

Kafkas, E., Koşar, M., Türemiş, N., and Başer, K. H. C. 2006. Analysis of sugars, organic acids and vitamin C contents of blackberry genotypes from Turkey. *Food Chemistry*, 97: 732–736.

Kaume, L., Howard, L. R., and Devareddy, L. 2012. The blackberry fruit: A review on its composition and chemistry, metabolism and bioavailability, and health benefits. *Journal of Agricultural and Food Chemistry*, 60: 5716–5727.

Kim, T. H., Jeon, E. J., Cheung, D. Y., Kim, C. W., Kim, S. S., et. al. 2013. Gastroprotective effects of grape seed proanthocyanidin extracts against nonsteroid anti-inflammatory drug-induced gastric injury in rats. *Gut and Liver*, 7: 282–289.

Knekt, P., Kumpulainen, J., Järvinen, R., Rissanen, H., Heliövaara, M., et al. 2002. Flavonoid intake and risk of chronic diseases. *American Journal of Clinical Nutrition*, 76: 560–568.

Kohalmy, K., and Vrzal, R. 2011. Regulation of Phase II Biotransformation Enzymes by Steroid Hormones. *Current Drug Metabolism*, 12: 104–123.

Lakra, P., and Gahlawat, I. N. 2022. Regular food chemicals as antioxidant towards prevention of diseases – An insight review. *Journal of Molecular Chemistry*, 2: 441.

Lee, H., Kwon, Y. J., and Chun, Y. J. 2025, November 18. Exploring the roles of cytochrome P450 enzymes and their inhibitors in cancers and non-neoplastic human diseases. *Archives of Pharmacal Research*, 48: 1224–1252.

Leibelt, D. A., Hedstrom, O. R., Fisher, K. A., Pereira, C. B., and Williams, D. E. 2003, July 1. Evaluation of chronic dietary exposure to indole-3-carbinol and absorption-enhanced 3,3′-diidolylmethane in Sprague-Dawley rats. *Toxicological Sciences*, 74(1): 10–21.

Lopez-Corona, A. V., Valencia-Espinosa, I., González-Sánchez, F. A., Sánchez-López, A. L., Garcia-Amezquita, L. E., and Garcia-Varela, R. 2022, June 1. Antioxidant, Anti-Inflammatory and Cytotoxic Activity of Phenolic Compound Family Extracted from Raspberries (Rubus idaeus): A General Review. *Antioxidants*, 11(6): 1192.

Lu, Y., Zhang, C., Bucheli, P., and Wei, D. 2006. Citrus flavonoids in fruit and Traditional Chinese Medicinal food ingredients in China. *Plant Foods for Human Nutrition*, 61: 57–65.

Lv, X., Zhao, S., Ning, Z., Zeng, H., Shu, Y., et al. 2015, December 24. Citrus fruits as a treasure trove of active natural metabolites that potentially provide benefits for human health. *Chemistry Central Journal*, 9(1): 1–14.

Lynch, C. F., and Cohen, M. B. 1995. Urinary system. *Cancer*, 75: 316–329.

Maldonado-Valderrama, J., Wilde, P., MacIerzanka, A., and MacKie, A. 2011. The role of bile salts in digestion. *Advances in Colloid and Interface Science*, 165(1): 36–46.

Manful, C. F., Fordjour, E., Subramaniam, D., Sey, A. A., Abbey, L., and Thomas, R. 2025, August 4. Antioxidants and reactive oxygen species: Shaping human health and disease outcomes. *International Journal of Molecular Sciences*, 26(15): 1–42.

Martí, N., Mena, P., Cánovas, J. A., Micol, V., and Saura, D. 2009. Vitamin C and the role of citrus juices as functional food. *Natural Product Communications*, 4(5): 677–700.

Martins, M. S., Gonçalves, A. C., Alves, G., and Silva, L. R. 2023, August 1. Blackberries and mulberries: Berries with significant health-promoting properties. *International Journal of Molecular Sciences*, 24(15): 12024–12034.

Masereeuw, R., Mutsaers, H. A. M., Toyohara, T., Abe, T., Jhawar, S., et al. 2014. The Kidney and Uremic Toxin Removal: Glomerulus or Tubule? *Seminars in Nephrology*, 34(2): 191–208.

Matthäus, B. 2008, July. Virgin grape seed oil: Is it really a nutritional highlight? *European Journal of Lipid Science and Technology*, 110(7): 645–650.

McFadden, S. A. 1996. Phenotypic variation in xenobiotic metabolism and adverse environmental response: Focus on sulfur-dependent detoxification pathways. *Toxicology*, 11(1–3): 43–65.

Mitra, V., and Metcalf, J. 2012. Metabolic functions of the liver. *Anaesthesia and Intensive Care Medicine*, 13(2): 54–55.

Moini, J., and Ferdowsi, K. 2024. Digestion, Absorption, and Metabolism. In: Moini, J., and Ferdowsi, K. (eds.). *Handbook of Nutritional Disorders*. CRC Press: pp. 18–42.

Moon, Y. J., Wang, X., and Morris, M. E. 2006. Dietary flavonoids: Effects on xenobiotic and carcinogen metabolism. *Toxicology in Vitro*, 20(2): 187–210.

Muir, J. G., Rose, R., Rosella, O., Liels, K., Barrett, J. S., et. al. 2009. Measurement of short-chain carbohydrates in common Australian vegetables and fruits by high-performance liquid chromatography (HPLC). *Journal of Agricultural and Food Chemistry*, 57: 554–565.

Mullen, W., McGinn, J., Lean, M. E. J., MacLean, M. R., Gardner, P., et al. 2002. Ellagitannins, flavonoids, and other phenolics in red raspberries and their contribution to antioxidant capacity and vasorelaxation properties. *Journal of Agricultural and Food Chemistry*, 50: 5191–5196.

Nair, A. R., Elks, C. M., Vila, J., Del Piero, F., Paulsen, D. B., and Francis, J. 2014. A blueberry-enriched diet improves renal function and reduces oxidative stress in metabolic syndrome animals: Potential mechanism of TLR4-MAPK signaling pathway. *PLoS ONE*, 9: e111976.

Nakhoul, N., and Batuman, V. 2011. Role of proximal tubules in the pathogenesis of kidney disease. *Contributions to Nephrology*, 169: 37–50.

Natochin Iu, V., 1973. The interrelationship between filtration-reabsorption and secretory processes in the evolution of the osmo- and ion-regulating functions of the kidney. *Zh Evol Biokhim Fiziol*, 9: 346–354.

Nemzer, B. V., Al-Taher, F., Yashin, A., Revelsky, I., and Yashin, Y. 2022, March 1. Cranberry: Chemical composition, antioxidant activity and impact on human health. Overview. *Molecules*, 27(5): 1–19.

Nissar, A. U., Farrukh, M. R., Kaiser, P. J., Rafiq, R. A., Afnan, Q., et al. 2013. Effect of N-acetyl cysteine (NAC), an organosulfur compound from Allium plants, on experimentally induced hepatic prefibrogenic events in Wistar rat. *Phytomedicine*, 20: 828–833.

Oszmiański, J., and Wojdyło, A. 2009. Comparative study of phenolic content and antioxidant activity of strawberry puree, clear, and cloudy juices. *European Food Research and Technology*, 228: 623–631.

Padayachee, A., Day, L., Howell, K., and Gidley, M. J. 2017. Complexity and health functionality of plant cell wall fibers from fruits and vegetables. *Critical Reviews in Food Science and Nutrition*, 57: 59–81.

Paredes-López, O., Cervantes-Ceja, M. L., Vigna-Pérez, M., and Hernández-Pérez, T. 2010. Berries: Improving human health and healthy aging, and promoting quality life—a review. *Plant Foods for Human Nutrition*, 65(3): 299–308.

Perez, J. L., Jayaprakasha, G. K., Cadena, A., Martinez, E., Ahmad, H., and Patil, B. S. 2010. In vivo induction of phase II detoxifying enzymes, glutathione transferase and quinone reductase by citrus triterpenoids. *BMC Complementary and Alternative Medicine*, 10: 51.

Petersen, C., Bharat, D., Wankhade, U. D., Kim, J. S., Cutler, B. R., et al. 2022. Dietary Blueberry Ameliorates Vascular Complications in Diabetic Mice Possibly through NOX4 and Modulates Composition and Functional Diversity of Gut Microbes. *Molecular Nutrition and Food Research*, 66(8): 2100784.

Pickard, J. M., Zeng, M. Y., Caruso, R., and Núñez, G. 2017, September 1. Gut microbiota: Role in pathogen colonization, immune responses, and inflammatory disease. *Immunological Reviews*, 279(1): 70–89.

Pizzino, G., Irrera, N., Cucinotta, M., Pallio, G., Mannino, F., et al. 2017. Oxidative stress: Harms and benefits for human health. *Oxidative Medicine and Cellular Longevity*, 2017(1): 1–13.

Pochi, P. E., Downing, D. T., and Strauss, J. S. 1970. Sebaceous gland response in man to prolonged total caloric deprivation. *The Journal of Investigative Dermatology*, 55: 303–309.

Rahm, C. 2024. Detoxing & Remediating Land, Air, and Water & Implications on Human and Animal Health. *PriMera Scientific Surgical Research and Practice*, 4(3): 77–81. DOI: 10.56831/PSSRP-04-133

Ramalho Mercé, A. L. 2023. Contaminated Waters and Depleted Soils: Impact in Nutrition. *Advances in Complementary & Alternative Medicine*, 819: 1–29.

Raz, R., Chazan, B., and Dan, M. 2004, May 15. Cranberry juice and urinary tract infection. *Clinical Infectious Diseases*, 38(10): 1419–1419.

Reinke, H., and Asher, G. 2016. Circadian clock control of liver metabolic functions. *Gastroenterology*, 150(3): 574–580.

Roca-Saavedra, P., Mendez-Vilabrille, V., Miranda, J. M., Nebot, C., Cardelle-Cobas, A., et al. 2018, February 1. Food additives, contaminants and other minor components: Effects on human gut microbiota—A review. *Journal of Physiology and Biochemistry*, 71(1): 69–83.

Ross, G. H., and Sternquist, M. C. 2012, September. Methamphetamine exposure and chronic illness in police officers: Significant improvement with sauna-based detoxification therapy. *Toxicology and Industrial Health*, 28(8): 758–768.

Rostom, H. 2018. Basic metabolism: Proteins. *Surgery (Oxford)*, 36(4): 153–158.

Rudrapal, M., Rakshit, G., Singh, R. P., Garse, S., Khan, J., and Chakraborty, S. 2024, March 30. Dietary polyphenols: Review on chemistry/sources, bioavailability/metabolism, antioxidant effects, and their role in disease management. *Antioxidants*, 13(4): 1–47.

Saha, L. 2014. Irritable bowel syndrome: Pathogenesis, diagnosis, treatment, and evidence-based medicine. *World Journal of Gastroenterology*, 20: 6759–6773.

Sainte-Marie, G. 2010, February 1. The lymph node revisited: Development, morphology, functioning, and role in triggering primary immune responses. *Anatomical Record*, 293(2): 320–337.

Saito, T., Miyake, M., Toba, M., Okamatsu, H., Shimizu, S., and Noda, M. 2002. Inhibition by apple polyphenols of ADP-ribosyltransferase activity of cholera toxin and toxin-induced fluid accumulation in mice. *Microbiology and Immunology*, 46: 249–255.

Sato, K. 1977. The physiology, pharmacology, and biochemistry of the eccrine sweat gland. *Reviews of Physiology, Biochemistry and Pharmacology*, 79: 51–131.

Sheweita, S. 2005. Drug-metabolizing enzymes: Mechanisms and functions. *Current Drug Metabolism*, 1: 107–132.

Shuai, Y., Sui, H., Tao, G., Huo, Q., Li, C., and Shao, N. 2022. Food Contaminants. In: Zhang, L. (ed.). *Nutritional Toxicology*, 107–166.

Singh, N., Toxicol, B. S.-E. P., and U. 2019. Role of toxicants in oxidative stress mediated DNA damage and protection by phytochemicals. *EC Pharmacol Toxicol*, 7(5): 325–330.

Singh, P. 2016. Lymphatic System. In: *Surgery Essence*. Jaypee Brothers Medical Publishers. 734–734. https://archive.org/details/OceanofPDF.comSURGERYESSENCESinghPriteshKumar

Slavin, J. 2013. Fiber and prebiotics: Mechanisms and health benefits. *Nutrients*, 5(4): 1417–1435.

Slavin, J. L., and Lloyd, B. 2012, July. Health benefits of fruits and vegetables. *Advances in Nutrition*, 3(4): 506–516.

Smith, K. R., and Thiboutot, D. M. 2008. Thematic review series: Skin lipids. Sebaceous gland lipids: Friend or foe? *Journal of Lipid Research*, 49(2): 271–281.

Standish, L. J. 2006. Textbook of Functional Medicine.*The Journal of Alternative and Complementary Medicine*, 12: 833–834.

Strassburg, C. P., Kneip, S., Topp, J., Obermayer-Straub, P., Barut, A., et. al. 2000. Polymorphic gene regulation and interindividual variation of UDP-glucuronosyltransferase activity in human small intestine. *Journal of Biological Chemistry*, 275: 36164–36171.

Sun, J., Chu, Y. F., Wu, X., and Liu, R. H. 2002. Antioxidant and antiproliferative activities of common fruits. *Journal of Agricultural and Food Chemistry*, 50: 7449–7454.

Surh, Y. J. 2003. Cancer chemoprevention with dietary phytochemicals. *Nature Reviews Cancer*, 3(10): 768–780.

Tomás-Barberán, F. A., and Andrés-Lacueva, C. 2012. Polyphenols and health: Current state and progress. *Journal of Agricultural and Food Chemistry*,; 60(36): 8773–8775.

Tsao, R. 2010. Chemistry and biochemistry of dietary polyphenols. *Nutrients*, 2(12): 1231–1246.

Ung, D., and Nagar, S. 2007. Variable sulfation of dietary polyphenols by recombinant human sulfotransferase (SULT) 1A1 genetic variants and SULT1E1. *Drug Metabolism and Disposition*, 35: 740–746.

Urugo, M. M., and Tringo, T. T. 2023, January 1. Naturally occurring plant food toxicants and the role of food processing methods in their detoxification. *International Journal of Food Science*, 1: 1–15.

Varzakas, T., Zakynthinos, G., and Verpoort, F. 2016. Plant food residues as a source of nutraceuticals and functional foods. *Foods*, 5: 1–32.

Vinson, J. A., Su, X., Zubik, L., and Bose, P. 2001. Phenol antioxidant quantity and quality in foods: Fruits. *Journal of Agricultural and Food Chemistry*. 49: 5315–5321.

Vistisen, K., Loft, S., Olsen, J. H., Vallentin, S., Ottesen, S., et al. 2004. Low CYP1A2 activity associated with testicular cancer. *Carcinogenesis.* 25: 923–929.

Wang, L.-Q., and James, M. 2005. Inhibition of sulfotransferases by xenobiotics. *Current Drug Metabolism.* 7: 83–104.

Wells, P. G., Mackenzie, P. I., Chowdhury, J. R., Guillemette, C., Gregory, P. A., et. al. 2004. Glucuronidation and the UDP-glucuronosyltransferases in health and disease. *Drug Metabolism and Disposition*, 32(3): 281–290.

Wiwanitkit, V. 2014. Oxidative stress and metabolic syndrome. *Korean Journal of Family Medicine*, 35(1): 44.

Wojtunik-Kulesza, K., Niziński, P., Krajewska, A., Oniszczuk, T., Combrzyński, M., and Oniszczuk, A. 2025, June 1. Therapeutic potential of ellagic acid in liver diseases. *Molecules*, 30(12): 1–26.

Wolfe, K., Wu, X., and Liu, R. H. 2003. Antioxidant activity of apple peels. *Journal of Agricultural and Food Chemistry*, 51: 609–614.

Xiao, T., Guo, Z., Fu, M., Huang, J., Wang, X., et al. 2022. Amelioration of alcoholic liver disease by activating PXR-Cytochrome P450s axis with blackberry extract. *Separations*, 99(10): 1–26.

Xu, C., Li, C. Y. T., and Kong, A. N. T. 2005, March 31. Induction of phase I, II and III drug metabolism/transport by xenobiotics. *Archives of Pharmacal Research*, 28(3): 249–268.

Yager, J. D. 2015, May 26. Mechanisms of estrogen carcinogenesis: The role of E2/E1-quinone metabolites suggests new approaches to preventive intervention—A review. *Steroids*, 99: 56–60.

Yang, M., Wang, D., Wang, X., Mei, J., and Gong, Q. 2024, June 14. Role of folate in liver diseases. *Nutrients*, 16(12): 1–18.

Zech, L. A., Gross, E. G., Peck, G. L., and Brewer, H. B. 1983. Changes in plasma cholesterol and triglyceride levels after treatment with oral isotretinoin: A prospective study. *Archives of Dermatology*, 119: 987–993.

Zhang, E. E., and Kay, S. A. 2010. Clocks not winding down: Unravelling circadian networks. *Nature Reviews Molecular Cell Biology*, 11: 764–776.

Zhang, X., Sandhu, A., Edirisinghe, I., and Burton-Freeman, B. 2018. An exploratory study of red raspberry (Rubus idaeus L.) (poly)phenols/metabolites in human biological samples. *Food and Function*, 9: 806–818.

Zhou, S. S., Li, D., Sun, W. P., Guo, M., Lun, Y. Z., et al. 2009. Nicotinamide overload may play a role in the development of type 2 diabetes. *World Journal of Gastroenterology*, 15: 5674–5684.

Zhou, S. S., Li, D., Zhou, Y. M., and Cao, J. M. 2012, April 26. The skin function: A factor of anti-metabolic syndrome. *Diabetology and Metabolic Syndrome*, 4(1): 15.

9 Enhancing Brain Health with Medicinal Fruits and Berries

H. G. Navyashree and Y. P. Naveen

9.1 INTRODUCTION

Humans are the unique organisms on Earth, and their uniqueness, in part or as a whole, is the result of their ability to analyse, comprehend, and plan their actions; this ability makes them the most powerful survivors on the globe. The ability is the result of the most advanced nervous system they bear. The nervous system is controlled by an advanced brain.

The brain is a vital organ that weighs about 1400 grams (in an adult weighing 70 kg), and is protected by one of the hardest parts of the body, the skull. The brain controls physical, physiological, and emotional coordination in the body. In fact, it consumes 40% of the daily energy consumed, and coordinates every essential activity of the body stated above. The function of the brain is manifested by intricate neural networks that can encode, store, and retrieve information, hence becoming the master control unit for memory and cognition (Sridhar et al., 2023). Cognitive capacities, including learning, identifying, and decision-making, are all shaped by synaptic connections and neuronal ensembles(Yuste et al., 2024). The functional unit of the nervous system is neurons. In the neurological system, neurons transmit chemical and electrical impulses. They are divided into three categories based on their functions: interneurons, which connect neurons within the central nervous system (CNS) for processing; motor neurons, which transmit commands from the CNS to muscles and glands; and sensory neurons, which convey information to the CNS. They are also classified by myelination and structural characteristics (unipolar, bipolar, multipolar, and anaxonic), all of which operate to transfer electrical signals for bodily processes like movement, sensation, and thinking.

The brain's main parts include the cerebellum (balance/coordination), brainstem (vital functions including breathing/heartbeat), and cerebrum (four lobes: frontal for planning/movement, parietal for senses/orientation, occipital for vision, and temporal for hearing/memory/speech), which are the main components of the brain. These are connected to the spinal cord by deep structures like the thalamus (sensory relay) and the limbic system (emotions/memory), which regulate everything from essential life support to thought and emotion. In addition, the hippocampus and prefrontal cortex carry different functions of the brain. The hippocampus, which has a similar appearance to a seahorse and is situated in the medial temporal lobe, is essential for memory formation, particularly for turning short-term memories into long-term ones. It is essential for both spatial navigation and episodic memory (Hughes, 2025).

Many studies propose that, along with the food, physical exercise capitalizes on the brain's inherent capacity for plasticity, which leads to less cortical atrophy, improved brain function, and enhanced cognitive function. Additionally, there is considerable specificity in the effects of physical activity on the brain, with prefrontal and hippocampal regions seeming to be more affected than other brain regions. Researchers propose that physical activity is another supporting intervention that can improve cognitive and emotional performance in late adulthood by influencing the brain's endogenous pharmacology (Erickson et al., 2013). A healthy brain is central to cognition, emotional

DOI: 10.1201/9781003688259-9

balance, and overall quality of life. Appropriate nutrition is undoubtedly the foundation of brain activity, and short-term changes in the amount and composition of nutrients consumed by healthy people have an impact on cognitive processes (Keservani et al., 2016).

It is said that a man is what he eats (a quote by Socrates, a great Greek philosopher); although the quote was said very early in Greek history, the scientists of the present time also agree with the concept based on various scientific observations. A good food builds a good body, and a good body yields a good mind. A good mind is nothing but a physiologically healthy brain system. Thus, food influences not just physiological health but also mental health.

Humans consume a variety of foods that consist of both natural and nature-derived components. Fruits and berries are the most widely used and most often used foods throughout history and also by various ethnicities on the globe. Medicinal fruits and berries act as functional foods that actively support brain resilience, memory, and emotional balance. The fruits, along with loaded macromolecules (proteins, carbohydrates, and a good amount of healthy fats (especially MUFA and PUFA)), contain micronutrients including vitamins (B-complex, like B6, B12, folate, vitamin C, and vitamin E) and minerals (magnesium, zinc, and iron). In addition to nutritional support, they have plant-derived phytochemicals, also termed phytoceuticals, which can enhance various physiological activities of the body and thus brain health. They also contain complex carbohydrates that can support memory, mood regulation, and protective age-related cognitive decline (Głąbska et al., 2020).

Humans consume a wide variety of fruits based on the geographical, economic, and social parameters. Fruits such as pomegranate, apple, red/purple grapes, pineapple, banana, mango, guava, java plum, lichi, cherries, avocado, plums, and prunes, and citrus fruits such as orange and lemon are some of the most consumed fruits around the globe.

The current book chapter sheds light on the most consumed fruits, their nutraceutical components, and their pharmacological value in enhancing brain health. The chapter also discusses the mechanism through which the nutraceuticals act to enhance and support brain health, and also to prevent/delay various disorders of the brain or mental health.

9.1.1 Disorders of Brain and their Pathogenesis

Most disorders of the brain arise because of any abnormal molecular issues in the neurons. In fact, neurodegenerative disorders mean the issue is with the functional units of the nervous system, that is, neurons. The brain's poor antioxidant defence system makes it more susceptible to oxidative stress than other organs for the following reasons. First, the brain requires a lot of oxygen, that is, 20% of the body's total oxygen intake. Second, the brain contains significant amounts of oxidation-reduction active metals like copper and iron, which actively catalyse the production of reactive oxygen species (ROS). Third, because of the increased extent of oxidative metabolism and the high levels of PUFA in the cell membrane, the brain's antioxidant defence mechanism is comparatively weak, and ROS production is extremely high. (Agussalim, 2025). Neurodegenerative disorders are caused by a complex interaction of variables, mainly protein misfolding and aggregation, which results in proteotoxicity, along with oxidative stress, mitochondrial dysfunction, and chronic neuroinflammation, which eventually causes progressive loss of neurons (Duranti & Villa, 2026). Genetics, environmental pollutants, and ageing are important risk factors that overpower cellular defences. ROS are produced more frequently during oxidative stress, which promotes DNA damage, lipid peroxidation, and neuroinflammation. Neurodegeneration is exacerbated by an imbalance in the protective functions of antioxidants and an excess of ROS. The most prevalent types of neurodegenerative disorders are multiple sclerosis (MS), Parkinson's disease (PD), Alzheimer's disease (AD), and Huntington's disease (HD). (Munteanu et al., 2024)

The pathophysiology of AD, PD, HD, and amyotrophic lateral sclerosis is aided by the production of misfolded proteins, mitochondrial dysfunction, oxidative stress, and cytokine imbalance in the neurons (Sheikh et al., 2013).

Inflammation is the immune system's reaction to difficulties like infections, injuries, or toxins (Chavda et al., 2024). It is an essential defence system to eradicate pathogens and restore brain health and damage. The inflammation involves the secretion of cytokines, recruitment of the immune cells (through chemokine cues), activation of phagocytosis, and other accessory immune responses that finally result in the clearance of the pathogens. However, cytokines produced in the systemic infection can cross the blood–brain barrier (BBB) and induce neuroinflammation in the brain. These cytokines can stimulate microglia (CNS immune cells) and astrocytes; for instance, IL-1βbinds to IL-1 receptor type I and activates NF-κb, MAPK, and JAK-STAT pathways. Activated NF-kB induces expression of pro-inflammatory cytokines (IL-6, IL-1β, and TNF-α) and components of the inflammasome, that is, NLRP3 and pro-IL-1β/pro-IL-18 (Chen et al., 2018). However, in long-term conditions, it starts the neuroinflammatory cascade and becomes invasive. Microglia often cannot cross a healthy, intact BBB, and glial cells live behind the BBB to keep an eye on the surroundings. When under stress, microglial cells release cytokines, ROS, and matrix metalloproteinases (MMPs; zinc-dependent proteases that cleave proteins in the extracellular matrix and on the cell surface) that weaken the tight junction components in the BBB, increasing permeability, which allows peripheral immune cells (monocytes and T cells) to infiltrate. Once these immune cells interact with microglia, this amplifies inflammation, the BBB gets compromised, and peripheral immune cells are enabled to worsen inflammation (Gullotta et al., 2023; Zhang et al., 2025).

Prolonged and unregulated release of cytokines further aggravates neuroinflammation. Neuroinflammation upregulates enzymes, including BACE1, thereby increasing β-amyloid production. Increased inflammation promotes new deposition of the plaques and worsens the condition. Pro-inflammatory cytokines activate GSK-3β, CDK5, and MAPKs. The activated kinases further phosphorylate tau proteins, leading them to dissociate from the actin filaments. The dissociated and hyperphosphorylated tau protein leads to the formation of neurofibrillary tangles and ultimately the death of the neurons (Rawat et al., 2022; Uddin et al., 2022).

The body's primary stress response mechanism is the neuroendocrine system: hypothalamic–pituitary–adrenal (HPA) axis, which coordinates a series of hormones (CRH, ACTH, and cortisol) to assist the body in adjusting to stressors and regulating metabolism, immunity, mood, and energy. Stress causes the hypothalamus to secrete CRH, which signals the pituitary to release ACTH. The pituitary then instructs the adrenal glands to release cortisol, the primary stress hormone, which then signals back to shut down the system when the stress has subsided. This process is known as negative feedback. Neuroinflammation and the HPA axis are closely related. Stress triggers the HPA axis, which releases cortisol. Chronic release of cortisol causes glucocorticoid resistance, mitochondrial dysfunction, and ROS release, which can cause inflammation, which is characterized by the activation of microglia and pro-inflammatory cytokines (Herman et al., 2016; Visentin et al., 2020).

Considering that oxidative stress plays a significant role in the pathophysiology of neurodegenerative diseases, and as the CNS is susceptible to oxidative stress, scavenging ROS or preventing their formation has emerged as a promising strategy to prevent and manage neurodegenerative diseases (Agussalim, 2025).

Current treatments for neurodegenerative diseases only address symptoms and target a particular portion of the brain; they do not change the underlying disease. (Hussain et al., 2018).

Finding therapeutic agents that can modulate illness causes may consequently prevent or delay disease progression, leading to the lifelong incapacity or death of people affected. Nonetheless, a body of research shows that dietary practices, exercise, bioactive chemicals from food, and nutritional conditions all help to mitigate the age-related decrease in memory and cognition (Bigford & Del Rossi, 2014). These days, antioxidants are appealing compounds to combat neurodegeneration. Excellent antioxidant qualities are exhibited by several vitamins (A, E, and C) and polyphenolic compounds (flavonoids). Antioxidants are mostly obtained through diet (Samanta et al., 2024).

9.2 FRUITS AND THEIR IMPORTANCE IN BRAIN HEALTH

9.2.1 Apples (*Malus domestica*)

Apples play a significant role in enhancing brain health; thanks to the rich content of flavonoids and antioxidant compounds, which also offer a neuroprotective role. Flavonoids found in apple peels include quercetin, catechin, epicatechin, and procyanidins. Studies have indicated that apple quercetin's anti-inflammatory, anti-amyloid, and antioxidant qualities may have neuroprotective effects (Chiang et al., 2023). Quercetin acts to modulate the inflammation in two ways: one is by directly scavenging the ROS and reactive nitrogen species, which are significant causes of oxidative stress-induced brain damage; in another way, it increases levels of natural antioxidant enzymes like glutathione (GSH), catalase (CAT), and superoxide dismutase (SOD) via activating the Nrf2-ARE signalling pathway (Debnath et al., 2025).

Catechins, also known as flavan-3-ols, are another type of flavonoid found in apples with potential neuroprotective qualities. They also function as strong antioxidants, reducing inflammation and promoting long-term brain health (Kopustinskiene et al., 2021). Catechins, especially epigallocatechin gallate (EGCG), show neuroprotective potential by countering oxidative stress, iron misregulation, and mitochondrial damage, which are also key drivers of ageing, stroke, and neurodegenerative diseases. They inhibit β-amyloid formation, memory decline in ageing models, and also protect neurons after ischaemia (Bernatoniene & Kopustinskiene, 2018). Additionally, apples contain minerals that support and boost brain function, such as potassium, a small amount of magnesium, and iron. Vitamin C and polyphenols of the apple provide antioxidant defence, preventing oxidative damage, which is believed to be a significant cause of dementia (Agussalim, 2025). Pectin and different soluble fibres promote a healthy gut flora, which is increasingly associated with enhanced mental health (Ribeiro et al., 2022).

Furthermore, apple polyphenols lower the production of pro-inflammatory cytokines, including TNF-α and IL-1β, by blocking the activation of NF-κB, a crucial regulator of inflammation (Zhang et al., 2024). When combined, apples' diversified phytochemical profile—which includes polyphenols, flavonoids, vitamins, and minerals—contributes to a multifactorial neuroprotective impact.

9.2.2 Pomegranate (*Punica granatum*)

Pomegranate is a fruit highly valued in traditional medicine; now it is recognized for its dense phytochemical profile. It is composed of high concentrations of polyphenols compared to other fruits and vegetables (Braidy et al., 2016). Their polyphenols (punicalagins and ellagic acid) reduce neuroinflammation, improve memory, and protect neurons from oxidative stress. The most common phytochemicals in pomegranate juice are polyphenols, encompassing the hydrolysable tannins known as ellagitannins(Seeram et al., 2006) (produced when ellagic acid and/or gallic acid attach with a carbohydrate to yield pomegranate ellagitannins), also called punicalagin (Keservani et al., 2016). Pomegranate contains anthocyanins, the subclass of flavonoids that provide characteristic colours, namely, purple, blue, and red pigments in many plant-based foods, which provide support to brain health (Bar-Ya'akov et al., 2019).

Anthocyanins are recognized for their pro-cognitive, anti-inflammatory, and neuroprotective qualities in observational and clinical trials. They may also enhance microvascular function. Additionally, there is preclinical evidence that foods high in anthocyanins can improve cerebellar physiology and motor learning, reverse age-related declines in neuronal signal transduction, and improve cognitive and motor behavioural deficits. This suggests that polyphenolic compounds can cross the BBB and localize in different brain regions crucial for memory and learning (do Rosario et al., 2024). Additionally, it contains vitamin K, which indirectly supports brain perfusion; vitamin C, which supports neurotransmitter synthesis; and urolithin, a pomegranate gut metabolite that enhances neuronal mitochondrial function, improves memory, and lessens dementia symptoms in an Alzheimer's model.

Pomegranate ellagitannins directly interact with misfolded proteins and restore redox equilibrium to have neuroprotective effects. While ellagic acid reduces α-synuclein aggregation, dissociates preformed aggregates, and lessens their neurotoxicity, pomegranate extract has been demonstrated to diminish Aβ accumulation through modulation of the β-secretase (BACE) pathway. These results show pomegranate polyphenols to be promising agents for preventing protein misfolding and maintaining the integrity of neurons (Aleksandrova et al., 2023, 76).

Pomegranate's potential as a multifactorial neuroprotective dietary intervention is highlighted by its antioxidant, anti-inflammatory, and protein-modulating effects that counteract misfolding and aggregation in the CNS. It also supports neuronal integrity and cognitive resilience.

9.2.3 Citrus Fruits

Oranges, grapefruit, lemons, and limes are examples of citrus fruits (Rutaceae) that are exceptionally rich in flavanones; for this reason, they are frequently referred to as citrus flavonoids. Oranges are packed with vitamin C, a powerful antioxidant that helps lessen oxidative stress caused by free radicals. Researchers have shown that high blood levels of vitamin C are linked to improved cognitive function, including memory and decision-making speed. Yet another significant component in oranges is folate, which is critical for brain function. It facilitates the production of neurotransmitters, which are essential for brain cell communication. Sufficient amounts of folate have been linked to better mood management and a decreased risk of cognitive decline.

Oranges are a great source of potassium, which is essential for nerve health and brain electrical signal transmission. Research studies suggest that the peel of the orange fruit shows antidepressant properties in mice by bringing back brain-derived neurotrophic factor (BDNF) (Ademosun, 2024).

One of the main citrus flavanones present in all citrus fruits is hesperidin (hesperetin-7-O-rutinoside). Hesperidin, also known as cirantin (hesperetin-7-O-rutinoside), is a flavonoid that is formed from the glycosylated derivative of hesperidin, a flavanone that is present in fruits including lemons, oranges, and tangerines. Hesperidin has been identified as having neuropharmacological effects, such as memory, nerve protection, and antidepressant effects. Hesperidin can readily pass through the BBB and offer neuroprotection mainly because it is lipophilic (Bellavite, 2023). Hesperidin exhibits neuroprotective potential by inhibiting β-amyloid aggregation, boosting endogenous antioxidant defences, reducing neuroinflammation and apoptosis, improving mitochondrial function, controlling autophagy, and encouraging neurogenesis. While a few clinical studies indicate hesperidin-rich supplementation may enhance cerebral blood flow, cognition, and memory, preclinical evidence supports its effectiveness in models of Alzheimer's disease by influencing molecular pathways such as RAGE/NF-κB, Akt/Nrf2, and AMPK/BDNF/CREB. However, further clinical trials are necessary to confirm its therapeutic value and prove human safety (Han et al., 2025).

Hesperidin may be a viable option for reducing oxidative stress and treating AD-type memory impairment. Hesperidin-enriched dietary supplements have been shown in clinical trials to substantially enhance retention, cognition, and cerebral blood flow (Hajialyani et al., 2019).

Following an episode of Huntington's illness, hesperidin-treated mice were shown to be less likely to show locomotor impairment and hippocampus function, suggesting that these flavanones show beneficial effects on neurons (Kean et al., 2015; Menze et al., 2012).

Naringenin-7-O-rutinoside (narirutin) is also present in oranges and orange juices, along with trace amounts of hesperetin-7-O-rutinoside-3′-O-glucoside, 4′-O-methyl-naringenin-7-O-rutinoside (didymin), and eriodictyol-7-O-rutinoside (eriocitrin) (Pontifex et al., 2021). The flavanone and aglycone of naringin, naringenin, are mostly present in citrus fruits. Because of their antioxidant properties, flavonoid dietary supplements have been demonstrated to ease the severity of NDs such as dementia, PD, and AD (Goyal et al., 2022, 78).

9.2.4 Lemons (*Citrus limon*)

Lemons are a natural resource for cognitive resilience since they are high in bioactive antioxidants, special pectins, vitamin C, folic acid, and potassium. Antioxidants, flavonoids, and special pectin molecules found in lemons work synergistically to improve neuroprotection, lower oxidative stress, and boost cognitive resilience. Through bio-actives like hesperidin, eriocitrin, naringenin, and integropectin. IntegroPectin, a novel citrus pectin produced by hydrodynamic cavitation, has better antioxidant and antibacterial action than traditional pectin because it retains high quantities of flavonoids and terpenes. Lemon integropectin exhibits strong neuroprotection against oxidative stress-induced neuronal damage *in vitro* by its cytoprotective effects on live cells and remarkable radical-scavenging ability, most likely due to mitochondrial damage protection induced by free radicals. The bioactives of the fruit enhance neurotransmitter balance, improve neuronal survival, and may help guard against neurodegenerative diseases (Nuzzo et al., 2021; Yadav et al., 2022).

9.2.5 Red/Purple Grapes (*Vitis vinifera*)

Grapes have been gaining attention for their ability to improve brain health, enhance memory, and provide neuroprotection. They contain vitamins and minerals such as vitamin K, copper, and potassium. They also contain a substantial quantity of polyphenolic agents, especially flavonoids (anthocyanins and proanthocyanidins) as well as stilbenes (resveratrol). Red or purple grapes (and their products, such as juice or extracts) have a significant therapeutic effect in improving brain health. Studies on humans and animals have demonstrated that red and purple grape products improve cognitive function, including memory, attention, learning, and motor skills. Purple grape juice sharply improved reaction time and increased tranquillity in young adults, while Concord grape juice enhanced memory in older adults with a mild decrease throughout a 12-week trial (Georgiev et al., 2014). Resveratrol's neuroprotective action extends beyond its antioxidant and anti-inflammatory roles. It has been shown to influence autophagy signalling by activating the AMPK–SIRT1 pathway. This activation promotes autophagy, thereby improving the efficiency of cellular protein degradation systems in removing misfolded and aggregated proteins (Ghavami et al., 2014). Foods high in anthocyanins, especially resveratrol of the grapes, show promise as neuroprotectors against neurodegenerative illnesses and ageing-related cognitive decline (Medina dos Santos et al., 2019).

9.2.6 Avocado (*Persea americana*)

Avocados are indeed an excellent source of nutrition for the brain because they contain oleic acid, which promotes healthy blood flow and cognition (Melo et al., 2019); dietary fibres that are beneficial for the gut–brain axis (Raskov et al., 2016); they enhance the microbial diversity (microbial diversity that potentiates gut–brain axis) as well as promote intestinal mobility (Maqsood et al., 2025) and reduce the risk of gastrointestinal disorders. The fruit also contains a significant amount of bioavailable vitamins C and E, which shield nerve cells from oxidative damage, and vitamins B and folate (Afzal et al., 2022), which help to regulate neurotransmitters like dopamine and serotonin (Essa et al., 2016), thereby helping to lower anxiety. In addition, the fruit contains lutein and carotenoids, the phytochemicals that can enhance memory, concentration, and processing speed (Johnson et al., 2015). Neurodegenerative disorders are associated with chronic inflammation. Avocados make up phytosterols like β-sitosterol that have been exhibited to have substantial benefits against inflammation by decreasing pro-inflammatory markers like TNF-α and IL-6, which engage in neuroinflammation (Bhuyan et al., 2019).

In addition to consuming more fruits and vegetables overall, regular Hass avocado eaters exhibit noticeably higher intakes of fibre, monounsaturated fats, antioxidant vitamins, folate, and other minerals. Avocado's contribution to improving nutritional density and promoting cardiometabolic and cognitive health is highlighted by this dietary profile (Dreher et al., 2021).

9.2.7 Plums (*Prunus domestica*) and Prunes

Prunes originate from dried plums, and plums are a varied group of stone fruit (drupes) that are purple or red in colour and rich in vitamins (A, C, K, and some B complex) and antioxidants. Plums are dehydrated in air at 85–90°C for 18 hours to lower the fruit's water content from 87% to about 33% in order to make prunes (Shukitt-Hale, Kalt, et al., 2009). Researchers believed it to have a diuretic effect due to high fibre components (Ayub et al., 2023). Plums have bioactive substances such as phenolic acids, anthocyanins, minerals (potassium, copper, and manganese), and pectins are abundant in plums. Fruits with insoluble fibres, like prunes, do not break down during digestion. They aid in preventing constipation and improving intestinal mobility through bowel movements (Maqsood et al., 2025). Polyphenols, such as anthocyanins of plums, have been shown to enhance cognitive performance. They also have anti-inflammatory and antioxidant features; hence, they can lower the risk of neurodegenerative disorders (Xu et al., 2025). Additionally, compounds like chlorogenic acid, neochlorogenic acid, and quercetin are prevalent in plums.

Prunes, even though they lack flavonoids, dried plums are high in hydroxycinnamate esters. An in vivo model of age-related cognitive impairment was used to compare their supplementation to flavonoid-rich products (Shukitt-Hale et al., 2009).

9.2.8 Mango (*Mangifera indica*)

Mangoes, which are members of the Anacardiaceae family, contain macronutrients including protein, lipids, fatty acids, and organic acids, and carbohydrates like sucrose, glucose, fructose, and certain plant-based structural components like cellulose and pectin, along with minerals including sodium, phosphorus, iron, calcium, magnesium, and potassium (Khan et al., 2017). Additionally, it contains micronutrients like vitamin B6, which help produce neurotransmitters like dopamine and serotonin that contribute to mood control and cognitive function. Mango leaves also contain

TABLE 9.1
Fruits, their Bioactives, and Effect on Brain Health

Fruits	Fruitaceuticals Present	Vitamins and Mineral	Effect on Brain Health
Apple	Flavonoid, phenolic acid	Vitamin C, potassium, fibre	Boost long-term cognitive resilience and lessen oxidative stress in brain cells
Pomegranate	Tannins, polyphenols	Vitamin C, vitamin K, folate	Enhanced memory retention, decreased neuroinflammation, and robust antioxidant protection
Orange	Flavonoids	Vitamin C, folate, potassium	Increase the production of neurotransmitters, guard against oxidative damage, and enhance attentiveness
Lemon	Flavonoids	Vitamin C, folate, potassium	Boosts neurotransmitter synthesis and protects against oxidative damage
Grape	Resveratrol, flavonoids	Vitamin K, copper, potassium	Improve memory, cognition, and protect from neurodegenerative diseases
Avocado	Phytosterols, carotenoids	Vitamin B, vitamin C, vitamin E, folate, potassium	Enhance memory, focus, and provide monounsaturated fats for cell membranes
Plum	Polyphenol, flavonoids	Vitamin A, vitamin C, vitamin B5, potassium, copper, manganese	Reduce neuroinflammation and enhance cognition
Mango	Carotenoids, flavonoid	Vitamin C, folate	Help produce neurotransmitters, enhance memory, and cognition

mangiferin, which prevents the diminishing of brain-derived neurotropic factor and lessens neuronal damage (López-Ríos et al., 2020). Indeed, support cognition by improving episodic and spatial memory through its anti-inflammatory and antioxidant qualities. It contains 25 different carotenoids (Lebaka et al., 2021) and flavonoids such as quercetin and kaempferol in pulp that contribute to a synergistic neuroprotective effect (Table 9.1).

9.2.9 Berries

Berries are often known because of their neuroprotective properties; they can also be referred to as "brain boosters." There are different types of these berries, namely, blueberries, strawberries, raspberries, and blackberries. Berries bear high bioactive molecules, especially polyphenols, flavonoids, and anthocyanins. Polyphenols in berries are very diverse, and they have the ability to provide neuroprotection because of their anti-inflammatory and antioxidant qualities (Singh et al., 2025).

Berries help to mitigate the neuroinflammatory state that frequently underlies neurodegenerative illnesses because of their capacity to both stop the spread of oxygen-free radicals (ROS) and cross the BBB. The flavonoids of the berries can interact with the neuronal structures involved in storing data and memory. They have shown the capacity to boost growth factors like BDNF, which control the growth and maintenance of neuronal functions and lead to neurogenesis and brain cell regeneration, as well as to expand and strengthen synaptic connections (De Amicis et al., 2022). Additionally, it has been demonstrated that flavonoids of berries boost blood flow in the brain, promoting vascularization (Martini et al., 2019) and angiogenesis throughout the brain, including the memory centre hippocampus. Furthermore, a study suggested berry fruit supplementation as a neuroprotector and the mitigation of symptoms in conditions like anxiety, depression, AD, and PD (Fernández Demeneghi et al., 2019).

Berries contain a good variety and quantity of anthocyanins, and these have optimal bioavailability and neuropharmacological potential. Anthocyanins form metabolites (protocatechuic acid, ferulic acid, and other phenolic derivatives) when the gut microbiota is present. which increases their solubility through phase 2 conjugation (glucuronidation, methylation, and sulfonation). Because they are tiny and lipophilic, these metabolites are better able to traverse the BBB. These compounds are transported via endothelial cells by tight junction proteins or carrier-mediated processes (glut transporters and p-glycoprotein). Consequently, endothelial cells' gene expression is modulated, supporting barrier function. These components' antioxidant capacity can neutralize superoxide and hydroxide radicals and directly scavenge ROS in the BBB. Additionally, by increasing Nrf2 signalling, endogenous antioxidant enzymes, including SOD, CAT, and glutathione peroxidase, are boosted. Furthermore, BNDF upregulation, NF-kappa B suppression, and neurotransmitter balance improve dopamine signalling and serotonin, which directly suppresses HPA axis activation.

9.2.10 Strawberries (*Fragaria × ananassa*)

Strawberries grow from a couple of ovaries to produce an aggregate fruit. They have high levels of folate, thiamine, riboflavin, niacinamide, and vitamins C, A, E, and K. Additionally, they are superfoods; they contain brain-supporting nutrients, including manganese, potassium, iodine, magnesium, copper, phosphate, and phenolic acids like hydroxycinnamic and hydroxybenzoic acid (Miller et al., 2019). It is a hub of antioxidants since it contains flavonoids such as anthocyanins, tannins, catechin, quercetin, kaempferol, and gallic acid in addition to antioxidants like caffeic acid and ellagic acid (Subash et al., 2014). A verbal learning test was implemented in certain investigations, including strawberries and a placebo, and the results revealed that strawberries improved cognition more than in the placebo group (Miller et al., 2021).

Strawberries' secondary metabolites have been shown in multiple laboratory and human studies to be able to modulate cellular oxidative stress, offer neuroprotection, improve endogenous detoxification systems, and alter metabolic and cognitive processes (Irshad et al., 2025). The various solvent extracts of strawberry have been documented in another investigation to enhance motor behavioural performance on the rod walking (Subash et al., 2014). Studies also show that supplementation of strawberries along with regular food can help to lower the likelihood of Alzheimer's disease by slightly less than a quarter (Agarwal et al., 2019).

9.2.11 Blueberries (*Vaccinium corymbosum*)

Blueberries belong to the Vaccinium genus within the Ericaceae family. Vitamins, minerals, dietary fibre, and polyphenols are abundant in blueberries. Berries are rich in nutrients and phytochemicals that have been evident in both *in vitro* and *in vivo* tests to enhance health and prevent a number of chronic diseases (Stull et al., 2024). Tannins, phenolic acids, and flavonoids (anthocyanins, flavonols, flavones, flavanols, and isoflavones) are the main phytochemicals found in berry fruits (Golovinskaia & Wang, 2021). It has 25 known anthocyanidins, which make up around 84% of the total antioxidant capacity, and are widely recognized as compounds in blueberries that contribute to the antioxidant capacity of active substances (Borges et al., 2010). Blueberry anthocyanins' potent antioxidant qualities have been thoroughly investigated and verified in both lab tests and living things (Wang et al., 2024). By releasing hydrogen atoms from the phenolic ring, anthocyanins aid in the neutralization of free radicals and the reduction of oxidative stress (Tena et al., 2020).

The 3-glucosides and 3-galactosides of delphinidin, malvidin, petunidin, cyanidin, and peonidin, as well as trace levels of their 3-arabinosides, were found to be anthocyanins in the unidentified lowbush blueberry variety (GAO & MAZZA, 1994). In addition to its neurocognitive benefits, blueberries offer anti-inflammatory and antioxidant properties (Cotman et al., 2000). As the flavanols in BBs communicate with nerve cells at the subcellular level, generating mechanisms of signalling that promote connections between the neurons, interactions between cells, and induce nerve cell regeneration, the abundance of antioxidants in BBs has been demonstrated to enhance and protect the brains of older adults and improve brain performance and memory in all age groups (Tran & Tran, 2021, 83).

The microbiota of the gastrointestinal tract extensively metabolizes blueberry anthocyanins alongside various polyphenols, suggesting that microbiota-induced polyphenol metabolites may modulate the health benefits of blueberries (Petersen et al., 2022). By controlling the gut microbiota, anthocyanins may ease neurodegeneration (Zhong et al., 2023).

9.2.12 Blackberries (*Rubus fruticosus*)

Blackberries are a member of the Rosaceae family found within the genus Rubus. Thanks to their unique flavour, exceptional nutritional value, and nutraceutical qualities, these have drawn focus. Essential vitamins (A, B, C, K, and E), minerals (manganese, copper, potassium, sodium, calcium, phosphorus, iron, and magnesium), sugars like glucose, fructose, and sucrose, and organic acids (the majority of it is malic acid with portions of ascorbic acid and galacturonic acid) are all present in them (Martins et al., 2023). The significant amounts of anthocyanins, ellagitannins (King et al., 1996), and other phenolic compounds contribute to the antioxidant ability of the berries based on their ability to absorb oxygen radicals by these phytochemicals, resulting in enhanced antioxidant ability of the berries relative to other fruits (King et al., 1996). Berries contain cyanidin 3-glucoside, cyanidin 3-galactoside, cyanidin 3-xyloside, cyanidin 3-rutinoside, cyanidin 3-sophoroside, cyanidin 3-glucosylrutinoside, cyanidin 3-arabinoside, malvidin 3-arabinoside, perlargonidin 3-glucoside, cyanidin 3-(3-malonyl)glucoside, and cyanidin 3-(6-malonyl)glucoside and are the only fruit that contains the zwitterionic anthocyanin cyanidin 3-dioxaloylglucoside (Wu & Prior, 2005).

The primary role of this fruit's neuroprotective potential is its antioxidant capacity, which can be improved by its incorporation of phenolic chemicals like anthocyanins. In fact, these substances have neuroprotective effects on the hippocampus and cortex, among other cerebral regions in the brain, and can cross the BBB that transfers these substances between the CNS and blood (Nowak et al., 2014). On top of that, flavonoids may have significant effects on mammalian cognitive function, possibly preventing losses of memory and learning associated with elderly age-grouped people (Subash et al., 2014). According to earlier research, rodents fed with blackberries performed noticeably better than control rats in terms of working memory, or short-term memory (Shukitt-Hale, Cheng, et al., 2009).

9.2.13 Cranberry (*Vaccinium macrocarpon*)

Cranberries, members of the Ericaceae family, bear antioxidants like polyphenols (flavonoids, phenolic acids, anthocyanins, and tannins). Hydroxybenzoic and hydroxycinnamic acids are the phenolic acids found in cranberries (Nemzer et al., 2022).

In addition to that, it has vitamin E and minerals, including potassium, phosphorus, iron, and manganese. It has been demonstrated that terpenoids in cranberries considerably reduce the expression of pro-inflammatory cytokines (IL-6, TNF-α) in THP-1 cells, a human monocyte cell line, blocking cellular pathways for anti-inflammatory actions.

Cranberries contain a significant quantity of proanthocyanidins (Nile & Park, 2014). The proanthocyanidins are potential scavengers of free radicals and thus can effectively reduce oxidative stress markers, making them promising therapeutic agents in the fight against the neuroinflammation-induced neurological complications (Yang et al., 2018, 136). Studies also have shown the acetylcholine esterase (the enzyme that breaks down acetylcholine) inhibitory potential of proanthocyanidins and thus can raise the levels of acetylcholine in the nerve junctions and improve cholinergic neuron signalling.

According to researchers, eating cranberries for three months can enhance one's episodic memory. Cranberries' ability to increase regional perfusion in areas of the brain, specifically in the temporal region (region responsible for memory and cognition), is one of the cognitive benefits of cranberries (Flanagan et al., 2022).

9.2.14 Raspberries (*Rubus occidentalis*)

Raspberries provide fat-soluble vitamins such as tocopherols and vitamin A, while the fruit's flesh is particularly rich in water-soluble vitamin C, along with thiamine, vitamin B6, and riboflavin. In addition, raspberries supply essential minerals such as potassium, calcium, magnesium, and iron. Their seeds also contain carotenoids such as β-carotene, zeaxanthin, and lutein.

Biological characteristics, including antioxidant and anti-inflammatory actions, were contributed by bioactive components such as anthocyanins, ellagitannins, and ellagic acid. Due to its high concentration of polyphenol, flavonoids, and phenolic acid content, it acts as a protective agent. (Mullen et al., 2003).

Especially hydrolysable tannins, which serve as an excellent source of cyanidin glycosides and are distinguished from many other berries by their notably high ellagitannin levels, which break down to yield ellagic acid (Havas et al., 1994).

Raspberries' polyphenols can help in lowering brain inflammation by preventing the activation of microglia, the immune cells that cause neuroinflammatory reactions. It has been shown that raspberry polyphenols can both increase the synthesis of anti-inflammatory cytokines like IL-10 and block important inflammatory signalling pathways like MAPK and NF-κB. The potential of raspberries to reduce the risk of cancer and neurodegenerative disorders linked to chronic neuroinflammation, such as AD and PD, is highlighted by this combined effect, which reduces pro-inflammatory mediators and increases anti-inflammatory and anticancer responses (Garcia et al., 2020).

9.2.15 Goji Berries (*Lysium barbarum*)

Goji berries are part of the Solanaceae family. They are also called Chinese wolfberries, which have been traditionally used in Asian countries as food as well as complementary medicine. They contain carotenoids (including lutein and zeaxanthin), flavonoids, and other phytochemicals, with potential antioxidant, anti-inflammatory, immunomodulatory, anticancer, and anti-ageing qualities (Noreen et al., 2025).

Goji berries are composed of about 1.5% fat, 16% dietary fibre, 13% protein, and 46% carbohydrates; micronutrients including copper, manganese, magnesium, and selenium; and vitamins such as nicotinic acid, thiamine, and riboflavin. Additionally, they are a rare and excellent source of zeaxanthin in the form of the highly concentrated carotenoid pigment, which is dipalmitin zeaxanthin (Ma et al., 2019).

Wolfberry has long been prized in Asian diets due to its anti-ageing qualities. According to supplementation studies, wolfberry polysaccharide fraction provides robust neuroprotection against β-amyloid-induced toxicity in rat cortical neurons (Ho et al., 2009).

In addition to that, using the Goji berry as an antioxidant extract has produced neuroplasticity and neuroprotection. Goji enhances neuronal structure through dendritic arborization and dendritic spine density, which promotes neuronal transmission and mitigates the development of neurodegenerative diseases (Ruíz-Salinas et al., 2020).

9.2.16 Acai Berries (*Euterpe oleracea*)

Acai berries are small, dark purple-coloured fruits from the Amazon rainforest. Now globally they are recognized as "superfoods" due to bioactive compounds such as phenolics, flavonoids, and anthocyanins, which are linked with its health benefits. Numerous nutrients have also been discovered in Euterpe oleracea fruits, including phytosterols, amino acids, soluble and insoluble fibres, vitamins, minerals, and an exceptionally high concentration of mono- and polyunsaturated fatty acids (MUFA and PUFA) (Schauss, et al., 2006).

The significant anthocyanins such as cyanidin 3-glucoside and cyanidin 3-rutinoside, and three minor anthocyanins, cyanidin 3-sambubioside, peonidin 3-glucoside, and peonidin 3-rutinoside, are present in these fruits (Schauss, et al., 2006); yet, compared to other deep-coloured berries, the overall anthocyanin concentration is lower quantitatively.

In vitro neuroprotection against the main neurologically damaging β-amyloid alloform, Aβ1, was demonstrated by acai berry extract. High concentrations of antioxidants found in berry pulp and acai extracts are typically thought to reduce the harmful effects of ROS in cell culture. However, a new discovery is that acai provides unique neuroprotection against β-amyloid loss of viability (Wong et al., 2013).

Acai's potential to mitigate and treat a variety of neurological conditions, such as dementia, PD, and AD, has been investigated. Although studies on acai's potential to improve brain function and health are still in their stage of development, more studies are necessary (Schauss, 2015).

9.2.17 Mulberry (*Morus nigra*)

Mulberries are high in plant proteins, carbs, and dietary fibre. They supply vital macronutrients and micronutrients (such as minerals, including iron, calcium, magnesium, zinc, nickel, potassium, and vitamins like vitamin C, vitamin K, and folate). Eighteen distinct amino acids were identified in mulberry fruit, nine of which were necessary for human health. Compared to monounsaturated and saturated fatty acids, mulberry fruit has more PUFAs. Furthermore, only mulberry fruits have been found to contain the important fatty acids behenic acid and palmitoleic acid.

Mulberries are a potential therapeutic food for neuroprotection because they contain bioactive substances that protect brain cells and enhance cognitive abilities. Anthocyanins, particularly cyanidin-3-glucoside derivatives, give mulberries their deep purple to black colour and operate as strong antioxidants. Phenolic acids (chlorogenic acid, gallic acid, protocatechuic acid, p-coumaric acid, O-coumaric acid, ferulic acid, caffeic acid, and vanillic acid), rutin, and quercetin have anti-inflammatory and pro-metabolic properties. Their antibacterial and antioxidant qualities are enhanced by tannins and ellagitannins.

Resveratrol, which is more prevalent in mulberries than other fruits, has been connected to neuroprotective and cardiovascular benefits. 1-Deoxynojirimycin (DNJ), an alkaloid, has a number of beneficial effects. First, it greatly improves the learning and spatial memory capacities of mice with cognitive impairments. Additionally, it reduces microglial activation, Aβ aggregation, and the levels of neuroinflammatory markers (TNF-α, IL-1β, and IL-6) in the hippocampus of rodents(Chen et al., 2018, 146).

In neuronal cells, anthocyanin-enriched *Morus alba* L. extract and vitamin C (MAC) have been shown to have strong synergistic neuroprotective effects against oxidative stress caused by H_2O_2. MAC efficiently suppresses apoptosis, restricts lipid peroxidation, increases endogenous antioxidant enzyme activity, and lowers ROS accumulation. This study shows that MAC is a viable nutraceutical option, intended to prevent or lessen neurological diseases linked to oxidative stress (Mairuae et al., 2025).

9.3 SYNERGISTIC EFFECT OF FRUITS OR FRUITACEUTICALS ON BRAIN HEALTH

Brain health has emerged as a prominent concern in recent times. Maintaining a balanced diet is essential for brain and cognitive function. According to another investigation, the antioxidant content of a mixture of wild bilberries, cranberries, raspberries, and strawberries was higher than that of single berries (Zafra-Stone et al., 2007).

Numerous studies have demonstrated that the cognitive impairment linked to neurodegenerative illnesses may be lessened by a nutritional strategy based on increased antioxidant consumption. According to a study, these polyphenol components enhance brain functions by directly influencing CNS cells. Both cerebral blood flow and blood pressure can be improved by polyphenolic substances by improving the integrity of the BBB and lowering inflammation. Polyphenols protect the brain from further damage following a stroke. (Grabska-Kobyłecka et al., 2023).

The MIND diet (Mediterranean DASH intervention for neurodegenerative delay), another newly popular healthy eating pattern, proposes the use of blueberries, a minimum of two servings per week (one serving means one cup of fresh berries). There is strong evidence that the consumption of this dietary pattern can decrease the risk of AD and slow rates of cognitive loss (Omar, 2019).

Polyphenols, B-vitamins, and omega-3 fatty acids are the main non-nutritional and nutritional components that affect neurogenesis, synaptic plasticity, and the reduction of oxidative stress. B vitamins and docosahexaenoic acid (DHA), a major component of neuronal membranes, can be involved in the methylation of DNA and the metabolism of homocysteine, both of which are critical for cognitive function. DHA also improves synaptic function and reduces neuroglial inflammation ("Brain Food: A Comprehensive Review of Nutritional Strategies for Cognitive Enhancement and Neuroprotection," 2025). And all the above components are found in the fruits in abundant and active form.

9.3.1 Bioavailability and Metabolism of Phytochemicals

The percentage of total phytochemicals consumed orally that enters the bloodstream and is used by tissue and organ is referred to as bioavailability. The dietary matrix, processing, and gut flora all have an impact on their bioavailability and are essential in converting complicated substances into molecules that can be absorbed. Due to their poor solubility, stability, and absorption properties, many phytochemicals have low bioavailability and bioactivity after intake. Interestingly, a lot of phytochemicals are included in food naturally in the form of glycosides, which are difficult for the upper gastrointestinal tract to absorb (Hu et al., 2023). Nevertheless, despite their health benefits, polyphenols' low bloodstream absorption rate (around 5–10%) and quick metabolism and excretion severely restrict their ability to reach target tissue (Saad et al., 2025).

Dietary flavonoids must be liberated from plant products by chewing, the action of the gastrointestinal tract's digestive fluids, and ultimately the microbes of the colon before they can be absorbed from the gut. The physicochemical characteristics of the flavonoid, including its molecular size and structure, lipophilicity, solubility, and pKa, will determine how well it is absorbed from the diet. With the exception of catechins, the majority of flavonoids are typically found in the diet as β-glycosides. Since glycosides were thought to be too hydrophilic to be absorbed through passive diffusion in the small intestine, only aglycones were anticipated to be absorbed. Only a little absorption of these glycosides is anticipated, as the colon's absorption capacity is significantly lower than the small intestine's (Hollman, 2004).

The polyphenols are found in ester, glycoside, and polymer forms that must be hydrolysed by the gut flora or intestinal enzymes in order to be absorbed. Although studies on this metabolic process are conflicting, hydrolysis may already take place in the oral cavity due to the activity of β-glucosidase that is either produced by the oral microbiota or present in saliva. The most commonly accepted mechanism occurs in the small intestine, where the molecules are hydrolysed, and aglycones are released by the lactase phlorizin hydrolase (LPH), which is found at the brush border of the epithelial cells. Through passive diffusion, the aglycones enter the intestinal cell. Before they reach the liver, they go through phase I biotransformation (oxidation, reduction, or hydrolysis) in the enterocyte. The colon receives 90–95% of the polyphenols that are not absorbed in the small intestine (Lavefve et al., 2019).

Berries share the pathway that yields their metabolites, including phenolic acids, terpenoids, and carotenoids. Berry phytochemicals like anthocyanins, flavonols, and ellagitannins exhibit restricted bioavailability because of poor intestinal absorption and extensive metabolism. However, because different berry species have variable regulation of the pathways that are crucial in mediating health effects, their bioactive contents and profiles differ. Their bioactivity is mostly driven by gut bacteria, which makes both parent chemicals and metabolites important for systemic and brain health (Baldassi et al., 2025).

These days, it's becoming more and more apparent that berry (poly)phenols are highly bioavailable because they travel from the small to the large intestine, where the local microbiota breaks them down into phenolic catabolites, which are then absorbed into the bloodstream in significant quantities (Di Pede et al., 2022).

By coordinating microbiota-brain transmission, which can contribute to neuroinflammation, a defining characteristic of neurodegenerative illnesses, the gut microbiota plays a significant role in these conditions. Numerous physiological and pathological processes are significantly influenced by the intricate network of communication between the gut and the brain, referred to as the gut-brain axis (Mayer et al., 2015). CNS and the gut are intimately related in both directions. While the sympathetic and parasympathetic nervous systems (known as the "Gut-Brain Axis") allow the gut to interact with the CNS, the CNS communicates with the gut's muscle and mucosal layer through both afferent and efferent autonomic pathways (ANS) (Roy Sarkar & Banerjee, 2019).

9.3.2 Strategies to Enhance the Bioavailability

Berry phytochemicals, especially anthocyanins, ellagic acid, quercetin, and resveratrol, often exhibit poor oral bioavailability because of low gastric pH stability, extensive liver metabolism, and limited intestinal epithelial permeability (Goktas et al., 2020).

These obstacles are solved via phytosome formulations. Phospholipids come together with plant bioactive substances to form phytosomes. Polyphenolic compounds are made more soluble, more bioavailable, and shielded from the body's premature breakdown by phytosomes or herbosomes.

To optimize the antioxidant potential, this tactic can also be employed in order to enhance their pharmacological effect (Kumar et al., 2024; Saritha et al., 2023).

Some researchers have taken benefit from substances known as "molecular enhancers" or "adjuvants," which alter step I of polyphenol metabolization by inhibiting the activity of enzymes, P-glycoprotein, or the multidrug resistance (MDR) efflux pump, which serve as obstacles to the absorption of polyphenols and other xenobiotic substances. Furthermore, by utilizing adenosine triphosphate (ATP) molecules, adenosine triphosphate-binding cassette (ABC) transporters are responsible for expelling foreign substances from human cells. Consequently, it has been demonstrated that co-administration of molecular enhancers with polyphenols increases their bioavailability by enhancing their intestinal absorption (Sahraeian et al., 2024).

9.4 CONCLUSION

Fruits and berries, as stated throughout the chapter, are important components of the culinary arts as they impart taste, appeal (sensory, psychological, and cultural), nutrition, and also complementary health support. The complementary health benefits are due to the presence of fruitaceuticals (bioactives of the plant with pharmacological benefits). The fruits apple, pomegranate, pineapple, guava, orange, mango, grapes, kiwi, avocado, plums, and apricots are rich in the various brain-supportive fruitaceuticals, namely, narirutin, didymin, eriocitrin, naringin, naringenin, integropectin, stilbenes, hydroxycinnamate esters, mangiferin, quercetin, kaempferol, protocatechuic acid, ferulic acid, delphinidin, malvidin, petunidin, cyanidin, cyanidin 3-glucoside, cyanidin 3-galactoside, cyanidin 3-xyloside, cyanidin 3-rutinoside, cyanidin 3-sophoroside, cyanidin 3-glucosylrutinoside, cyanidin 3-arabinoside, cyanidin 3-sambubioside, peonidin 3-glucoside, and peonidin 3-rutinoside and peonidin. The bioactives of the fruit have a special characteristic of being present in an easily bioavailable form, in addition to being present in adequate amounts. These bioactives have potential antioxidant, anti-inflammatory, antineuroinflammatory, neuromodulatory, and neuroplasticizing abilities. All these properties will help in improving brain health and psychological well-being (Figure 9.1).

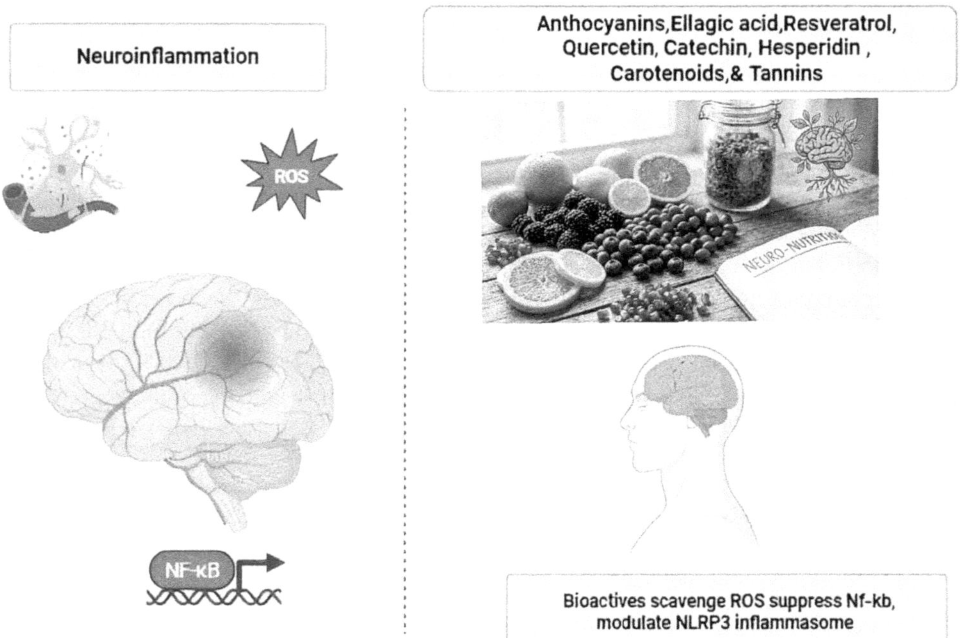

FIGURE 9.1 The fruitaceuticals scavenge free radicals, ameliorate neuroinflammation, balance the neurotransmitters, and thus improve brain health.

Fruits are also rich in dietary fibre (soluble forms that are loved by micro-organisms) that can act as a prebiotic for the growth of beneficial microbiota and thus can improve the gut–brain axis. With all of the above beneficial packages towards brain health, there are very few toxic or non-toxic foods, thus making them a special part of the culinary. Though they have been consumed since time immemorial and have been accepted by the various cultures as an indispensable part of the diet, not just in the view of nutrition but also as a complementary medication in various illnesses, the scientific validation for the health-beneficial properties is still in the back seat. There is an imminent need for systematically planned studies to validate the brain health benefits of these affordable and valuable medicines.

Abbreviation

Sl.no	Abbreviations	Definition
1	BBB	Blood brain barrier
2	ROS	Reactive oxygen species
3	NLRP3	NLR Family Pyrin Domain Containing 3
4	Nf-kappa-b	Nuclear factor kappa B
5	Nrf2	Nuclear factor erythroid 2-related factor 2
6	CNS	Central nervous system
7	PUFA	Polyunsaturated fatty acid
8	MUFA	Monounsaturated fatty acid
9	DNA	Deoxyribonucleic acid
10	MS	Multiple sclerosis
11	AD	Alzheimer's disease
12	HD	Huntington's disease
13	IL-1ß	Interleukin-beta
14	IL-6	Interleukin-6
15	Pro IL-18	Pro interleukin-8
16	TNF-α	Tumour Necrosis Factor-alpha
17	MMPs	Matrix metalloproteinases
18	BACE-1	Beta-site Amyloid Precursor Protein Cleaving Enzyme 1
19	GSK-3B	Glycogen synthase kinase 3 beta
20	CDK-5	Cyclin-dependent kinase 5
21	MAPKs	Mitogen-activated protein kinases
22	HPA axis	Hypothalamus-pituitary-adrenal axis
23	CRH	Corticotropin-releasing hormone
24	ACTH	Adrenocorticotropic hormone
25	GSH	Glutathione
26	CAT	Catalase
27	SOD	Superoxide dismutase
28	Nrf2-AND	Nrf2 Antioxidant Response Element
29	EGCG	Epigallocatechin gallate
30	Aβ	Amyloid beta
31	BDNF	Brain-derived neurotrophic factor
32	RAGE	Receptor for advanced glycation end products
33	AMPK	AMP-activated protein kinase

(Continued)

Sl.no	Abbreviations	Definition
34	CREB	cAMP response element-binding protein
35	SIRT1	Sirtuin 1
36	THP-1	Tohoku Hospital Pediatrics-1
37	DNJ	1-Deoxynojirimycin
38	MAC	*Morus alba* and Vitamin C
39	MIND	Mediterranean-DASH Intervention for Neurodegenerative Delay
40	DHA	Docosahexaenoic acid
41	LPH	Lactase phlorizin hydrolase
42	ANS	Autonomic pathways
43	MDR	Multiple drug resistance
44	ATP	Adenosine triphosphate
45	ATC	Adenosine triphosphate-binding cassette

REFERENCES

Ademosun, A. O. (2024). Enrichment with citrus peels as a strategy for improving the health benefits and nutritional value of breakfast cereals: a review. *Human Nutrition & Metabolism*, 36, 200239. https://doi.org/10.1016/j.hnm.2024.200239

Afzal, M., Akhtar, A., Bukhari, R., & Ul Hasan, S. Z. (2022). A review on avocado fruit: description, morphological characteristics, composition, nutritional benefits and propagation technique. *Plant Cell Biotechnology and Molecular Biology*, 23(29–30), 32–41. https://doi.org/10.56557/pcbmb/2022/v23i29-307772

Agarwal, P., Holland, T. M., Wang, Y., Bennett, D. A., & Morris, M. C. (2019). Association of strawberries and anthocyanidin intake with Alzheimer's dementia risk. *Nutrients*, 11(12), 3060. https://www.mdpi.com/2072-6643/11/12/3060

Agussalim, H. M. S. (2025). The effect of white-colored fruits' micronutrients on brain cell function in adults with early memory loss. *International Neurology and Neurological Research*, 1(1), 03.

Aleksandrova, S., Alexova, R., Dragomanova, S., Kalfin, R., Nicoletti, F., Fagone, P., Petralia, M. C., Mangano, K., & Tancheva, L. (2023). Preventive and therapeutic effects of *Punica granatum* L. polyphenols in neurological conditions. *International Journal of Molecular Sciences*, 24(3). https://doi.org/10.3390/ijms24031856

Ayub, H., Nadeem, M., Mohsin, M., Ambreen, S., Khan, F. A., Oranab, S., Rahim, M. A., Zubair Khalid, M., Zongo, E., Zarlasht, M., & Ullah, S. (2023). A comprehensive review on the availability of bioactive compounds, phytochemicals, and antioxidant potential of plum (*Prunus domestica*). *International Journal of Food Properties*, 26(1), 2388–2406. https://doi.org/10.1080/10942912.2023.2249254

Baldassi, C., Kolniak-Ostek, J., Castellarin, S. D., & Pico, J. (2025). Chemistry, biosynthesis, content, and bioavailability of berry bioactives. In A. C. Karaca & E. Capanoglu (Eds.), *Berry fruits* (pp. 67–130). Academic Press. https://doi.org/10.1016/B978-0-443-16011-0.00003-1

Bar-Ya'akov, I., Tian, L., Amir, R., & Holland, D. (2019). Primary metabolites, anthocyanins, and hydrolyzable tannins in the pomegranate fruit. *Frontiers in Plant Science*, 10. https://doi.org/10.3389/fpls.2019.00620

Bellavite, P. (2023). Neuroprotective potentials of flavonoids: experimental studies and mechanisms of action. *Antioxidants*, 12(2), 280. https://www.mdpi.com/2076-3921/12/2/280

Bernatoniene, J., & Kopustinskiene, D. M. (2018). The role of catechins in cellular responses to oxidative stress. *Molecules*, 23(4), 965. https://www.mdpi.com/1420-3049/23/4/965

Bhuyan, D. J., Alsherbiny, M. A., Perera, S., Low, M., Basu, A., Devi, O. A., Barooah, M. S., Li, C. G., & Papoutsis, K. (2019). The odyssey of bioactive compounds in avocado (*Persea americana*) and their health benefits. *Antioxidants*, 8(10), 426. https://www.mdpi.com/2076-3921/8/10/426

Bigford, G. E., & Del Rossi, G. (2014). Supplemental substances derived from foods as adjunctive therapeutic agents for treatment of neurodegenerative diseases and disorders. *Advances in Nutrition*, 5(4), 394–403.

Borges, G., Degeneve, A., Mullen, W., & Crozier, A. (2010). Identification of flavonoid and phenolic antioxidants in black currants, blueberries, raspberries, red currants, and cranberries. *Journal of Agricultural and Food Chemistry*, 58(7), 3901–3909. https://doi.org/10.1021/jf902263n

Braidy, N., Essa, M. M., Poljak, A., Selvaraju, S., Al-Adawi, S., Manivasagam, T., Thenmozhi, A. J., Ooi, L., Sachdev, P., & Guillemin, G. J. (2016). Consumption of pomegranates improves synaptic function in a transgenic mice model of Alzheimer's disease. *Oncotarget*, 7(40), 64589.

Brain food: a comprehensive review of nutritional strategies for cognitive enhancement and neuroprotection. (2025). *African Journal of Biomedical Research*, 28(3S), 229–236. https://doi.org/10.53555/AJBR.v28i3S.7602

Chavda, V. P., Feehan, J., & Apostolopoulos, V. (2024). Inflammation: The cause of all Diseases. *Cells*, 13(22). https://doi.org/10.3390/cells13221906

Chen, L., Deng, H., Cui, H., Fang, J., Zuo, Z., Deng, J., Li, Y., Wang, X., & Zhao, L. (2018). Inflammatory responses and inflammation-associated diseases in organs. *Oncotarget*, 9(6), 7204–7218. https://doi.org/10.18632/oncotarget.23208

Chiang, M.-C., Tsai, T.-Y., & Wang, C.-J. (2023). The potential benefits of quercetin for brain health: a review of anti-inflammatory and neuroprotective mechanisms. *International Journal of Molecular Sciences*, 24(7), 6328. https://www.mdpi.com/1422-0067/24/7/6328

Cotman, C. W., Stern, P., & Carstensen, L. (2000). *The aging mind: opportunities in cognitive research*. National Academies Press (US).

De Amicis, R., Mambrini, S. P., Pellizzari, M., Foppiani, A., Bertoli, S., Battezzati, A., & Leone, A. (2022). Systematic review on the potential effect of berry intake in the cognitive functions of healthy people. *Nutrients*, 14(14), 2977. https://www.mdpi.com/2072-6643/14/14/2977

Debnath, I., Ghosh, S., Bhunia, S., Nayak, A., Basak, S., Nandi, S., & Bhattacharjee, S. (2025). Mechanistic insights and therapeutic potential of quercetin in neuroprotection: a comprehensive review of pathways and clinical perspectives. *BIO Integration*, 6(1), 978.

Di Pede, G., Favari, C., Bresciani, L., Almutairi, T. M., Del Rio, D., & Crozier, A. (2022). Occurrence, bioavailability and metabolism of berry (poly)phenols. In D. Klimis-Zacas & A. Rodriguez-Mateos (Eds.), *Berries and berry bioactive compounds in promoting health*. The Royal Society of Chemistry. https://doi.org/10.1039/9781839165450-00041

do Rosario, V., Lorzadeh, E., Brodaty, H., Anstey, K. J., Chan, K., Roodenrys, S., Kent, K., Bliokas, V., Phillipson, L., Weston-Green, K., Francois, M. E., Jiang, X., George, J., Potter, J., Batterham, M. J., & Charlton, K. (2024). Assessing the effect of anthocyanins through diet and supplementation on cognitive function in older adults at risk for dementia: protocol for a randomised controlled trial. *BMJ Open*, 14(9), e086435. https://doi.org/10.1136/bmjopen-2024-086435

Dreher, M. L., Cheng, F. W., & Ford, N. A. (2021). A comprehensive review of Hass avocado clinical trials, observational studies, and biological mechanisms. *Nutrients*, 13(12), 4376. https://www.mdpi.com/2072-6643/13/12/4376

Duranti, E., & Villa, C. (2026). Misfolded proteins and cognitive decline: Mechanistic insights into neurodegenerative disorders. *Neurology International*, 18(3). https://doi.org/10.3390/neurolint18030048

Erickson, K. I., Gildengers, A. G., & Butters, M. A. (2013). Physical activity and brain plasticity in late adulthood. *Dialogues in Clinical Neuroscience*, 15(1), 99–108.

Essa, M. M., Akbar, M., & Guillemin, G. (2016). Erratum. In M. M. Essa, M. Akbar, & G. Guillemin (Eds.), *The benefits of natural products for neurodegenerative diseases* (pp. E1–E1). Springer International Publishing. https://doi.org/10.1007/978-3-319-28383-8_24

Fernández Demeneghi, R., Vargas Moreno, I., Héctor-Gabriel, A.-M., Puga-Olguín, A., Herrera-Meza, S., Romo-González, T., Uscanga, Y., Patraca, L., & Guzman-Geronimo, R. (2019). Berry supplementation and their beneficial effects on some central nervous system disorders. https://doi.org/10.5772/intechopen.90428

Flanagan, E., Cameron, D., Sobhan, R., Wong, C., Pontifex, M. G., Tosi, N., Mena, P., Del Rio, D., Sami, S., Narbad, A., Müller, M., Hornberger, M., & Vauzour, D. (2022). Chronic consumption of cranberries (*Vaccinium macrocarpon*) for 12 weeks improves episodic memory and regional brain perfusion in healthy older adults: a randomised, placebo-controlled, parallel-groups feasibility study. *Frontiers in Nutrition*, 9. https://doi.org/10.3389/fnut.2022.849902

Gao, L., & Mazza, G. (1994). Quantitation and distribution of simple and acylated anthocyanins and other phenolics in blueberries. *Journal of Food Science*, 59(5), 1057–1059. https://doi.org/10.1111/j.1365-2621.1994.tb08189.x

Garcia, G., Pais, T. F., Pinto, P., Dobson, G., McDougall, G. J., Stewart, D., & Santos, C. N. (2020). Bioaccessible raspberry extracts enriched in ellagitannins and ellagic acid derivatives have anti-neuroinflammatory properties. *Antioxidants*, 9(10), 970. https://www.mdpi.com/2076-3921/9/10/970

Georgiev, V., Ananga, A., & Tsolova, V. (2014). Recent advances and uses of grape flavonoids as nutraceuticals. *Nutrients*, 6(1), 391–415. https://doi.org/10.3390/nu6010391

Ghavami, S., Shojaei, S., Yeganeh, B., Ande, S. R., Jangamreddy, J. R., Mehrpour, M., Christoffersson, J., Chaabane, W., Moghadam, A. R., Kashani, H. H., Hashemi, M., Owji, A. A., & Łos, M. J. (2014). Autophagy and apoptosis dysfunction in neurodegenerative disorders. *Progress in Neurobiology*, 112, 24–49. https://doi.org/10.1016/j.pneurobio.2013.10.004

Głąbska, D., Guzek, D., Groele, B., & Gutkowska, K. (2020). Fruit and vegetable intake and mental health in adults: a systematic review. *Nutrients*, 12(1), 115. https://www.mdpi.com/2072-6643/12/1/115

Goktas, Z., Zu, Y., Abbasi, M., Galyean, S., Wu, D., Fan, Z., & Wang, S. (2020). Recent advances in nanoencapsulation of phytochemicals to combat obesity and its comorbidities. *Journal of Agricultural and Food Chemistry*, 68(31), 8119–8131. https://doi.org/10.1021/acs.jafc.0c00131

Golovinskaia, O., & Wang, C.-K. (2021). Review of functional and pharmacological activities of berries. *Molecules*, 26(13), 3904. https://www.mdpi.com/1420-3049/26/13/3904

Goyal, A., Verma, A., Dubey, N., Raghav, J., & Agrawal, A. (2022). Naringenin: A prospective therapeutic agent for Alzheimer's and Parkinson's disease. *Journal of Food Biochemistry*, 46(12), e14415. https://doi.org/10.1111/jfbc.14415

Grabska-Kobyłecka, I., Szpakowski, P., Król, A., Książek-Winiarek, D., Kobyłecki, A., Głąbiński, A., & Nowak, D. (2023). Polyphenols and their impact on the prevention of neurodegenerative diseases and development. *Nutrients*, 15(15), 3454. https://www.mdpi.com/2072-6643/15/15/3454

Gullotta, G. S., Costantino, G., Sortino, M. A., & Spampinato, S. F. (2023). Microglia and the blood-brain barrier: An external player in acute and chronic neuroinflammatory conditions. *International Journal of Molecular Sciences*, 24(11). https://doi.org/10.3390/ijms24119144

Hajialyani, M., Hosein Farzaei, M., Echeverría, J., Nabavi, S. M., Uriarte, E., & Sobarzo-Sánchez, E. (2019). Hesperidin as a neuroprotective agent: a review of animal and clinical evidence. *Molecules*, 24(3), 648. https://www.mdpi.com/1420-3049/24/3/648

Han, X., Zhang, Y., Zhang, L., Zhuang, Y., & Wang, Y. (2025). Efficacy and molecular mechanisms of hesperidin in mitigating Alzheimer's disease: a systematic review. *European Journal of Medicinal Chemistry*, 283, 117144. https://doi.org/10.1016/j.ejmech.2024.117144

Havas, S., Heimendinger, J., Reynolds, K., Baranowski, T., Nicklas, T. A., Bishop, D., Buller, D., Sorensen, G., Beresford, S. A. A., Cowan, A., & Damron, D. (1994). 5 a day for better health: a new research initiative. *Journal of the American Dietetic Association*, 94, 32.

Herman, J. P., McKlveen, J. M., Ghosal, S., Kopp, B., Wulsin, A., Makinson, R., Scheimann, J., & Myers, B. (2016). Regulation of the hypothalamic-pituitary-adrenocortical stress response. *Comprehensive Physiology*, 6(2), 603–621. https://doi.org/10.1002/cphy.c150015

Ho, Y.-S., Yu, M.-S., Yik, S.-Y., So, K.-F., Yuen, W.-H., & Chang, R. C.-C. (2009). Polysaccharides from wolfberry antagonizes glutamate excitotoxicity in rat cortical neurons. *Cellular and Molecular Neurobiology*, 29(8), 1233–1244. https://doi.org/10.1007/s10571-009-9419-x

Hollman, P. C. H. (2004). Absorption, bioavailability, and metabolism of flavonoids. *Pharmaceutical Biology*, 42(sup1), 74–83. https://doi.org/10.3109/13880200490893492

Hu, Y., Lin, Q., Zhao, H., Li, X., Sang, S., McClements, D. J., Long, J., Jin, Z., Wang, J., & Qiu, C. (2023). Bioaccessibility and bioavailability of phytochemicals: influencing factors, improvements, and evaluations. *Food Hydrocolloids*, 135, 108165. https://doi.org/10.1016/j.foodhyd.2022.108165

Hughes, D. C. (2025). *Characterizing the effects of age on autistic adults using network analysis*. Arizona State University.

Hussain, R., Zubair, H., Pursell, S., & Shahab, M. (2018). Neurodegenerative diseases: regenerative mechanisms and novel therapeutic approaches. *Brain Sciences*, 8(9), 177.

Irshad, R., Amir, R. M., Azam, M. T., Abbasi, F. A., Fatima, E., Muhammad, Z., Kausar, R., Razzaq, S., Chaudhary, Z., & Afzal, S. (2025). Nutraceutical potential of strawberries, its phytochemicals and health benefits: a comprehensive review. *Indus Journal of Bioscience Research*, 3(12), 36–45. https://doi.org/10.70749/ijbr.v3i12.2711

Johnson, E., Vishwanathan, R., Mohn, E., Haddock, J., Rasmussen, H., & Scott, T. (2015). Avocado consumption increases neural lutein and improves cognitive function. *The FASEB Journal*, 29(S1), 32.8. https://doi.org/10.1096/fasebj.29.1_supplement.32.8

Kean, R. J., Lamport, D. J., Dodd, G. F., Freeman, J. E., Williams, C. M., Ellis, J. A., Butler, L. T., & Spencer, J. P. E. (2015). Chronic consumption of flavanone-rich orange juice is associated with cognitive benefits: an 8-week, randomized, double-blind, placebo-controlled trial in healthy older adults. *The American Journal of Clinical Nutrition*, 101(3), 506–514. https://doi.org/10.3945/ajcn.114.088518

Keservani, R. K., Sharma, A. K., & Kesharwani, R. K. (2016). Medicinal effect of nutraceutical fruits for cognition and brain health. *Scientifica*, 2016(1), 3109254. https://doi.org/10.1155/2016/3109254

Khan, N., Khushtar, M., Ahmad, N., Hasan, N., Khan, Z., Idris, S., Ahmad, A., & Zishan, M. (2017). Nutritional importance and pharmacological activity of *Mangifera indica*. *World Journal of Pharmacy and Pharmaceutical Sciences*, 6, 258–273. https://doi.org/10.20959/wjpps20175-8974

King, R. A., Broadbent, J. L., & Head, R. J. (1996). Absorption and excretion of the soy isoflavone genistein in rats. *The Journal of Nutrition*, 126, 176.

Kopustinskiene, D. M., Bernatoniene, J., Jakstas, V., & Morkuniene, R. (2021). Catechins, neuroprotection, and brain mitochondria. In Marcos Roberto de Oliveira (Ed.), *Mitochondrial physiology and vegetal molecules* (pp. 455–470). Academic Press.

Kumar, M., Dhivya, K., Lokesh, D., Nivethitha, S., Kumar, M., Sarathi, M., & Astalakshmi, N. (2024). Phytosomes: nature's secret to enhanced bioavailability. *Journal of Applied Pharmaceutical Research*, 12, 88–99. https://doi.org/10.69857/joapr.v12i6.702

Lavefve, L., Howard, L., & Carbonero, F. (2019). Berry polyphenols metabolism and impact on human gut microbiota and health. *Food & Function*, 11. https://doi.org/10.1039/C9FO01634A

Lebaka, V. R., Wee, Y.-J., Ye, W., & Korivi, M. (2021). Nutritional composition and bioactive compounds in three different parts of mango fruit. *International Journal of Environmental Research and Public Health*, 18(2), 741. https://www.mdpi.com/1660-4601/18/2/741

López-Ríos, L., Wiebe, J. C., Vega-Morales, T., & Gericke, N. (2020). Central nervous system activities of extract *Mangifera indica* L. *Journal of Ethnopharmacology*, 260, 112996. https://doi.org/10.1016/j.jep.2020.112996

Ma, Z. F., Zhang, H., Teh, S. S., Wang, C. W., Zhang, Y., Hayford, F., Wang, L., Ma, T., Dong, Z., Zhang, Y., & Zhu, Y. (2019). Goji berries as a potential natural antioxidant medicine: an insight into their molecular mechanisms of action. *Oxidative Medicine and Cellular Longevity*, 2019(1), 2437397. https://doi.org/10.1155/2019/2437397

Mairuae, N., Jittiwat, J., Apaijit, K., Noisa, P., Bai, G., Hou, Y., & Palachai, N. (2025). Synergistic effects of anthocyanin-enriched *Morus alba* L. extract and vitamin C: promising nutraceutical ingredients in functional food development for neuroprotection. *Foods*, 14(21), 3630. https://www.mdpi.com/2304-8158/14/21/3630

Maqsood, S., Arshad, M. T., Ikram, A., & Gnedeka, K. T. (2025). Fruit-based diet and gut health: a review. *Food Science & Nutrition*, 13(5), e70159. https://doi.org/10.1002/fsn3.70159

Martini, D., Marino, M., Angelino, D., Del Bo', C., Del Rio, D., Riso, P., & Porrini, M. (2019). Role of berries in vascular function: a systematic review of human intervention studies. *Nutrition Reviews*, 78(3), 189–206. https://doi.org/10.1093/nutrit/nuz053

Martins, M. S., Gonçalves, A. C., Alves, G., & Silva, L. R. (2023). Blackberries and mulberries: berries with significant health-promoting properties. *International Journal of Molecular Sciences*, 24(15), 12024. https://www.mdpi.com/1422-0067/24/15/12024

Mayer, E. A., Tillisch, K., & Gupta, A. (2015). Gut–brain axis and the microbiota. *The Journal of Clinical Investigation*, 125(3), 926–938. https://doi.org/10.1172/jci76304

Medina dos Santos, N., Berilli Batista, P., Batista, Â. G., & Maróstica Júnior, M. R. (2019). Current evidence on cognitive improvement and neuroprotection promoted by anthocyanins. *Current Opinion in Food Science*, 26, 71–78. https://doi.org/10.1016/j.cofs.2019.03.008

Melo, M. F. F. T. de, Pereira, D. E., Moura, R. de L., Silva, E. B. da, Melo, F. A. L. T. de, Dias, C. de C. Q., Silva, M. de C. A., Oliveira, M. E. G. de, Viera, V. B., Pintado, M. M. E., Santos, S. G. de, & Soares, J. K. B. (2019). Maternal supplementation with avocado (*Persea americana* Mill.) pulp and oil alters reflex maturation, physical development, and offspring memory in rats. *Frontiers in Neuroscience*, 13. https://doi.org/10.3389/fnins.2019.00009

Menze, E. T., Tadros, M. G., Abdel-Tawab, A. M., & Khalifa, A. E. (2012). Potential neuroprotective effects of hesperidin on 3-nitropropionic acid-induced neurotoxicity in rats. *NeuroToxicology*, 33(5), 1265–1275. https://doi.org/10.1016/j.neuro.2012.07.007

Miller, K., Feucht, W., & Schmid, M. (2019). Bioactive compounds of strawberry and blueberry and their potential health effects based on human intervention studies: a brief overview. *Nutrients*, 11(7), 1510. https://www.mdpi.com/2072-6643/11/7/1510

Miller, M. G., Thangthaeng, N., Rutledge, G. A., Scott, T. M., & Shukitt-Hale, B. (2021). Dietary strawberry improves cognition in a randomised, double-blind, placebo-controlled trial in older adults. *British Journal of Nutrition*, 126(2), 253–263. https://doi.org/10.1017/S0007114521000222

Mullen, W., Yokota, T., Lean, M. E., & Crozier, A. (2003). Analysis of ellagitannins and conjugates of ellagic acid and quercetin in raspberry fruits by LC-MSn. *Phytochemistry*, 64, 617.

Munteanu, C., Galaction, A. I., Turnea, M., Blendea, C. D., Rotariu, M., & Poştaru, M. (2024). Redox homeostasis, gut microbiota, and epigenetics in neurodegenerative diseases: A systematic review. *Antioxidants*, *13*(9), 1062. https://www.mdpi.com/2076-3921/13/9/1062

Nemzer, B. V., Al-Taher, F., Yashin, A., Revelsky, I., & Yashin, Y. (2022). Cranberry: chemical composition, antioxidant activity and impact on human health: overview. *Molecules*, 27(5), 1503. https://www.mdpi.com/1420-3049/27/5/1503

Nile, S. H., & Park, S. W. (2014). Edible berries: bioactive components and their effect on human health. *Nutrition*, 30(2), 134–144. https://doi.org/10.1016/j.nut.2013.04.007

Noreen, S., Niazi, M. K., Arshad, M. T., Ikram, A., & Gnedeka, K. T. (2025). Goji berries: a review of their bioactive components and health-promoting properties. *International Journal of Food Science and Technology*, 60(2). https://doi.org/10.1093/ijfood/vvaf232

Nowak, R., Olech, M., & Nowacka, N. (2014). Plant polyphenols as chemopreventive agents. In R. R. Watson, V. R. Preedy, & S. Zibadi (Eds.), *Polyphenols in human health and disease* (pp. 1289–1307). Academic Press. https://doi.org/10.1016/B978-0-12-398456-2.00086-4

Nuzzo, D., Picone, P., Giardina, C., Scordino, M., Mudò, G., Pagliaro, M., Scurria, A., Meneguzzo, F., Ilharco, L. M., Fidalgo, A., Alduina, R., Presentato, A., Ciriminna, R., & Di Liberto, V. (2021). New neuroprotective effect of lemon IntegroPectin on neuronal cellular model. *Antioxidants*, 10(5), 669. https://www.mdpi.com/2076-3921/10/5/669

Omar, S. H. (2019). Mediterranean and MIND diets containing olive biophenols reduces the prevalence of Alzheimer's disease. *International Journal of Molecular Sciences*, 20(11), 2797. https://www.mdpi.com/1422-0067/20/11/2797

Petersen, C., Bharat, D., Wankhade, U. D., Kim, J.-S., Cutler, B. R., Denetso, C., Gholami, S., Nelson, S., Bigley, J., Johnson, A., Chintapalli, S. V., Piccolo, B. D., Satheesh Babu, A. K., Paz, H. A., Shankar, K., Symons, J. D., & Anandh Babu, P. V. (2022). Dietary blueberry ameliorates vascular complications in diabetic mice possibly through NOX4 and modulates composition and functional diversity of gut microbes. *Molecular Nutrition & Food Research*, 66(8), 2100784. https://doi.org/10.1002/mnfr.202100784

Pontifex, M. G., Malik, M. M. A. H., Connell, E., Müller, M., & Vauzour, D. (2021). Citrus polyphenols in brain health and disease: current perspectives. *Frontiers in Neuroscience*, 15. https://doi.org/10.3389/fnins.2021.640648

Raskov, H., Burcharth, J., Pommergaard, H. C., & Rosenberg, J. (2016). Irritable bowel syndrome, the microbiota and the gut–brain axis. *Gut Microbes*, 7(5), 365–383. https://doi.org/10.1080/19490976.2016.1218585

Rawat, P., Sehar, U., Bisht, J., Selman, A., Culberson, J., & Reddy, P. H. (2022). Phosphorylated tau in Alzheimer's disease and other tauopathies. *International Journal of Molecular Sciences*, 23(21). https://doi.org/10.3390/ijms232112841

Ribeiro, J. A., dos Santos Pereira, E., de Oliveira Raphaelli, C., Radünz, M., Camargo, T. M., da Rocha Concenço, F. I. G., Cantillano, R. F. F., Fiorentini, Â. M., & Nora, L. (2022). Application of prebiotics in apple products and potential health benefits. *Journal of Food Science and Technology*, 59(4), 1249–1262. https://doi.org/10.1007/s13197-021-05062-z

Roy Sarkar, S., & Banerjee, S. (2019). Gut microbiota in neurodegenerative disorders. *Journal of Neuroimmunology*, 328, 98–104. https://doi.org/10.1016/j.jneuroim.2019.01.004

Ruíz-Salinas, A. K., Vázquez-Roque, R. A., Díaz, A., Pulido, G., Treviño, S., Floran, B., & Flores, G. (2020). The treatment of goji berry (*Lycium barbarum*) improves the neuroplasticity of the prefrontal cortex and hippocampus in aged rats. *The Journal of Nutritional Biochemistry*, 83, 108416. https://doi.org/10.1016/j.jnutbio.2020.108416

Saad, A. M., Mohammed, D. M., Alkafaas, S. S., Ghosh, S., Negm, S. H., Salem, H. M., Fahmy, M. A., Semary, H. E., Ibrahim, E. H., AbuQamar, S. F., El-Tarabily, K. A., & El-Saadony, M. T. (2025). Dietary polyphenols and human health: sources, biological activities, nutritional and immunological aspects, and bioavailability: a comprehensive review. *Frontiers in Immunology*, 16, 1653378. https://doi.org/10.3389/fimmu.2025.1653378

Sahraeian, S., Rashidinejad, A., & Golmakani, M.-T. (2024). Recent advances in the conjugation approaches for enhancing the bioavailability of polyphenols. *Food Hydrocolloids*, 146, 109221. https://doi.org/10.1016/j.foodhyd.2023.109221

Samanta, S., Chakraborty, S., & Bagchi, D. (2024). Pathogenesis of neurodegenerative diseases and the protective role of natural bioactive components. *Journal of the American Nutrition Association*, 43(1), 20–32.

Saritha, M., Ramya, A., Sonia, A., Monika, S., & Sowmya, K. (2023). Phytosomes: a novel approach for improving the efficacy of herbal extracts. *Research Journal of Pharmacy and Technology*, 6028–6031. https://doi.org/10.52711/0974-360X.2023.00978

Schauss, A. G. (2015). The effect of acai (*Euterpe* spp.) fruit pulp on brain health and performance. In R. R. Watson & V. R. Preedy (Eds.), *Bioactive nutraceuticals and dietary supplements in neurological and brain disease* (pp. 179–186). Academic Press. https://doi.org/10.1016/B978-0-12-411462-3.00019-9

Schauss, A. G., Wu, X., Prior, R. L., Ou, B., Huang, D., Owens, J., Agarwal, A., Jensen, G. S., Hart, A. N., & Shanbrom, E. (2006). Antioxidant capacity and other bioactivities of the freeze-dried Amazonian palm berry, *Euterpe oleracea* Mart. (acai). *Journal of Agricultural and Food Chemistry*, 54(22), 8604–8610. https://doi.org/10.1021/jf0609779

Schauss, A. G., Wu, X., Prior, R. L., Ou, B., Patel, D., Huang, D., & Kababick, J. P. (2006). Phytochemical and nutrient composition of the freeze-dried Amazonian palm berry, *Euterpe oleracea* Mart. (acai). *Journal of Agricultural and Food Chemistry*, 54(22), 8598–8603.

Seeram, N. P., Henning, S. M., Zhang, Y., Suchard, M., Li, Z., & Heber, D. (2006). Pomegranate juice ellagitannin metabolites are present in human plasma and some persist in urine for up to 48 Hours12. *The Journal of Nutrition*, 136(10), 2481–2485. https://doi.org/10.1093/jn/136.10.2481

Sheikh, S., Safia, Haque, E., & Mir, S. S. (2013). Neurodegenerative diseases: multifactorial conformational diseases and their therapeutic interventions. *Journal of Neurodegenerative Diseases*, 2013, 563481. https://doi.org/10.1155/2013/563481

Shukitt-Hale, B., Cheng, V., & Joseph, J. A. (2009). Effects of blackberries on motor and cognitive function in aged rats. *Nutritional Neuroscience*, 12(3), 135–140. https://doi.org/10.1179/147683009X423292

Shukitt-Hale, B., Kalt, W., Carey, A. N., Vinqvist-Tymchuk, M., McDonald, J., & Joseph, J. A. (2009). Plum juice, but not dried plum powder, is effective in mitigating cognitive deficits in aged rats. *Nutrition*, 25(5), 567–573. https://doi.org/10.1016/j.nut.2008.10.018

Singh, L., Wani, A. W., Shaifali, N., Sadawarti, R. K., Kaur, H., Bashir, O., Majeed, J., Mirza, A. A., Sharma, J., Ali, S., Alsalman, A. J., Al Hawaj, M. A., Almalak, A. E., Zengin, G., & Farid, A. (2025). A comprehensive review on neurotrophic receptors and their implications in brain health: exploring the neuroprotective potential of berries. *Journal of Berry Research*, 15(1), 48–64. https://doi.org/10.1177/18785093241301881

Sridhar, S., Khamaj, A., & Asthana, M. K. (2023). Cognitive neuroscience perspective on memory: overview and summary. *Frontiers in Human Neuroscience*, 17. https://doi.org/10.3389/fnhum.2023.1217093

Stull, A. J., Cassidy, A., Djoussé, L., Johnson, S. A., Krikorian, R., Lampe, J. W., Mukamal, K. J., Nieman, D. C., Porter Starr, K. N., Rasmussen, H., Rimm, E. B., Stote, K. S., & Tangney, C. (2024). The state of the science on the health benefits of blueberries: a perspective. *Frontiers in Nutrition*, 11. https://doi.org/10.3389/fnut.2024.1415737

Subash, S., Essa, M. M., Al-Adawi, S., Memon, M. A., Manivasagam, T., & Akbar, M. (2014). Neuroprotective effects of berry fruits on neurodegenerative diseases. *Neural Regeneration Research*, 9(16), 1557–1566. https://doi.org/10.4103/1673-5374.139483

Tena, N., Martín, J., & Asuero, A. G. (2020). State of the art of anthocyanins: antioxidant activity, sources, bioavailability, and therapeutic effect in human health. *Antioxidants*, 9(5), 451. https://www.mdpi.com/2076-3921/9/5/451

Tran, P. H. L., & Tran, T. T. D. (2021). Blueberry supplementation in neuronal health and protective technologies for efficient delivery of blueberry anthocyanins. *Biomolecules*, 11(1). https://doi.org/10.3390/biom11010102

Uddin, M. S., Kabir, M. T., Jalouli, M., Rahman, M. A., Jeandet, P., Behl, T., Alexiou, A., Albadrani, G. M., Abdel-Daim, M. M., Perveen, A., & Ashraf, G. M. (2022). Neuroinflammatory signaling in the pathogenesis of alzheimer's disease. *Current Neuropharmacology*, 20(1), 126–146. https://doi.org/10.2174/1570159x19666210826130210

Visentin, A. P. V., Colombo, R., Scotton, E., Fracasso, D. S., da Rosa, A. R., Branco, C. S., & Salvador, M. (2020). Targeting inflammatory-mitochondrial response in major depression: current evidence and further challenges. *Oxidative Medicine and Cellular Longevity*, 2020, 2972968. https://doi.org/10.1155/2020/2972968

Wang, L., Lan, W., & Chen, D. (2024). Blueberry (*Vaccinium* spp.) anthocyanins and their functions, stability, bioavailability, and applications. *Foods*, 13(17). https://doi.org/10.3390/foods13172851

Wong, D. Y. S., Musgrave, I. F., Harvey, B. S., & Smid, S. D. (2013). Açaí (*Euterpe oleracea* Mart.) berry extract exerts neuroprotective effects against β-amyloid exposure in vitro. *Neuroscience Letters*, 556, 221–226. https://doi.org/10.1016/j.neulet.2013.10.027

Wu, X., & Prior, R. L. (2005). Systematic identification and characterization of anthocyanins by HPLC-ESI-MS/MS in common foods in the United States: fruits and berries. *Journal of Agricultural and Food Chemistry*, 53(7), 2589–2599. https://doi.org/10.1021/jf048068b

Xu, J., Li, C., Wu, W., Kong, L., Xiao, L., Ma, W., & Zhang, L. (2025). Cranberry research progress: A systematic review of chemical composition, pharmacological mechanisms, clinical applications, and nutritional significance. *International Journal of Molecular Sciences*, 26(19), 9707. https://www.mdpi.com/1422-0067/26/19/9707

Yadav, Y. C., Singh, A., Kannaujia, S. K., & Yadav, R. (2022). Neuroprotective effect of *Citrus limon* juice against scopolamine-induced amnesia in Wistar rats: role of cholinergic neurotransmission monitoring and beta-actin signaling. *Pharmacological Research – Modern Chinese Medicine*, 5, 100191. https://doi.org/10.1016/j.prmcm.2022.100191

Yang, L., Xian, D., Xiong, X., Lai, R., Song, J., & Zhong, J. (2018). Proanthocyanidins against oxidative stress: From molecular mechanisms to clinical applications. *BioMed Research International*, 2018, 8584136. https://doi.org/10.1155/2018/8584136

Yuste, R., Cossart, R., & Yaksi, E. (2024). Neuronal ensembles: Building blocks of neural circuits. *Neuron*, 112(6), 875–892. https://doi.org/10.1016/j.neuron.2023.12.008

Zafra-Stone, S., Yasmin, T., Bagchi, M., Chatterjee, A., Vinson, J. A., & Bagchi, D. (2007). Berry anthocyanins as novel antioxidants in human health and disease prevention. *Molecular Nutrition & Food Research*, 51(6), 675–683. https://doi.org/10.1002/mnfr.200700002

Zhang, L., Tang, M., Falsaperla, R., & Zhang, K. (2025). The role of microglia in sepsis-associated encephalopathy: a narrative review. *Translational Pediatrics*, 14(9), 2363–2373. https://doi.org/10.21037/tp-2025-496

Zhang, W., Zhong, Y., Wang, Z., Tang, F., & Zheng, C. (2024). Apple polysaccharide improves age-matched cognitive impairment and intestinal aging through microbiota–gut–brain axis. *Scientific Reports*, 14(1), 16215. https://doi.org/10.1038/s41598-024-67132-4

Zhong, H., Xu, J., Yang, M., Hussain, M., Liu, X., Feng, F., & Guan, R. (2023). Protective effect of anthocyanins against neurodegenerative diseases through the microbial–intestinal–brain axis: a critical review. *Nutrients*, 15(3), 496. https://www.mdpi.com/2072-6643/15/3/496

10 Anti-Cancer Potential of Medicinal Fruits and Berries

Noureddine Chaachouay, Abdellatif Rahim, and Azamal Husen

10.1 INTRODUCTION

The burden of cancer as one of the causes of morbidity and mortality in the world requires a multifaceted approach that goes beyond the conventional treatment plan and the measures to include active prevention interventions (Thun et al., 2010). Although other developments are unfolding regarding targeted therapies and diagnostics efforts, the hunt toward complementary and available risk-reducing measures is of primary concern (Daca-Alvarez et al., 2025). In this scenario, nutritional modification has become an important and adjustable fortification. This chapter initiates the urgency of such preventive tools by establishing the high demand for such actions, as the regular intake of bioactive-rich foods is a promising, sustainable, and affordable aspect of the public health programs that can alleviate the rates of cancer incidence. The therapeutic use of fruits and berries has a long historical origin and is centuries old, being central to the empirical information of traditional medical systems in cultures (Gurib-Fakim, 2006; Chaachouay et al., 2021). Retrospectively, these natural products were culturally used in treating symptoms and conditions that today are heralded as indicative of chronic disease and malignancy, such as the Ayurvedic use of the amla (*Emblica officinalis*), used by the Native Americans as blueberries, the *Vaccinium* species (Murugan et al., 2024). This ancient ethnobotanical tradition has not been limited to a historical appendix; it is a crucial source of human knowledge and offers a legitimate starting point by which the scientific inquiry into the chemopreventive benefits of such foods can be addressed in the modern and contemporary world (Dean, 2024; Mishra & Kumar, 2025). The shift into evidence-based science translation of the traditional wisdom is an important paradigm shift, enabled by the so-called phytoceutical revolution. The isolation and separation of particular, powerful phytochemicals in their fruit matrices, i.e., anthocyanins, ellagitannins, and resveratrol, drives this movement (Rai & Tzima, 2021). Contemporary analytical and molecular studies have allowed investigators to go beyond correlation and make accurate investigations into the nature of the interaction of these compounds with cellular signaling pathways that are the cornerstone of carcinogenesis (Ahmad et al., 2024). This part examines the emergence of strict scientific investigation that aims to certify mechanisms and determine efficacy. Together, the preclinical researchers have shed light on an attractive multitarget profile of most of the fruit and berry phytochemicals (Iqubal et al., 2021). They seem to exert their anti-cancer activity predominantly not of one silver-bullet action but rather of an orchestration of actions. These are free radical scavenging activity, suppression of chronic inflammation, selective induction of apoptosis in malignant cells, and tumor invasion and angiogenic inhibition (Hamed et al., 2012). This complex pharmacology profile makes them special agents that can regulate various phases of cancer progression, including initiation and progression (Gialeli et al., 2011). This explores critical and overall existing scientific evidence on the anti-cancer potential of the target medicinal fruits and berries. The range of it includes an individual analysis of their major bioactive components, reported pharmacokinetics, and intricate mechanisms of action at the molecular, cellular, and organismal levels. Taken together, the main objectives are (1) to overview the epidemiological, preclinical, and

DOI: 10.1201/9781003688259-10

clinical data on the chemopreventive and therapeutic uses of key examples, such as pomegranate, *Vaccinium* berries, citrus, and others; (2) to discuss the key issue on translation, such as the bioavailability, metabolism, and design of effective human trials; and (3) to speculate on the future of this field, including the possible integration of these natural compounds into personalized nutrition interventions in cancer prevention and adjuvant.

10.2 MAJOR BIOACTIVE COMPOUNDS IN ANTI-CANCER FRUITS AND BERRIES

10.2.1 Polyphenols

Polyphenolic compounds, which form a wide phytochemical group that is famous due to its multifunctional anti-cancer property, form important reservoirs in fruits and berries (Wan et al., 2025). The bright anthocyanins that are found in blueberries and blackberries serve as effective antioxidants, and at the same time, they also regulate intracellular signaling pathways to hinder inflammation and tumor angiogenesis (Ma et al., 2021). Examples of flavonoids that cause apoptosis and arrest the cell cycle of a cancer cell are quercetin, derived from apple skin, whereas condensed proanthocyanidins (PACs) in cranberries and grapes inhibit essential enzymatic pathways responsible for the metastasis of cancer. Of particular interest are the ellagitannins of pomegranate and raspberries; the substrates pass through microbial biotransformation in the gut to form the urolithins, which are powerful anti-inflammatory and anti-proliferative substances in colonic epithelium (Tao et al., 2026). Similarly, stilbenes such as resveratrol, isolated in grape skins, stimulate the cell survival proteins, i.e., sirtuins, and induce widespread chemopreventive actions across the carcinogenesis spectrum, i.e., from inhibition of early DNA damage to inhibition of the spread of known tumors (Duta-Bratu et al., 2023).

Fruit matrices do have other essential bioactive components other than polyphenols. An example is terpenoids and carotenoids, lycopene in watermelon and β-cryptoxanthin in citrus, which provide strong antioxidant protection and directly interrupt growth-factor signaling that promotes malignant progression (Chaachouay et al., 2025; Verma et al., 2025). The necessary vitamins, including ascorbic acid, may, in large doses, induce selective oxidative stress in cancerous cells, but folate performs a dual role in ensuring genomic stability and controlling gene expression. In addition, organosulfur compounds—which are mostly found in bitter melon—and unique proteolytic enzymes, including bromelain extracted from pineapple, exhibit specific anti-cancer effects, which span induction of detoxifying enzymes to protection of stromal matrices, which protect tumors (Rajan et al., 2022). Notably, the indigestible dietary fiber of whole fruits has a significant role in the prevention of cancer, both by entrapment of putative carcinogens in the gastrointestinal environment and by fermentation to short-chain fatty acids (SCFAs) to strengthen the colonic mucosa and extend systemic anti-inflammatory effects (Azcárate-Peril et al., 2011). The multifunctional arrangement of this heterogeneous blend of phytochemicals has a synergistic effect within the intact fruit matrix, thus providing an effective dietary approach to the prevention and treatment of malignancy.

10.2.2 Terpenoids and Carotenoids

Carotenoids and terpenoids form a crucial family of fat-soluble phytochemicals that have strong anti-cancer properties (Amponsah et al., 2024). Lycopene, which is red in color and is common in watermelon and pink guava, is an exceptionally powerful antioxidant that removes singlet oxygen, thus avoiding oxidative destruction of cellular macromolecules (Dutta & Dutta, 2022). It also lowers insulin-like growth factor signaling and enhances gap-junctional intercellular communication, resulting in inhibition of the proliferation of prostate cancer models. Similarly, the oxygenated carotenoids, β-cryptoxanthin (rich in papaya and citrus) and lutein, not only serve as antioxidants but also impact cellular signaling: 8-cryptoxanthin induces apoptosis and activates retinoic acid receptors, leading to cell differentiation, reducing cancer risk (Vitamin, 2022); lutein suppresses light-induced oxidative stress and inhibits angiogenic signaling. Taken together, such a combination of inhibitory effects of these compounds on tumor development is a synergistic mixture of direct antioxidant data, alteration of growth-regulatory processes, and tumor angiogenesis.

10.2.3 Vitamins and Organosulfur Compounds

The essential micronutrients and hard-to-find organosulfur compounds have a strong potential to promote the anti-neoplastic effect of fruit-derived constituents, and their action occurs via complementary mechanistic pathways (Paliyath et al., 2011; Chaachouay et al., 2020). The citrus fruits and berries contain vitamin C (ascorbic acid), which is a dual-purpose free radical scavenger, replenishing other cellular antioxidants and, in large doses, acts as a preferential pro-oxidant to form hydrogen peroxide to trigger the oxidative stress response and apoptosis in vulnerable cancer cells (Martí et al., 2009). It is also a vital cofactor of enzymes that are used in the production of connective tissues as well as in the regulation of the expression of epigenetic genes. As a B vitamin, citrus and avocado contain folate, which plays a paradoxical but central role in stabilizing DNA synthesis and methylation, thus inhibiting the occurrence of genetic mutation and enhancing stable silencing of oncogenes during the early prevention of neoplasms (Abd El-Azim, 2021). The use of folate in the prevention of neoplasms must be weighed with caution in the therapy of established neoplasms. Lastly, glucosinolates, present in fruits like bitter melon (*Momordica charantia*), are transformed into bioactive isothiocyanates (e.g., charantin-related compounds) (Gayathry & John, 2022). These isothiocyanates strongly activate phase 2 detoxification, activate the apoptotic pathway via several signaling cascades, and inhibit the activity of histone deacetylase (HDACs), so they affect the destruction of tumor cells and also increase the endogenous hepatic detoxification mechanisms in the human body.

10.2.4 Dietary Fiber and Unique Compounds

A combination of dietary fibers and an assortment of biologically distinct bioactive compounds completes the anti-cancer nutrient repertoire, doing so indirectly through physiological mechanisms and directly by interfering with molecular processes (Tao et al., 2020). Dietary fiber, including insoluble fiber, fecal bulk, and soluble fiber, forms a basis of cancer prevention; the insoluble fiber is augmented by fecal bulk and fermentation to generate SCFAs like butyrate, which not only fulfills the primary role of energy in the colonocytes of the colonic epithelium but also has the potent anti-inflammatory and pro-apoptotic properties of an HDAC inhibitor (Dahiya & Nigam, 2023; Chaachouay et al., 2024). Systemic immunity is regulated along with fiber; specific enzymes and triterpenoids do also confer specific effects; bromelain, a proteolytic complex of enzymes obtained in pineapples, has anti-metastatic effects by degrading the extracellular matrix, immune-modulatory effects through the effects of a single compound, bromelain, and anti-angiogenic effects, such as ursolic acid, a pentacyclic triterpenoid, which can be found in apple peels and cranberry skins, leading to apoptosis via the mitochondria.

10.3 MECHANISMS OF ANTI-CANCER ACTION

10.3.1 Antioxidant and Anti-Inflammatory Activities

The antioxidant and anti-inflammatory activities of fruit and berry bioactives are the key protective roles against carcinogenesis. These practices exist at two interconnections with each other (Folmer et al., 2014). To begin with, direct scavenging of reactive oxygen species (ROS) and subsequent reduction of oxidative stress are performed by such compounds as anthocyanins, vitamin C, and lycopene, which neutralize ROS and chelate pro-oxidant metals (Ozougwu, 2016). This will help avoid any oxidative harm to DNA, lipids, and proteins, which otherwise would trigger oncogenic mutations. Second, and probably the most critical point, these bioactives have a strong effect on important pro-inflammatory signaling pathways, primarily by inhibiting the central transcription factor NF-κB and down-regulating cyclooxygenase-2 (COX-2). The synthesis of inflammatory cytokines (e.g., TNF-α, IL-6), expression of the anti-apoptotic proteins, and synthesis of pro-inflammatory prostaglandins through inhibition of COX-2 phytonutrients such as quercetin, resveratrol, and ursolic acid suppress the NF-κB, which is a key mediator of inflammation, cell survival, and proliferation (Upadhyay, 2025). This bi-fold action interferes with the long-lasting inflammatory microenvironment that is an established promoter of tumor progression, angiogenesis, and metastasis, eliminating a vital source of fuel to cancer evolution.

10.3.2 Modulation of Carcinogen Metabolism and Detoxification

The chemopreventive active effect of fruits and berry phytochemicals is significantly promoted by their subtle regulating effects on carcinogen metabolism and detoxification, which are mechanisms that strategically modifying the enzymatic processing of procarcinogens by the body (Swetha et al., 2022). This modification is done through a dual-phase enzymatic approach, which works in both extra-hepatic and hepatic systems. Many bioactive components are also important in this effect, including the isolated isothiocyanates of bitter melon and various flavonoids, which are potent induction of phase II detoxifying enzymes, including glutathione S-transferase (GST) and the Nicotinamide Adenine Dinucleotide (NADH) quinone oxidoreductase (Sur & Ray, 2020). These enzymes add intermediates, potentially DNA-damaging reactive species, to glutathione, thus transforming them into hydrophilic excretable products. At the same time, the same phytochemicals selectively inhibit an activation enzyme of phase I, especially the isoforms of enzymes of the cytochrome P450 (CYP450) family involved in the transformation of non-genotoxic procarcinogens into genotoxic ones (Das et al., 2017). These dietary compounds modulate the activity of the potentially carcinogenic phase I and the protective phase II pathway, resulting in a metabolic remodeling toward the detoxification pathway, decreasing the concentration of hepatogens, and thus having a solid, systemic approach to cancer prevention.

10.3.3 Inhibition of Tumor Cell Proliferation and Induction of Apoptosis

One common feature of the direct anti-tumor anti-microbial properties of fruit- and berry-derived bioactive components is that they can inhibit the increase in tumor cell population and induce the programmed death of cells (apoptosis), and hence, they are able to attack the tumor cell characteristics of uncontrolled growth and immortality of cells (Lin, 2024). The latter effect is accomplished through an orchestrated perturbation of the cell-cycle apparatus and apoptotic signaling pathways. Firstly, phytochemicals including quercetin, resveratrol, and ellagic acid impose cell cycle arrest at essential checkpoints, mostly the G1/S and G2/M progressions, by modulating the transcriptional activity and enzymatic activity of key regulators, such as cyclins, cyclin-dependent kinases (CDKs), and their inhibitors (e.g., p21Cip1 and p27Kip1), and interrupt the uncontrolled division of malignant cells (Singh & Agarwal, 2006). Later on, these compounds induce apoptosis through intrinsic (mitochondrial) and extrinsic (death-receptor) pathways. They trigger the intrinsic pathway, which is controlled by the Bcl-2 protein family; Bax and Bak are pro-apoptotic, and Bcl-2 and Bcl-xL are anti-apoptotic and are up-regulated and down-regulated, respectively. This perturbation enhances the release of cytochrome c, the assemblage of apoptosomes, and the sequential activation of a caspase cascade, which are apoptotic enzymes. At the same time, the compounds have the ability to activate the extrinsic pathway by increasing the cell surface expression of death receptors, such as CD95 and DR5 (Green, 2022). Combined, these processes aim to selectively eliminate malignant cells and leave healthy and normal tissue intact.

10.3.4 Anti-Angiogenic and Anti-Metastatic Effects

In addition to attacking cancerous cells directly, the bioactive compounds of fruits and berries can have an indirect effect on the surroundings of the tumor by exerting their strong anti-angiogenic and anti-metastatic actions, the factors that are absolutely inevitable in slowing tumor growth and spreading. One of the key processes is the suppression of VEGF and HIF-1α signaling, which is the key regulatory pathway of de novo vascularization (Guo & Mo, 2020). The constituents include anthocyanins, resveratrol, and ursolic acid, which reduce the activity and expression of the vasodilator endothelial growth factor (VEGF) and the receptors often through inhibition of hypoxia-inducible factor-1alpha (HIF-1alpha), a transcriptional factor that is highly expressed in the hypoxic central region of neoplasms (Abbaszadeh et al., 2019). This intervention halts tumor growth and limits tumor size by blocking the supply of nutrients and oxygen to the tumor.

At the same time, they break the metastatic cascade by inhibiting matrix metalloproteins (MMPs) and invasion by cells through these phytochemicals. They inhibit the production and enzymatic functions of major MMPs such as MMP-2 and MMP-9, which break down the extracellular matrix and basement membrane, thus enabling the malignancies not to detach themselves in the original location, penetrate tissues, and form another colony that is fatal (Sarwar et al., 2018). This two-fold mechanistic intervention is an effective containment of the tumor, and it abrogates the maintenance as well as the propagation of the tumor.

10.3.5 Epigenetic Modulation and Effects on the Tumor Microenvironment

Recent studies highlight that the chemopreventive effects of fruit and berry compounds are expanded to control the expression of genes without modifying the DNA sequence by epigenetic regulation and the rearrangement of the systemic and local tumor microenvironment (Gerhauser, 2012; Jiang et al., 2021). The main bioactive metabolites, including folate, urolithins (built off of ellagitannins), and sulforaphane, appear to directly interact with DNA methylation and histone modification; both are capable of donating methyl groups required to silence oncogenes (proper DNA methylation), or act as an inhibitor of DNA methyltransferase (DNMTs) and HDACs to activate once-silenced tumor suppressor genes in cancer cells (Kocabas & Sanlier, 2023).

At the same time, these nutrients have intense immunomodulation and communication with the intestinal microbiota. It is a fermentation of dietary fiber to generate SCFAs (e.g., butyrate) that positively modulate the responsiveness of immune cells (such as T-cells and macrophages) to control pro-inflammatory cytokines and boost anti-tumor surveillance (Yao et al., 2022). Moreover, phytochemicals and their microbial conjugates have the potential to alter the gut microbiome composition to a preferential configuration, which, in process, regulates systemic inflammation, enhances gut barrier protection against microbial translocation, and develops other bioactive molecules that circulate to affect distal tumor stations, creating an entire, body-wide defense system against carcinogenesis (Santhiravel et al., 2022) (Figure 10.1).

10.4 PROMINENT MEDICINAL FRUITS AND BERRIES: EVIDENCE FROM PRECLINICAL AND CLINICAL STUDIES

10.4.1 Pomegranate (*Punica granatum*)

Pomegranate (*Punica granatum*) is one of the brightest examples of such medicinal fruits that have significant scientific factors that prove its anti-cancerous properties, which may be largely explained by the unique features of its bioactive components (Ranjha et al., 2023). A large body of preclinical and clinical evidence supports its effectiveness, especially in prostate, breast, and colon cancer models. Various in vitro or in vivo studies indicate that pomegranate extract induces apoptosis, inhibits proliferation, and angiogenesis in prostate cancer; several human clinical trials have shown a

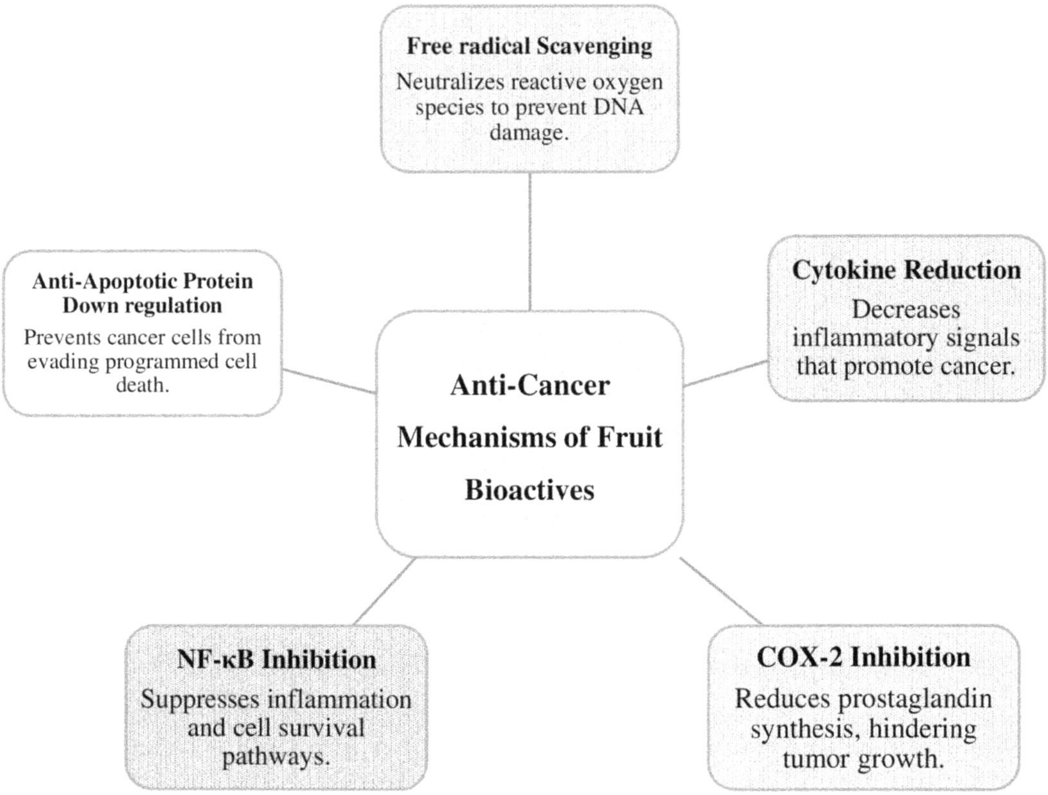

FIGURE 10.1 Mechanisms of anti-cancer action in medicinal fruits and berries.

considerable prolongation of prostate-specific antigen (PSA) doubling time, which is likely to slow the progression of recurrent disease (Amri et al., 2020; Deng et al., 2017; Sartippour et al., 2008). In line with this, experimental studies in breast cancer models have shown significant anti-proliferative activity and inhibition of estrogen-sensitive tumors, but studies in colorectal cancer have shown that it can suppress inflammation, induce cell-cycle arrest, and decrease the incidence of precancerous lesions (Bhat et al., 2001; Liang et al., 2006). Together, these outcomes put pomegranate as a potent dietary controller of several cancer hallmarks at different locations in the human organism.

10.4.2 Berries (Blueberry, Raspberry, Blackberry, and Strawberry)

Fruits such as blueberries, raspberries, blackberries, and strawberries are common berries and form some of the strongest biofunctional foods, with anti-cancer capacity that is strongly supported by scientific studies (Baby et al., 2018; Yadav, 2021). Their high levels of critical phytochemicals, such as anthocyanins and ellagic acid, are greatly correlated with their high efficacy, whose effect is synergistic (Chaachouay et al., 2023; Yu et al., 2024). An extensive body of preclinical and clinical studies has shown their usefulness in the treatment of a range of malignancies. Through research findings, it has been established that the berry extracts have strong anti-tumor effects, more so against gastrointestinal cancers like esophageal and colorectal carcinoma, where they inhibit cell formation, induce cell death, and suppress inflammatory reactions in animal models with cancer caused by carcinogens (Bishayee et al., 2016). The evidence provided in literature investigating oral malignancies shows that oral lesions at precancerous stages can be reversed by berry preparations in human clinical trials and that the growth of oral cancer cells can be inhibited. In addition, the bioactive components of these berries exhibit notable activity in estrogen and androgen receptor-mediated cancers such as breast and prostate cancer through the regulation of the estrogen and androgen receptor-mediated signaling pathways, the inhibition of tumor growth, and the sensitization of the malignancy cells to conventional therapies (De Amicis et al., 2019). This set of results supports the idea in favor of berries as broad and effective dietary agents in cancer chemoprevention in various organ systems.

10.4.3 Citrus Fruits (Orange, Grapefruit, and Lemon)

Citrus fruits, including oranges, grapefruits, and lemons, have significant chemopreventive capacity, much of which could be due to an arsenal of central bioactive products, especially the flavonoid glycoside hesperidin and naringenin, as well as D-limonene monoterpenes (Duarte et al., 2016). Their preclinical evidence confirms their wide range of anti-neoplastic effects, including induction of apoptosis, inhibition of tumor cell growth and invasion, and their amplification of detoxification activities and a wide interest in their protective effect against gastric, breast, and cutaneous cancer (Tajaldini & Asadi, 2021). However, an outstanding clinical issue that sets in a salient clinical concern is with regard to drug interactions, which are naringenin and associated furanocoumarins that provide strong and non-reversible inhibition against the intestinal cytochrome P450 3A4 (CYP3A4) enzyme complex (Dayyih et al., 2024). This kind of inhibition may significantly increase the bioavailability and plasma levels of countless numbers of co-administered drugs with specific types of chemotherapeutic agents, statins, and calcium-blockers, increasing the risk of severe toxicity. In line with this, even though the fresh intake of citrus fruits is widely promoted on the grounds of cancer prevention, the simultaneous intake of grapefruit or the juice thereof and many prescription drugs requires careful medical monitoring to prevent negative interactions.

10.4.4 Grapes and Red Wine Constituents

The frequently cited example of dense reservoirs of essential bioactive constituents is the grape, as well as red wine, with a significant portion being the stilbene resveratrol and oligomeric PAC, which provide a greater contribution to the chemopreventive effects of the substance (Teixeira et al., 2014). Literature presents comprehensive and manifold evidence as to the part played by these compounds in chemoprevention and as adjunctive therapies, which is gained after long credentials of preclinical studies. Regular observations are that resveratrol regulates every stage of carcinogenesis by acting as an antioxidant, providing anti-inflammatory effects through NF-KB inhibition, inducing cell-cycle arrest and apoptosis, and having specific anti-angiogenic effects (Bejenaru et al., 2024). PACs are also important because they inhibit matrix metalloproteinase and, according to this, they prevent invasion and metastasis (Katiyar, 2006). Though human clinical trials have had these challenges in recreating the high-dose effects seen in the laboratory due to bioavailability issues, epidemiological and smaller-scale intervention studies indicate that there is a protective relationship, especially against cardiovascular disease and specific cancers. Significantly, as of late, studies point to their possible use as adjunctive agents since resveratrol has been shown to sensitize cancer cells to traditional chemotherapy and radiotherapy to the extent that it will lead these compounds to a better future in the collective treatment schemes in the future.

10.4.5 Cranberry (*Vaccinium macrocarpon*)

Cranberry (*Vaccinium macrocarpon*) is characterized by a unique phytochemical composition, and the anti-cancer effect can be explained by the presence of specific key bioactives: type A PACs, which possess a unique double A-type linkage, and pentacyclic triterpenoid ursolic acid (Kowalczyk et al., 2025). The support of its chemopreventive ability is especially strong in urinary tract malignancies as well as gastrointestinal cancer. In the case of urinary tract cancers, the proven capacity of cranberries in preventing the colonization of the uropathogenic bacteria, like *Escherichia coli*, on the epithelium of the bladder is essential, as it suppresses chronic inflammation and recurrent infections, which are both put forward to causally influence the development of bladder malignancy (González de Llano et al., 2021). In addition to this indirect effect, cranberry extracts have direct anti-neoplastic effects, as revealed in gastrointestinal systems; cranberry extracts induce apoptosis in esophageal, colorectal, and gastric cancer cells and inhibit the growth of these cells in vitro and in rodents (Weh et al., 2016). This can be attributed to the fact that these effects are linked to the various signaling pathways being modulated, such as the inhibition of NF-KB and COX-2 by ursolic acid and the induction of an oxidative stress that is localized to explicitly malignant cells by the unique PACs, indicating that cranberry can be viewed as a multitargeted functional food against inflammatory etiology-related and infectious etiology-related human cancers.

10.4.6 Goji Berry (*Lycium barbarum*)

Lycium barbarum, the Goji berry, which is also used as a traditional Chinese herbal medicine, is currently receiving recognition as an oncological agent in modern medicine thanks to its unique and multipotent bioactive compounds, in particular *Lycium barbarum* polysaccharides (LBPs), a type of bioactive glycoconjugated polysaccharide, and the carotenoid zeaxanthin (Jiang et al., 2024). The main factual evidence that led to the development of its anti-neoplastic potential lies in its strong immunomodulatory effect and its ability to act as an adjunctive support instead of using direct cytotoxic deterrents (Bernitsa et al., 2023). It has been repeatedly found through preclinical studies that LBPs enhance the innate immunity as well as adaptive immunity by stimulating macrophages and complementing the actions of natural killer (NK) cells and T-lymphocytes to increase their proliferation and the production of cytokines, further enhancing natural tumor defense mechanisms. In addition, there is evidence that extracts of *L. barbarum* could protect normal tissues against chemotherapy-induced adverse effects, such as myelosuppression and hepatotoxicity, and even help decrease the fatigue caused by cancer and improve general living conditions (Almukram et al., 2024). Therefore, this research projects the Goji berry as a complementary drug but not a monotherapy, which is able to modulate the tumor microenvironment through immune upkeep and possibly enhance therapeutic tolerance and therapeutic outcomes in the conventional adjunctive cancer treatments.

10.4.7 Other Notable Fruits

Outside of the fruits that are generally characterized as targets of study, there is a range of other botanicals with an elaborate history in traditional medicinal systems with promising anti-cancer properties. Bitter melon (*Momordica charantia*) contains bioactive compounds like charantin and momordicin, which have been shown to induce apoptosis, inhibit malignant cell proliferation, and inhibit tumor glucose metabolism, a highly important process, bearing in mind that cancer is dependent on glycolysis (Sur & Ray, 2020). The mangosteen (*Garcinia mangostana*) is famous because it has a purple pericarp that contains a high amount of xanthones, especially α-mangostin (Kaur et al., 2020). Empirical evidence suggests that these compounds are potent anti-inflammatory, pro-apoptotic, and anti-metastatic in nature, and promising activity is demonstrated in xenograft as well as cell culture systems of breast, colon, and prostate cancer. The two fruits show how the use of ethnobotanical leads can be used to produce highly targeted malignancy-centered molecular mechanisms.

In the same way, Indian gooseberry (*Emblica officinalis*) and avocado (*Persea americana*) provide unique chemopreventive profiles (Shafi & Iqbal, 2025). Amla is also one of the richest natural sources of vitamin C, and it also possesses highly potent tannins, including emblicanin A and B (Dasaroju & Gottumukkala, 2014), which have remarkable antioxidant and radioprotective activity and have demonstrated efficiency in preventing carcinogenesis caused by chemicals in animal models, with a particular effect being exerted in the hepatic and cutaneous systems. Instead, avocado provides bioactive lipids, folate, and a distinct group of compounds known as avocatin B, which has been demonstrated to induce apoptosis specifically on leukemia stem cells while sparing healthy

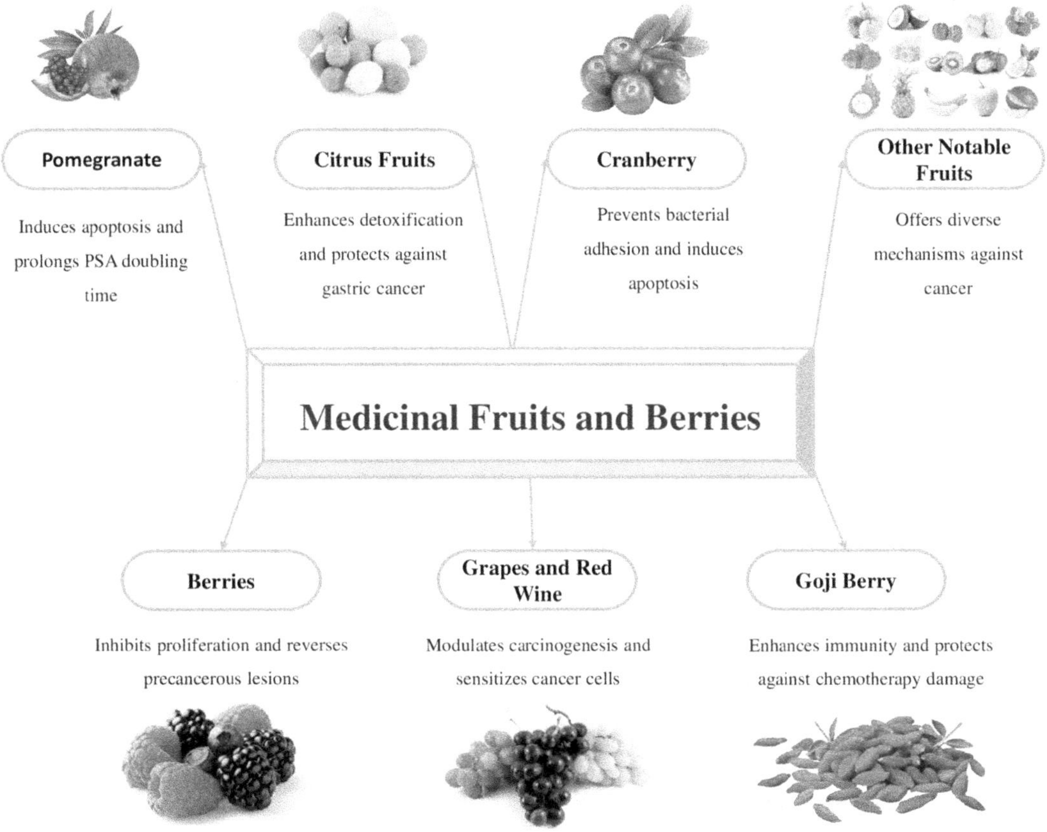

FIGURE 10.2 Summary of key anti-cancer effects by medicinal fruits and berries.

hematopoietic cells (Collignon et al., 2024; Walker, 2021; Chaachouay et al., 2025). All these fruits clearly highlight the enormous and diverse pharmacopoeia of the plant kingdom, with each member having different bioactive compounds and mechanisms that add up to the dietary staple of cancer prevention and integrative care (Figure 10.2).

10.5 SYNERGISTIC EFFECTS AND DELIVERY CHALLENGES

The anti-cancer activity of medicinal fruits and berries cannot be properly valued by looking at the constituents separately; it is significantly increased by the principle of a whole food benefit (Baby et al., 2018). This idea can explain the relationship between the interaction of phytochemicals, in which a complex combination of compounds found in an intact fruit, including polyphenols, vitamins, dietary fibers, and organic acids, acts together to provide a better biological effect than any single compound (Liu, 2013). As an example, the vitamin C present in berries has the ability to repair the tracked oxidized flavonoids, therefore, restoring their antioxidant property, and endogenous lipid in avocado has the ability to significantly boost the absorption of fat-soluble carotenoids associated with other fruit sources (Stephen & Radhakrishnan, 2022; Chaachouay et al., 2025). The complex interactions of such networks enable multitarget activity in modulating cancer hallmarks, such as all three inhibiting inflammation, inducing apoptosis, and preventing angiogenesis, and are thus a more effective and overall defense mechanism compared to single-agent supplements.

Even though this synergy exists, significant translational barriers still exist in terms of the high bioavailability issues controllable with polyphenols that are among the most widely investigated anti-cancer agents in these foods. Table bioactive compounds, such as anthocyanins, ellagitannins, and resveratrol, are highly insoluble in aqueous solutions, are labile in the gastrointestinal environment, are immediately metabolized by gut microorganisms and liver enzymes, and are poorly absorbed by cells (Ayvaz et al., 2022). As a result, an insignificant proportion of the dose applied is carried to the systemic circulation in its active form, thus significantly limiting its therapeutic potential at oncological application sites, including solid tumors (Najafi et al., 2021). This poor absorption feature has traditionally limited the therapeutic potential of fruit-derived polyphenols, creating a disaffiliating gap between the stimulation of in-vitro activity and quantifiable health benefits during human studies.

To beat such daunting challenges, modern research has been vigorously seeking new systems of delivery that will help to protect, carry, and deliver these valuable phytochemicals to the location. Such advanced systems as polymeric nanoparticles, lipid-based liposomes, and phospholipid complexes are designed to promote the encapsulation of sensitive polyphenols, which protects them against degradation and increases their solubility as well as transmembrane delivery (Ozkan et al., 2020). Also, these systems can be functionalized to deliver detrimental effects to specific tumor tissues, which may target the bioactive cargo being delivered to the tumor tissue with minimal off-target effects (Zhao et al., 2020). These innovative approaches to enhancing the clinical potential of fruit and berry bioactives by improving their stability, bioavailability, and tissue-specific accumulation will turn fruit and berry bioactives from being clinical dietary chemopreventatives to effective and dependable adjuncts in cancer treatment.

10.6 CLINICAL APPLICATIONS AND PUBLIC HEALTH PERSPECTIVES

The strategic point at which the clinical application of medicinal fruits and berries happens is their use in the prevention of cancer, both at the primary level and secondary levels (Huang et al., 2009). Primary prevention encompasses long-term habitual intake of these foods in a well-balanced diet to decrease the possibility of cancer beginning in healthy populations. The synergizing combination of antioxidants, anti-inflammatory substances, and fiber is active to reduce the damage of DNA, extinguish chronic inflammation, and maintain healthy metabolism and immune activity (Oyegoke et al., 2024). Secondary prevention aims at those with pre-cancerous lesions or who have had cancer, who have been shown to have a response to specific bioactive compounds that help to arrest or reverse the development of lesions, as well as prevent recurrence, as seen with the same on berry extracts with oral leukoplakia or pomegranate with rising levels of PSA.

Other than prevention, there is an increasing interest in the use of such fruits as an adjunct to other traditional treatment measures like chemotherapy and radiation. There is a solid argument from preclinical research, which shows that certain phytochemicals are capable of making cancer cells resistant to treatment. As an example, resveratrol and curcumin have been demonstrated to augment the activity of some chemotherapeutic agents acting on drug-resistance mechanisms and the prevention of survival signals in tumors (Maleki Dana et al., 2022). Such an adjunctive method should not be used to substitute traditional treatment; rather, it forms a synergistic effect, which may lead to reduced drug dosage or even treatment resistance, which may positively affect overall treatment outcomes, though a strict timing and dosing schedule must be observed to ensure there is no interference.

The reduction in the side effects of treatments and the quality of life are much-treasured patient-centered appeals. The severe adverse effects of the chemo- and radiotherapy, such as mucositis, neuropathy, fatigue, and immunosuppression, are significant challenges (Naidu et al., 2004). There is evidence to support the idea that some fruit compounds have supportive effects, such as anti-inflammatory effects of berry anthocyanins, which could be effective in protecting oral and gastrointestinal mucosa, and immunomodulatory polysaccharides in Goji berries may have an effect of maintaining the level of white blood cells (Sun et al., 2023). By reducing these impairing side effects, the incorporation of these functional foods in the diet of a patient can help to keep up the treatment schedules and boost the daily health and strength of a patient in a trying stage tremendously.

Acquiring such scientific evidence into practice needs a clear guideline on dietary prescriptions and community health guidelines. Every existing recommendation has a similar suggestion of a high intake of different fruits and vegetables as the foundation of cancer prevention. The next step to make is no longer generic prescriptions but more focused, evidence-based prescriptions based on particular types of fruits, preparations, and serving frequency that are associated with certain health consequences. The public health programs should prioritize education, availability, and socioeconomic factors of healthy food. Moreover, in the case of therapeutic and adjunctive applications, there is a need to develop clinically validated and standardized extracts and concrete guidelines on how to safely combine them with conventional care to ascertain efficacy, safety, and responsible application by both the caregivers and the patients.

10.7 FUTURE DIRECTIONS AND CHALLENGES

Medicinal fruits and berries serve as promising future approaches to the treatment of cancer, but this is inevitable to occur only when significant scientific and translational challenges are resolved (Seeram, 2008). The most prominent one is the creation of standardized extracts and clinical trials that should be large and of high quality. Existing evidence is often constrained by the research based on non-standardized extracts that have imprecise phytochemical composition and thus impede the comparison and reproducibility of results. In order to develop clinical credibility, future studies should utilize well-characterized and standardized preparations in randomized controlled trials with well-defined prespecified endpoints. This approach of such rigor in their methods will provide a high level of evidence that these natural compounds can be widely accepted and recommended by mainstream oncology, thus turning them into a part of evidence-based adjunctive medicine, not complementary nutrition anymore.

In the future, the trend of individualized nutrition provides a game-changing opportunity, and it is directed by the progress of the fields of pharmacogenomics and microbiome. A person is likely to respond to particular phytochemicals depending on genetic polymorphisms, where the specific make-up of the gut microbiota dictates the bioconversion of the compound in the diet to active or inactive metabolites (Karn et al., 2021). In turn, the research of the future should outline the way a genomic composition and microbial profile of an individual attenuate the effectiveness of interventions based on fruits. Such information may form the basis of the creation of specific dietary prescriptions, where certain fruits or their extracts are prescribed based on the genetic and microbiome profile of an individual to generate the greatest cancer-preventive and therapeutic advantages and synergy.

Finally, the efficient implementation of these interventions into the oncology practice requires a deeper understanding of understanding dose–response relations, safety, and logistic issues thereof. It is necessary to ascertain the optimal and effective dose during preventive and adjunctive treatment, not to be confused with the usual dietary intake, as well as to have a thorough evaluation of possible herb-drug interactions and safety in the long run. The insight is found in the development of collaborative models whereby oncologists, dietitians, and pharmacologists are collaborators in the development of integrative care plans (Schoenfeld et al., 2020). The major problem will consist of patient and healthcare provider education, product quality and consistency, and insurance coverage on clinically validated nutraceutical intervention to decrease the general chasm between nutritional science and standardized clinical practice in an attempt to enhance patient outcomes.

10.8 CONCLUSION

To sum up, medicinal fruits and berries have a huge potential in cancer treatment, which is confirmed by the amount of evidence provided in this chapter. A voluminous array of bioactive compounds, comprising polyphenols like anthocyanins and resveratrol, carotenoids like lycopene, and some distinctive molecules like bromelain, perform their actions based on an intricate system of action. These include antioxidant action and anti-inflammatory action, control of carcinogen metabolism, direct induction of apoptosis in cancerous cells, inhibition of tumor angiogenesis and metastasis, and epigenetic control. Repeated preclinical and progressively strong clinical research results of particular fruits, consisting of pomegranate, berries, citrus, and goji berry, are strong demonstrations of their efficiency in models of prostate, breast, gastrointestinal, and other cancer types. Importantly, this potential is not maximized by individual supplements but rather by the synergistic effect of whole foods, whereby a complex natural matrix is found to increase total bioactivity. Although problems of bioavailability and more standardized clinical trials are still to be overcome, the incorporation of these fruits into a balanced diet is an influential, available, and risk-free part of a synergistic public health approach. As such, a high intake of fruits and berries of various colors and types must be strongly encouraged as one of the pillars in a holistic form of preventive measures against cancer, as a supportive measure to conventional treatment to reduce side effects, and, finally, as a quality-of-life enhancer during and after cancer management.

REFERENCES

Abbaszadeh, H., Keikhaei, B., & Mottaghi, S. (2019). A review of molecular mechanisms involved in anticancer and antiangiogenic effects of natural polyphenolic compounds. *Phytotherapy Research, 33*(8), 2002–2014. https://doi.org/10.1002/ptr.6403

Abd El-Azim, A. O. (2021). The effects of B-vitamins, dietary bioactive agents and functional foods on hyperhomocysteinemia. In M. I. Waly (Ed.), *Nutritional management and metabolic aspects of hyperhomocysteinemia* (pp. 225–241). Springer International Publishing. https://doi.org/10.1007/978-3-030-57839-8_18

Ahmad, H. A., Seemab, K., Wahab, F., & Khan, M. I. (2024). Signaling pathways in drug development. *Drug Development and Safety.* https://www.intechopen.com/chapters/89167

Almukram, A. M. A., Al-Mamoori, F., Sameer, H. N., Farhan, D. M., Al-Shammari, A. H., Al-Hussaniy, H. A., & Oraibi, A. I. (2024). Strategies for attenuating chemotherapy side effects: A review. *Research Journal of Pharmacy and Technology, 17*(11), 5624–5631.

Amponsah, S. K., Amoafo, E. B., Ofori, E. K., & Bugyei, K. A. (2024). Anticancer potentials of food phytochemicals. In *Plant food phytochemicals and bioactive compounds in nutrition and health* (pp. 279–297). CRC Press. https://www.taylorfrancis.com/chapters/edit/10.1201/9781003340201-11/anticancer-potentials-food-phytochemicals-seth-kwabena-amponsah-emmanuel-boadi-amoafo-emmanuel-kwaku-ofori-kwasi-agyei-bugyei

Amri, Z., Kharroubi, W., Fidanzi-Dugas, C., Leger, D. Y., Hammami, M., & Liagre, B. (2020). Growth inhibitory and pro-apoptotic effects of ornamental pomegranate extracts in DU145 human prostate cancer cells. *Nutrition and Cancer, 72*(6), 932–938. https://doi.org/10.1080/01635581.2019.1660385

Ayvaz, H., Cabaroglu, T., Akyildiz, A., Pala, C. U., Temizkan, R., Ağçam, E., Ayvaz, Z., Durazzo, A., Lucarini, M., & Direito, R. (2022). Anthocyanins: Metabolic digestion, bioavailability, therapeutic effects, current pharmaceutical/industrial use, and innovation potential. *Antioxidants, 12*(1), 48.

Azcárate-Peril, M. A., Sikes, M., & Bruno-Bárcena, J. M. (2011). The intestinal microbiota, gastrointestinal environment and colorectal cancer: A putative role for probiotics in prevention of colorectal cancer? *American Journal of Physiology-Gastrointestinal and Liver Physiology, 301*(3), G401–G424. https://doi.org/10.1152/ajpgi.00110.2011

Baby, B., Antony, P., & Vijayan, R. (2018). Antioxidant and anticancer properties of berries. *Critical Reviews in Food Science and Nutrition, 58*(15), 2491–2507. https://doi.org/10.1080/10408398.2017.1329198

Bejenaru, L. E., Biţă, A., Belu, I., Segneanu, A.-E., Radu, A., Dumitru, A., Ciocîlteu, M. V., Mogoşanu, G. D., & Bejenaru, C. (2024). Resveratrol: A review on the biological activity and applications. *Applied Sciences, 14*(11), 4534.

Bernitsa, S., Dayan, R., Stephanou, A., Tzvetanova, I. D., & Patrikios, I. S. (2023). Natural biomolecules and derivatives as anticancer immunomodulatory agents. *Frontiers in Immunology, 13*, 1070367.

Bhat, K. P., Lantvit, D., Christov, K., Mehta, R. G., Moon, R. C., & Pezzuto, J. M. (2001). Estrogenic and antiestrogenic properties of resveratrol in mammary tumor models. *Cancer Research, 61*(20), 7456–7463.

Bishayee, A., Haskell, Y., Do, C., Siveen, K. S., Mohandas, N., Sethi, G., & Stoner, G. D. (2016). Potential benefits of edible berries in the management of aerodigestive and gastrointestinal tract cancers: Preclinical and clinical evidence. *Critical Reviews in Food Science and Nutrition*, *56*(10), 1753–1775. https://doi.org/10.1080/10408398.2014.982243

Chaachouay, N., Azeroual, A., & Zidane, L. (2025). *Comprehensive Guide to Hallucinogenic Plants*. 9781003460336, 455.

Chaachouay, N. (2025). Synergy, additive effects, and antagonism of drugs with plant bioactive compounds. *Drugs and Drug Candidates*, *4*(1), 4. https://doi.org/10.3390/ddc4010004

Chaachouay, N., Azeroual, N., Bencherki, B., Douira, A., & Zidane, L. (2023). Various Metabolites and or Bioactive Compounds from Vegetables, and Their Use Nanoparticles Synthesis, and Applications. In A. Husen (Éd.), *Nanomaterials from Agricultural and Horticultural Products* (pp. 187–209). Springer Nature. https://doi.org/10.1007/978-981-99-3435-5_10

Chaachouay, N., Douira, A., & Zidane, L. (2021). COVID-19, prevention and treatment with herbal medicine in the herbal markets of Salé Prefecture, North-Western Morocco. *European journal of integrative medicine*, *42*, 101285. https://doi.org/10.1080/10408398.2023.2260474

Chaachouay, N., Benkhnigue, O., Douira, A., & Zidane, L. (2024). Avocado (*Persea americana* Mill) and its phytoconstituents Poisonous medicinal plants used in the popular pharmacopoeia of the Rif, northern Morocco. *Toxicon*, *189*, 24–32. https://doi.org/10.1016/j.toxicon.2020.10.028

Chaachouay, N., Douira, A., Hassikou, R., Brhadda, N., Dahmani, J., Belahbib, N., Ziri, R., & Zidane, L. (2020). Étude floristique et ethnomédicinale des plantes aromatiques et médicinales dans le Rif (Nord du Maroc). *PhD Thesis, Département de Biologie-Université Ibn Tofail-Kénitra.* tel-03376377

Collignon, T. E., Webber, K., Piasecki, J., Rahman, A. S. W., Mondal, A., Barbalho, S. M., & Bishayee, A. (2024). Avocado (*Persea americana* Mill) and its phytoconstituents: Potential for cancer prevention and intervention. *Critical Reviews in Food Science and Nutrition*, *64*(33), 13072–13092. https://doi.org/10.1016/j.eujim.2021.101285

Daca-Álvarez, M., Brunori, A., Carbone, A., Carbonell, C., Tangen, C. M., Unger, J. M., Lucia, M. S., Oliva, M., De Censi, A., Brenner, D. R., Thompson, I. M., & Balaguer, F. (2025). Emerging strategies for drug-based cancer risk reduction. *American Society of Clinical Oncology Educational Book*, *45*(3), e473708. https://doi.org/10.1200/EDBK-25-473708

Dahiya, D., & Nigam, P. S.-N. (2023). Inclusion of dietary fibers in nutrition provides prebiotic substrates to probiotics for the synthesis of beneficial metabolites SCFA to sustain gut health minimizing risk of IBS, IBD, CRC. *Recent Progress in Nutrition*, *3*(3), doi-10.

Das, D. N., Panda, P. K., Naik, P. P., Mukhopadhyay, S., Sinha, N., & Bhutia, S. K. (2017). Phytotherapeutic approach: A new hope for polycyclic aromatic hydrocarbons induced cellular disorders, autophagic and apoptotic cell death. *Toxicology Mechanisms and Methods*, *27*(1), 1–17. https://doi.org/10.1080/15376516.2016.1268228

Dasaroju, S., & Gottumukkala, K. M. (2014). Current trends in the research of *Emblica officinalis* (amla): A pharmacological perspective. *International Journal of Pharmaceutical Sciences Review and Research*, *24*(2), 150–159.

Dayyih, W. A., Al-Ani, I., Hailat, M., Alarman, S. M., Zakaraya, Z., Assab, M. A., & Alkhader, E. (2024). Review of grapefruit juice-drug interactions mediated by intestinal CYP3A4 inhibition. *Journal of Applied Pharmaceutical Science*, *14*(5), 059–068.

De Amicis, F., Chimento, A., Montalto, F. I., Casaburi, I., Sirianni, R., & Pezzi, V. (2019). Steroid receptor signaling as targets for resveratrol actions in breast and prostate cancer. *International Journal of Molecular Sciences*, *20*(5), 1087.

Dean, M. (2024). Exploring ethnobotanical knowledge: Qualitative insights into the therapeutic potential of medicinal plants. *Golden Ratio of Data in Summary*, *4*(2), 154–166.

Deng, Y., Li, Y., Yang, F., Zeng, A., Yang, S., Luo, Y., Zhang, Y., Xie, Y., Ye, T., & Xia, Y. (2017). The extract from *Punica granatum* (pomegranate) peel induces apoptosis and impairs metastasis in prostate cancer cells. *Biomedicine & Pharmacotherapy*, *93*, 976–984.

Duarte, A., Carvalho, C., & Miguel, G. (2016). Bioactive compounds of citrus as health promoters. In *Natural bioactive compounds from fruits and vegetables as health promoters: Part I* (pp. 29–97). Bentham Science Publishers. https://www.benthamdirect.com/content/books/9781681082394.chapter-2

Duta-Bratu, C.-G., Nitulescu, G. M., Mihai, D. P., & Olaru, O. T. (2023). Resveratrol and other natural oligomeric stilbenoid compounds and their therapeutic applications. *Plants*, *12*(16). https://doi.org/10.3390/plants12162935

Dutta, D., & Dutta, D. (2022). Lycopene as nutraceuticals. In P. Balakrishnan & S. Gopi (Eds.), *Handbook of nutraceuticals and natural products* (1st ed., pp. 205–243). Wiley. https://doi.org/10.1002/9781119746843.ch11

Folmer, F., Basavaraju, U., Jaspars, M., Hold, G., El-Omar, E., Dicato, M., & Diederich, M. (2014). Anticancer effects of bioactive berry compounds. *Phytochemistry Reviews*, *13*(1), 295–322. https://doi.org/10.1007/s11101-013-9319-z

Gayathry, K. S., & John, J. A. (2022). A comprehensive review on bitter gourd (*Momordica charantia* L.) as a gold mine of functional bioactive components for therapeutic foods. *Food Production, Processing and Nutrition*, *4*(1), 10. https://doi.org/10.1186/s43014-022-00089-x

Gerhauser, C. (2012). Cancer chemoprevention and nutri-epigenetics: State of the art and future challenges. In J. M. Pezzuto & N. Suh (Eds.), *Natural products in cancer prevention and therapy* (Vol. 329, pp. 73–132). Springer, Berlin Heidelberg. https://doi.org/10.1007/128_2012_360

Gialeli, C., Theocharis, A. D., & Karamanos, N. K. (2011). Roles of matrix metalloproteinases in cancer progression and their pharmacological targeting. *The FEBS Journal*, *278*(1), 16–27. https://doi.org/10.1111/j.1742-4658.2010.07919.x

González de Llano, D., Moreno-Arribas, M., & Bartolomé, B. (2021). Cranberry polyphenols and prevention against urinary tract infections: A brief review. https://digital.csic.es/handle/10261/253304

Green, D. R. (2022). The death receptor pathway of apoptosis. *Cold Spring Harbor Perspectives in Biology*, *14*(2), a041053.

Guo, Z., & Mo, Z. (2020). Keap1–Nrf2 signaling pathway in angiogenesis and vascular diseases. *Journal of Tissue Engineering and Regenerative Medicine*, *14*(6), 869–883. https://doi.org/10.1002/term.3053

Gurib-Fakim, A. (2006). Medicinal plants: Traditions of yesterday and drugs of tomorrow. *Molecular Aspects of Medicine*, *27*(1), 1–93. https://doi.org/10.1016/j.mam.2005.07.008

Hamed, E. A., Zakhary, M. M., & Maximous, D. W. (2012). Apoptosis, angiogenesis, inflammation, and oxidative stress: Basic interactions in patients with early and metastatic breast cancer. *Journal of Cancer Research and Clinical Oncology*, *138*(6), 999–1009. https://doi.org/10.1007/s00432-012-1176-4

Huang, W.-Y., Cai, Y.-Z., & Zhang, Y. (2009). Natural phenolic compounds from medicinal herbs and dietary plants: Potential use for cancer prevention. *Nutrition and Cancer*, *62*(1), 1–20. https://doi.org/10.1080/01635580903191585

Iqubal, A., Rahman, S. O., Ahmed, M., Bansal, P., Haider, M. R., Iqubal, M. K., Najmi, A. K., Pottoo, F. H., & Haque, S. E. (2021). Current quest in natural bioactive compounds for Alzheimer's disease: Multi-targeted-designed-ligand based approach with preclinical and clinical based evidence. *Current Drug Targets*, *22*(6), 685–720. https://doi.org/10.2174/1389450121999201209201004

Jiang, C., Chen, Z., Liao, W., Zhang, R., Chen, G., Ma, L., & Yu, H. (2024). The medicinal species of the *Lycium* genus (goji berries) in East Asia: A review of its effect on cell signal transduction pathways. *Plants*, *13*(11), 1531.

Jiang, W., Xia, T., Liu, C., Li, J., Zhang, W., & Sun, C. (2021). Remodeling the epigenetic landscape of cancer—Application potential of flavonoids in the prevention and treatment of cancer. *Frontiers in Oncology*, *11*, 705903.

Karn, A., Zhao, C., Yang, F., Cui, J., Gao, Z., Wang, M., Wang, F., Xiao, H., & Zheng, J. (2021). In vivo biotransformation of citrus functional components and their effects on health. *Critical Reviews in Food Science and Nutrition*, *61*(5), 756–776. https://doi.org/10.1080/10408398.2020.1746234

Katiyar, S. K. (2006). Matrix metalloproteinases in cancer metastasis: Molecular targets for prostate cancer prevention by green tea polyphenols and grape seed proanthocyanidins. *Endocrine, Metabolic & Immune Disorders Drug Targets*, *6*(1), 17–24.

Kaur, G., Singh, A., & Dar, B. N. (2020). Mangosteen (*Garcinia mangostana* L.). In G. A. Nayik & A. Gull (Eds.), *Antioxidants in fruits: Properties and health benefits* (pp. 83–101). Springer Singapore. https://doi.org/10.1007/978-981-15-7285-2_5

Kocabas, Ş., & Sanlier, N. (2023). A comprehensive overview of the complex relationship between epigenetics, bioactive components, cancer, and aging. *Critical Reviews in Food Science and Nutrition*, *63*(19), 3333–3345. https://doi.org/10.1080/10408398.2021.1986803

Kowalczyk, T., Merecz-Sadowska, A., Gładys, A., Osicka, W., Dudzic, M., Picot, L., Piekarski, J., Szemraj, J., Rijo, P., & Sitarek, P. (2025). The chemopreventive and antitumor potential of cranberry (*Vaccinium macrocarpon* Aiton) in gastrointestinal cancers: A review of molecular mechanisms, preclinical evidence, and commercial perspectives. *Food Reviews International*, 1–29. https://doi.org/10.1080/87559129.2025.2597422

Liang, Y., Brekken, R. A., & Hyder, S. M. (2006). Vascular endothelial growth factor induces proliferation of breast cancer cells and inhibits the anti-proliferative activity of anti-hormones. *Endocrine-Related Cancer*, *13*(3), 905–919.

Lin, X. (2024). The importance of bioactive compounds and plant extracts in the treatment of pancreatic cancer. https://www.preprints.org/frontend/manuscript/666a882dcc401f7f767c0007736f560b/download_pub

Liu, R. H. (2013). Dietary bioactive compounds and their health implications. *Journal of Food Science*, *78*(S1). https://doi.org/10.1111/1750-3841.12101

Ma, Z., Du, B., Li, J., Yang, Y., & Zhu, F. (2021). An insight into anti-inflammatory activities and inflammation-related diseases of anthocyanins: A review of both in vivo and in vitro investigations. *International Journal of Molecular Sciences*, *22*(20). https://doi.org/10.3390/ijms222011076

Maleki Dana, P., Sadoughi, F., Asemi, Z., & Yousefi, B. (2022). The role of polyphenols in overcoming cancer drug resistance: A comprehensive review. *Cellular & Molecular Biology Letters*, *27*(1), 1. https://doi.org/10.1186/s11658-021-00301-9

Martí, N., Mena, P., Cánovas, J. A., Micol, V., & Saura, D. (2009). Vitamin C and the role of citrus juices as functional food. *Natural Product Communications*, *4*(5), 1934578X0900400506. https://doi.org/10.1177/1934578X0900400506

Mishra, P., & Kumar, J. (2025). Exploring ethnobotany: Understanding the dynamic relationship between humans and plants. *Diverse Aspects of Ethnobotany*, 196, 196–221.

Murugan, S. P., Murugesan, K., Muthusamy, Y., Basalingappa, K. M., Murugan, M., Mahendran, N., Rajendran, P., Kuppannan, G., & Arthanari, M. (2024). The role of bioactive principles from traditional medicinal plants in metabolic disorder. In *Plant Derived Bioactive Compounds in Human Health and Disease* (pp. 8–32). CRC Press.

Naidu, M. U. R., Ramana, G. V., Rani, P. U., Suman, A., & Roy, P. (2004). Chemotherapy-induced and/or radiation therapy-induced oral mucositis—complicating the treatment of cancer. *Neoplasia*, *6*(5), 423–431.

Najafi, M., Majidpoor, J., Toolee, H., & Mortezaee, K. (2021). The current knowledge concerning solid cancer and therapy. *Journal of Biochemical and Molecular Toxicology*, *35*(11), e22900. https://doi.org/10.1002/jbt.22900

Oyegoke, R. A., Oladele, J. O., Oladele, O. T., & Oladiji, A. T. (2024). Role of dietary fibre and phytonutrients in human health and nutrition. In *Nutrition and diet in health* (pp. 121–147). CRC Press. https://www.taylorfrancis.com/chapters/edit/10.1201/9781003361497-12/role-dietary-fibre-phytonutrients-human-health-nutrition-rukayat-oyegoke-johnson-olaleye-oladele-oluwaseun-titilope-oladele-adenike-temidayo-oladiji

Ozkan, G., Kostka, T., Esatbeyoglu, T., & Capanoglu, E. (2020). Effects of lipid-based encapsulation on the bioaccessibility and bioavailability of phenolic compounds. *Molecules*, *25*(23), 5545.

Ozougwu, J. C. (2016). The role of reactive oxygen species and antioxidants in oxidative stress. *International Journal of Research*, *1*(8), 1–8.

Paliyath, G., Bakovic, M., & Shetty, K. (2011). *Functional foods, nutraceuticals, and degenerative disease prevention*. John Wiley & Sons. https://books.google.fr/books?hl=fr&lr=&id=bGpuyrFr9M8C

Rai, D. K., & Tzima, K. (2021). A review on chromatography–mass spectrometry applications on anthocyanin and ellagitannin metabolites of blackberries and raspberries. *Foods*, *10*(9). https://doi.org/10.3390/foods10092150

Rajan, P. K., Dunna, N. R., & Venkatabalasubramanian, S. (2022). A comprehensive overview on the anti-inflammatory, antitumor, and ferroptosis functions of bromelain: An emerging cysteine protease. *Expert Opinion on Biological Therapy*, *22*(5), 615–625. https://doi.org/10.1080/14712598.2022.2042250

Ranjha, M. M. A. N., Shafique, B., Wang, L., Irfan, S., Safdar, M. N., Murtaza, M. A., Nadeem, M., Mahmood, S., Mueen-ud-Din, G., & Nadeem, H. R. (2023). A comprehensive review on phytochemistry, bioactivity, and medicinal value of bioactive compounds of pomegranate (*Punica granatum*). *Advances in Traditional Medicine*, *23*(1), 37–57. https://doi.org/10.1007/s13596-021-00566-7

Santhiravel, S., Bekhit, A. E.-D. A., Mendis, E., Jacobs, J. L., Dunshea, F. R., Rajapakse, N., & Ponnampalam, E. N. (2022). The impact of plant phytochemicals on the gut microbiota of humans for a balanced life. *International Journal of Molecular Sciences*, *23*(15), 8124. https://doi.org/10.3390/ijms23158124

Sartippour, M., Seeram, N., Rao, J., Moro, A., Harris, D., Henning, S., Firouzi, A., Rettig, M., Aronson, W., Pantuck, A., & Heber, D. (2008). Ellagitannin-rich pomegranate extract inhibits angiogenesis in prostate cancer in vitro and in vivo. *International Journal of Oncology*. https://doi.org/10.3892/ijo.32.2.475

Sarwar, M. S., Zhang, H.-J., & Tsang, S. W. (2018). Perspectives of plant natural products in inhibition of cancer invasion and metastasis by regulating multiple signaling pathways. *Current Medicinal Chemistry*, *25*(38), 5057–5087.

Schoenfeld, R., Nguyen, G. C., & Bernstein, C. N. (2020). Integrated care models: Optimizing adult ambulatory care in inflammatory bowel disease. *Journal of the Canadian Association of Gastroenterology*, *3*(1), 44–53.

Seeram, N. P. (2008). Berry fruits for cancer prevention: Current status and future prospects. *Journal of Agricultural and Food Chemistry*, *56*(3), 630–635. https://doi.org/10.1021/jf072504n

Shafi, L., & Iqbal, P. (2025). Indian gooseberry: Role in cancer prevention and treatment. In *Neglected and underutilized crops* (pp. 94–115). CRC Press. https://www.taylorfrancis.com/chapters/edit/10.1201/9781003477792-7/indian-gooseberry-lubna-shafi-pervaiz-iqbal

Singh, R., & Agarwal, R. (2006). Natural flavonoids targeting deregulated cell cycle progression in cancer cells. *Current Drug Targets*, *7*(3), 345–354. https://doi.org/10.2174/138945006776055004

Stephen, J., & Radhakrishnan, M. (2022). Avocado (*Persea americana* Mill.) fruit: Nutritional value, handling and processing techniques, and health benefits. *Journal of Food Processing and Preservation*, *46*(12), e17207. https://doi.org/10.1111/jfpp.17207

Sun, Q., Du, M., Kang, Y., & Zhu, M.-J. (2023). Prebiotic effects of goji berry in protection against inflammatory bowel disease. *Critical Reviews in Food Science and Nutrition*, *63*(21), 5206–5230. https://doi.org/10.1080/10408398.2021.2015680

Sur, S., & Ray, R. B. (2020). Bitter melon (*Momordica charantia*), a nutraceutical approach for cancer prevention and therapy. *Cancers*, *12*(8), 2064.

Swetha, M., Keerthana, C. K., Rayginia, T. P., & Anto, R. J. (2022). Cancer chemoprevention: A strategic approach using phytochemicals. *Frontiers in Pharmacology*, *12*, 809308.

Tajaldini, M., & Asadi, J. (2021). The use of bio-active compounds of citrus fruits as chemopreventive agents and inhibitor of cancer cells viability. *Anti-Cancer Agents in Medicinal Chemistry*, *21*(9), 1058–1068.

Tao, J., Li, S., Gan, R.-Y., Zhao, C.-N., Meng, X., & Li, H.-B. (2020). Targeting gut microbiota with dietary components on cancer: Effects and potential mechanisms of action. *Critical Reviews in Food Science and Nutrition*, *60*(6), 1025–1037. https://doi.org/10.1080/10408398.2018.1555789

Tao, M., Mao, J., Wang, G., Jiang, Y., Li, G., Zhang, K., Sun, L., & Yang, G. (2026). Ellagitannins from red raspberry (*Rubus idaeus* L.): A comprehensive review on chemistry characteristics and beneficial effects. *Food & Function*. https://doi.org/10.1039/D5FO04185F

Teixeira, A., Baenas, N., Dominguez-Perles, R., Barros, A., Rosa, E., Moreno, D. A., & Garcia-Viguera, C. (2014). Natural bioactive compounds from winery by-products as health promoters: A review. *International Journal of Molecular Sciences*, *15*(9), 15638–15678.

Thun, M. J., DeLancey, J. O., Center, M. M., Jemal, A., & Ward, E. M. (2010). The global burden of cancer: Priorities for prevention. *Carcinogenesis*, *31*(1), 100–110. https://doi.org/10.1093/carcin/bgp263

Upadhyay, R. K. (2025). Plant origin anti-inflammatory and immunomodulatory agents and their mechanism of action: A review. *International Journal of Green Pharmacy (IJGP)*, *19*(01). http://www.greenpharmacy.info/index.php/ijgp/article/download/3650/1354

Verma, H., Gangwar, P., Yadav, A., Yadav, B., Rao, R., Kumari, S., Kaur, S., Dhiman, M., & Mantha, A. K. (2025). Phytochemicals in lifestyle-associated human diseases. In V. Rani, B. Saxena, & N. Atale (Eds.), *Advancing biotechnology: From science to therapeutics and informatics* (pp. 223–250). Springer Nature Switzerland. https://doi.org/10.1007/978-3-031-80973-6_19

Vitamin, A. (2022). Carotenoids. *Professionals, NIH fact sheet for health*. https://www.researchgate.net/profile/Suzanne-Adams-4/publication/341051931_VITAMIN_A_Functions_and_Deficiency/links/5eaaffab45851592d6ae45a6/VITAMIN-A-Functions-and-Deficiency.pdf

Walker, S. (2021). *Impact of an avocado-containing diet on progression and relapse of acute myeloid leukemia in mouse models* (PhD thesis, University of Guelph). https://atrium.lib.uoguelph.ca/bitstream/10214/26653/3/Walker_Sarah_202201_MSc.pdf

Wan, C., Langyan, S., Echeverría, J., Devkota, H. P., Tewari, D., Moosavi, M. A., Ezzat, S. M., Perez-Vazquez, A., Fraga-Corral, M., Cravotto, G., Prieto, M. A., Belwal, T., & Li, M. (2025). Edible fruits and berries as a source of functional polyphenols: Current scene and future perspectives. *Phytochemistry Reviews*, *24*(2), 1197–1225. https://doi.org/10.1007/s11101-023-09892-x

Weh, K. M., Clarke, J., & Kresty, L. A. (2016). Cranberries and cancer: An update of preclinical studies evaluating the cancer inhibitory potential of cranberry and cranberry-derived constituents. *Antioxidants*, *5*(3), 27.

Yadav, R. B. (2021). Potential benefits of berries and their bioactive compounds as functional food component and immune boosting food. In *Immunity boosting functional foods to combat COVID-19* (pp. 75–90). CRC Press. https://www.taylorfrancis.com/chapters/edit/10.1201/9781003242604-7/potential-benefits-berries-bioactive-compounds-functional-food-component-immune-boosting-food-ritika-yadav

Yao, Y., Cai, X., Fei, W., Ye, Y., Zhao, M., & Zheng, C. (2022). The role of short-chain fatty acids in immunity, inflammation, and metabolism. *Critical Reviews in Food Science and Nutrition*, *62*(1), 1–12. https://doi.org/10.1080/10408398.2020.1854675

Yu, J., Xiao, X., Chen, B., Deng, Z., Chen, X., Fan, Y., & Li, H. (2024). Synergistic and antagonistic activity of selected dietary phytochemicals against oxidative stress-induced injury in cardiac H9c2 cells via the Nrf2 signaling pathway. *Foods*, *13*(15), 2440.

Zhao, Z., Ukidve, A., Kim, J., & Mitragotri, S. (2020). Targeting strategies for tissue-specific drug delivery. *Cell*, *181*(1), 151–167.

11 Managing Diabetes through Medicinal Fruits and Berries

Nazim Husain and Jamal Akhtar

11.1 INTRODUCTION

Diabetes mellitus is defined by the American Diabetes Association (ADA) as a group of metabolic disorders of carbohydrate metabolism in which glucose is both underutilized as an energy source and overproduced due to inappropriate gluconeogenesis and glycogenolysis, resulting in hyperglycemia (American Diabetes Association Professional Practice Committee for Diabetes, 2026). It represents one of the most pressing global health challenges of the 21st century. According to the latest 11th edition of the International Diabetes Federation (IDF) Diabetes Atlas, diabetes affected 11.11% of the global adult population in 2024, corresponding to 589 million adults worldwide, with projections indicating that this number will rise to 853 million by 2050 (Genitsaridi et al., 2026). This epidemic encompasses both Type 1 diabetes mellitus (T1DM), characterized by autoimmune destruction of pancreatic β cells, and Type 2 diabetes mellitus (T2DM), which accounts for approximately 90–95% of all diabetes cases and is primarily characterized by insulin resistance and progressive β-cell dysfunction (Lu et al., 2024).

The global prevalence of diabetes is increasing rapidly, driven by population growth and an aging demographic ((NCD-RisC, 2016). In 2000, the global age-standardized prevalence of diabetes was 4.6%, rising to an estimated 10.5% by 2021 (Magliano et al., 2021). T2DM accounts for over 90% of all diabetes cases, significantly exceeding the prevalence of T1DM and gestational diabetes (DeFronzo et al., 2015). The rise in type 2 diabetes has been particularly notable since the mid-20th century, largely fueled by the global increase in obesity and overweight populations (Kahn et al., 2014). In India, a recent report indicated a diabetes prevalence of 9.3%, with only 45.8% of those affected being aware of their condition, 36.1% receiving treatment, and 15.7% achieving adequate glycemic control (Mathur et al., 2022).

Individuals with T2DM are at heightened risk of developing concurrent conditions such as hypertension (affecting 80–90%), lipid disorders (affecting 70–80%), and obesity or overweight (affecting 60–70%) (Preis et al., 2009). Diabetes diagnosed at 40 years of age is associated with an average reduction in life expectancy of 5.8 years for men and 6.8 years for women, emphasizing the importance of early prevention and intervention (Gregg et al., 2014). Early-onset T2DM (<40 years) is linked to poorer metabolic control and a greater likelihood of chronic complications and premature mortality (Chen et al., 2020).

The pathophysiology of T2DM involves insulin resistance and impaired insulin secretion due to pancreatic β-cell dysfunction, resulting in relative insulin deficiency (Guthrie & Guthrie, 2004; Petersen & Shulman, 2006). The condition has a strong genetic predisposition and is strongly associated with obesity and a sedentary lifestyle (Ali, 2013; Kolb & Martin, 2017). Insulin resistance is exacerbated by aging, physical inactivity, and excess body weight, with a body mass index (BMI) between 25 and 29.9 kg/m² classified as overweight and a BMI >30 kg/m² as obese (Cappola et al., 2023).

Screening for hyperglycemia is recommended for individuals with classic symptoms of diabetes, while asymptomatic individuals at high risk (e.g., those with obesity and additional risk factors)

DOI: 10.1201/9781003688259-11

should also be screened (American Diabetes Association Professional Practice Committee, 2024b). Diabetes is diagnosed based on elevated blood glucose or hemoglobin A1c (HbA1c) levels. Prediabetes is characterized by a fasting plasma glucose (FPG) of 100–125 mg/dL (5.6–6.9 mmol/L), a 2-hour post-oral glucose test result of 140–199 mg/dL (7.8–11.0 mmol/L), or an HbA1c of 5.7–6.4% (39–47 mmol/mol). A definitive diabetes diagnosis requires two confirmed values of FPG ≥126 mg/dL (≥7.0 mmol/L), 2-hour plasma glucose ≥200 mg/dL (≥11.1 mmol/L) following a 75 g oral glucose test, or HbA1c ≥6.5% (≥48 mmol/mol). Alternatively, diabetes can be diagnosed with a random plasma glucose ≥200 mg/dL (≥11.1 mmol/L) accompanied by symptoms of hyperglycemia or hyperglycemic crisis. A single blood sample may suffice for diagnosis if both HbA1c and FPG meet the diagnostic thresholds (American Diabetes Association Professional Practice Committee, 2024b).

The cornerstone of T2DM management is a personalized treatment plan that incorporates pharmacotherapy and continuous self-management education, often facilitated by diabetes educators or dietitians (American Diabetes Association Professional Practice Committee, 2024c; Powers et al., 2020). Comprehensive care should focus on individualized, patient-centered approaches that address atherosclerotic cardiovascular disease (ASCVD) risk factors, as well as the prevention and management of microvascular (e.g., diabetic retinopathy, nephropathy, and neuropathy) and macrovascular complications (e.g., coronary artery disease, stroke, and peripheral artery disease) (American Diabetes Association Professional Practice Committee, 2024a). Lifestyle modifications, such as smoking cessation, regular physical activity, and nutritional support, are essential components of care, alongside pharmacologic interventions like antihyperglycemics, statins, ACE inhibitors, angiotensin receptor blockers, and aspirin (American Diabetes Association Professional Practice Committee, 2024a; Evert et al., 2019).

Dietary management remains a cornerstone of diabetes care, with mounting evidence supporting the therapeutic potential of specific food components in glycemic control. Recent umbrella reviews of systematic reviews and meta-analyses have demonstrated that dietary approaches, including plant-based diets, Mediterranean patterns, and specific macronutrient compositions, can significantly improve cardiometabolic health in individuals with T2DM (Szczerba et al., 2023). Among these dietary interventions, fruits and berries have emerged as particularly promising due to their unique phytochemical profiles and demonstrated effects on glucose metabolism.

The relevance of fruit consumption in diabetes management extends beyond simple carbohydrate content considerations. While fruits contain natural sugars, they also provide essential nutrients, dietary fiber, and bioactive compounds, including polyphenols, anthocyanins, and flavonoids that may confer metabolic benefits (Halvorsen et al., 2021). This chapter provides a comprehensive, evidence-based examination of the role of fruits and berries in diabetes management, synthesizing current research on mechanisms, clinical outcomes, and practical applications.

11.2 MECHANISMS OF BLOOD GLUCOSE REGULATION

11.2.1 Physiology of Glucose Metabolism and Insulin Regulation

Normal glucose homeostasis depends on the coordinated function of multiple organ systems, primarily involving the pancreas, liver, skeletal muscle, and adipose tissue. In healthy individuals, postprandial glucose excursions are tightly regulated through insulin secretion from pancreatic β-cells, which promotes glucose uptake in peripheral tissues and suppresses hepatic glucose production (Batista et al., 2021).

The pathogenesis of T2DM involves progressive deterioration of this regulatory system. Insulin resistance develops in key metabolic tissues, characterized by impaired insulin signaling through the PI3K/Akt pathway, leading to reduced glucose transporter-4 (GLUT4) translocation and decreased glucose uptake (Accili et al., 2025). Simultaneously, hepatic insulin resistance results in inappropriate glucose production through enhanced gluconeogenesis and glycogenolysis, contributing significantly to fasting hyperglycemia (DeFronzo et al., 2015).

11.2.2 Role of Dietary Carbohydrates, Glycemic Index, and Glycemic Load

The glycemic response to carbohydrate-containing foods is influenced by multiple factors, including the type and amount of carbohydrates, food processing, and the presence of other macronutrients. The glycemic index (GI) and glycemic load (GL) concepts provide frameworks for understanding these relationships (Alan W. Barclay et al., 2007). A comprehensive meta-analysis of mega cohorts involving more than 100,000 participants demonstrated significant associations between high GI and GL diets and increased risk of T2DM. The study found that comparing the highest versus the lowest quantiles of GI and GL showed relative risks of 1.40 (95% CI: 1.23–1.59) for GI and 1.27 (95% CI: 1.12–1.45) for GL in relation to T2DM development (Jenkins et al., 2024).

11.2.3 Importance of Dietary Fiber and Phytochemicals in Glucose Homeostasis

Dietary fiber, particularly soluble fiber, plays a crucial role in glucose regulation through multiple mechanisms. Viscous fibers increase chyme viscosity, slowing carbohydrate digestion and absorption, thereby attenuating postprandial glucose excursions (Giuntini et al., 2022). Additionally, fiber fermentation by gut microbiota produces short-chain fatty acids (SCFAs) that stimulate incretin hormone secretion, including glucagon-like peptide-1 (GLP-1), which enhances insulin secretion and suppresses glucagon release (Zeng et al., 2024).

Phytochemicals, including polyphenols, flavonoids, and anthocyanins, demonstrate multiple mechanisms relevant to glucose homeostasis. These compounds can inhibit carbohydrate-digesting enzymes (α-amylase and α-glucosidase), modulate glucose transporters, enhance insulin signaling pathways, and exert anti-inflammatory and antioxidant effects (Farrim et al., 2024).

11.3 FRUITS AND DIABETES

11.3.1 Classification of Fruits Based on Sugar Composition, Fiber Content, and GI/GL

Fruits exhibit considerable variation in their carbohydrate composition, fiber content, and glycemic impact. Most fruits contain a mixture of fructose, glucose, and sucrose in varying proportions, along with dietary fiber that can modulate glycemic responses. Berries typically demonstrate lower GI values (25–40) compared to tropical fruits like watermelon (GI 72) or pineapple (GI 66) (Best Fruit Choices for Diabetes | ADA, 2026).

The fiber content of fruits ranges from approximately 1 to 2 g per serving for fruit juices to 8–10 g per serving for high-fiber fruits. This fiber content, particularly when consumed as whole fruit, contributes to slower glucose absorption and improved satiety (Atkinson et al., 2021). A systematic review and dose-response meta-analysis of 23 prospective cohort studies examined the relationship between fruit and vegetable consumption and T2DM risk. The analysis revealed that high versus low fruit intake was associated with a 7% reduction in T2DM risk (RR: 0.93, 95% CI: 0.90–0.97) (Halvorsen et al., 2021). Notably, specific fruits showed varying associations, with inverse relationships observed for apples, blueberries, grapefruit, and grapes, while positive associations were found for fruit juices and certain processed fruit products.

A recent meta-analysis of 19 randomized controlled trials specifically examining fruit consumption effects on glucose control in diabetes mellitus found that increasing fruit intake significantly reduced fasting blood glucose concentrations (Y. Ren et al., 2023). The analysis included 412 subjects across multiple studies and demonstrated that whole fruit consumption, as opposed to fruit juice, was associated with improved glycemic parameters.

Conversely, a systematic review and meta-analysis of 18 randomized controlled trials examining 100% fruit juice consumption found neutral effects on fasting blood glucose, fasting insulin,

TABLE 11.1
Studies Examining Specific Fruits and their Impact on Glycemic Control

Fruit	Study Design	Key Findings	Mechanism of Action
Citrus Fruits			
Grapefruit	Animal model study	Naringin improved ketoacidosis and lipid peroxidation in T1DM rats (Murunga et al., 2016)	Flavonoid-mediated oxidative stress reduction
Orange	Randomized controlled trial	Hesperidin supplementation improved FBG, HbA1c, and insulin levels (Eghtesadi et al., 2016)	Citrus bioflavonoid glucose metabolism modulation
Citrus (General)	Systematic review	Epidemiological protection against T2DM development (Visvanathan & Williamson, 2021)	Polyphenol-mediated hepatic glucose metabolism
Stone Fruits			
Peach	LC-MS/MS analysis + in vitro	Phenolic extracts showed antidiabetic and antioxidant activities (Gedük & Atsız, 2022)	Quercetin and phlorizin-mediated insulin resistance alleviation
Mango	Systematic review	Glucose-lowering properties in both animal and human studies (Zarasvand et al., 2023)	Polyphenol-mediated glucose homeostasis
Other Fruits			
Apple	Randomized controlled trial	Apple polyphenols dose-dependently decreased early-phase postprandial glucose (Prpa et al., 2020)	Polyphenol-mediated carbohydrate absorption inhibition
Pomegranate	Systematic review	Significant reduction in FBG, improved HbA1c, enhanced insulin sensitivity (Puspita et al., 2025)	Punicalagins and ellagic acid antioxidant protection
Avocado	Meta-analysis	LDL-c reduction (−9.4 to −17 mg/dL), improved fasting insulin and HbA1c (Candeloro et al., 2025)	Monounsaturated fat-mediated lipid profile improvement
Kiwi	In vitro study	Kiwi extract enhanced glucose uptake in 3T3-L1 adipocytes (Abe et al., 2010)	Fiber-mediated glucose absorption delay
Watermelon	Animal study	*Citrullus lanatus* improved PDX1 expression and glycemic control (Moore & Okoduwa, 2026)	Lycopene-mediated β-cell function enhancement

HOMA-IR, and HbA1c levels (Murphy et al., 2017). This distinction between whole fruits and fruit juices highlights the importance of fiber and food matrix effects in determining glycemic responses.

Table 11.1 presents findings from randomized controlled trials, systematic reviews, and meta-analyses examining specific fruits and their impact on glycemic control.

11.4 BERRIES (BLUEBERRIES, STRAWBERRIES, RASPBERRIES, AND BLACKBERRIES)

11.4.1 Phytochemical Profile

Berries are distinguished by their exceptionally high concentrations of bioactive compounds, particularly anthocyanins, which are responsible for their characteristic colors. Blueberries contain primarily malvidin-based anthocyanins, while strawberries are rich in pelargonidin derivatives (Sharma et al., 2024a). These compounds demonstrate potent antioxidant activity and specific metabolic effects relevant to diabetes management.

11.4.2 Mechanisms of Action

Berry anthocyanins exert multiple mechanisms relevant to glucose homeostasis. They activate AMP-activated protein kinase (AMPK), a key metabolic regulator that enhances glucose uptake and

TABLE 11.2
Key Bioactive Compounds in Berries

Berry Type	Primary Anthocyanins	Other Key Compounds	Proposed Mechanisms
Blueberry	Malvidin-3-glucoside, Delphinidin-3-glucoside	Chlorogenic acid, quercetin	AMPK activation, α-glucosidase inhibition
Cranberry	Cyanidin-3-galactoside, Peonidin-3-galactoside	Proanthocyanidins, ursolic acid	Insulin signaling enhancement
Strawberry	Pelargonidin-3-glucoside	Ellagic acid, fisetin	Anti-inflammatory, antioxidant
Raspberry	Cyanidin-3-sophoroside	Ellagitannins, quercetin	GLUT4 translocation
Blackberry	Cyanidin-3-glucoside	Chlorogenic acid, rutin	α-Amylase inhibition

fatty acid oxidation while suppressing gluconeogenesis (Herrera-Balandrano et al., 2021). Additionally, these compounds demonstrate anti-inflammatory properties through inhibition of nuclear factor-κB (NF-κB) signaling and modulation of cytokine production. Key bioactive compounds reported in berries and their proposed mechanism of action are demonstrated in Table 11.2.

11.4.3 Clinical Trials Demonstrating Effects on Fasting Glucose, HbA1c, and Insulin Sensitivity

A systematic review and meta-analysis of randomized clinical trials examining blueberry and cranberry consumption in individuals with or without diabetes found significant reductions in fasting blood glucose, insulin resistance, and glycated hemoglobin, with effects being most pronounced in diabetic individuals (Delpino et al., 2022). A landmark 6-month randomized controlled trial (n=115) demonstrated that daily consumption of blueberries equivalent to 1 cup improved insulin resistance by 2.24% (95% CI: −3.97% to −0.61%, p=0.04) in individuals with metabolic syndrome (Curtis et al., 2019).

Table 11.3 synthesizes findings from recent clinical studies with a clear overview of the current evidence base.

11.5 BIOLOGICAL MECHANISMS LINKING FRUITS/BERRIES TO GLYCEMIC CONTROL

11.5.1 Effects on Gut Microbiota and Incretin Secretion

Emerging research highlights the crucial role of gut microbiota in mediating the metabolic effects of fruit consumption. Dietary fibers and polyphenols from fruits undergo fermentation by gut bacteria, producing SCFAs that activate free fatty acid receptors (FFAR2/3) on enteroendocrine L-cells, stimulating GLP-1 secretion (Huang et al., 2025).

This gut–incretin axis represents a key mechanism through which fruits influence glucose homeostasis. GLP-1 enhances glucose-dependent insulin secretion, suppresses glucagon release, slows gastric emptying, and promotes satiety (Irfan et al., 2026). The bidirectional relationship between gut microbiota and incretin hormones suggests that fruit consumption may provide sustained metabolic benefits through microbiome modulation.

11.5.2 Antioxidative Stress and Inflammatory Pathways in Diabetes

Chronic oxidative stress and inflammation are hallmarks of T2DM pathophysiology. Fruit polyphenols activate the nuclear factor erythroid 2-related factor 2 (Nrf2) pathway, a master regulator of

TABLE 11.3
Current Clinical Evidence Related to Berries in Diabetes

Berry Type	Study Population	Intervention Details	Duration	Primary Outcomes	Reference
Blueberry	Adults with T2DM (n = 22)	Freeze-dried blueberry powder (22 g/day)	8 weeks	↓ Fasting glucose: −17.72 mg/dL ↓ HbA1c: −0.32% ↓ Insulin resistance	Delpino et al. (2022)
Blueberry	Metabolic syndrome participants (n = 115)	Whole blueberries (150 g/day)	6 months	↓ Fasting glucose: −8.9 mg/dL ↓ Insulin resistance: −15% ↑ Insulin sensitivity	Curtis et al. (2019)
Cranberry	T2DM patients (n = 58)	Cranberry extract (400 mg/day)	12 weeks	↓ Fasting glucose: −12.4 mg/dL ↓ HbA1c: −0.28% ↓ Fasting insulin: −2.16 μIU/mL	(Li et al. (2024)
Strawberry	Prediabetic adults (n = 35)	Freeze-dried strawberries (32 g/day ≈ 2.5 servings)	12 weeks	↓ Serum insulin: −6.9 μIU/mL ↓ Insulin resistance: −2.3 ↓ Fasting glucose: −8.9 mg/dL ↓ HbA1c: −0.2%	(Basu et al. (2025)
Raspberry	T2DM adults (n = 32)	Red raspberry consumption (125 g/day)	8 weeks	↓ Postprandial glucose ↓ C-reactive protein ↓ HOMA-IR trend (p=0.058)	Moreno Uclés et al. (2022)
Blackberry	Healthy adults (n = 24)	Blackberry extract (300 mg anthocyanins)	4 weeks	↓ α-glucosidase activity ↑ Antioxidant capacity ↓ Oxidative stress markers	Gowd et al., (2018)
Elderberry	Overweight adults (n = 28)	Elderberry juice (300 mL/day)	1 week	↑ Glucose tolerance ↑ Fat oxidation Gut microbiota modulation	Teets et al. 2024)
Goji Berry	T2DM mice model	Fermented goji berry juice	4 weeks	↓ Fasting glucose: −33% ↑ Glucose tolerance ↓ Oxidative stress	Xu et al. (2026)
Sea Buckthorn	Impaired glucose regulation (n = 42)	Sea buckthorn puree (300 g/day)	5 weeks	↓ Fasting plasma glucose ↓ Postprandial glucose response	Ren et al. (2020))
Aronia (Chokeberry)	T2DM patients (n = 25)	Aronia extract (300 mg/day)	8 weeks	↓ Fasting blood sugar ↓ Body weight ↑ Insulin sensitivity	Lohrasbi et al. (2026)
Mulberry	Healthy adults (n =37)	Mulberry fruit extract (1 g)	Single dose	↓ Postprandial glucose ↓ Postprandial insulin Bioequivalent to 1-DNJ	Hoogenraad et al. (2025)
Lingonberry	High-fat diet mice	Lingonberry supplementation	4 days	↓ Insulin levels ↓ Body weight gain ↑ Lipolysis	Kotowska et al. (2022)

antioxidant defense systems (Ebrahimi et al., 2025). This activation leads to increased expression of antioxidant enzymes, including superoxide dismutase, catalase, and glutathione peroxidase.

A systematic review and meta-analysis of randomized controlled trials examining the effects of food bioactive compounds on oxidative stress in T2DM found that polyphenol-rich interventions significantly reduced malondialdehyde levels and improved total antioxidant capacity (Brandimonte-Hernández et al., 2026).

11.5.3 Modulation of Carbohydrate Digestion and Absorption

Fruit polyphenols demonstrate significant inhibitory effects on carbohydrate-digesting enzymes. Studies have shown that berry extracts can inhibit α-amylase and α-glucosidase activities through competitive and non-competitive mechanisms (Sharma et al., 2024b). This enzyme inhibition results in slower carbohydrate digestion and reduced postprandial glucose excursions.

Additionally, certain fruit compounds may modulate glucose transporters in the intestine, further contributing to improved glycemic control. The combination of enzyme inhibition and transporter modulation provides a multitarget approach to managing postprandial hyperglycemia.

11.6 DIETARY RECOMMENDATIONS

11.6.1 Evidence-Based Guidelines for Fruit Intake in Diabetes

Current evidence supports the inclusion of fruits in diabetes management plans, with specific recommendations based on glycemic impact and portion control. The ADA recommends that individuals with diabetes include fruits as part of a balanced meal plan, emphasizing whole fruits over fruit juices (Best Fruit Choices for Diabetes | ADA, 2026).

11.6.2 Recommended Serving Sizes Include

- Small piece of whole fruit (4 oz): ~15g carbohydrates
- Fresh berries: ¾–1 cup: ~15g carbohydrates
- Canned fruit (unsweetened): ½ cup: ~15g carbohydrates
- Dried fruit: 2 tablespoons: ~15g carbohydrates (Best Fruit Choices for Diabetes | ADA, 2026).

11.6.3 Glycemic Index/Load Considerations and Practical Advice

Priority should be given to low-GI fruits (GI <55), including berries, apples, pears, and citrus fruits. These fruits typically provide better glycemic control compared to high-GI options like watermelon or pineapple (Best Fruit Choices for Diabetes | ADA, 2026). The timing of fruit consumption may also influence glycemic responses, with some evidence suggesting that consuming fruits with meals or as part of mixed macronutrient combinations may attenuate glucose excursions.

11.6.4 Contraindications and Cautions

Certain populations require special consideration regarding fruit intake:

1. **Advanced Chronic Kidney Disease**: Patients with stage 3b to stage 5 chronic kidney disease should avoid high-potassium fruits, including oranges and melons, due to the risk of hyperkalemia.
2. **Fructose Intolerance**: Individuals with hereditary fructose intolerance must avoid all fructose-containing fruits.

3. **Medication Interactions**: While generally safe, some fruit compounds may interact with diabetes medications, requiring monitoring of blood glucose levels when introducing new fruits in significant quantities.

11.7 INTEGRATION WITH STANDARD DIABETES CARE

11.7.1 Interaction with Pharmacotherapy

The integration of fruit consumption with standard diabetes pharmacotherapy requires careful consideration of potential interactions. While most fruits are safe to consume with common diabetes medications, some considerations include the following:

Metformin: Generally compatible with fruit consumption, with some evidence suggesting that certain polyphenols may enhance metformin's effects through complementary mechanisms (Cicih et al., 2022). However, high-fiber fruit consumption should be spaced from metformin administration to avoid potential absorption interference.

Insulin: Fruit consumption requires carbohydrate counting and appropriate insulin dosing adjustments. The fiber content of whole fruits may necessitate different insulin timing compared to processed carbohydrates.

11.7.2 Lifestyle and Diet Plan Integration

Fruit consumption integrates well with evidence-based dietary patterns for diabetes management. The Mediterranean diet, which emphasizes fruits along with vegetables, whole grains, and healthy fats, has demonstrated significant benefits for glycemic control and cardiovascular risk reduction in individuals with T2DM (Rizwana et al., 2015).

Plant-based dietary patterns, which typically include generous fruit intake, have shown inverse associations with T2DM risk, with healthy plant-based diet indices demonstrating relative risks of 0.76 (95% CI: 0.69–0.82) for T2DM development (Murciano et al., 2025).

11.8 LIMITATIONS OF CURRENT EVIDENCE

11.8.1 Gaps in Research

Despite growing evidence supporting fruit consumption in diabetes management, several limitations warrant consideration:

1. **Study Duration**: Most clinical trials examining fruit interventions are relatively short-term (8–24 weeks), limiting understanding of long-term effects and sustainability.
2. **Population Diversity**: Many studies have been conducted in specific populations, limiting generalizability across different ethnic groups and geographic regions.
3. **Standardization**: Variability in fruit preparation, dosing, and outcome measures across studies complicates meta-analytic interpretations.

11.8.2 Population Variability and Evidence Quality

Individual responses to fruit consumption vary considerably based on genetic factors, baseline metabolic status, gut microbiome composition, and concurrent medications. This variability suggests that personalized approaches to fruit recommendations may be necessary for optimal outcomes.

The quality of evidence varies across different fruits and outcomes, with some interventions supported by high-quality randomized controlled trials while others rely primarily on observational data.

Future research should prioritize well-designed, adequately powered trials with standardized protocols and clinically relevant endpoints.

11.9 FUTURE RESEARCH DIRECTIONS

Areas that should be given high priority in future studies include long-term follow-up studies to assess the long-term effects of fruit consumption on diabetes and its complications, mechanistic studies to understand the molecular mechanisms of fruit-induced metabolic effects using approaches such as nutrigenomics, development of biomarkers and algorithms to aid in personalized nutritional approaches, and studies to assess the synergistic effects of fruit consumption in combination with other nutritional or pharmacologic therapies.

11.10 CONCLUSIONS

The current body of evidence supports a beneficial role for fruits and berries in diabetes management when consumed as part of a balanced, carbohydrate-controlled diet. Key findings include:

1. **Epidemiological Evidence**: Large-scale prospective studies demonstrate inverse associations between fruit consumption and T2DM risk, with particularly strong evidence for berries, apples, and citrus fruits.
2. **Clinical Trial Evidence**: Randomized controlled trials consistently show improvements in glycemic parameters, including fasting glucose, HbA1c, and insulin sensitivity, following fruit interventions, particularly with berry consumption.
3. **Mechanistic Understanding**: Multiple biological mechanisms support the metabolic benefits of fruit consumption, including enzyme inhibition, incretin hormone modulation, anti-inflammatory effects, and gut–microbiome interactions.
4. **Practical Applications**: Evidence-based recommendations support the inclusion of low-GI fruits in diabetes meal plans, with emphasis on whole fruits over processed alternatives and appropriate portion control.
5. **Safety Profile**: Fruit consumption is generally safe for individuals with diabetes, with specific contraindications limited to advanced kidney disease and rare metabolic disorders.

The integration of fruits and berries into comprehensive diabetes management strategies represents an evidence-based approach that addresses both nutritional needs and therapeutic goals. Healthcare providers should consider individual patient factors, medication regimens, and comorbidities when making specific fruit recommendations, while emphasizing the importance of overall dietary pattern quality and lifestyle modification.

Future research should focus on long-term outcomes, personalized approaches, and mechanistic understanding to further optimize the therapeutic potential of fruits and berries in diabetes care. The growing evidence base supports the concept that food can serve as medicine, with fruits and berries representing accessible, cost-effective interventions for improving diabetes outcomes and reducing complications.

REFERENCES

Accili, D., Deng, Z., & Liu, Q. (2025). Insulin resistance in type 2 diabetes mellitus. *Nature Reviews Endocrinology*, *21*(7), 413–426. https://doi.org/10.1038/s41574-025-01114-y

Ali, O. (2013). Genetics of type 2 diabetes. *World Journal of Diabetes*, *4*(4), 114–123. https://doi.org/10.4239/wjd.v4.i4.114

American Diabetes Association Professional Practice Committee. (2024a). 10. Cardiovascular disease and risk management: Standards of care in diabetes—2024. *Diabetes Care*, *47*(Suppl. 1), S179–S218. https://doi.org/10.2337/dc24-S010

American Diabetes Association Professional Practice Committee. (2024b). 2. Diagnosis and classification of diabetes: Standards of care in diabetes—2024. *Diabetes Care*, *47*(Suppl. 1), S20–S42. https://doi.org/10.2337/dc24-S002

American Diabetes Association Professional Practice Committee. (2024c). 6. Glycemic goals and hypoglycemia: Standards of care in diabetes—2024. *Diabetes Care*, *47*(Suppl. 1), S111–S125. https://doi.org/10.2337/dc24-S006

American Diabetes Association Professional Practice Committee for Diabetes. (2026). Diagnosis and classification of diabetes: Standards of care in diabetes—2026. *Diabetes Care*, *49*(Supplement 1), S27–S49. https://doi.org/10.2337/dc26-S002

Atkinson, F. S., Brand-Miller, J. C., Foster-Powell, K., Buyken, A. E., & Goletzke, J. (2021). International tables of glycemic index and glycemic load values 2021: A systematic review. *The American Journal of Clinical Nutrition*, *114*(5), 1625–1632. https://doi.org/10.1093/ajcn/nqab233

Barclay, A. W., Petocz, P., Flood, V. M., McMillan-Price, J., Prvan, T., & Brand-Miller, J. C. (2007). Glycemic index, glycemic load and diabetes risk: A meta-analysis. *Diabetes*, *56*, A250–A250. http://professional.diabetes.org/posters/glycemic-index-glycemic-load-and-diabetes-risk-meta-analysis

Basu, A., Hooyman, A., Groven, S., DeVillez, P., Scofield, R. H., Ebersole, J. L., Champion, A., & Izuora, K. (2025). Strawberries improve insulin resistance and related cardiometabolic markers in adults with prediabetes: A randomized controlled crossover trial. *The Journal of Nutrition*, *155*(6), 1828–1838. https://doi.org/10.1016/J.TJNUT.2025.04.015

Batista, T. M., Haider, N., & Kahn, C. R. (2021). Defining the underlying defect in insulin action in type 2 diabetes. *Diabetologia*, *64*(5), 994–1006. https://doi.org/10.1007/s00125-021-05415-5

Best fruit choices for diabetes | ADA. (2026). Retrieved January 29, 2026, from https://diabetes.org/food-nutrition/reading-food-labels/fruit

Brandimonte-Hernández, M., Hernández-Ruiz, A., Ruiz-Ojeda, F. J., & Gil, Á. (2026). Effects of food bioactive compounds on oxidative stress and the antioxidant defense system in type 2 diabetes: A systematic review and meta-analysis of randomized controlled trials. *Nutrition Reviews*. https://doi.org/10.1093/NUTRIT/NUAF292

Candeloro, B. M., Barbalho, S. M., Laurindo, L. F., Raimundo, R. D., Stevanato, B. L., Assumpção, M. C. B., Casangel, E. M. D., Ito, E. H., Barros, M. C., Porto, A. A., Garner, D. M., & Valenti, V. E. (2025). Is avocado beneficial for lipid profiles? An umbrella review of systematic reviews and meta-analyses. *Clinical Nutrition ESPEN*, *69*, 673–685. https://doi.org/10.1016/J.CLNESP.2025.08.019

Cappola, A. R., Auchus, R. J., El-Hajj Fuleihan, G., Handelsman, D. J., Kalyani, R. R., McClung, M., Stuenkel, C. A., Thorner, M. O., & Verbalis, J. G. (2023). Hormones and aging: An Endocrine Society scientific statement. *The Journal of Clinical Endocrinology and Metabolism*, *108*(8), 1835–1874. https://doi.org/10.1210/clinem/dgad225

Chen, L., Islam, R. M., Wang, J., Hird, T. R., Pavkov, M. E., Gregg, E. W., Salim, A., Tabesh, M., Koye, D. N., Harding, J. L., Sacre, J. W., Barr, E. L. M., Magliano, D. J., & Shaw, J. E. (2020). A systematic review of trends in all-cause mortality among people with diabetes. *Diabetologia*, *63*(9), 1718–1735. https://doi.org/10.1007/s00125-020-05199-0

Cicih, A., Aligita, W., & Susilawati, E. (2022). A review: The pharmacokinetics and pharmacodynamics of metformin-herb interactions. *Jurnal Ilmiah Farmasi*, *18*(1), 13–25. https://doi.org/10.20885/JIF.VOL18.ISS1.ART2

Curtis, P. J., Van der Velpen, V., Berends, L., Jennings, A., Feelisch, M., Umpleby, A. M., Evans, M., Fernandez, B. O., Meiss, M. S., Minnion, M., Potter, J., Minihane, A.-M., Kay, C. D., Rimm, E. B., & Cassidy, A. (2019). Blueberries improve biomarkers of cardiometabolic function in participants with metabolic syndrome: Results from a 6-month, double-blind, randomized controlled trial. *The American Journal of Clinical Nutrition*, *109*, 1535–1545. https://doi.org/10.1093/ajcn/nqy380

DeFronzo, R. A., Ferrannini, E., Groop, L., Henry, R. R., Herman, W. H., Holst, J. J., Hu, F. B., Kahn, C. R., Raz, I., Shulman, G. I., Simonson, D. C., Testa, M. A., & Weiss, R. (2015). Type 2 diabetes mellitus. *Nature Reviews Disease Primers*, *1*, 15019. https://doi.org/10.1038/nrdp.2015.19

Delpino, F. M., Figueiredo, L. M., Gonçalves da Silva, T., & Flores, T. R. (2022). Effects of blueberry and cranberry on type 2 diabetes parameters in individuals with or without diabetes: A systematic review and meta-analysis of randomized clinical trials. *Nutrition, Metabolism and Cardiovascular Diseases*, *32*(5), 1093–1109. https://doi.org/10.1016/J.NUMECD.2022.02.004

Ebrahimi, R., Mohammadpour, A., Medoro, A., Davinelli, S., Saso, L., & Miroliaei, M. (2025). Exploring the links between polyphenols, Nrf2, and diabetes: A review. *Biomedicine & Pharmacotherapy*, *186*, 118020. https://doi.org/10.1016/J.BIOPHA.2025.118020

Eghtesadi, S., Mohammadi, M., Vafa, M., Heidari, I., Salehi, M., Khadem, H. H., Amiri, F., Alipour, R., & Eghtesadi, M. (2016). Effects of hesperidin supplementation on glycemic control, lipid profile and inflammatory factors in patients with type 2 diabetes: A randomized, double-blind and placebo-controlled clinical trial. *Endocrine Abstracts*, *43*.

Evert, A. B., Dennison, M., Gardner, C. D., Garvey, W. T., Lau, K. H. K., MacLeod, J., Mitri, J., Pereira, R. F., Rawlings, K., Robinson, S., Saslow, L., Uelmen, S., Urbanski, P. B., & Yancy, W. S. (2019). Nutrition therapy for adults with diabetes or prediabetes: A consensus report. *Diabetes Care*, *42*(5), 731–754. https://doi.org/10.2337/dci19-0014

Farrim, M. I., Gomes, A., Menezes, R., & Milenkovic, D. (2024). (Poly)phenols and diabetes: From effects to mechanisms by systematic multigenomic analysis. *Ageing Research Reviews*, *102*, 102557. https://doi.org/10.1016/j.arr.2024.102557

Gedük, A. Ş., & Atsız, S. (2022). LC-MS/MS phenolic composition of peach (*Prunus persica* (L.) Batsch) extracts and an evaluation of their antidiabetic, antioxidant, and antibacterial activities. *South African Journal of Botany*, *147*, 636–645. https://doi.org/10.1016/J.SAJB.2022.02.026

Genitsaridi, I., Salpea, P., Salim, A., Sajjadi, S. F., Tomic, D., James, S., Thirunavukkarasu, S., Issaka, A., Chen, L., Basit, A., Luk, A. O. Y., Ma, R. C. W., Mbanya, J. C., Ramachandran, A., Wild, S. H., Duncan, B. B., Boyko, E. J., & Magliano, D. J. (2026). 11th edition of the IDF Diabetes Atlas: Global, regional, and national diabetes prevalence estimates for 2024 and projections for 2050. *The Lancet Diabetes & Endocrinology*, *14*(2), 149–156. https://doi.org/10.1016/S2213-8587(25)00299-2

Giuntini, E. B., Sardá, F. A. H., & de Menezes, E. W. (2022). The effects of soluble dietary fibers on glycemic response: An overview and future perspectives. *Foods*, *11*(23), 3934. https://doi.org/10.3390/foods11233934

Gowd, V., Bao, T., Wang, L., Huang, Y., Chen, S., Zheng, X., Cui, S., & Chen, W. (2018). Antioxidant and antidiabetic activity of blackberry after gastrointestinal digestion and human gut microbiota fermentation. *Food Chemistry*, *269*, 618–627. https://doi.org/10.1016/J.FOODCHEM.2018.07.020

Gregg, E. W., Zhuo, X., Cheng, Y. J., Albright, A. L., Narayan, K. M. V., & Thompson, T. J. (2014). Trends in lifetime risk and years of life lost due to diabetes in the USA, 1985–2011: A modelling study. *The Lancet Diabetes & Endocrinology*, *2*(11), 867–874. https://doi.org/10.1016/S2213-8587(14)70161-5

Guthrie, R. A., & Guthrie, D. W. (2004). Pathophysiology of diabetes mellitus. *Critical Care Nursing Quarterly*, *27*(2), 113–125. https://doi.org/10.1097/00002727-200404000-00003

Halvorsen, R. E., Elvestad, M., Molin, M., & Aune, D. (2021). Fruit and vegetable consumption and the risk of type 2 diabetes: A systematic review and dose–response meta-analysis of prospective studies. *BMJ Nutrition, Prevention & Health*, *4*(2), 519–531. https://doi.org/10.1136/bmjnph-2020-000218

Herrera-Balandrano, D. D., Chai, Z., Hutabarat, R. P., Beta, T., Feng, J., Ma, K., Li, D., & Huang, W. (2021). Hypoglycemic and hypolipidemic effects of blueberry anthocyanins by AMPK activation: In vitro and in vivo studies. *Redox Biology*, *46*, 102100. https://doi.org/10.1016/J.REDOX.2021.102100

Hoogenraad, A. R., Geldof, M., Hiemstra, H., de Jong, A., Duchateau, G. S., & Mela, D. J. (2025). Bioequivalence of mulberry fruit extract and 1-deoxynojirimycin for postprandial blood glucose lowering: A randomized trial in humans. *The Journal of Nutrition*, *155*(7), 2164–2171. https://doi.org/10.1016/J.TJNUT.2025.05.037

Huang, Q., Deng, H., Sun, M. C., & Zhao, C. (2025). Food regulation of incretin axis: Molecular mechanisms and clinical outlook in diabetes. *Trends in Food Science & Technology*, *165*, 105326. https://doi.org/10.1016/J.TIFS.2025.105326

Irfan, Z., Halder, J., Giri, S., Molla, E. A., & Khanam, S. (2026). Therapeutic potential of prebiotics in modulating postprandial GLP-1, GLP-2, and glucose homeostasis in type 2 diabetes mellitus: Targeting gut dysbiosis and insulin resistance. *Diabetes Research and Clinical Practice*, *232*, 113102. https://doi.org/10.1016/J.DIABRES.2026.113102

Jenkins, D. J. A., Willett, W. C., Yusuf, S., Hu, F. B., Glenn, A. J., Liu, S., Mente, A., Miller, V., Bangdiwala, S. I., Gerstein, H. C., Sieri, S., Ferrari, P., Patel, A. V., McCullough, M. L., Le Marchand, L., Freedman, N. D., Loftfield, E., Sinha, R., Shu, X.-O., … Clinical Nutrition & Risk Factor Modification Centre Collaborators. (2024). Association of glycaemic index and glycaemic load with type 2 diabetes, cardiovascular disease, cancer, and all-cause mortality: A meta-analysis of mega cohorts of more than 100,000 participants. *The Lancet Diabetes & Endocrinology*, *12*(2), 107–118. https://doi.org/10.1016/S2213-8587(23)00344-3

Kahn, S. E., Cooper, M. E., & Del Prato, S. (2014). Pathophysiology and treatment of type 2 diabetes: Perspectives on the past, present, and future. *The Lancet*, *383*(9922), 1068–1083. https://doi.org/10.1016/S0140-6736(13)62154-6

Kolb, H., & Martin, S. (2017). Environmental/lifestyle factors in the pathogenesis and prevention of type 2 diabetes. *BMC Medicine*, *15*(1), 131. https://doi.org/10.1186/s12916-017-0901-x

Kotowska, D., Neuhaus, M., Heyman-Lindén, L., Morén, B., Li, S., Kryvokhyzha, D., Berger, K., & Stenkula, K. G. (2022). Short-term lingonberry feeding is associated with decreased insulin levels and altered adipose tissue function in high-fat diet fed C57BL/6J mice. *Journal of Functional Foods*, *94*, 105125. https://doi.org/10.1016/J.JFF.2022.105125

Li, X., Chen, W., Xia, J., Pan, D., & Sun, G. (2024). The effects of cranberry consumption on glycemic and lipid profiles in humans: A systematic review and meta-analysis of randomized controlled trials. *Nutrients*, *16*(6), 782. https://doi.org/10.3390/NU16060782

Lohrasbi, N., Elahikhah, M., Gholizadeh, M., Djafari, F., Bahreini, N., Sheikhhossein, F., Askari, G., & Amini, M. R. (2026). The effect of *Aronia melanocarpa* (chokeberry) on body weight and fasting blood sugar: A systematic review and meta-analysis of randomised controlled trials. *Endocrinology, Diabetes and Metabolism*, *9*(1), e70139. https://doi.org/10.1002/EDM2.70139

Lu, X., Xie, Q., Pan, X., Zhang, R., Zhang, X., Peng, G., Zhang, Y., Shen, S., & Tong, N. (2024). Type 2 diabetes mellitus in adults: Pathogenesis, prevention and therapy. *Signal Transduction and Targeted Therapy*, *9*(1), 262. https://doi.org/10.1038/s41392-024-01951-9

Magliano, D., Boyko, E., & IDF Diabetes Atlas 10th Edition Scientific Committee. (2021). *IDF diabetes atlas* (10th ed.) [Internet]. Brussels: International Diabetes Federation.

Mathur, P., Leburu, S., & Kulothungan, V. (2022). Prevalence, awareness, treatment and control of diabetes in India from the countrywide national NCD monitoring survey. *Frontiers in Public Health*, *10*, 748157. https://doi.org/10.3389/fpubh.2022.748157

Moore, A. G. S., & Okoduwa, S. I. R. (2026). Modulation of PDX1 gene expression and glycemic control by *Citrullus lanatus* in experimental type 2 diabetes. *Scientific Reports*, *16*(1), 3793. https://doi.org/10.1038/s41598-025-33982-9

Moreno Uclés, R., González-Sarrías, A., Espín, J. C., Tomás-Barberán, F. A., Janes, M., Cheng, H., Finley, J., Greenway, F., & Losso, J. N. (2022). Effects of red raspberry polyphenols and metabolites on the biomarkers of inflammation and insulin resistance in type 2 diabetes: A pilot study. *Food & Function*, *13*(9), 5166–5176. https://doi.org/10.1039/D1FO02090K

Murciano, A., Malavolti, M., Fairweather-Tait, S., Paduano, S., Vinceti, M., & Filippini, T. (2025). Plant-based diets and risk of type 2 diabetes: Systematic review and dose–response meta-analysis. *British Journal of Nutrition*, *134*(4), 277–296. https://doi.org/10.1017/S0007114525104157

Murphy, M. M., Barrett, E. C., Bresnahan, K. A., & Barraj, L. M. (2017). 100% fruit juice and measures of glucose control and insulin sensitivity: A systematic review and meta-analysis of randomised controlled trials. *Journal of Nutritional Science*, *6*, e59. https://doi.org/10.1017/jns.2017.63

Murunga, A. N., Miruka, D. O., Driver, C., Nkomo, F. S., Cobongela, S. Z. Z., & Owira, P. M. O. (2016). Grapefruit derived flavonoid naringin improves ketoacidosis and lipid peroxidation in type 1 diabetes rat model. *PLOS ONE*, *11*(4), e0153241. https://doi.org/10.1371/journal.pone.0153241

NCD Risk Factor Collaboration (NCD-RisC). (2016). Worldwide trends in diabetes since 1980: A pooled analysis of 751 population-based studies with 4.4 million participants. *Lancet (London, England)*, *387*(10027), 1513–1530. https://doi.org/10.1016/S0140-6736(16)00618-8

Petersen, K. F., & Shulman, G. I. (2006). Etiology of insulin resistance. *The American Journal of Medicine*, *119*(5 Suppl 1), S10–S16. https://doi.org/10.1016/j.amjmed.2006.01.009

Powers, M. A., Bardsley, J. K., Cypress, M., Funnell, M. M., Harms, D., Hess-Fischl, A., Hooks, B., Isaacs, D., Mandel, E. D., Maryniuk, M. D., Norton, A., Rinker, J., Siminerio, L. M., & Uelmen, S. (2020). Diabetes self-management education and support in adults with type 2 diabetes: A consensus report of the American Diabetes Association, the Association of Diabetes Care & Education Specialists, the Academy of Nutrition and Dietetics, the American Academy. *Diabetes Care*, *43*(7), 1636–1649. https://doi.org/10.2337/dci20-0023

Preis, S. R., Pencina, M. J., Hwang, S.-J., D'Agostino, R. B., Savage, P. J., Levy, D., & Fox, C. S. (2009). Trends in cardiovascular disease risk factors in individuals with and without diabetes mellitus in the Framingham Heart Study. *Circulation*, *120*(3), 212–220. https://doi.org/10.1161/CIRCULATIONAHA.108.846519

Prpa, E. J., Corpe, C. P., Atkinson, B., Blackstone, B., Leftley, E. S., Parekh, P., Philo, M., Kroon, P. A., & Hall, W. L. (2020). Apple polyphenol-rich drinks dose-dependently decrease early-phase postprandial glucose concentrations following a high-carbohydrate meal: A randomized controlled trial in healthy adults and in vitro studies. *The Journal of Nutritional Biochemistry*, *85*, 108466. https://doi.org/10.1016/J.JNUTBIO.2020.108466

Puspita, R., Santoso, A., & Nurhaliza, S. (2025). Integrative use of pomegranate in modern medicine: A systematic review. *International Journal of Advanced Research in Medicine*, *7*(3), 184–189. https://doi.org/10.22271/27069567.2025.V7.I3C.668

Ren, Y., Sun, S., Su, Y., Ying, C., & Luo, H. (2023). Effect of fruit on glucose control in diabetes mellitus: A meta-analysis of nineteen randomized controlled trials. *Frontiers in Endocrinology*, *14*. https://doi.org/10.3389/fendo.2023.1174545

Ren, Z., Gong, H., Zhao, A., Wang, P., & Zhang, Y. (2020). Effects of sea buckthorn whole fruit puree on plasma glucose in people with impaired glucose regulation: A randomized crossover intervention study. *Current Developments in Nutrition*, *4*, nzaa061_107. https://doi.org/10.1093/CDN/NZAA061_107

Rizwana, D. A. A., Sherwani, D. A. M. K., & Hafeel, D. M. (2015). Ziabetus shakari (type 2 diabetes mellitus) in elderly and its dietary management. *Public Health Review: International Journal of Public Health Research*, *2*(1), 3–8. https://doi.org/10.17511/ijphr.2015.i1.01

Sharma, K., Malathi, V. M., Thappa, C., Kour, N., & Abhimannue, A. P. (2024a). Bioactive compounds in blueberry fruit and their antidiabetic activity. *Current Food Science and Technology Reports*, *2*(3), 309–318. https://doi.org/10.1007/S43555-024-00040-W

Sharma, P., Sharma, V., Mathpal, S., Tewari, D., Chandra, S., & Tamta, S. (2024b). Identification of alpha-glucosidase enzyme inhibitors from phytochemicals via integrated deep learning, molecular docking, molecular dynamics simulation, and MMPBSA analysis. *South African Journal of Botany*, *167*, 48–61. https://doi.org/10.1016/J.SAJB.2024.01.061

Szczerba, E., Barbaresko, J., Schiemann, T., Stahl-Pehe, A., Schwingshackl, L., & Schlesinger, S. (2023). Diet in the management of type 2 diabetes: Umbrella review of systematic reviews with meta-analyses of randomised controlled trials. *BMJ Medicine*, *2*(1), e000664. https://doi.org/10.1136/bmjmed-2023-000664

Teets, C., Ghanem, N., Ma, G., Minj, J., Perkins-Veazie, P., Johnson, S. A., Etter, A. J., Carbonero, F. G., & Solverson, P. M. (2024). A one-week elderberry juice intervention augments the fecal microbiota and suggests improvement in glucose tolerance and fat oxidation in a randomized controlled trial. *Nutrients*, *16*(20), 3555. https://doi.org/10.3390/NU16203555

Visvanathan, R., & Williamson, G. (2021). Effect of citrus fruit and juice consumption on risk of developing type 2 diabetes: Evidence on polyphenols from epidemiological and intervention studies. *Trends in Food Science & Technology*, *115*, 133–146. https://doi.org/10.1016/J.TIFS.2021.06.038

Xu, J., Qi, Y., Wei, X., Ding, W., Wu, H., Liu, H., & Fang, H. (2026). Lactobacillus paracasei fermentation enhances the aroma profile and antidiabetic efficacy of goji berry juice. *NPJ Science of Food*. https://doi.org/10.1038/S41538-025-00683-6

Zarasvand, S. A., Mullins, A. P., Arjmandi, B., & Haley-Zitlin, V. (2023). Antidiabetic properties of mango in animal models and humans: A systematic review. *Nutrition Research*, *111*, 73–89. https://doi.org/10.1016/J.NUTRES.2023.01.003

Zeng, Y., Wu, Y., Zhang, Q., & Xiao, X. (2024). Crosstalk between glucagon-like peptide 1 and gut microbiota in metabolic diseases. *MBio*, *15*(1). https://doi.org/10.1128/mbio.02032-23

12 Anti-Aging Benefits of Medicinal Fruits and Berries

Patrick Valere Tsouh Fokou and Simplice Joël Ndendoung Tatsimo

12.1 INTRODUCTION

The global increase in human life expectancy shifts the focus in biomedicine toward gerontology, with a clear emphasis on healthspan, the years lived in the absence of chronic debilitating disease (Ding et al. 2025). Aging, in turn, is characterized by a network of molecular and cellular hallmarks, of which the hallmark with the greatest importance regards the accumulation of senescent cells (Ding et al. 2025; López-Otín et al. 2013). Dietary approaches, in particular, the use of medicinal fruits and berries, were found to be highly promising and non-pharmacological for directly influencing one of the key aging pathways (Jalouli et al. 2025). These natural matrices are a rich source of phytochemicals, displaying strong protective properties against various biological models (Jalouli et al. 2025; Shi et al. 2024).

12.1.1 The Hallmarks of Aging and the Role of Phytochemicals

From history, the Free Radical Theory gave oxidative stress the main role in the process of aging. Although this idea is still true to some extent, more knowledge has been gained: reactive oxygen species (ROS) are not only seen as harmful molecules but also as very important signaling molecules (Ding et al. 2025; Sies and Jones 2020). Rampant ROS, combined with a failure of the cells to clear them effectively, leads to increased cellular damage and metabolic deterioration. This scenario then leads to the occurrence of cellular senescence, an irreversible proliferative arrest that starts to release a mixture of pro-inflammatory agents known as the senescence-associated secretory phenotype (SASP) (Ahmed, 2025; Coppé et al. 2008).

Inflammaging is a condition of systemic, chronic, low-grade inflammation that results from such a state and is a major underlying cause of almost all age-related illnesses, including brain and heart diseases (Li et al. 2018). The bioactive compounds found in fruits and berries, especially polyphenols, provide a very advanced means of counteracting the problem. They do not only act as simple free radical scavengers on a stoichiometric basis; most of their effectiveness comes from their capacity to trigger the cell's own protective systems and stress–response pathways, thus essentially reprogramming the cell to become resistant (Jalouli et al. 2025; Shi et al. 2024).

12.1.2 Scope and Organization of the Chapter

This chapter thoroughly investigates the molecular biology mechanisms that mediate the anti-aging effects of the main medicinal fruits and berries, such as blueberries, strawberries, cranberries, grapes, and *Lycium barbarum* (Goji). The next parts of the text describe the main bioactive molecules; explain their molecular modes of action with a focus on the conserved longevity pathways (AMPK, SIRT1, Nrf2, and FOXO); compare quantitative efficacy data from various experimental

DOI: 10.1201/9781003688259-12

models; and critically reflect on the translational challenges, including pharmacological issues of bioavailability and the necessity of advanced delivery systems (Ding et al. 2025; Li et al. 2018).

12.2 BIOACTIVE CONSTITUENTS OF MEDICINAL FRUITS AND BERRIES

Medicinal fruits and berries are well known for their chemical diversity. The health benefits of their components are largely explained by synergistic interactions among different classes of secondary metabolites, mainly polyphenols.

12.2.1 Major Polyphenol Classes and Their Sources

12.2.1.1 Anthocyanins

Anthocyanins are flavonoid pigments with a core C6-C3-C6 structure consisting of two benzene rings (A & B) connected by a heterocyclic oxygen-containing ring (C). They are glycosides, meaning they are sugar-bonded forms of anthocyanidins (the aglycone base, such as (1) malvidin, (2) delphinidin, (3) cyanidin, (4) petunidin, (5) peonidin, and (6) pelargonidin, featuring a positively charged flavylium cation (Figure 12.1). Glycosides typically affiliate with the 3-position on the C-ring (resulting in 3-monoglycosides) or the 5-position on the A-ring (resulting in 3,5-diglycosides). Different glycosidic units (mono-, di-, and tri-glycosides are present in anthocyanin structures. The predominant monoglycosides are hexoses (galactose, glucose, or rhamnose) or pentoses (arabinose and xylose) (Wang et al. 2024; Yang et al. 2022).

This category of compounds imparts the various vibrant colors—such as deep red, blue, and purple—of berries like blueberry (*Vaccinium* spp.), strawberry (*Fragaria* × *ananassa*), and bilberry, and has been shown to effectively reduce oxidative stress and protect against age-related functional decline in animal models. Their effectiveness was mainly conferred by direct free radical scavenging and cellular signaling pathways modulation, including AMPK, SIRT1, and Nrf2 (Del Bo et al. 2015; Giampieri et al. 2017; Wilson et al. 2006). One of the key benefits of anthocyanins that has been highlighted in various studies is their ability to serve as potent neuroprotective and anti-inflammatory agents (Ma et al. 2025). A study released in 2024 further supports this point by showing how these pigments are significantly involved in the control of longevity pathways conserved through evolution. They demonstrated that blueberry anthocyanins induce the expression of the most important anti-aging genes, such as daf-16 (the equivalent of the FOXO transcription factor), in the nematode model *C. elegans* (Ding et al. 2025). The mechanism by which these metabolites facilitate the

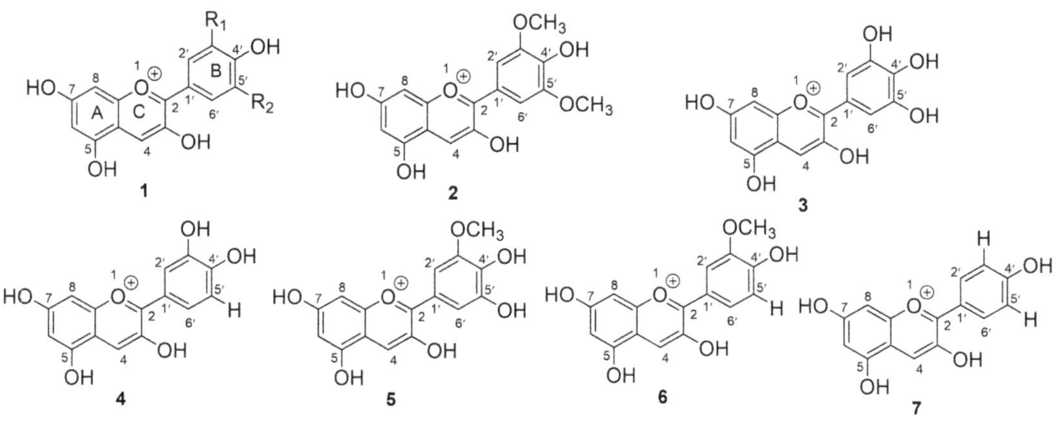

FIGURE 12.1 Chemical structures of the aglycone base of different types of anthocyanidins.

maintenance of cognitive functions during aging has been shown to include the reduction of neuroinflammation, enhancement of neuronal communication, and an increase in neuroplasticity (i.e., the ability of the brain to change and adapt—its most basic property) (Lautrup et al. 2019).

12.2.1.2 Proanthocyanidins (PCDs)

Proanthocyanidins are oligo- or polymers of monomeric flavan-3-ols (e.g., catechin, 8) produced as an end product of the flavonoid biosynthetic pathway (Figure 12.2). The degree of polymerization can range between 3 and 11 (Rauf et al. 2019).

Proanthocyanidins occur in very high amounts in cranberries and blueberries. Oligomeric proanthocyanidins (OPC) show a remarkable anti-senescence effect as they can alleviate the deteriorating effect of aging on, e.g., skeletal health by additionally reducing oxidative stress and inflammatory response in osteocytes (Yu et al. 2024).

A major point that marked these compounds as different was their mode of action. Using the *C. elegans* model, it was shown that proanthocyanidin-enriched fractions from blueberries were the most potent fraction, increasing lifespan and thermotolerance, rather than anthocyanins or chlorogenic acid (Wilson et al. 2006). Notably, this response was through the calcium/calmodulin-dependent protein kinase II (CaMKII) pathway and the osmotic stress resistance protein-1 (OSR-1) pathway and was independent of classic antioxidant resistance (Wilson et al. 2006). The difference in the action of two closely related structures (anthocyanins vs. proanthocyanidins) from the same fruit highlights a key point in nutritional geroscience: effectiveness may be very specific to the chemical structure, and this may be a result of factors such as membrane permeability or binding affinity of the target; therefore, highly standardized, fraction-specific extracts should be used for translational studies.

12.2.1.3 Stilbenes

Stilbenes are polyphenolic allelochemicals synthesized by plants, especially grapes, peanuts, rhubarb, berries, etc., to defend themselves under stressful conditions. They are characterized by a carbon skeleton of 1,2-diphenylethylene (C6–C2–C6), consisting of an ethylene moiety in the middle of two benzene rings (Figure 12.3). Resveratrol (RSV) is the most well-known of the stilbenes, especially because of its role in the French paradox (Al-Khayri et al. 2023).

Stilbenes such as RSV and its highly bioavailable derivative pterostilbene are most often linked with grapes, grape products, and certain berries. These compounds are well-known activators of the sirtuin pathways. Pterostilbene, a compound besides blueberries, can also be found in other fruits, and is the subject of ongoing research due to its potentially better bioavailability than RSV

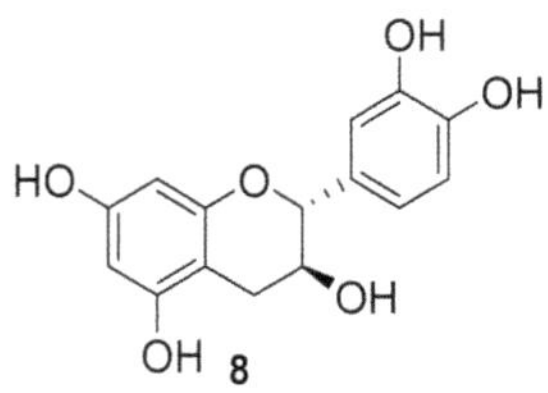

8

FIGURE 12.2 Structure of (+)-catechin.

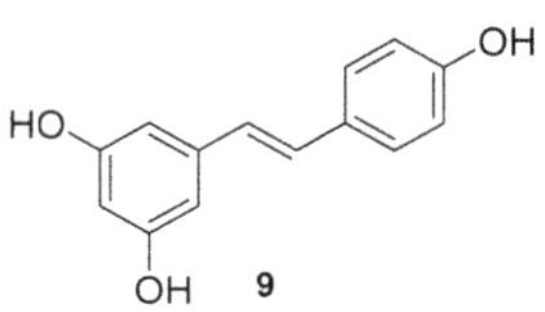

9

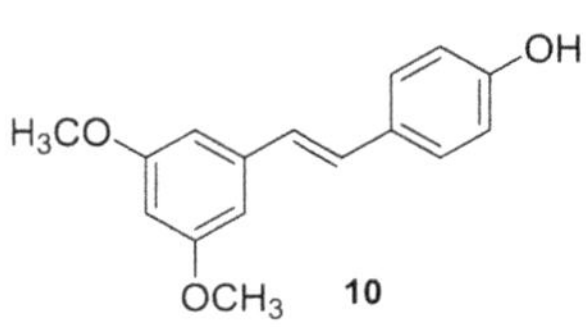

10

FIGURE 12.3 Chemical structure of resveratrol and pterostilbene.

(Muralitharan et al. 2025). In 2025, a scoping review reported that, while RSV can significantly improve some features of skin photoaging, early studies indicate that pterostilbene may have some benefits (Ahmed, 2025). However, more human trials are necessary to verify its effectiveness in the prevention and treatment of photoaging.

12.2.2 Non-Phenolic Bioactives: Polysaccharides

Polyphenols aside, some medicinal fruits that are not well known may contain complex carbohydrates with robust biological activities. Goji berry (*L. barbarum*) is the fruit of the plant, of which the polysaccharides (*Lycium barbarum* polysaccharides, LBP) form the main active ingredients. LBP has a good deal of relevance to anti-aging studies, as its immunomodulatory, antioxidant, and anti-inflammatory properties have been well established. It was in 2025 that the latest findings pointed out the molecular mechanism responsible for LBP's anti-neuroinflammation effect: LBP reduces microglial inflammation caused by radiation through the blockade of the IKK-β/IκBα/NF-κB signaling pathway. As a result, microglia switched their behavior away from the destructive pro-inflammatory M1 phenotype toward the protective, anti-inflammatory M2 phenotype (Zhang et al. 2025). The fact that a bioactive compound (LBP) can change the phenotype reveals a quite elaborate senomorphic mechanism, thus opening up perspectives for researchers beyond mere scavenging radicals to the real reprogramming of resident immune cells responsible for neuroinflammation and age-related cognitive decline.

12.3 MECHANISTIC MODULATION OF AGING PATHWAYS

Medicinal fruits and berries have clinical relevance because they can act as pleiotropic agents and therefore are able to interact with complex, highly conserved metabolic and stress response signaling networks that are essential for cellular resilience and longevity (Figure 12.4).

12.3.1 Transcriptional Regulation of Antioxidant Defense: The Nrf2 Pathway

Flavonoids and stilbenes are not just direct antioxidants but also secondary modulators that increase the cellular antioxidant capacity by activating the nuclear factor erythroid 2-related factor 2 (Nrf2) pathway. The process is that Nrf2, which is normally attached to the Keap1 protein in the cytoplasm, gets released. It is the exposure to electrophilic compounds, which are often the metabolic transformation products of polyphenols, that causes the release of Nrf2. By moving into the nucleus, Nrf2 binds to the antioxidant response element (ARE) and triggers the transcription of essential endogenous phase II detoxifying and antioxidant enzymes such as superoxide dismutase, catalase, and glutathione peroxidase. The primary compound being metabolized, the effect still persists due to this induction. The initiation of the Nrf2 pathway is proven to be effective. Experiments on old Wistar rats given strawberry polyphenols showed a remarkable increase in antioxidant enzyme levels by 16.9to 55.9%, along with a 34.2 to 56.8% reduction in oxidative damage markers (e.g., lipid peroxidation), thus providing quantitative evidence of Nrf2 pathway activation in a mammalian system. Besides, the grape/berry-derived stilbene RSV demonstrates antioxidative properties via the co-activation of the Nrf2 and SIRT1 pathways in aged kidney models, thus underlining the compound's capability to mobilize multiple protective mechanisms simultaneously (Giampieri et al. 2017; Li et al. 2018; Wilson et al. 2006).

12.3.2 Metabolic Reprogramming via Energy-Sensing Pathways

Aging is generally associated with decreasing energy metabolism and mitochondrial dysfunction. Berry bioactive components activate pathways that mimic the beneficial effects of caloric restriction (CR), which is a very effective method for increasing longevity.

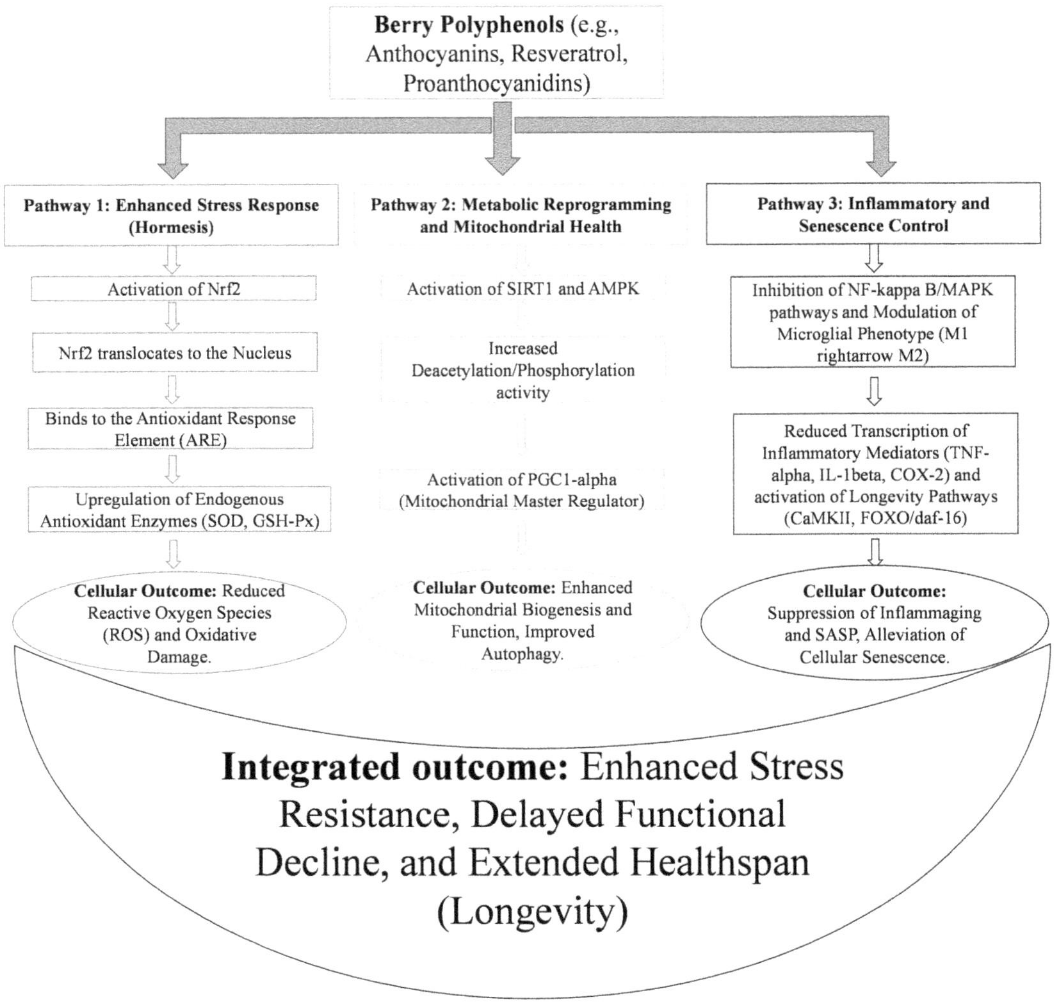

FIGURE 12.4 Schematic representation of the pleiotropic anti-aging mechanisms modulated by medicinal fruit and berry polyphenols. This diagram shows how polyphenols obtained from berries can simultaneously affect three major aging-related pathways: oxidative stress, metabolic imbalance, and inflammatory/senescence.

12.3.2.1 SIRT1 and AMPK Activation

Sirtuin-1 (SIRT1) enzyme and AMP-activated protein kinase (AMPK) are fundamental metabolic sensors. Stilbenes (RSV and pterostilbene) and strawberry polyphenols are known activators of these pathways. The activation of this axis results in enhanced cellular energy levels and the promotion of peroxisome proliferator-activated receptor gamma co-activator-1α (PGC-1α) expression, which is the master regulator of mitochondrial biogenesis. This metabolic reprogramming leads to an increase in mitochondrial turnover and the overall health of mitochondria, which in turn results in the optimization of cellular energy production, reduction in ROS generation, and eventually, the delay of cellular senescence (Giampieri et al. 2017; Li et al. 2018). One of the ways RSV acts in the kidneys of old mice to reverse oxidative stress and mitochondrial dysfunction is through activation of the SIRT1/AMPK signaling, which is an important part of systemic protection (Kim et al. 2018).

12.3.3 Mitigating Chronic Inflammation

Chronic, systemic inflammation is characteristic of aging. Phytochemicals suppress senescence-associated features by releasing less inflammatory molecules in this chronic inflammatory milieu.

12.3.3.1 NF-κB Suppression and Cytokine Reduction

Berry polyphenols and LBP mainly work via inhibition of the central transcriptional regulator nuclear factor kappa B (NF-κB) (Zhang et al. 2025). NF-κB suppression breaks down the genetic transcription and subsequent release of pro-inflammatory cytokines that make up the SASP; thus, it regulates the communication between senescent cells and healthy cells. Research indicates that berry bioactives, such as extracts from blueberry and cranberry, downregulate inflammatory mediators such as interleukin-8 (IL-8), monocyte chemoattractant protein-1 (MCP-1), cyclooxygenase-2 (COX-2), and tumor necrosis factor alpha (TNF-α). Various serum metabolites derived from blueberries and strawberries consumed by subjects were demonstrated in vitro (using human microglial cells) to lower the expression of iNOS, COX-2, and TNF-α among other markers (Rutledge et al. 2019; Zhang et al. 2025). Moreover, strong quantitative clinical evidence confirms the reality of these observations. A meta-analysis on acai berry consumption published in 2024 provided strong validation of the research, demonstrating that humans on average, had a 35% drop in inflammatory markers after just eight weeks of supplementation, thus forming a strong quantitative link (Dias et al. 2024).

12.3.4 Direct Effects on Cellular Senescence

Cellular senescence is one of the main pathological mechanisms leading to tissue aging. Berries have compounds with a range of properties that could help to lower the senescent load. OPCs exhibited a protective effect against Cd-induced osteocyte senescence both in vitro and in vivo. This protective effect against senescence-associated cell cycle arrest, downregulation of p53/p21, and ROS accumulation explains the role of OPCs in delaying skeletal aging(Yu et al. 2024). On the other hand, procyanidin A1 (PC A1), one of the OPCs, was reported to delay aging and reduce cellular senescence by promoting oxidative stress tolerance and activating autophagy (Li et al. 2025). In terms of neuroprotection, anthocyanins have been shown to promote brain health by dampening neuroinflammation (e.g., TNF-α and IL-1) and improving synaptic plasticity (Lautrup et al. 2019; Rutledge et al. 2019). They play a key role in reducing age-related memory loss in animal models, probably by facilitating neuronal communication and restoring the calcium buffering capacity, which is impaired in old age.

12.3.5 Non-Antioxidant Longevity Pathways

Identifying mechanisms that work independently of classical antioxidant activity is a major theoretical shift in geroscience research of phytochemicals, thus firmly establishing their role as signaling molecules. By using *C. elegans* as a model, researchers have shown that the lifespan-enhancing effect of proanthocyanidin-rich fractions from blueberries is dependent on the CaMKII and OSR-1 pathways, with no measurable improvement in oxidative stress resistance (Wilson et al. 2006). The discovery reveals that certain polyphenols function as selective neuromodulators or calcium regulators that trigger intrinsic stress survival mechanisms rather than just acting as radical scavengers. This non-antioxidant role highlights the structural specificity necessary to turn on highly conserved survival pathways. Moreover, recent transcriptomic analysis (Ding et al. 2025) has revealed the molecular intricacy of the mechanism, showing that *C. elegans*' lifespan extension by blueberry anthocyanins is achieved through the activation of highly conserved insulin/insulin-like signaling (IIS), and PI3K/Akt pathways, and as a result, the master longevity regulator, FOXO (daf-16), is upregulated (Ding et al. 2025). The different polyphenolic fractions (PCDs vs. anthocyanins) from the same source (blueberry) have anti-aging effects through different, non-overlapping regulatory networks (CaMKII vs. FOXO/IIS), which is a strong indication that eating the complex whole food matrix protects you through multiple, parallel pathways, thus giving a robust and synergistic defense against a variety of aging stressors.

12.4 EFFICACY ACROSS DIVERSE EXPERIMENTAL MODELS

Empirical evidence of the anti-aging effects has been corroborated in different biological systems, which show that the mechanisms are conserved throughout the phyla and, therefore, the results can be generalized.

12.4.1 Invertebrate Models: Lifespan Extension and Genetic Pathways

Invertebrate models such as *C. elegans* and *Drosophila melanogaster* are used for lifespan evaluation and delineation of conserved aging pathways, as they allow the tests to be performed rapidly and lifespans are assessed directly (Phelps et al. 2025). Blueberry anthocyanins were found to very effectively extend the lifespan and improve healthy aging parameters (e.g., locomotion and pharyngeal pumping rate) in a dose-dependent manner in *C. elegans* (Ding et al. 2025). Although the berry polyphenols' exact percentage in the core dataset was qualitative, a comparison of the berry polyphenols' effects on lifespan with the literature suggested that some natural compounds can extend the nematode lifespan by over 27% (Bei et al. 2025). The mechanistic basis revealed by these models and the engaged pathways, including the CaMKII/OSR-1 and FOXO/IIS/PI3K/Akt, being identified as mediators, illustrate that these fruit-derived bioactives interact with crucial survival and longevity genes (Neri, 2012; Wilson et al. 2006).

12.4.2 Mammalian Models: Organ-Specific Neuroprotection and Systemic Health

Mammalian research is essential in confirming tissue- and organ-specific protection that applies to physiological human aging. In a study, chronic supplementation with blueberry extracts substantially reversed age-related behavioral and neuronal impairments in aging F344 rats (Galli et al. 2002; Joseph et al. 2000). These neuroprotective effects are highly associated with enhanced neuronal signaling, decreased susceptibility to oxidative stress, and improved calcium regulation in the brain (Joseph et al. 2000). Administration of strawberry polyphenols to aged Wistar rats yielded whole-body benefits, especially in connection with liver function and mitochondrial health. This was corroborated by the quantitative data previously mentioned, revealing an increase in antioxidant enzyme activity by as much as 55.9% (Galli et al. 2002). In the same manner, RSV was able to protect aged mouse kidneys via activation of Nrf2 and SIRT1 signaling, thus relieving oxidative stress and mitochondrial dysfunction (Kim et al. 2018).

The findings presented here depict the compounds as being systemic and pleiotropic. They do not merely focus on one disease but rather enhance the overall resilience of several key organs by fine-tuning basic metabolic and stress–response pathways.

12.4.3 Human Intervention Trials and Biomarker Validation

Although there is still a scarcity of human clinical data that conclusively connect molecular mechanisms and long-term lifespan outcomes, intervention trials indirectly indicate the translation of these preclinical mechanisms to human health benefits that can be measured. Human intervention trials with different berries (blueberry, cranberry, and strawberry) have constantly demonstrated enhancements of plasma and serum antioxidant levels and a significant decrease in protein and lipid oxidation markers in the body after both single and repeated consumption (Del Bo et al. 2015). Apart from the general antioxidant capability being increased, the anti-inflammatory effect also comes through

TABLE 12.1
Quantitative Efficacy of Medicinal Fruit and Berry Bioactives on Key Aging Biomarkers

Berry/Fruit Source	Bioactive Compound Class	Model System	Key Anti-Aging Endpoint	Quantitative Effect Size Reported	Key Reference
Strawberry (*F.* × *ananassa*)	Polyphenols/ Anthocyanins	Old Wistar rats	Increase in antioxidant enzymes (SOD, GSH-Px)	16.9–55.9% increase ($p<0.05$)	Giampieri et al. (2017)
Strawberry (*F.* × *ananassa*)	Polyphenols/ Anthocyanins	Old Wistar rats	Reduction in oxidative damage (lipid/protein oxidation)	34.2–56.8% decrease ($p<0.05$)	Giampieri et al. (2017)
Blueberry (*V. angustifolium*)	Proanthocyanidin fraction	*C. elegans*	Lifespan extension	Significant, dose-dependent extension (CaMKII-dependent)	Wilson et al. (2006)
Açai Berry (*E. oleracea*)	Anthocyanins	Human intervention (Meta-analysis, 2025)	Reduction in inflammatory markers	Average 35% reduction after 8 weeks	Dias et al. (2024)
Lycium barbarum	Polysaccharide (LBP)	BV2 Microglial cells	Reduction in pro-inflammatory cytokines (IL-1β, TNF-α)	Significantly decreased release (NF-κ B inhibition)	Zhang et al. (2025)

very clearly in humans. Metabolites of blueberries and strawberries, when tested ex vivo, were found to downregulate the expression of key inflammatory markers in microglial cells. Furthermore, the meta-analysis of the aforementioned 2025 trial that was carried out both in vitro and in vivo confirmed that acai consumption can lead to a 35% decrease, on average, in the levels of inflammatory markers in human subjects, thus providing a strong quantitative indication of the anti-inflammatory effect (Dias et al. 2024). Table 12.1 integrates the available quantitative efficacy data, illustrating the remarkable influence of specific fruit bioactives on essential aging biomarkers in varied models.

The positive results observed in mammalian models often require chronic intervention (e.g., months of supplementation). This necessity underscores that the beneficial mechanisms are not based on acute, transient radical scavenging but rather on the sustained activation of gene expression programs (Nrf2, SIRT1, and AMPK) and the gradual, beneficial modification of cellular phenotypes (such as the M1-to-M2 shift in microglia). Chronic, low-dose exposure is therefore essential to induce the cellular adaptations (hormesis) necessary for robust anti-aging effects.

12.5 TRANSLATIONAL CHALLENGES AND BIOAVAILABILITY

Even though there is strong mechanistic and preclinical evidence, medicinal fruit and berry bioactives still have a limited impact on the execution of trustworthy human trials, which are mainly hindered by their complex pharmacokinetics.

12.5.1 Limitations of Bioavailability and Metabolism

One of the major limitations for several key polyphenols, such as anthocyanins and RSV, is their very low oral bioavailability inherent to their nature, along with their fast metabolism and high clearance rates. Take anthocyanins, for example; they can be easily broken down by heat, changes in pH inside the digestive tract, and chemical degradation, hence the amount of the intact parent compound that enters the systemic circulation is very little (Noman et al. 2025). In addition, the polyphenols, once ingested, are exposed to a great extent of metabolic conversion by the gut microbiota as well as by the liver (through phase I and phase II metabolism). This generates circulating metabolites (e.g., glucuronides and sulfates) that may have biological activities that are different from, or even stronger than, those of the original compound. Due to the very fast elimination and high metabolism of the parent compounds, it is recognized that the systemic anti-aging effects observed (e.g., reduced inflammation and increased Nrf2 activation) are mainly resulting from these circulating metabolites. Therefore, the focus of research has swung from the measurement of intact parent compounds in plasma to the identification, quantification, and investigation of specific, stable, and bioavailable metabolites' natural signaling capabilities. The reason that the human serum/microglial cell model demonstrated efficacy is that it reflects the biological potency of these circulating conversion products (Rutledge et al. 2019).

12.5.2 Strategies for Enhancing Bioefficacy

Various scientific studies are currently examining strategies for overcoming the chemical instability and poor absorption of these compounds in order to maximize their therapeutic potential.

12.5.2.1 Advanced Delivery Systems

To slow down degradation and facilitate systemic delivery, state-of-the-art encapsulation techniques are being devised. Such drug delivery systems comprise microencapsulation, liposomes, and nanoparticles. Nanoparticles are particularly designed to offer protection from thermal damage and to act as a barrier from gastric acidity, which can be crucial for the most efficacious targeting of sensitive organs like the brain (Jalouli et al. 2025). Such developments are vital to set forth definitive dosage guidelines and unravel the long-term benefits of using human-optimized formulations (Noman et al. 2025).

12.5.2.2 Whole Food Matrix Versus Isolated Compounds

One of the ongoing controversies in nutritional science is whether highly purified, isolated compounds work better than a whole food matrix that offers complex synergy (Kumkum et al. 2024). Studies have shown that combinations of berry polyphenols and their metabolites in the blood are capable of scavenging ROS and thereby reducing oxidative stress and inflammation. It could be that the whole food matrix provides protective synergy, which may be due to the presence of co-factors (e.g., proteins or co-pigments) that stabilize the main bioactives, a phenomenon that may be lost with highly purified supplements. Meanwhile, studying isolated compounds (like, for instance, RSV) is crucial for delineating particular mechanistic routes.

12.5.2.3 The Role of the Microbiome and Personalized Nutrition

Bioavailability is quite affected by one's gut microbiota composition, as this basically controls how effectively parent polyphenols are changed into active metabolites (e.g., the conversion of ellagitannins into urolithin A). Because of such differences in individuals, the supplementation with berries for anti-aging is a matter of very personalized effectiveness. Hence, it is essential for upcoming researchers to combine studies of polyphenol interventions with microbiota profiling to develop personalized nutritional strategies, especially in the case of age-related neurological disorders (Jalouli et al. 2025; Leng et al. 2025).

12.6 CONCLUSION AND FUTURE REMARKS

Medicinal fruits and berries are potent functional foods that can help alleviate the symptoms of aging. Their bioactive-rich composition, especially the anthocyanins, proanthocyanidins, and stilbenes, deeply mediates an anti-aging response via hormesis. The anti-aging effect, which is based on hormesis, refers to the simultaneous activation of internal cellular defense mechanisms

(Nrf2, SIRT1, and AMPK) and the alleviation of inflammaging (NF-κB, pro-inflammatory cytokines). Quantitative evidence demonstrates that these natural products are highly effective at the molecular level, as shown by the impressive upregulation of endogenous antioxidant enzymes and the downregulation of systemic inflammatory markers observed in both animal and human models. Analyses of specific compound fractions that can direct special conserved longevity effects, such as a CaMKII-dependent lifespan extension induced by blueberry proanthocyanidins, reveal their role as very powerful modulators of cellular longevity programs.

Despite the robust mechanistic foundation, the successful clinical translation of berry bioactives requires addressing several critical research gaps. Pharmacological optimization is paramount: innovative delivery systems (nanoparticles, liposomes) are essential to overcome poor bioavailability and instability, thereby ensuring adequate concentration of active metabolites at target tissues. Furthermore, future research must move beyond associative studies to establish definitive causality, particularly regarding the ability of these chemicals to modulate epigenetic modifications that affect gene expression patterns pertinent to health and aging. Specifically, long-term, double-blind randomized controlled trials (RCTs) in older human populations must be prioritized to:

1. Establish clear dosage recommendations for standardized, metabolite-producing extracts.
2. Directly link improvements in molecular biomarkers (e.g., sustained Nrf2 activation or reduction of cellular senescence markers) to measurable human healthspan outcomes, such as improved cognitive function and reduced frailty index.
3. Determine whether the natural complexity of the whole food matrix offers a superior, synergistic efficacy compared to highly purified, single-compound nutraceuticals.

Only through this rigorous, translational research focused on pharmacology and long-term human outcomes can medicinal fruits and berries transition from promising natural remedies to validated, evidence-based anti-aging therapeutics.

REFERENCES

Ahmed, M. 2025. Targeting aging pathways with natural compounds: a review of curcumin, epigallocatechin gallate, thymoquinone, and resveratrol. *Immun. Ageing* 22(1): 28.

Al-Khayri, J.M., R. Mascarenhas, H.M. Harish, Y. Gowda, V.V. Lakshmaiah and P. Nagella. 2023. Stilbenes, a versatile class of natural metabolites for inflammation—an overview. *Molecules* 28(9): 3786.

Bei, Y., T. Wang and S. Guan. 2025. Berberine extends lifespan in C. elegans through multi-target synergistic antioxidant effects. *Antioxidants* 14(4): 450.

Coppé, J.-P., C.K. Patil, F. Rodier, Y. Sun, D.P. Muñoz and J. Goldstein. 2008. Senescence-associated secretory phenotypes reveal cell-nonautonomous functions of oncogenic RAS and the p53 tumor suppressor. *PLoS Biol.* 6(12): e301.

Del Bo, C., D. Martini, M. Porrini, D. Klimis-Zacas and P. Riso. 2015. Berries and oxidative stress markers: an overview of human intervention studies. *Food Funct.* 6(9): 2890–2917.

Dias, K.K.B., G.A. de Jesus, A.A.F. da Costa, F.F. Costa, G.N. da Rocha Filho and R.J. Oliveira. 2024. Biological activities from açaí (Euterpe spp. Mart.) seeds and their pharmacological aspects: a systematic review and meta-analysis. *PharmaNutrition* 29: 100405.

Ding, J., J. Liu, Q. Guo and N. Zhang. 2025. Transcriptomic approaches to investigate the anti-aging effects of blueberry anthocyanins in a Caenorhabditis elegans aging model. *Antioxidants* 14(1): 35.

Galli, R.L., B. Shukitt-Hale, K.A. Youdim and J.A. Joseph. 2002. Fruit polyphenolics and brain aging. *Ann. N. Y. Acad. Sci.* 959(1): 128–132.

Giampieri, F., J.M. Alvarez-Suarez, M.D. Cordero, M. Gasparrini, T.Y. Forbes-Hernandez and S. Afrin. 2017. Strawberry consumption improves aging-associated impairments, mitochondrial biogenesis and functionality through the AMP-activated protein kinase signaling cascade. *Food Chem.* 234: 464–471.

Jalouli, M., M.A. Rahman, P. Biswas, H. Rahman, A.H. Harrath and I.S. Lee. 2025. Targeting natural antioxidant polyphenols to protect neuroinflammation and neurodegenerative diseases: a comprehensive review. *Front. Pharmacol.* 16: 1492517.

Joseph, J.A., N.A. Denisova, D. Bielinski, D.R. Fisher and B. Shukitt-Hale. 2000. Oxidative stress protection and vulnerability in aging: putative nutritional implications for intervention. *Mech. Ageing Dev.* 116(2–3): 141–153.

Kim, E.N., J.H. Lim, M.Y. Kim, T.H. Ban, I.A. Jang and H.E. Yoon. 2018. Resveratrol, an Nrf2 activator, ameliorates aging-related progressive renal injury. *Aging* 10(1): 83–99.

Kumkum, R., K. Aston-Mourney, B.A. McNeill, D. Hernández and L.R. Rivera. 2024. Bioavailability of anthocyanins: whole foods versus extracts. *Nutrients* 16(10): 1403.

Lautrup, S., D.A. Sinclair, M.P. Mattson and E.F. Fang. 2019. NAD(+) in brain aging and neurodegenerative disorders. *Cell Metab.* 30(4): 630–655.

Leng, P., Y. Wang and M. Xie. 2025. Ellagic acid and gut microbiota: interactions, and implications for health. *Food Sci. Nutr.* 13(4): e70133.

Li, Y., L. Xiang and J. Qi. 2025. Procyanidin A1 from peanut skin exerts anti-aging effects and attenuates senescence via antioxidative stress and autophagy induction. *Antioxidants* 14(3): 322.

Li, Y.R., S. Li and C.C. Lin. 2018. Effect of resveratrol and pterostilbene on aging and longevity. *Biofactors* 44(1): 69–82.

López-Otín, C., M.A. Blasco, L. Partridge, M. Serrano and G. Kroemer. 2013. The hallmarks of aging. *Cell* 153(6): 1194–1217.

Ma, X., Z. Jin, Z. Rao and L. Zheng. 2025. Health benefits of anthocyanins against age-related diseases. *Front. Nutr.* 12: 1618072.

Muralitharan, R.V., S.F. Masre, D.F. Basri and A.R. Ghazali. 2025. Pterostilbene and resveratrol: exploring their protective mechanisms against skin photoaging - a scoping review. *Biochem. Biophys. Rep.* 42: 102011.

Neri, C. 2012. Role and therapeutic potential of the pro-longevity factor FOXO and its regulators in neurodegenerative disease. *Front. Pharmacol.* 3: 15.

Noman, A.M., M.T. Sultan, M. Maaz, A. Mazhar, N. Tariq and M. Imran et al. 2025. Nutraceutical potential of anthocyanins: a comprehensive treatise. *Food Sci. Nutr.* 13(5): e70164.

Phelps, G.B., J. Morin, C. Pinto, L. Schoenfeldt, S. Guilmot and A. Ocampo. 2025. Comprehensive evaluation of lifespan-extending molecules in C. elegans. *Aging Cell* 24(4): e14424.

Rauf, A., M. Imran, T. Abu-Izneid, H. Iahtisham Ul, S. Patel and X. Pan. 2019. Proanthocyanidins: a comprehensive review. *Biomed. Pharmacother.* 116: 108999.

Rutledge, G.A., D.R. Fisher, M.G. Miller, M.E. Kelly, D.F. Bielinski and B. Shukitt-Hale. 2019. The effects of blueberry and strawberry serum metabolites on age-related oxidative and inflammatory signaling in vitro. *Food Funct.* 10(12): 7707–7713.

Shi, R., D. Gao, R. Stoika, K. Liu, A. Sik and M. Jin. 2024. Potential implications of polyphenolic compounds in neurodegenerative diseases. *Crit. Rev. Food Sci. Nutr.* 64(16): 5491–5514.

Sies, H. and D.P. Jones. 2020. Reactive oxygen species (ROS) as pleiotropic physiological signalling agents. *Nat. Rev. Mol. Cell Biol.* 21(7): 363–383.

Wang, L., W. Lan and D. Chen. 2024. Blueberry (Vaccinium spp.) anthocyanins and their functions, stability, bioavailability, and applications. *Foods* 13(17): 2851.

Wilson, M.A., B. Shukitt-Hale, W. Kalt, D.K. Ingram, J.A. Joseph and C.A. Wolkow. 2006. Blueberry polyphenols increase lifespan and thermotolerance in Caenorhabditis elegans. *Aging Cell* 5(1): 59–68.

Yang, W., Y. Guo, M. Liu, X. Chen, X. Xiao and S. Wang. 2022. Structure and function of blueberry anthocyanins: a review of recent advances. *J. Funct. Foods* 88: 104864.

Yu, G., Z. Wang, A. Gong, X. Fu, N. Chen and D. Zhou. 2024. Oligomeric proanthocyanidins ameliorate cadmium-induced senescence of osteocytes through combating oxidative stress and inflammation. *Antioxidants* 13(12): 1515.

Zhang, X., Q. Liu, L. Liu, F. Qian, B.X. Ren and F.R. Tang. 2025. Lycium barbarum polysaccharides suppress radiation-induced microglial inflammation by inhibiting the IKKβ/IκBα/NF-κB pathway. *Dose Response* 23(4): 15593258251382964.

13 Medicinal Fruits and Berries in Skin Care

Patcharaporn Saengratwatchara, Kamonwan Chaikhong, and Anchalee Prasansuklab

13.1 INTRODUCTION

Skin is a crucial barrier between the external environment and the internal body components. It is the largest and most complex organ because it has three major layers: epidermal, dermal, and subcutaneous layers. These layers contain collagen, connective tissue, extracellular matrix (ECM), and other supporting proteins (Venus et al. 2010). Skin serves as a first line of defense against invading microorganisms and a physical protective barrier from environmental threats such as ultraviolet (UV) radiation and pollution. It also plays a vital role in the maintenance of the body's internal balance, including water, temperature, sensory perception, and immunity (Proksch et al. 2008). Furthermore, skin serves as a first impression in an individual's physical appearance. That is why the appearance of skin also greatly affects mental health and quality of life (Puizina-Ivic 2008). When aging changes the skin's structure and function, it can harm the overall health and well-being of the individual.

Aging is a natural process that occurs all the time in human life. It involves various factors altered in the body, such as the generation of oxidative stress, mitochondrial dysfunction, and internal inflammation. These factors can contribute to cellular aging. In addition, external stimuli, including food, chemical agents, and human activities in their daily lives, can markedly accelerate this biological aging. The characteristics of skin aging are fine lines, wrinkles, thinning of the epidermis, skin dehydration, a rough and coarse skin texture, loss of elasticity, and hyperpigmentation (Naharro-Rodriguez et al. 2025). Skin aging, in the form of cellular aging, is a major concern today because it is particularly noticeable and shows gradual damage over time. The appearance of skin greatly affects mental well-being, self-esteem, social interaction, and quality of life. Although various current therapies, cosmetic procedures, and clinical treatments can improve skin appearance, there are also some side effects from chemical ingredients and aesthetic procedures, such as skin rash, skin redness, skin irritation, laser sensitivity, and high cost (Zhang and Duan 2018). These weaknesses have inspired the search for new and safer compounds from nature as sustainable alternatives.

In recent years, many studies have been interested in bioactive compounds from plants and natural products, especially fruits and berries, because they are rich in phytochemical constituents, including polyphenols, flavonoids, anthocyanins, vitamin A, ascorbic acid, resveratrol, and other compounds. These compounds exhibit potential alternatives for the treatment of various skin conditions. Moreover, they are widely known to have strong antioxidants, anti-inflammatory, and photoprotective properties, which are relevant for skin protection and anti-skin aging (Ribeiro et al. 2015). Interestingly, many fruits and berries have been traditionally used in medicinal recipes worldwide, supporting their long-term safety and efficacy (Nichols and Katiyar 2010). Using medicinal fruits and berries in skin care products matches the growing desire for natural and environmentally friendly ingredients. This highlights their potential role in developing safe, effective, and sustainable solutions for anti-aging. In this chapter, we review biological activities and research related to skin benefits, in order to provide basic information on interesting fruits and berries for future product development.

DOI: 10.1201/9781003688259-13

13.2 SKIN AGING

13.2.1 Pathophysiology of Skin Aging

Skin aging is a natural biological event in human life that cannot be avoided. It generally occurs with the passage of time from birth, commonly known as chronological aging or intrinsic aging. This natural aging process is associated with changes in metabolic activity, immune function, hormonal balance, and general regulation. The skin characteristics from these factors exhibit fine wrinkles, skin dryness, thinness, minor pigmentation, loss of elasticity, increased sagging, compromised skin barrier function, epidermal layer thinning, and decreased cell turnover (Naharro-Rodriguez et al. 2025). However, it is undeniable that the skin aging process can also be accelerated by extrinsic factors related to lifestyle and environment, termed as extrinsic aging. Examples include unhealthy food diet, smoking, particulate matter and air pollution, UV irradiation, or chemicals used in daily life. These factors predominantly enhanced the characteristics of skin aging appearance, such as coarse or deep wrinkles, rough skin, thickened texture, age spots or dark spots, prominent sagging, reduced elasticity, skin dehydration, and epidermal hyperplasia (Hussein et al. 2025; Sriram and Gopal 2025). Previous studies have shown that all skin layers can be influenced during intrinsic and extrinsic aging. Keratinocytes in the epidermal skin layer are disturbed, causing skin barrier malfunction (Russell-Goldman and Murphy 2020). Likewise, there was a decreased number of melanocytes, which can cause the reduction of melanin pigment production, thereby making it more sensitive to UV exposure and, in turn, leading to increased dark spots on the skin (Hughes and Bishop 2022). In addition, there was a negative impact on dermis, the layer of skin comprised of fibroblasts that provide strength and structural support to the epidermis by forming extracellular matrix and connective tissues. Damage to this skin layer results in decreased fibroblast activity, loss of collagen bundles and elastic fibers, and reduced skin firmness and elasticity. The flattening of the epidermal and dermal junction also disrupts the exchange of water, nutrients, and waste products, causing the skin to be unhealthy (Russell-Goldman and Murphy 2020).

A central mechanism underlying both intrinsic and extrinsic aging involves exacerbated reactive oxygen species (ROS) generation, mitochondrial dysfunction, and damage to biological macromolecules supporting skin's structural integrity (Papaccio et al. 2022). Excessive ROS can further stimulate chronic inflammatory responses through activation of the nuclear transcription factor kappa B (NF-κB) and subsequently increased production of primary pro-inflammatory mediators such as interleukin (IL)-1, IL-6, and tumor necrosis factor alpha (TNF-α). Additionally, ROS formation could induce the activator protein 1 (AP-1) transcriptional activity through mitogen-activated protein kinase (MAPK) pathways and sequentially upregulate the expressions of Smad7 and matrix metalloproteinases (MMPs), a group of proteolytic enzymes involved in ECM degradation and the skin's structural integrity. The function of these enzymes is normally counteracted by tissue inhibitors of metalloproteinases (TIMPs); however, this regulation gradually declines with age (Feng et al. 2024). Simultaneously, upregulation of Smad7 expression impairs the transforming growth factor beta (TGF-β) signaling cascade, further limiting collagen synthesis and remodeling (Lee et al. 2021). Also, the accumulation of ROS-induced DNA strand breaks triggers the DNA damage response (DDR) pathway via activating the ATM (ataxia telangiectasia mutated) and ATR (ATM and Rad3-related) kinases, which in turn activate cell cycle regulatory proteins, including tumor suppressors p53 and p21. These proteins then inhibit cyclin-dependent kinases (CDKs), ultimately leading to cell cycle arrest and cellular senescence of the skin. Senescence cells can secrete a complex mixture of factors called senescence-associated secretory phenotype (SASP), majorly comprises of cytokines, proteases, and growth factors, which further promote chronic inflammation, cumulative damage, and impaired function of skin cells (Stojanovic et al. 2025).

ROS also plays a crucial role in skin pigmentation in melanocytes by regulating α-melanocyte-stimulating hormone (α-MSH) secretion from keratinocytes (Jia and He 2025). When keratinocytes are induced by factors of intrinsic or extrinsic aging, the p53 is activated and leads to the transcriptional upregulation of the pro-opiomelanocortin (POMC) gene, which serves as a precursor of α-MSH. The secreted α-MSH then binds to melanocortin-1 receptors (MC1R) on melanocytes. This binding triggers a cyclic adenosine monophosphate (cAMP)-dependent signaling pathway, which involves cascade activation of adenyl cyclase, cAMP, protein kinase A, and cAMP response element-binding protein (CREB). Further, CREB promotes melanogenesis through activating transcription of microphthalmia-associated transcription factor (MITF) and other melanogenesis-related enzymes, including tyrosinase-related proteins and tyrosinase. Dysregulation of pigment cells results in the accumulation of melanin pigment, leading to hyperpigmentation and age spots (Ahuja et al. 2025). Figure 13.1 summarizes the molecular mechanisms underlying intrinsic and extrinsic skin aging, as well as the treatment targets.

13.2.2 Current Treatment Strategies for Anti-Skin Aging

Delaying and preventing skin aging can simply be achieved by avoiding external triggers, such as protecting skin from UV radiation using sunscreen, staying in shaded areas, or reducing exposure to environmental pollutants. However, in everyday life, complete avoidance of these stimuli is not feasible (Gilchrest 2013). Consequently, vitamins and dietary supplements, including retinoic acid (vitamin A), ascorbic acid (vitamin C), niacinamide (vitamin B3), antioxidants, estrogens, growth

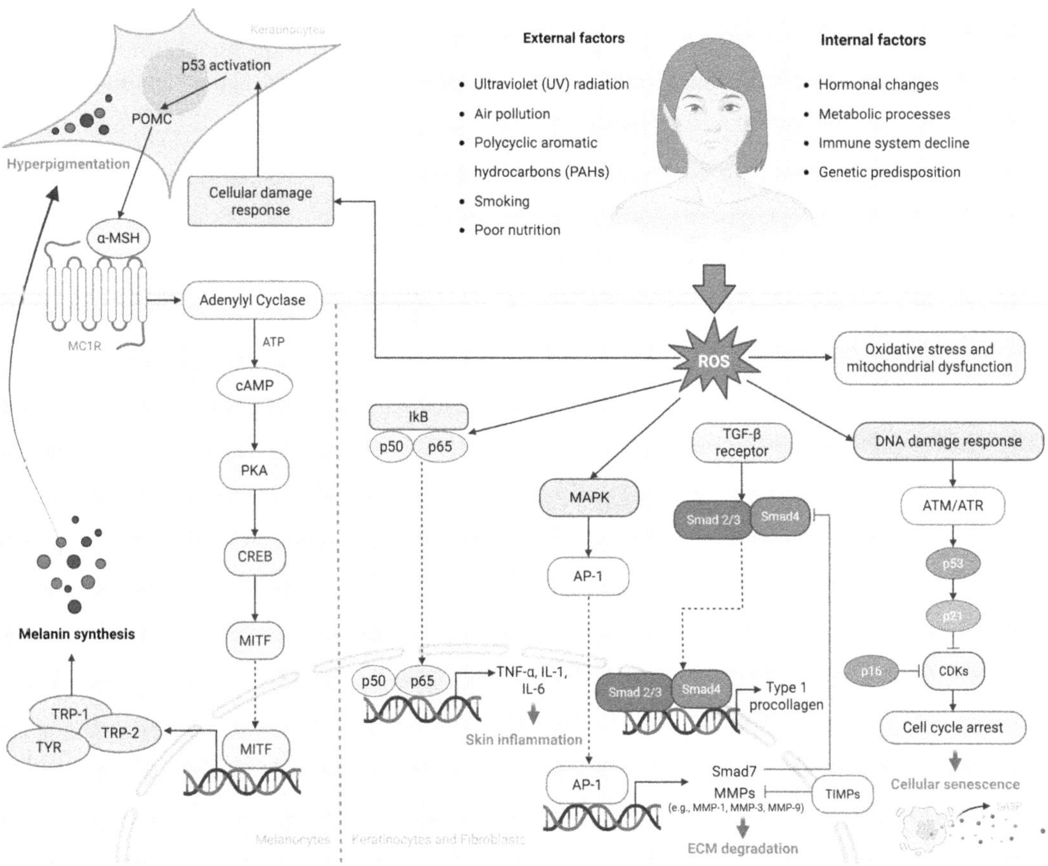

FIGURE 13.1 Molecular mechanisms underlying skin aging progression and its potential therapeutic targets. A regular arrow (→) represents activation or promotion; a bar-headed line (⊣) represents inhibition; and a dashed arrow (⇢) represents multistep processes and nuclear translocation.

factors, and cytokines, are commonly employed with the aim of protecting, delaying, nourishing, and restoring the skin by mitigating the deterioration processes that may contribute to intrinsic cellular aging (Mohiuddin 2019).

Recently, different therapeutic and cosmetic methods have been developed to counteract those harmful effects. Dermatologists and specialists have explored different ways to improve skin regeneration and decrease wrinkles. There are various treatment levels, from topical treatments to minimally invasive procedures, and positive outcomes have been obtained in recent years. Among non-surgical interventions, the injectable treatments remain necessary. For example, a pivotal phase 3 through a double-blind and randomized-controlled trial on botulinum toxin type A formulation in Chinese patients showed significant improvement in moderate-to-severe glabellar wrinkles (Wu et al. 2023). These injections offer rapid improvements but also carry limitations in cost, the risk of bruising, and the need for repeated maintenance in the future.

In the topical treatment, using active compounds like retinoids has been shown in a randomized controlled trial to improve signs of naturally aged skin by enhancing collagen and glycosaminoglycan levels, and reducing wrinkles (Kafi et al. 2007). In addition, other active ingredients, such as antioxidants, peptides, or growth-factor derivatives, have often been used. However, many topical products still face problems related to the stability of active compounds, limited skin penetration, and difficulty in maintaining long-term use (McDaniel et al. 2019).

Nowadays, energy-based technologies such as lasers, high-intensity focused ultrasound (HIFU), radiofrequency (RF), and intense pulsed light (IPL) are increasingly used in treatment procedures to specifically target multiple layers of the skin in a more complementary manner (Jia and Feng 2025). Despite remarkable progress, a key challenge remains that most current treatments primarily focus on visible symptoms such as wrinkles, skin laxity, or pigmentation, rather than an approach to fundamental biological processes of aging, including cellular senescence, ECM degradation, and chronic inflammation (Shin et al. 2023). Therefore, developing integrated treatment strategies is necessary through combining topical formulations, injectables, and procedural techniques, which eventually can tailor the biological characteristics of individual skin.

13.3 NATURAL ANTI-AGING SKIN CARE PRODUCTS

Skin care and skin-nourishing products are currently considered essential for daily personal care. Nowadays, topical skin care products are sold in a wide variety of formulation types, i.e., serums, moisturizers, lotions, creams, sunscreens, and special essences. Each formulation differs according to its properties and the intended performance, such as fine texture, rapid absorption into deep skin layers, moisturization, water retention, UV protection, wrinkle reduction, or treatment of specific skin concerns. In addition, the product types can be classified based on functional objectives; e.g., cleansing products, hydrating formulations, whitening or brightening products, acne prevention and protection, anti-aging products, photoprotective sunscreens, barrier-repair therapeutics, and skin-rejuvenating products (Surber et al. 2018; Zhang et al. 2024).

However, certain limitations remain unsolved regarding the handling of skin care products, even when they have been certified or approved under the existing standards process. The application of synthetic formulations on human skin may result in adverse reactions ranging from mild to severe, such as allergic responses, skin rashes, pain, or burning sensations. In response to these concerns, many researchers and the cosmetics industry have focused on the concept of clean and eco-friendly beauty, concentrating on reducing or avoiding the use of certain synthetic ingredients to enhance consumer safety (Rubin and Brod 2019). Various biological constituents have been commonly found in skin care products, including polyphenols, flavonoids, carotenoids, polysaccharides, anthocyanins, and vitamin C. These substances exerted their antioxidant, anti-inflammatory, and anti-skin aging effects. After UV exposure, they could enhance collagen production and maintain the skin microbiome. However, these natural compounds have some limitations. They are unstable, very sensitive to temperature, and have inconsistent phytocompound concentrations, which affect the standardized quality in production. Additionally, some individuals might be allergic to the skin caused by these active ingredients. Compared to synthetic substances, natural constituents are beneficial for long-term administration due to minimal adverse effects (Wirtu 2024).

Several berries and fruits, as well as their phytocompounds, have gained attention in skin care products due to their antioxidants and anti-inflammation (Azevedo et al. 2025). As a result, they are increasingly popular as bioactive ingredients for cosmetic development and skin care products. Because fruits and berries are generally edible, the extracts are regarded as having a certain level of safety for use in cosmetic and pharmaceutical applications. In addition, they serve as a natural reservoir of antioxidants and can be further developed into cosmeceuticals with high commercial potential, representing a sustainable alternative to synthetic ingredients.

13.4 POTENTIAL FRUITS AND BERRIES AS SKIN CARE INGREDIENTS

As previously mentioned, many studies have documented natural plant-derived products as potential antioxidants, anti-inflammatories, and anti-aging contributors that can protect cells from oxidative damage and delay physiological deterioration (Nowak-Perlak et al. 2025). Fruits and berries are examples of the most consumed plant products abundant in bioactive molecules, making them of great nutritional and biomedical interest (Bharadvaja et al. 2023). Moreover, fruits refer to the edible reproductive parts of flowering plants, while berries represent a subgroup of small, fleshy fruits particularly rich in pigment ingredients, including anthocyanins, flavonoids, and vitamins (Giménez Martínez et al. 2025). In recent years, fruits and berries have attracted our attention as promising and valuable sources of skin care ingredients such as polyphenolic compounds and vitamins that can neutralize ROS or other forms of free radicals, prevent photoaging, promote defense oxidant enzyme, and stimulate collagen synthesis (Nowak-Perlak et al. 2025). This section summarizes the biological properties and underlying mechanisms of potential fruits and berries and their derived bioactive components that are relevant to skin health promotion and anti-skin aging.

13.4.1 Goji Berry

Lycium barbarum L., or goji berry, is a small and sweet red fruit belonging to the Solanaceae family (Hu et al. 2022). This berry has been cultivated in several regions in the Eastern parts of Asia, including China. It has also been utilized in both internal and external remedies of Traditional Chinese Medicine (TCM). Goji berry is generally processed into dried fruit (known as *Lycii Fructus*), food and beverage products, such as tea, juice, and soup for health purposes (Bora et al. 2019). In addition, this berry has gained attention in scientific research due to their bioactive ingredients, i.e., polysaccharides, flavonoids, carotenoids, anthocyanins, and vitamins. Previous studies have reported various beneficial properties of this berry, such as antioxidants, anti-inflammation, anti-aging, hypoglycemic and lipid-lowering effects, immunotherapy, and retinoprotective and neuroprotective effects (Kulczyński and Gramza-Michałowska 2016). High levels of carotenoids within the berry were found to exhibit skin photoprotection by attenuating the effects of ultraviolet and visible-light damage (Li et al. 2017). Berry containing flavonoids and phenolic compounds, particularly rutin and quercetin, has been reported for its anti-inflammatory and free radical-scavenging activities through suppressing ROS signaling and inflammatory mediators (Liang et al. 2018). Other bioactive compounds of goji berry, including vitamins C and E, betaine, and trace minerals, are well known to maintain cellular redox homeostasis and enhance tissue protection under environmental stress (Wu et al. 2024). Therefore, this berry is beneficial for skin protection and rejuvenation.

Moreover, *L. barbarum* polysaccharides (LBPs), a water-soluble glycoconjugate, have shown their antioxidant and anti-cell senescence activities. LBPs could enhance antioxidant defense mechanisms after UVB exposure by activating the nuclear factor E2–related factor 2 (Nrf2)/antioxidant response element (ARE) pathway, which resulted in increased activities of glutathione peroxidase and superoxide dismutase, as well as inhibition of p38/MAPK signaling and MMP-9 expression in human keratinocytes (HaCaT) and fibroblasts. These ultimately resulted in decreasing collagen deconstruction (Li et al. 2017; Liang et al. 2018). Oral administration of zeaxanthin-rich *L. barbarum* extract mitigated UVB-induced epidermal thickening and photoaging features in BALB/c mice (Zhu et al., 2025). Interestingly, *Lycii Fructus* extract exerted its photoprotection in a randomized and placebo-controlled clinical trial by reducing UVB-mediated erythema formation and enhancing the minimal erythema dose (MED) (Tsunenaga et al., 2022). Taken together, the collected data suggest that *L. barbarum* has multiple beneficial roles, including antioxidant, photoprotective, as well as anti-inflammation, which delay skin aging. However, further large-scale clinical studies are still required to verify long-term efficacy, optimal dosage, and standardized formulations.

13.4.2 Acai

Euterpe oleracea Mart., or Acai palm, belongs to the Arecaceae family grown in the floodplains of the Amazonian area of Central and South America. It is a small, dark purple berry about 1–2 centimeters wide with a large seed and a thin layer of edible pulp. For thousands of years, people living by the Amazon River have relied on acai as a main part of their diet, usually consuming acai pulp. Traditionally, different parts of the palm were used in medicine. The fruit pulp is put on ulcers, and the oil is taken to help with diarrhea (Mi Hyun Kang et al. 2017). Recently, acai has gained worldwide popularity, largely because it is considered a superfood with strong nutritional and antioxidant properties.

Currently, acai consumption is rising in popularity worldwide due to its status as a superfood containing an exceptional nutritional profile and health benefits. The antioxidant capacity of acai pulp is ranked highest among fruits due to high content of antioxidant compounds such as polyphenols, fatty acids, and a large amount of anthocyanins, which contribute to its dark purple color. The main polyphenolic compounds, anthocyanins, are responsible for the color and its strong antioxidant effect. Notable examples include cyanidin-3-O-rutinoside and cyanidin-3-O-glucoside (Sanches et al. 2024). The water extract of acai fruit also contains other important plant chemicals like flavonoids and proanthocyanidins, especially quercetin, orientin, and vitexin. In addition, acai pulp is also rich in unsaturated fatty acids, predominantly oleic acid and linoleic acid (Mi Hyun Kang et al. 2017).

Acai fruit extract and oil possess potent anti-aging effects on the skin linked to their synergistic antioxidant and anti-inflammatory actions, as shown by many studies. Acai's polyphenols, especially anthocyanins, neutralize harmful ROS. Previous studies have shown that anthocyanin-rich plant extracts possessed strong free radical scavenging properties and were able to attenuate oxidative stress in human cells, a mechanism also relevant to the bioactive compounds found in Acai (Remigante et al. 2022). Anti-inflammatory effects have also been reported, as plant-derived polyphenolic extracts were able to attenuate pro-inflammatory signaling by reducing TNF-α production and modulating cyclooxygenase-2 (COX-2) expression (Dias et al. 2015). It also offers enzyme inhibition and structural protection. It can inhibit MMP-1 activity, which helps prevent early collagen breakdown linked to photoaging (Zimmer et al. 2025). Furthermore, acai extracts can enhance the migration of human fibroblasts in cell culture, helping tissue regeneration and repair (Mi Hyun Kang et al. 2017; Kang et al. 2017). This suggests the acai's potential benefits for wound healing and structural repair on sun-damaged skin.

Recently, acai oil and extracts have been widely used as active ingredients in skin care products. The oil acts as a strong emollient and is high in lipids. It moisturizes and protects the skin while helping restore the skin barrier. Chemical analysis found that acai oil contains phenolic acid and procyanidins that have strong antioxidant activity, making it well-suited for topical use (Pacheco-Palencia et al. 2008). Recent formulation studies have also demonstrated its cosmetic benefits. An acai oil–based organogel containing hyaluronic acid improved skin hydration. It also demonstrated good retention on ex vivo skin while remaining biocompatible, showing promise for moisturizing and barrier-supporting cosmetics (Sanches et al. 2024). Acai seed extract from manufacturing by-products was found to have potent antioxidant, photoprotective, calming, and firming activities in vitro. It was also regarded as safe from mutagenicity and irritation (Zimmer et al. 2025). The use of nanoencapsulation technology has further improved stability, texture, and active ingredient retention. These advancements lead to longer-lasting moisturization and protection. To determine whether acai-based skin care can improve skin suppleness or minimize wrinkles, further comprehensive clinical research is still required.

13.4.3 Blueberry

Blueberry (*Vaccinium* spp.), a tropical fruit with a diameter ranging from 0.4 to 0.8 cm, originates from Canada and the northeast of the United States of America. The bioactive compounds in blueberries have been linked to various therapeutic effects, i.e., anti-cardiovascular disease, anti-diabetes, anti-aging, and anti-inflammation (Pambianchi et al. 2021; Piechowiak et al. 2021). A powerful antioxidant property of blueberry extract has been demonstrated in a number of previous studies. The in vitro analysis of antioxidant properties revealed that the hydroethanolic extract of *V. angustifolium* fruits contained high levels of total phenolic and total flavonoid contents, as well as promising scavenging activity against DPPH free radicals. Phytochemical profiling of this extract, analyzed using ultra-high performance liquid chromatography-tandem mass spectrometry technique, has identified the presence of polyphenolic and flavonoid compounds, including procyanidin B1 and B2, gallic acid, quercetin, cyanidin, trans-caffeic acid, and ascorbic acid (Omachi et al. 2025). The study in the HaCaT cell line has shown that the ethanol peel extract of blueberry (*V. corymbosum* L.) could inhibit the formation of ROS induced by hydrogen peroxide (H_2O_2), thereby suppressing oxidative stress (Piechowiak et al. 2021).

Besides antioxidants, blueberries are also a great source of effective anti-aging and skin protective agents. Blueberry fruit extract has demonstrated antioxidative and anti-inflammatory effects in different skin cell models. It was found that exacerbated ROS production and inflammasome activation stimulated by ozone could be reduced in HaCaT keratinocytes, human 3D skin tissues, and human skin explants when pre-treated with the extract of *V. angustifolium* (lowbush blueberry) (Pambianchi et al. 2020). Furthermore, the indigenous Alaskan blueberry extract has exhibited a protective effect on UV-irradiated human ex vivo skin cell model via inhibiting oxidative stress and suppressing aberrant expression of key mediators in inflammatory pathways, such as 4-hydroxynonenal (4-HNE) and COX-2 (Pambianchi et al. 2021). The anti-skin aging of blueberry extract was also clearly demonstrated in the recent randomized-controlled trial study by Jiang and colleagues. The A.G.E. interrupter advanced anti-wrinkle cream containing 4% blueberry extract as a primary ingredient has been tested in Chinese patients, with topical application on one side of their faces for a period of 180 days following micro-focused ultrasound (MFUS) treatment. At the end of the study, most patients showed significant improvement in skin elasticity, fine lines, and dermal thickness on the side of the face treated with A.G.E. cream (Jiang et al. 2025).

13.4.4 Strawberry

Strawberry (*Fragaria ananassa*) has been previously reported for various bioactive phytocompounds, such as polyphenols, flavonoids, vitamin C, β-carotene, and anthocyanins. Strawberry extract containing these compounds exhibited their antioxidant properties by lowering DNA damage and maintaining mitochondrial activity in UVA-induced fibroblasts (Giampieri et al. 2014; Giampieri et al. 2012). In addition, strawberry extract has been applied to study its photoprotective effect in skin care products. UVA-stimulated cell death in human fibroblasts was suppressed by strawberry, and this inhibitory effect was also enhanced by strawberry-derived formation with CoQ_{10}, indicating a synergistic interaction (Gasparrini et al. 2015). Moreover, hydrolyzed strawberry extract, which contains a high content of phenolics, bioflavonoids, and anthocyanins, has shown its anti-aging property on the skin. The study has used the microneedle mesotherapy technique for applying strawberry hydrolysate enriched with L-ascorbic acid formulation on the facial skin of healthy woman volunteers. The formulation could increase skin hydration and elasticity, accompanied by improving skin brightness and texture (Markiewicz et al. 2019). Overall, strawberries and its bioactive ingredients possess anti-skin aging and photoprotective properties that are beneficial in skin care products.

13.4.5 Pomegranate

Pomegranate (*Punica granatum*) is among the popular plants used as an ingredient in cosmeceutical and dermatological industries (Baumann 2007). This fruit contains a variety of phytoconstituents such as flavonoids, polyphenols, anthocyanidins, hydrolyzable tannins, with ellagitanins as major components (Hayouni et al. 2011). Studies on crude extract of pomegranate peel have supported its notable anti-aging activity with strong antioxidant and protective effect against UV exposure. It was found that the pomegranate rind extract reduced UVB-induced cell damage in human fibroblasts. This treatment effectively increased the production of procollagen type I and reduced the expression of MMP-1, an enzyme responsible for collagen degradation, suggesting a mechanism for protecting the ECM from solar damage. The protective effects were primarily attributed to the major polyphenols present, especially catechin (Park et al. 2010). In addition, this extract has shown the potential to alleviate inflammaging, the process of age-associated dysregulated inflammation. It was found that the expressions of pro-inflammatory mediators, including TNF-α, IL-1β, and IL-17α, have substantially decreased after topical application of the extract in gel formulation (5% w/w) (Karim et al. 2023).

The benefits of *P. granatum* extend to oral supplementation as part of a "beauty from within" approach. After 60 days of continuous consumption, participants in a clinical study assessing a proprietary oral extract showed significant improvements in various skin characteristics. Along with improvements in skin elasticity, firmness, moisturization, and tone evenness, there was a noticeable decrease in fine lines and crow's feet wrinkles. Furthermore, the extract showed potential in regulating pigmentation through the decreasing of tyrosinase levels (Krishnakumar et al. 2024). This collective evidence firmly establishes pomegranate as a crucial botanical resource for both topical and systemic dermatological applications.

13.4.6 Grape

Grapes (*Vitis vinifera* L.) are among the oldest cultivated plants, first grown in Western Asia. While many people think of grapes mainly for winemaking, they can also be enjoyed fresh, dried as raisins, or squeezed into juice. In the past, people in Europe and Asia used grape leaves and unripe fruit in folk medicine to help with inflammation, pain, bleeding, and diarrhea. Today, grape seeds, once thrown away, are valued for their bioactive compounds and are used to make grape seed extract and grape seed oil. Grapes and their byproducts, e.g., peels and seeds, provide beneficial effects for health and beauty care linked to their polyphenolic content. Anthocyanins, which give grapes their rich color and potent antioxidant properties, are found in the skin and pulp, particularly in red or black grapes. Mostly found in wine and grape skins, resveratrol is a potent antioxidant that helps activate sirtuins. Proanthocyanidins, one of the most potent natural antioxidants, are abundant in grape seed extract (Mantena and Katiyar 2006). The high concentration of the vital fatty acid—linoleic acid, a modulator of the integrity of the skin barrier—makes grape seed oil valuable.

Grape-derived products, namely resveratrol, grape seed extract, and grape seed oil, have been reported to perform synergistic anti-aging benefits. For instance, proanthocyanidin-containing grape seed extract scavenged free radicals and down-regulated MAPK and NF-κB expression, resulting in protection of keratinocytes from UV-induced oxidative stress and inflammation (Mantena and Katiyar 2006). Furthermore, clinical data of a sunscreen formulated by 3% of grape seed extract supported the beneficial effects by improving skin hydration and elasticity and reducing erythema in human subjects (Yarovaya et al. 2022). Additionally, a topical drug delivery system for grape seed oil has improved its distribution and effectiveness. For example, a nanoemulsion utilizing an ionic liquid enhanced skin penetration, resulting in upregulated expression of type I collagen together with MMP-1 suppression (Yang et al. 2024). Furthermore, resveratrol activated sirtuin 1 (SIRT1), which inhibits NF-κB and reduces TNF-α in fibroblasts, thereby alleviating chronic inflammation of human skin (Meng et al. 2021; Zhu et al. 2011). Proanthocyanidins produced from grapes suppressed the activity of collagenase and elastase, so stabilizing the ECM and protecting against structural degradation (Wittenauer et al. 2015). Oligomeric procyanidins (OPCs) may regulate collagen homeostasis rather than simply increasing production, even though they decrease procollagen I secretion in fibroblasts, suggesting involvement in ECM remodeling (Kim et al. 2017). Additionally, under oxidative stress or inflammation, proanthocyanidins increase VEGF expression in keratinocytes, promoting angiogenesis and healing (Khanna et al. 2001).

Cosmetic research supports the use of grape-derived compounds. For instance, resveratrol has been demonstrated for its inhibitory activities on MMP-1 and -3 expression in dermal fibroblasts by reducing NF-κB acetylation, thereby helping preserve collagen (Lu and Chen 2021). Furthermore, a study on 3D human skin models reported advantageous impacts by resveratrol derivatives on the dermal structure through increasing collagen expression and decreasing advanced glycation end products (AGEs) (Markiewicz et al. 2022). Additionally, resveratrol reduced UVA-induced ROS and stopped collagen fiber breakdown by activating autophagy via AMPK signaling in human skin fibroblasts and in vivo (Xia et al. 2024). A topical 2% of grape seed extract cream considerably sped up dermal wound healing in human subjects after minor surgical procedures, according to a randomized clinical experiment (Hemmati et al. 2014). A systemic photoprotection has been exhibited by proanthocyanidin-containing grape seed extract by protecting human skin from UV radiation, resulting in enhanced flexibility and moisture content (Mantena and Katiyar 2006). Lastly, grape seed oil is highly valued as a lightweight, non-comedogenic moisturizer because of its high linoleic acid concentration. It is a great option for a variety of skin types because it minimizes trans-epidermal water loss, strengthens the skin barrier, and is quickly absorbed. *V. vinifera* is a key component of many high-performance cosmetic products because its constituents collectively offer strong evidence for its all-encompassing anti-aging, anti-inflammatory, and protective properties.

13.4.7 Annurca Apple

Annurca apple (*Malus pumila* Miller cv. Annurca), originally cultivated in Southern Italy, is distinguished as a unique functional food due to the exceptionally high levels of major phytoconstituents, especially procyanidin B2 and the triterpenoid ursolic acid (UA). Previous clinical and mechanism research studies have demonstrated the substantial dermatological benefits of this fruit's extract and bioactive substances, especially for aging skin (Di Lorenzo et al. 2024; Tenore et al. 2018). The ECM degradation in the dermis layer is a well-known primary cause of skin wrinkles as well as the loss of skin integrity and firmness through the breakdown of key structural proteins, primarily collagen and elastin fibers. Interestingly, the recent in vitro study showed that UA-rich annurca apple extract in sunflower oil, termed annurca apple oleolite (AAO), possessed strong inhibitory activity toward elastase, an enzyme involved in the breakdown of elastin. Moreover, the anti-skin-aging efficacy of this fruit was supported by the results of the randomized, placebo-controlled, double-blind, two-arm clinical study. The participants demonstrated a noticeable decrease in wrinkles on the nasolabial fold and forehead area and improvement of skin moisture in their facial skin, after being topically applied with an AAO-based formulation for 28 days continuously (Di Lorenzo et al. 2024).

In addition to elastin and collagen, keratin is a key structural protein in the epidermis, which is essential for maintaining skin's structural integrity, skin barrier function, and overall skin health. Polyphenolic-rich annurca apple extract in the form of a nutraceutical supplement has been shown to promote the upregulation of keratin biosynthesis in HaCaT keratinocytes (Tenore et al. 2018).

The underlying mechanisms may be through the activation of mitochondrial respiration and fatty acid β-oxidation, thereby increasing ATP synthesis, and ultimately driving the production of keratin (Badolati et al. 2018). These findings have supported the potential benefits of the annurca apple as a source of anti-aging agents in skin care products.

13.4.8 Camu-camu

Myrciaria dubia (Kunth) McVaugh (camu-camu) is a tropical fruit from the Amazon rainforest, ranging from 1.0 to 3.2 cm in diameter. It is a natural source rich in phenolic acid and polyphenols such as gallic acid, ellagic acid, and quercetin. The compounds exhibit a variety of health benefits, i.e., anti-inflammatory, immunomodulatory, antiallergic, antioxidative, anti-aging, anti-diabetes, and anti-cancer properties (Do et al. 2021). An in vitro investigation of the antioxidant properties of camu-camu extract revealed that the parameters, including scavenging values against both $ABTS^{\bullet+}$ and $DPPH^{\bullet}$ free radicals, total polyphenol and phenolic content, and total flavonoid content, were remarkably three- to sixfold higher than those of other fruits (*Averrhoa carambola* L. and *Cyphomandra betacea*) (Nowak et al. 2023).

Currently, the camu-camu fruit is extensively investigated through in vitro and in vivo assays for its pro-oxidant inhibitory activities. The focus of the studies includes the reduction of ROS, anti-inflammatory, and anti-aging activities. For example, camu-camu hydroalcoholic extract has been explored in a study using HaCaT cells exposed to high glucose levels. It is worth noting that the extract exhibited the inhibition of ROS overproduction and the downregulation of two key pathways associated with the reduction of inflammatory cytokines and chemokines, the calcium-dependent nuclear factor of activated T cells (NFAT), and the MAPK/AP-1/NF-κB signaling. Additionally, the extract involved in the modulated expression of transcription factor Nrf2 and NAD(P)H quinone oxidoreductase 1, offering protection to keratinocyte cells from oxidative stress induced by high glucose levels (Do et al. 2021). Furthermore, the human skin barrier was enhanced by the use of camu-camu fruit as a primary ingredient in cosmetic creams that contain the SUPCERAT™ complex and epidermal lipids (stained by Nile red). In the treatment of metastatic melanoma, prebiotics from camu-camu fruit were combined with immune checkpoint inhibitors (nivolumab). The treatment resulted in a significantly increased amount of hyaluronic acid and ceramide in the epidermis and inhibitory activity against the main causes of skin's restricted flexibility and facial wrinkles, including elastase and collagenase enzymes (Landau et al. 2024). In addition, camu-camu fruit, when given together with the nivolumab, helped the body respond to treatment, reduced the spread of metastatic melanoma on the human skin by modifying the body's microbiome, and reduced immune-related adverse events (irAEs) (Pang et al. 2023).

13.4.9 Star Fruit

Star fruit (*Averrhoa carambola* L.) or carambola is a tropical fruit, cultivated worldwide, especially in Southeast Asia. It is classified as a member of the Oxalidaceae family. The name star fruit comes from its resemblance to a star in yellow-green color. The whole fruit, including its peel, is edible. The taste varies from sweet to sour flavors, and feels like a mix of tropical and citrus fruits such as pear, orange, and green grape (Luan et al. 2021). Star fruit plays an important role in traditional medicine recipes, including Ayurveda and TCM. According to its pharmaceutical activities, it is used to treat fever, cough, headache, and skin conditions such as eczema. In addition, some evidence suggests that star fruit juice exhibits the diuretic property and relieves the side effects of excessive alcohol consumption (Shui and Leong 2006). It is also rich in vitamins and minerals such as vitamin B, ascorbic acid, potassium, and dietary fiber, which help the human body to control blood sugar and cholesterol levels (Luan et al. 2021). Furthermore, the fruit contains plenty of phytochemical constituents, including phenolic acids and flavonoids, which provide antioxidant, anti-inflammatory activities, blood sugar-lowering, and immune-enhancing properties for health benefits (Luan et al. 2021; Shui and Leong 2006).

Both in vitro and in vivo works have documented the notable anti-aging potential of star fruit. The ethanolic extract and its derived flavonoid glycosides, including isovitexin 2″-O-α-L-rhamnoside, carambolasides, and carambolaflavone B, significantly inhibited oxidative stress-induced carbonylated proteins in the skin's outermost layer, from which carambolasides P exerted the strongest inhibition. Additionally, the extract and its compounds (isovitexin 2″-O-α-L-rhamnoside, carambolaflavone B, and carambolasides J and P) reduced protein carbonylation induced by engine exhaust, and the extract, along with carambolasides P, markedly suppressed the expression of COX-2 and IL-1α inflammatory mediators in particulate matter-exposed human epidermis keratinocytes (Wu et al. 2022). By performing in vivo evaluation in a skin inflammation model, the star fruit ethanol extract and its three fractions (butanol, ethyl acetate, and hexane) were found to significantly reduce croton oil-induced ear edema formation in male Swiss mice and the activity of tissue myeloperoxidase (MPO). The strongest inhibition was demonstrated by the ethyl acetate fraction with 75% and 54% inhibition of ear thickness and MPO activity, respectively (Cabrini et al. 2011). This extract also improved dermal and epidermal structure in hypertrophic scars, with effects comparable to triamcinolone acetate (Córdova-Aguilar et al. 2024). Overall, the findings indicate that star fruit extract and its compounds are effective in preventing premature skin aging, particularly via environmental induction. However, evidence from clinical studies remains limited, which should be emphasized to confirm its cosmeceutical benefits.

13.4.10 Mango

Mangifera indica L., or mango, is a tropical fruit of a medium- to large-sized evergreen tree belonging to the Anacardiaceae family. The edible part is the pulp or the mesocarp, which has a unique taste and aroma. While the unripe pulp is firm and compact, it becomes gradually softer during ripening, forming yellow–orange fibrous tissues (Mirza et al. 2021). In traditional medicines, mango has been used for varied disorders, such as animal bites, hypertension, anemia, asthma, bronchitis, diarrhea, hemorrhage, leucorrhea, and rheumatism (Ediriweera et al. 2017). Both unripe and ripe mango fruits serve as valuable sources of polyphenols and carotenoids. Mango peel contains notable phenolic constituents (Ajila et al. 2010), such as mangiferin, syringic acid, ellagic acid, protocatechuic acid, gentisic acid, and quercetin. Some previous studies indicated that raw *M. indica* fruit peel ethanol extract possesses high levels of total phenolics and antioxidant activities by DPPH•, ABTS•+, and FRAP tests (Musika et al. 2021). For anti-aging of the skin, the ethanolic extract of *M. indica* fruit peel demonstrated high anti-hyaluronidase activity compared to the control (gallic acid) (García-Villegas et al. 2023). Extracts from unripe mango fruits were evaluated for anti-aging activity on skin cells. Extracts from unripe mango fruits have been evaluated in skin cell models, where they protected dermal fibroblasts from UVA-induced damage and enhanced recovery post-irradiation effects attributed to their antiglycation and anti-melanogenic properties (Yu et al., 2019). Moreover, dried mango extract has shown anti-photoaging effects, inhibiting UVB-induced skin thickening, wrinkle formation, and loss of collagen fibers in HR-1 hairless mice exposed to UVB radiation (Song et al. 2013). Conclusively, these data represent the potential of mango-derived products as natural antioxidant sources and as a photoprotective agent, suitable for skin care formulations, therapeutic applications, and cosmetic product development.

13.5 CONCLUSION AND FUTURE REMARKS

As of now, the trend of using natural extracts for cosmeceuticals is gaining significant attention, particularly in anti-aging treatments. Many natural-derived ingredients from fruits or berries have demonstrated their abilities to effectively slow down the deterioration of the skin while maintaining its vitality and functionality (Ribeiro et al. 2015). In this chapter, ten common fruits and berries, along with their bioactive components, have shown promise in reducing age-related skin changes in animal research and human studies. Table 13.1 summarizes the targeted skin concerns,

TABLE 13.1
Summary of Anti-Aging Fruits and Berries with Targeted Skin Concerns and Related Mechanisms

Plant Name	Target Skin Concern	Anti-Skin Aging Mechanism	Potential Type of Skin Care Product Application	Key References
Goji berry	Skin barrier damage, collagen fiber breakdown, and disarrangement	- Preventing basal layer vacuolization and collagen breakdown, and increasing the uniformity of collagen fibers - Inhibiting oxidative damage via promoting SOD activity while reducing MDA level - Reducing excessive autophagy and mitophagy through decreasing LC3B, parkin, and PINK1 markers	- Firming products - Anti-photoaging products	Wu et al. (2024))
Goji berry	Skin eschar, swelling, hyperpigmentation, collagen, and elastin degradation	- Alleviating appearance of UV-damaged skin - Reducing oxidative stress via Nrf2 signaling activation - Preventing collagen and elastin breakdown by downregulating the expression of MMP-3 and MMP-9	- Firming products - Dietary supplements for skin health	Zhu et al. (2025)
Goji berry	Skin erythema due to UVB exposure	- Inhibiting UVB-induced erythema formation by reducing the expression of oxidative stress markers (total GSSH, carbonyl protein, lipid peroxidase, 8-OHdG) while enhancing glutathione levels	- Anti-redness products - Anti-photoaging product - Dietary supplements for skin health	Tsunenaga et al. (2022)
Acai	Skin dehydration and loss of water retention	- Increasing water retention in the stratum by elevating hyaluronic acid, controlling thermal skin, and increasing skin hydration	- Moisturizing products	Sanches et al. (2024)
Blueberry	Skin redness	- Improving skin redness through inhibiting ROS production, inflammasome activation, and inhibiting transcription of key players (caspase-1/IL-18) in the inflammasome pathway	- Anti-redness products	Pambianchi et al. (2020)
Blueberry	Impaired skin barrier	- Preserving epidermal structural proteins by increasing filaggrin and involucrin	- Barrier repairing products	Pambianchi et al. (2021)
Blueberry	Loss of elasticity, fine lines, and dermal thinning	- Improving skin elasticity - Improving dermal thickness by activating glycosaminoglycan synthesis and increasing the arrangement of ECM components	- Firming product - Barrier repairing product - Anti-aging products	Jiang et al. (2025)
Strawberry	Skin irritation, loss of firmness and elasticity, wrinkles, skin dehydration, and hyperpigmentation	- Improving skin sensitization, elasticity, and hydration - Improving skin tone by reducing melanin synthesis	- Firming products - Moisturizing products - Whitening products	Markiewicz et al. (2019)

Pomegranate	Skin wrinkles, fine lines, tactile roughness, dull skin, and hyperpigmentation	- Enhancing elasticity and firmness - Improving skin tone by inhibiting tyrosinase activity	- Firming products - Whitening products	Krishnakumar et al. (2024)
Grape	Skin dehydration, loss of elasticity, skin erythema, and hyperpigmentation	- Improving skin hydration - Improving overall skin tone by reducing melanin and erythema index - Improving skin elasticity	- Moisturizing products - Whitening products - Firming products - Anti-aging products	Yarovaya et al. (2022)
Grape	Skin dehydration, wrinkles, loss of elasticity, and dull skin	- Improving water retention in the skin - Enhancing firmness and elasticity - Improving skin tone by reducing melanin index	- Moisturizing products - Firming products - Anti-aging products	Yang et al. (2024)
Grape	Collagen degradation, skin redness, and epidermal thickness due to UVA exposure	- Increasing collagen level by reducing MMP-1 expression through AMPK activation - Decreasing epidermal layer thickness by increasing the arrangement of collagen fibers - Improving skin redness by suppressing infiltration of inflammatory cells	- Firming products - Anti-photoaging products	Xia et al. (2024)
Annurca apple	Wrinkles, fine lines, and nasolabial fold	- Reducing skin aging signs by inhibiting elastase activity and improving skin viscoelasticity	- Firming products	Di Lorenzo et al. (2024)
Star fruit	Skin edema	- Suppressing inflammation-induced edema by inhibiting MPO activity	- Anti-edema products	Cabrini et al. (2011)
Star fruit	Hypertrophic scar	- Increasing papillary dermis layer while decreasing the thickness of reticular dermis	- Anti-scar products	Córdova-Aguilar et al. (2024)
Camu-camu	Skin dehydration and skin barrier integrity	- Enhancing skin hydration by increasing hyaluronic acid production - Improving skin barrier integrity by increasing epidermal lipid content	- Moisturizing products - Barrier repairing products	Landau et al. (2024)
Mango	Collagen degradation, wrinkles, and epidermal thickness due to UVB exposure	- Increasing collagen bundles - Reducing wrinkle length and epidermal thickness	- Firming products - Anti-photoaging products - Dietary supplements for skin health	Song et al. (2013)

Abbreviation: AMPK, AMP-activated protein kinase; ECM, Extracellular matrix; GSSG, Glutathione disulfide; IL-18, Interleukin-18; LC3B, Microtubule-associated protein 1 light-chain 3 beta; MDA, Malondialdehyde; MMP, Matrix metalloproteinase; MPO, myeloperoxidase; Nrf2, Nuclear factor E2-related factor 2; 8-OHdG, 8-Hydroxy-2'-deoxyguanosine; PINK1, PTEN-induced kinase 1; ROS, Reactive oxygen species; SOD, Superoxide dismutase; UV, Ultraviolet.

mechanisms underlying anti-skin aging, and potential types of skin care product application for each fruit or berry.

In conclusion, a number of studies suggest that several fruits and berries, either in the form of crude extract or formulation, possess strong anti-aging abilities such as protecting the skin against free radicals, reducing skin inflammation, enhancing collagen production, stabilizing the extracellular matrix, and preserving skin structure and function. These findings highlight potential applications of fruits and berries and related compounds as promising anti-aging ingredients of skin care products to promote skin health and alleviate changes associated with skin aging. Further research focusing on the optimization of plant extraction methods for large-scale production, together with the development of formulation techniques with nanotechnology, is indispensable.

REFERENCES

Ahuja, K., S. Raju, S. Dahiya and R. K. Motiani. 2025. ROS and calcium signaling are critical determinant of skin pigmentation. *Cell Calcium* 125: 102987.

Ajila, C. M., L. Jaganmohan Rao and U. J. S. Prasada Rao. 2010. Characterization of bioactive compounds from raw and ripe *Mangifera indica* L. peel extracts. *Food Chem. Toxicol.* 48: 3406–3411.

Azevedo, F. S., A. R. Pires, M. A. Lila, G. Valacchi, R. T. Hoskin, M. M. Sarandy et al. 2025. Berry fruit extracts as topical cosmeceuticals for skin health applications: a systematic review. *Cosmetics* 12: 87.

Badolati, N., E. Sommella, G. Riccio, E. Salviati, D. Heintz, S. Bottone et al. 2018. Annurca apple polyphenols ignite keratin production in hair follicles by inhibiting the pentose phosphate pathway and amino acid oxidation. *Nutrients* 10: 1406.

Baumann, L. S. 2007. Less-known botanical cosmeceuticals. *Dermatol. Ther.* 20: 330–342.

Bharadvaja, N., S. Gautam and H. Singh. 2023. Natural polyphenols: a promising bioactive compounds for skin care and cosmetics. *Mol. Biol. Rep.* 50: 1817–1828.

Bora, P., S. Ragaee and E.-S. M. Abdel-Aal. 2019. Effect of incorporation of goji berry by-product on biochemical, physical and sensory properties of selected bakery products. *LWT* 112: 108225.

Cabrini, D. A., H. H. Moresco, P. Imazu, C. D. Silva, E. F. Pietrovski, D. A. G. B. Mendes et al. 2011. Analysis of the potential topical anti-inflammatory activity of *Averrhoa carambola* L. in mice. *Evid. Based Complement. Alternat. Med.* 2011: 908059.

Córdova-Aguilar, A., E. Mujica-Albán, J. M. Ortiz-Sánchez, S. Suárez-Cunza, R. Iglesias-Bustamante and D. J. Blanco-Victorio. 2024. Effect of *Averrhoa carambola* L. on the skin in an animal model of hypertrophic scarring: efecto de la *Averrhoa carambola* L. en la piel en un modelo animal de cicatrización hipertrófica. *Rev. Fac. Med. Hum.* 24: 45–51.

Di Lorenzo, R., M. Maisto, L. Ricci, V. Piccolo, A. Marzocchi, G. Greco et al. 2024. Annurca apple oleolite as functional ingredient for the formulation of cosmetics with skin-antiaging activity. *Int. J. Mol. Sci.* 25: 1677.

Dias, M. M., H. S. Martino, G. Noratto, A. Roque-Andrade, P. C. Stringheta, S. Talcott et al. 2015. Anti-inflammatory activity of polyphenolics from açai (*Euterpe oleracea* Martius) in intestinal myofibroblasts CCD-18Co cells. *Food Funct.* 6: 3249–3256.

Do, N. Q., S. Zheng, B. Park, Q. T. N. Nguyen, B.-R. Choi, M. Fang et al. 2021. Camu-camu fruit extract inhibits oxidative stress and inflammatory responses by regulating NFAT and Nrf2 signaling pathways in high glucose-induced human keratinocytes. *Molecules* 26: 3174.

Ediriweera, M. K., K. H. Tennekoon and S. R. Samarakoon. 2017. A review on ethnopharmacological applications, pharmacological activities, and bioactive compounds of *Mangifera indica* (mango). *Evid. Based Complement. Alternat. Med.* 2017: 6949835.

Feng, C., X. Chen, X. Yin, Y. Jiang and C. Zhao. 2024. Matrix metalloproteinases on skin photoaging. *J. Cosmet. Dermatol.* 23: 3847–3862.

García-Villegas, A., Á. Fernández-Ochoa, A. Rojas-García, M. E. Alañón, D. Arráez-Román, M. l. L. Cádiz-Gurrea et al. 2023. The potential of *Mangifera indica* L. peel extract to be revalued in cosmetic applications. *Antioxidants* 12: 1892.

Gasparrini, M., T. Y. Forbes-Hernandez, S. Afrin, J. M. Alvarez-Suarez, A. M. Gonzàlez-Paramàs, C. Santos-Buelga et al. 2015. A pilot study of the photoprotective effects of strawberry-based cosmetic formulations on human dermal fibroblasts. *Int. J. Mol. Sci.* 16: 17870–17884.

Giampieri, F., J. M. Alvarez-Suarez, L. Mazzoni, T. Y. Forbes-Hernandez, M. Gasparrini, A. M. Gonzàlez-Paramàs et al. 2014. An anthocyanin-rich strawberry extract protects against oxidative stress damage and

improves mitochondrial functionality in human dermal fibroblasts exposed to an oxidizing agent. *Food Funct.* 5: 1939–1948.
Giampieri, F., J. M. Alvarez-Suarez, S. Tulipani, A. M. Gonzàles-Paramàs, C. Santos-Buelga, S. Bompadre et al. 2012. Photoprotective potential of strawberry (*Fragaria* × *ananassa*) extract against UV-A irradiation damage on human fibroblasts. *J. Agric. Food Chem.* 60: 2322–2327.
Gilchrest, B. A. 2013. Photoaging. *J. Invest. Dermatol.* 133: E2–E6.
Giménez Martínez, R. J., F. Rivas García, J. C. March Cerdá, Á. Hernández-Ruíz, M. I. González Castro, M.-I. Valverde-Merino et al. 2025. Bioactive substances and skin health: an integrative review from a pharmacy and nutrition perspective. *Pharmaceuticals* 18: 373.
Hayouni, E. A., K. Miled, S. Boubaker, Z. Bellasfar, M. Abedrabba, H. Iwaski et al. 2011. Hydroalcoholic extract based-ointment from *Punica granatum* L. peels with enhanced in vivo healing potential on dermal wounds. *Phytomedicine* 18: 976–984.
Hemmati, A. A., G. Houshmand, Z. B. Moosavi, M. Bahadoram and N. S. Maram. 2014. The topical effect of grape seed extract 2% cream on surgery wound healing. *Glob. J. Health Sci.* 7: 52.
Hu, Z., Y. Ma, J. Liu, Y. Fan, A. Zheng, P. Gao et al. 2022. Assessment of the bioaccessibility of carotenoids in goji berry (*Lycium barbarum* L.) in three forms: in vitro digestion model and metabolomics approach. *Foods* 11: 3731.
Hughes, B. K. and C. L. Bishop. 2022. Current understanding of the role of senescent melanocytes in skin ageing. *Biomedicines* 10: 3111.
Hussein, R. S., S. Bin Dayel, O. Abahussein and A. A. El-Sherbiny. 2025. Influences on skin and intrinsic aging: biological, environmental, and therapeutic insights. *J. Cosmet. Dermatol.* 24: e16688.
Jia, X. and Y. Feng. 2025. Energy-based skin rejuvenation: a review of mechanisms and thermal effects. *J. Cosmet. Dermatol.* 24: e16657.
Jia, X. and L. He. 2025. Unveiling the dual-edged roles of melanocytes: key insights into skin photoaging and therapeutic avenues. *Pharmacol. Res.* 218: 107848.
Jiang, X., S. Peng, Y. Chen, Y. Nan and W. Song. 2025. A randomized, investigator-blinded, split-face, controlled trial evaluating the efficacy and satisfaction of a topical product containing blueberry extract and Pro-Xylane combined with micro-focused ultrasound for anti-aging. *J. Cosmet. Dermatol.* 24: e70281.
Kafi, R., H. S. R. Kwak, W. E. Schumacher, S. Cho, V. N. Hanft, T. A. Hamilton et al. 2007. Improvement of naturally aged skin with vitamin A (retinol). *Arch. Dermatol.* 143: 606–612.
Kang, M. H., S. Choi and B. H. Kim. 2017. Skin wound healing effects and action mechanism of acai berry water extracts. *Toxicol. Res.* 33: 149–156.
Karim, S., M. I. Khan and A. Ahmad. 2023. A mechanistic insight into cytokines and growth factors in Saudi pomegranate peel extract (*Punica granatum* L.) mediated diabetic wound healing in alloxan induced diabetic rats. *J. King Saud Univ. Sci.* 35: 102952.
Khanna, S., S. Roy, D. Bagchi, M. Bagchi and C. K. Sen. 2001. Upregulation of oxidant-induced VEGF expression in cultured keratinocytes by a grape seed proanthocyanidin extract. *Free Radic. Biol. Med.* 31: 38–42.
Kim, B. J., J. K. Park, B. K. Kim, S. J. Park, M. K. Kim, C. W. Lee et al. 2017. Oligomeric procyanidins (OPCs) inhibit procollagen type I secretion of fibroblasts. *Tissue Eng. Regen. Med.* 14: 297–306.
Krishnakumar, A., D. K. Chellappan, S. Jeyakodi, M. Dalal and S. Shetty. 2024. Efficacy and safety of a proprietary *Punica granatum* extract in skin health — a randomized, placebo-controlled clinical study in healthy volunteers. *Am. J. Transl. Res.* 16: 8043.
Kulczyński, B. and A. Gramza-Michałowska. 2016. Goji berry (*Lycium barbarum*): composition and health effects–a review. *Pol. J. Food Nutr. Sci.* 66: 67–75.
Landau, M., S. Landau-Blum, K. Samuel, L. M. Audrey, P. Armelle, V. Julien et al. 2024. The effect of a novel complex, composed of ceramide, energizing peptide and camu camu extract, on epidermal barrier function and dermal antiaging properties in ex vivo human skin small live cohort. *J. Cosmet. Dermatol. Sci. Appl.* 14: 1–11.
Lee, H., Y. Hong and M. Kim. 2021. Structural and functional changes and possible molecular mechanisms in aged skin. *Int. J. Mol. Sci.* 22: 12489.
Li, H., Z. Li, L. Peng, N. Jiang, Q. Liu, E. Zhang et al. 2017. *Lycium barbarum* polysaccharide protects human keratinocytes against UVB-induced photo-damage. *Free Radic. Res.* 51: 200–210.
Liang, B., L. Peng, R. Li, H. Li, Z. Mo, X. Dai et al. 2018. *Lycium barbarum* polysaccharide protects HSF cells against ultraviolet-induced damage through the activation of Nrf2. *Cell. Mol. Biol. Lett.* 23: 18.
Lu, Y.-E. and Y.-J. Chen. 2021. Resveratrol inhibits matrix metalloproteinase-1 and -3 expression by suppressing of p300/NFκB acetylation in TNF-α-treated human dermal fibroblasts. *Chem. Biol. Interact.* 337: 109395.
Luan, F., L. Peng, Z. Lei, X. Jia, J. Zou, Y. Yang et al. 2021. Traditional uses, phytochemical constituents and pharmacological properties of *Averrhoa carambola* L.: a review. *Front. Pharmacol.* 12: 699899.

Mantena, S. K. and S. K. Katiyar. 2006. Grape seed proanthocyanidins inhibit UV-radiation-induced oxidative stress and activation of MAPK and NF-κB signaling in human epidermal keratinocytes. *Free Radic. Biol. Med.* 40: 1603–1614.
Markiewicz, A., M. Zasada, A. Erkiert-Polguj, M. Wieckowska-Szakiel and E. Budzisz. 2019. An evaluation of the antiaging properties of strawberry hydrolysate treatment enriched with L-ascorbic acid applied with microneedle mesotherapy. *J. Cosmet. Dermatol.* 18: 129–135.
Markiewicz, E., J. Jerome, T. Mammone and O. C. Idowu. 2022. Anti-glycation and anti-aging properties of resveratrol derivatives in the in-vitro 3D models of human skin. *Clin. Cosmet. Investig. Dermatol.* 15: 911–927.
McDaniel, D. H., J. M. Waugh, L. I. Jiang, T. J. Stephens, A. Yaroshinsky, C. Mazur et al. 2019. Evaluation of the antioxidant capacity and protective effects of a comprehensive topical antioxidant containing water-soluble, enzymatic, and lipid-soluble antioxidants. *J. Clin. Aesthet. Dermatol.* 12: 46.
Meng, T., D. Xiao, A. Muhammed, J. Deng, L. Chen and J. He. 2021. Anti-inflammatory action and mechanisms of resveratrol. *Molecules* 26: 229.
Mirza, B., C. R. Croley, M. Ahmad, J. Pumarol, N. Das, G. Sethi et al. 2021. Mango (*Mangifera indica* L.): a magnificent plant with cancer preventive and anticancer therapeutic potential. *Crit. Rev. Food Sci. Nutr.* 61: 2125–2151.
Mohiuddin, A. K. 2019. Skin aging & modern age anti-aging strategies. *Int. J. Clin. Dermatol. Res.* 7: 209–240.
Musika, S., N. Pokratok, J. Pliankratoke, C. Khongla, C. Kupradit, A. Ranok et al. 2021. Antioxidant, antityrosinase and antibacterial activities of fruit peel extracts. *IJAT* 14: 1447–1460.
Naharro-Rodriguez, J., S. Bacci, M. L. Hernandez-Bule, A. Perez-Gonzalez and M. Fernandez-Guarino. 2025. Decoding skin aging: a review of mechanisms, markers, and modern therapies. *Cosmetics* 12: 144.
Nichols, J. A. and S. K. Katiyar. 2010. Skin photoprotection by natural polyphenols: anti-inflammatory, antioxidant and DNA repair mechanisms. *Arch. Dermatol. Res.* 302: 71–83.
Nowak, D., M. Gośliński, K. Przygoński and E. Wojtowicz. 2023. *Averrhoa carambola* L., *Cyphomandra betacea*, *Myrciaria dubia* as a source of bioactive compounds of antioxidant properties. *Foods* 12: 753.
Nowak-Perlak, M., M. Olszowy and M. Woźniak. 2025. The natural defense: anti-aging potential of plant-derived substances and technological solutions against photoaging. *Int. J. Mol. Sci.* 26: 8061.
Omachi, D. O., T. Shahnaz, B. Gines, N. Dawkins and J. O. Onuh. 2025. Phytochemical profiles, in vitro antioxidants and antihypertensive properties of wild blueberries (*Vaccinium angustifolium*). *Foods* 14: 4281.
Pacheco-Palencia, L. A., S. Mertens-Talcott and S. T. Talcott. 2008. Chemical composition, antioxidant properties, and thermal stability of a phytochemical enriched oil from açai (*Euterpe oleracea* Mart.). *J. Agric. Food Chem.* 56: 4631–4636.
Pambianchi, E., F. Ferrara, A. Pecorelli, B. Woodby, M. Grace, J.-P. Therrien et al. 2020. Blueberry extracts as a novel approach to prevent ozone-induced cutaneous inflammasome activation. *Oxid. Med. Cell. Longev.* 2020: 9571490.
Pambianchi, E., Z. Hagenberg, A. Pecorelli, M. Grace, J.-P. Therrien, M. A. Lila et al. 2021. Alaskan bog blueberry (*Vaccinium uliginosum*) extract as an innovative topical approach to prevent UV-induced skin damage. *Cosmetics* 8: 112.
Pang, S. A., A. Elkrief, M. P. Capella and W. H. Miller. 2023. Two cases of durable and deep responses to immune checkpoint inhibition-refractory metastatic melanoma after addition of camu camu prebiotic. *Curr. Oncol.* 30: 7852–7859.
Papaccio, F., S. Caputo and B. Bellei. 2022. Focus on the contribution of oxidative stress in skin aging. *Antioxidants* 11: 1121.
Park, H. M., E. Moon, A. J. Kim, M. H. Kim, S. Lee, J. B. Lee et al. 2010. Extract of *Punica granatum* inhibits skin photoaging induced by UVB irradiation. *Int. J. Dermatol.* 49: 276–282.
Piechowiak, T., B. Skóra, K. Grzelak-Błaszczyk and M. Sójka. 2021. Extraction of antioxidant compounds from blueberry fruit waste and evaluation of their in vitro biological activity in human keratinocytes (HaCaT). *Food Anal. Methods* 14: 2317–2327.
Proksch, E., J. M. Brandner and J. M. Jensen. 2008. The skin: an indispensable barrier. *Exp. Dermatol.* 17: 1063–1072.
Puizina-Ivic, N. 2008. Skin aging. *Acta Dermatovenerol. APA* 17: 47.
Remigante, A., S. Spinelli, E. Straface, L. Gambardella, D. Caruso, G. Falliti et al. 2022. Açaì (*Euterpe oleracea*) extract protects human erythrocytes from age-related oxidative stress. *Cells* 11: 2391.
Ribeiro, A. S., M. Estanqueiro, M. B. Oliveira and J. M. Sousa Lobo. 2015. Main benefits and applicability of plant extracts in skin care products. *Cosmetics* 2: 48–65.

Rubin, C. B. and B. Brod. 2019. Natural does not mean safe—the dirt on clean beauty products. *JAMA Dermatol.* 155: 1344–1345.

Russell-Goldman, E. and G. F. Murphy. 2020. The pathobiology of skin aging: new insights into an old dilemma. *Am. J. Pathol.* 190: 1356–1369.

Sanches, S. C., L. M. M. C. Ferreira, R. R. Pereira, D. G. Lynch, I. N. Ramos, A. S. Khayat et al. 2024. Acai oil-based organogel containing hyaluronic acid for topical cosmetic: in vitro and ex vivo assessment. *Pharmaceutics* 16: 1195.

Shin, S. H., Y. H. Lee, N.-K. Rho and K. Y. Park. 2023. Skin aging from mechanisms to interventions: focusing on dermal aging. *Front. Physiol.* 14: 1195272.

Shui, G. and L. P. Leong. 2006. Residue from star fruit as valuable source for functional food ingredients and antioxidant nutraceuticals. *Food Chem.* 97: 277–284.

Song, J. H., E. Y. Bae, G. Choi, J. W. Hyun, M. Y. Lee, H. W. Lee et al. 2013. Protective effect of mango (*Mangifera indica* L.) against UVB-induced skin aging in hairless mice. *Photodermatol. Photoimmunol. Photomed.* 29: 84–89.

Sriram, R. and V. Gopal. 2025. Mechanistic insights on skin ageing and dermatologic interventions to slow ageing process. *Indian J. Dermatol.* 70: 135–145.

Stojanovic, B., I. Jovanovic, M. Dimitrijevic Stojanovic, B. S. Stojanovic, V. Kovacevic, I. Radosavljevic et al. 2025. Oxidative stress-driven cellular senescence: mechanistic crosstalk and therapeutic horizons. *Antioxidants* 14: 987.

Surber, C., N. Dragicevic and J. Kottner. 2018. Skin care products for healthy and diseased skin. *Curr. Probl. Dermatol.* 54: 183–200.

Tenore, G. C., D. Caruso, G. Buonomo, M. D'Avino, R. Santamaria, C. Irace et al. 2018. Annurca apple nutraceutical formulation enhances keratin expression in a human model of skin and promotes hair growth and tropism in a randomized clinical trial. *J. Med. Food* 21: 90–103.

Tsunenaga, M., W. Xu, T. Myojin, T. Nakamura, T. Kon, Y. Nakamura et al. 2022. Modulating effects of oral administration of *Lycii fructus* extracts on UVB-induced skin erythema: a randomized, placebo-controlled study. *Biomed. Rep.* 17: 62.

Venus, M., J. Waterman and I. McNab. 2010. Basic physiology of the skin. *Surgery* 28: 469–472.

Wirtu, Y. D. 2024. A review of environmental and health effects of synthetic cosmetics. *Front. Environ. Sci.* 12: 1402893.

Wittenauer, J., S. Mäckle, D. Sußmann, U. Schweiggert-Weisz and R. Carle. 2015. Inhibitory effects of polyphenols from grape pomace extract on collagenase and elastase activity. *Fitoterapia* 101: 179–187.

Wu, F., B. Dang, L. Hu, S. Zhu, Z. Liu, X. Cao et al. 2024. *Lycium barbarum* polysaccharide inhibits blue-light-induced skin oxidative damage with the involvement of mitophagy. *Photochem. Photobiol.* 100: 604–621.

Wu, P., H. Iwahashi, H. H. Xie, Y. Wang, Y. Y. Zhou, A. Kiso et al. 2022. Star fruit extract and C-glycosylated flavonoid components have potential to prevent air pollutant-induced skin inflammation and premature aging. *Nat. Prod. Bioprospect.* 12: 13.

Wu, Y., F. Fang, W. Lai, C. Li, L. Li, Q. Liu et al. 2023. Efficacy and safety of abobotulinumtoxina for the treatment of glabellar lines in Chinese patients: a pivotal, phase 3, randomized, double-blind and open-label phase study. *Aesthetic Plast. Surg.* 47: 351–364.

Xia, Y., H. Zhang, X. Wu, Y. Xu and Q. Tan. 2024. Resveratrol activates autophagy and protects from UVA-induced photoaging in human skin fibroblasts and the skin of male mice by regulating the AMPK pathway. *Biogerontology* 25: 649–664.

Yang, B., X. Zhang, L. Zhang, J. Wu, W. Wang, Q. Huang et al. 2024. Ionic liquid-based grapeseed oil emulsion for enhanced anti-wrinkle treatment. *Pharmaceuticals* 17: 1273.

Yarovaya, L., N. Waranuch, W. Wisuitiprot and W. Khunkitti. 2022. Clinical study of Asian skin changes after application of a sunscreen formulation containing grape seed extract. *J. Cosmet. Dermatol.* 21: 4523–4535.

Yu, C.-H., Y. C. Lin, S. H. Su, K. K. Kan, L. F. Liu, L. S. Lin et al. 2019. Unripe fruit extracts of *Mangifera indica* L. protect against AGEs formation, melanogenesis and UVA-induced cell damage. *FNS* 10: 188–197.

Zhang, C., W. Song, B. Yu, X. Chen, W. Fan, L. Gao et al. 2024. Expert consensus on perioperative integrated skincare for noninvasive energy-based device aesthetic procedures in clinical practice in China. *J. Eur. Acad. Dermatol. Venereol.* 38 Suppl. 6: 26–36.

Zhang, S. and E. Duan. 2018. Fighting against skin aging: the way from bench to bedside. *Cell Transplant.* 27: 729–738.

Zhu, L., Q. Shen, Y. Xu and C. Li. 2025. Protective effect of zeaxanthin from *Lycium barbarum* L. on ultraviolet B-induced skin photodamage in mice through Nrf2-related pathway. *Antioxidants* 14: 632.

Zhu, X., Q. Liu, M. Wang, M. Liang, X. Yang, X. Xu et al. 2011. Activation of Sirt1 by resveratrol inhibits TNF-α induced inflammation in fibroblasts. *PLoS One* 6: e27081.

Zimmer, C. G. M., M. Heller, E. C. Santos, C. S. Freitas, P. O. Benedet, J. M. Siqueira et al. 2025. The potential use of the standardized *Euterpe oleracea* (açaí) seed extract in cosmetic products. *Int. J. Cosmet. Sci.* 47: 563–584.

14 Supporting Gastrointestinal Health With Medicinal Fruits and Berries

Nayomi Deshani De Silva, Anoja Priyadarshani Attanayake, and Thushara Indika Sampath

14.1 INTRODUCTION

Gastrointestinal health is fundamental to overall human well-being, as the digestive system ensures nutrient absorption and energy homeostasis while playing a pivotal role in immunity, neurological signaling, and metabolic regulation (Shang et al., 2024). A bustling ecosystem of microbes, mostly bacteria, yet complemented by viruses and eukaryotic organisms, resides within the gut, contributing to human health in remarkable ways (Shang et al., 2024). Consequences of disruption of the gastrointestinal system extend from local discomfort to systemic ailments such as gastroesophageal reflux disease, irritable bowel syndrome, peptic ulcers, and colorectal cancer. The global prevalence of gastrointestinal disorders contributes significantly to both rates of illness and death. For long-standing health conditions, gastrointestinal disorders are often accompanied by multiple complications, which could profoundly affect public health and increase the economic burden (Suri et al., 2024; Ansari et al., 2025). Though many conventional pharmacological therapies provide symptomatic relief, they are accompanied by several limitations, such as adverse side effects, reduced efficacy in long-term management, increased risk of medication interactions, and hypersensitivity reactions (Ansari et al., 2025). Therefore, adopting a dietary or therapeutic intervention enriched with medicinal fruits and berries emerges as a highly effective approach for supporting gastrointestinal health.

Medicinal fruits and berries have long been valued in ethnomedicine as healthy, affordable, and accessible alternatives to synthetic drugs. Fruits and berries with medicinal value are increasingly recognized as crucial contributors to gastrointestinal health due to their distinctive blend of nutrients and bioactive compounds (Ansari et al., 2025). Indeed, medicinal fruits and berries are rich in numerous phytoconstituents, including polyphenols and flavonoids, along with vitamins and dietary fiber. The availability of diverse phytoconstituents enables them to modulate various biological targets simultaneously by enhancing their therapeutic efficacy. Medicinal fruits, being high in fiber, vitamins, antioxidants, and polyphenols, contribute to enhancing immune function, reducing gastric inflammation, and fostering microbial diversity (Maqsood et al., 2025). Dietary fibers in fruits and berries act as prebiotic substrates for advantageous microbiota, thereby promoting the generation of short-chain fatty acids (SCFAs). Accordingly, these attributes emphasize the important role of the consumption of medicinal fruits and berries in nurturing a balanced gut ecosystem and supporting gastrointestinal health (Maqsood et al., 2025).

Emerging evidence highlights that citrus fruits (Family: Rutaceae), such as *Citrus sinensis* L. Osbeck (orange), *Citrus limon* L. Osbeck (lemon), and *Citrus paradisi* Macfadyen (grapefruit), *Malus domestica* Borkh (apple, Family: Rosaceae), *Pyrus communis* L. (pears, Family: Rosaceae), *Musa acuminata* Colla (banana, Family: Musaceae), *Punica granatum* L. (pomegranate, Family: Punicaceae), and berries such as *Vaccinium sect.* Cyanococcus (blueberries, Family: Ericaceae),

DOI: 10.1201/9781003688259-14

Fragaria ananassa Duchesne (strawberries, Family: Rosaceae), and *Rubus idaeus* L. (raspberries, Family: Rosaceae) contribute to gastrointestinal well-being through the positive effects of fibers, polyphenols, and antioxidants on gut health and the immune system (Maqsood et al., 2025). Therefore, integrating medicinal fruits and berries into the daily diet offers a promising approach to preventing and managing gastrointestinal ailments. Accordingly, it is important to highlight the consolidated mechanistic insights into medicinal fruits and berries as potent modulators of gut microbiota and intestinal physiology, preclinical evidence from scientific investigations on fruits and berries supporting gastrointestinal health, and research gaps to guide future dietary and therapeutic applications of fruits and berries.

14.2 OVERVIEW OF GASTROINTESTINAL PHYSIOLOGY AND REGULATION

The gastrointestinal system is a highly integrated network of organs that ensures the digestion, absorption, and assimilation of nutrients required for human survival (Greenwood-Van Meerveld et al., 2017). The gastrointestinal system integrates with the vascular, lymphatic, and nervous systems to facilitate the regulation of digestive activity, efficient circulation of absorbed nutrients to organs, and the regulation of food ingestion (Greenwood-Van Meerveld et al., 2017). It performs mechanical and chemical processing of food through coordinated motility, secretion, enzymatic breakdown, and absorption, extending from the mouth to the anus. The small intestine serves as the center for nutrient and mineral absorption. Water and electrolyte reabsorption and the formation of feces are mainly performed in the large intestine. Further, it supports a diverse microbial ecosystem that contributes to metabolism, immune modulation, and overall health maintenance.

The gastrointestinal tract harbors an extensively diverse and metabolically active community of microorganisms, including bacteria, yeast, viruses, and parasites. Among these dynamic microbial ecosystems, *Firmicutes* and *Bacteroidetes* are the foremost bacterial phyla represented in the human gut microbiota, accounting for 90% of the total microbial population in a healthy intestine. Phyla *Verrucomicrobia* and *Fusobacteria* are present in minor proportions in the gastrointestinal tract (Kanmani et al., 2020). Age, genetics, lifestyle, environment, and diet are the factors that can affect the makeup of the gut microbiota. As dietary practices tend to vary according to geographical factors, this variability could also be seen reflected in the microbiota of various populations. Gut-residing microbiota have coevolved with humans and are recognized as part of a symbiotic system that plays an integral role in maintaining physiological balance and supporting human health. The gut microbiota performs various crucial functions in maintaining host health through multiple mechanisms. The intestine helps to maintain its barrier integrity through the generation of SCFAs with anti-inflammatory effects (Figure 14.1), epithelial renewal, metabolite production, protection against enteric pathogens, and modulation of mucosal and systemic immune responses are also mediated by microbial populations (Van Hul et al., 2024). The homeostatic balance and composition of the gut microbiota are profoundly modulated by diet, antimicrobial interventions, and diverse environmental factors, making the microbiota a key mediator in the immunological defense and physiological resilience of the host cells.

14.3 GASTROINTESTINAL HEALTH AND DISEASES

Gastrointestinal health is the state in which the digestive tract supports efficient digestion and absorption, maintains mucosal barrier functions, preserves a balanced microbiota, and sustains tightly coordinated immune-neuroendocrine interactions to support overall homeostasis. Further, it is defined as a state of physical and mental well-being without any gastrointestinal complaints. Immune cells, goblet cells, and Paneth cells are induced by the gut microbiota to assemble functionally active components such as anti-inflammatory cytokines and immunoglobulins. This contributes to the maintenance of gut homeostasis preventing increased intestinal permeability and inflammation, thereby promoting leading to a healthy gut (Figure 14.1) (Kanmani et al., 2020).

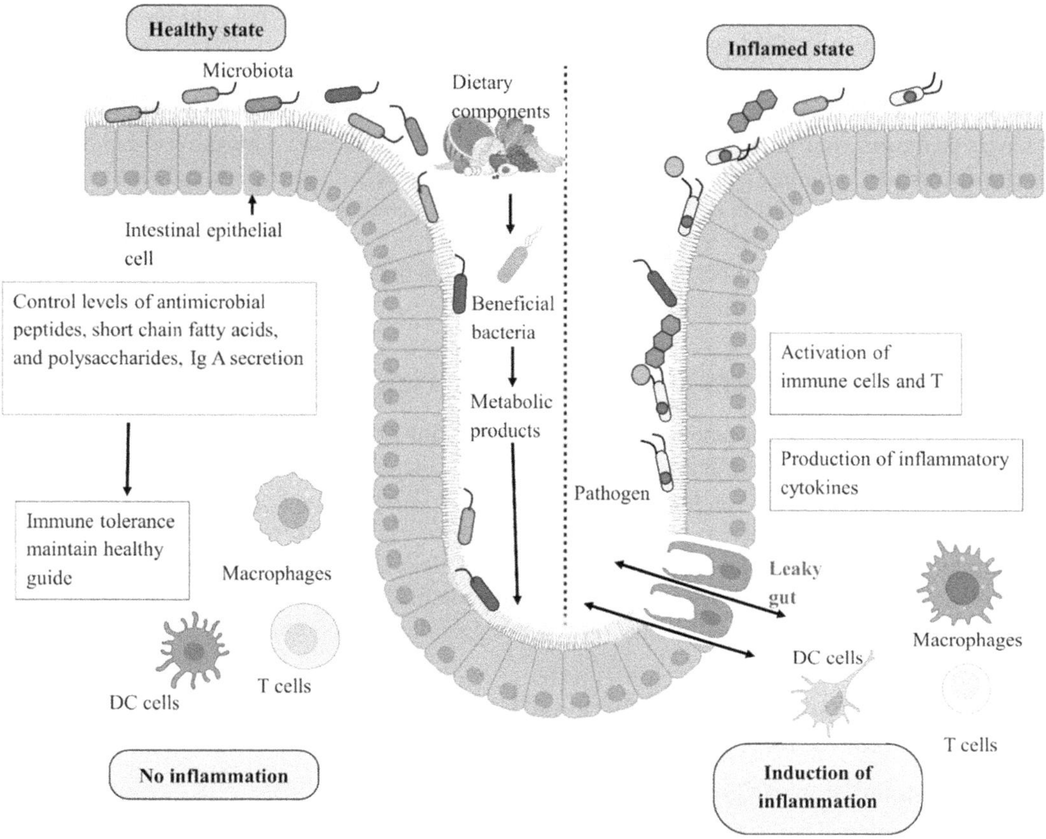

FIGURE 14.1 Interaction between host and gut microbiota maintaining gut health and triggering inflammation.

The combinations of abnormal gut movement, alterations in gut lining and immune function, visceral hypersensitivity, and imbalance in gut microbiota lead to gastrointestinal disorders. Abnormalities in gastrointestinal function could lead to life-threatening diseases. Antibiotics, obesity, dietary modifications, and various environmental factors can lead to the initiation of gut dysbiosis. This condition enhances the loss of gut homeostasis, bacterial overgrowth, and impairs the strength and protection of the gut lining. The penetrability of pathogens and their associated molecular patterns is enhanced with the disturbance of the gut barrier integrity, and induces intestinal inflammation (Kanmani et al., 2020).

14.4 HOW DIET SHAPES GASTROINTESTINAL HEALTH

The optimal functioning of the gastrointestinal system is extremely dependent on a balanced and adequate diet. Nutrient levels at either extreme, excess or deficiency, could compromise immune system function and are linked to inflammation-driven diseases, including gastrointestinal diseases. In contrast, balanced and adequate nutrients significantly impact gastrointestinal health by reshaping the arrangement and activity of gut microbiota, mucosal barrier, and motility. Fibers, proteins, fats, and polyphenols derived from the diet act as substrates for gut microbiota, influencing the human metabolome and physiology (Manske, 2024). Trends in gastrointestinal health accentuate the utilization of a fiber-rich, herb-based diet rich in fruits, vegetables, whole grains, and fermented foods to nurture a healthy gut community. A diet enriched in fiber, particularly prebiotic fibers, exerts a favorable effect on the metabolome characterized by elevated SCFA production and reduced concentrations of harmful metabolites such as indoxyl sulfate and p-cresyl sulfate associated with pathological states.

There is an intricate crosstalk among nutrition, immunological regulation, and the microbiome, with investigations on fruit-derived dietary patterns and their effects on gastrointestinal well-being and the immune system. Among the several dietary strategies, the consumption of medicinal fruits and berries is recommended for supporting gastrointestinal health due to their potential consequences on the gut microbiota and gut motility (Manske, 2024). The interconnection of polyphenols, fibers, flavonoids, vitamins, and antioxidants in medicinal fruits and berries further fosters mutually beneficial relationships within the gut microbiota, lessens intestinal inflammation, enhances bacterial populations that generate SCFAs, and promotes diversity in microbes. Accordingly, dietary intake of medicinal fruits and berries would have wide-ranging implications for supporting gastrointestinal health while preventing disorders.

14.5 MEDICINAL FRUITS AND BERRIES IN GASTROINTESTINAL HEALTH

Fruits are the seed-bearing, palatable products arising from the mature ovary of flowering plants. They are central to the human diet, providing minerals, vitamins, dietary fiber, polyphenols, flavonoids, and antioxidants in dried, fresh, or juiced forms. Medicinal fruits supporting gastrointestinal health encompass diverse groups such as berries, citrus fruits, stone fruits, pome fruits, grapes (*Vitis vinifera* L., Family; Vitaceae), kiwi fruits (*Actinidia chinensis* var. deliciosa, Family; Actinidiaceae), and cactus fruits. The beneficial health effects of consuming medicinal fruits and berries include modulation of gut microbiota, modulation of immune function, and modulation of oxidative stress (Figure 14.2).

14.5.1 Mechanisms Underlying Medicinal Fruits and Berries in Gastrointestinal Health

14.5.1.1 Modulation of Gut Microbiota

Consuming medicinal fruits and berries helps regulate gut microbes and their function, resulting in beneficial effects on host gastrointestinal health. Fruits and berries contain several active phytochemicals, and they exert valuable effects on human beings, capable of regulating the bacterial population growth in the intestine. Phenolic compounds are capable of regulating gut microbiota, favoring the enhancement of helpful bacteria species like *Akkermansia* spp., *Lactobacillus* spp., and *Bifidobacterium* spp. Inhibition of pathogen multiplication by phenolic compounds promotes the

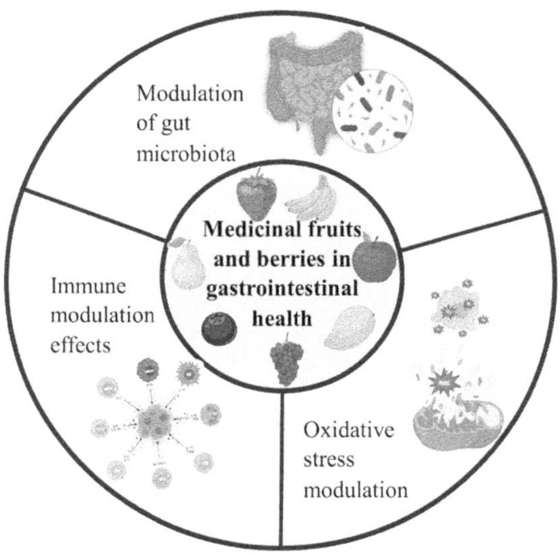

FIGURE 14.2 Mechanisms underlying medicinal fruits and berries in gastrointestinal health.

expansion of beneficial bacteria (Pinheiro et al., 2024). Crosstalk among the gut microbiota and antioxidant compounds present in medicinal fruits and berries is decisive for homeostasis, as they help expand bacterial assortment and stimulate microbiome-derived metabolite production. As a case in point, administration of strawberries has the potential to enhance *Bifidobacterium* abundance in diabetic mice, highlighting the defensive roles of berry-derived polyphenols in the gut (Petersen et al., 2019). In addition, fruits such as apples, oranges, and avocados (*Persea americana* Mill., Family; Lauraceae) contain high dietary fiber content and are a vital energy source for anaerobic bacteria whose fermentation generates SCFAs. Gallic acid from fruits and berries acts as a substrate for bacterial fermentation or is decomposed into protocatechuic acid. Flavonoids are metabolized by some *Lactobacilli* and *Bifidobacteria* species and release sugars, which can serve as growth substrates for the strains. Intestinal epithelial cells could absorb aglycones, benefiting the host by strengthening gut barrier integrity and providing fuel for beneficial microbes (Petersen et al., 2019).

14.5.1.2 Immune Modulation Effects

Inflammation occurs under the condition of dysbiosis, leading to disrupted immune regulation and overgeneration of cytokines that drive inflammation. Consumption of fruits could confer potent anti-inflammatory properties and reinstate microbial balance. It was revealed that polyphenols in grapes, berries, and apples attenuate intestinal inflammation by hindering nuclear factor kappa-B. Recent work spotlights the association between medicinal fruits/berries and immune function, shaping immunological responses through improved intestinal health. Accordingly, medicinal fruits and berries diminish gut inflammation via converging actions along the microbiota–gut–immune axis. Regulation of T cell activity, promotion of immunological tolerance, and prevention of overactivation occur with the improved production of SCFAs (Maqsood et al., 2025). Fruit-derived vitamins, polyphenols, fibers, and flavonoids jointly foster eubiosis. This gives sustenance to gut-associated lymphoid tissue function and strengthens mucosal immunity. Heat-induced gastrointestinal disorders are averted by improving microbial assortment and gut barrier function with dietary intake of fruits. Flavonoids and carotenoids derived from fruits and berries have anti-inflammatory properties. Collectively, these mechanisms position fruit-based dietary patterns as immunomodulatory strategies to reduce gastrointestinal inflammation.

14.5.1.3 Oxidative Stress Modulation

Flavonoids, vitamin C, and carotenoids are common antioxidants derived from medicinal fruits and berries. They are well capable of improving immune competence and protecting tissues from oxidative injury, supporting gastrointestinal health. Effective immunity supporting gastrointestinal health requires a balance between the defensive system and oxidative stress. Vitamin C from citrus fruits, kiwi fruits, and berries could directly scavenge free radicals and support the production of other antioxidants such as vitamin E and glutathione, in turn supporting redox homeostasis and the smooth function of macrophages, lymphocytes, and neutrophils (Hyży et al., 2025). Flavonoids and carotenoids support the protection and stability of the mucosal surface in the gastrointestinal system.

14.5.2 Commonly Reported Medicinal Fruits in Gastrointestinal Care

14.5.2.1 Citrus Fruits

Oranges, grapefruits, lemons, limes, and pomelos are categorized under citrus fruits and contain hesperidin, naringin, and apigenin as flavonoids, terpenes, polymethoxylated flavones, vitamin C, limonene as essential oils, pectin as dietary fiber, and marmin, merazin, and scopoletin as coumarins (He et al., 2018). Citrus fruits are extensively valued as an essential part of the diet due to their support of immune function through strong anti-inflammatory activity. Citrus fruits are widely regarded as an important dietary constituent for immunological functions with anti-inflammatory and immune-enhancing properties supporting gastrointestinal health (He et al., 2018). In addition, improvement of immune cell actions, skin strength, and permeability is achieved by citrus fruits enriched in vitamin C. Practically, dietary fibers and flavonoids in citrus fruits often escape digestion in the upper gut and make their way to the colon. Bioactive derivatives are made when these flavonoids and fibers interact with gut microbes (Zayed et al., 2025). This interaction helps keep both gut and systemic health. It encourages the maturation of valuable microbes and stimulates the SCFA generation, which is responsible for modulating oxidative stress and the immune response.

Quercetin and hesperidin are well-known flavonoids found in citrus fruits and exert potent anti-inflammatory properties. Attenuation of intestinal inflammation is promoted through suppression of pro-inflammatory cytokines with quercetin and hesperidin. In addition, citrus fruit flavonoids could decrease intestinal permeability while modulating the intestinal microbiota. This facilitates the reduction of inflammatory bowel diseases and various other digestive disorders. Beta-carotene, being an antioxidant in citrus fruits, activates immune responses while lowering oxidative stress and intestinal inflammation (He et al., 2018).

The synergy among carotenoids, flavonoids, vitamin C, and terpenes in citrus fruits is largely responsible for strengthening the immune system and gastrointestinal health (Wang et al., 2024). Moreover, it helps to diminish inflammation and increase resistance to infection. Dietary flavonoids from citrus species are among the most important naturally occurring phytochemicals with several promising effects on intestinal barrier function. After consumption, citrus flavonoids are released from the lumen and are subjected to biotransformation by host epithelial cells. This leads to the generation of various downstream metabolites. Citrus flavonoid glycosides are resistant to digestion by stomach enzymes and pass into the small intestine (Wang et al., 2024). A small proportion may penetrate through the intestinal lining and enter the portal blood circulation. Then, aglycones are released through direct hydrolysis in the presence of lactase phloridzin hydrolase and/or taken up into the enterocytes via sodium–glucose cotransporter 1. These formed aglycones can either cross into the portal blood or undergo further phase II conjugation in the enterocytes to afford glucuronidated/sulfated metabolites (Wang et al., 2024). It is reported that active metabolites of citrus fruits could support gastrointestinal health through the maintenance of the mucus layer by promoting mucin-2 production, controlling the growth and maturation of immune cells, and increasing secretory immunoglobulin A secretion. Further, active metabolites in citrus fruits are capable of inhibiting oxidative damage (Wang et al., 2024).

14.5.2.2 Pome Fruits

A pome is a fleshy fruit consisting of an outer, thickened fleshy layer. Pome fruits, such as pears, apples, and quince (*Cydonia oblonga* Mill. Family; Rosaceae) contain a synergistic combination of soluble and insoluble fibers, and polyphenols that contribute to intestinal health.

The dietary fiber content in apples and pears ranges from 1.4–2.7% and 1.6–3.6%, respectively, compared to fresh weight (Hoogeveen et al., 2025). Pectin is an active compound that exists in the skins of apples and pears and is recognized as the key soluble fiber. Apples and pears have attracted attention since the availability of pectin is needed to control the immune system and alter the profile of gut microbiota (Hoogeveen et al., 2025). SCFAs such as butyrate, propionate, and acetate are generated during pectin fermentation in the large intestine, serving as a prime energy source for colonocytes. Accordingly, the formation of SCFAs plays a momentous role in perpetuating normal colonic function. Improved blood flow, stimulation of electrolyte fluid absorption, activation of regulatory T cells, and reduction of intestinal pH are facilitated by the generation of SCFAs. Activation of T cells causes the restriction of excessively active immune responses and prolonged inflammatory responses. Further, pectin is abundant in pome fruits as a fermentable fiber that can act as a prebiotic that selectively enriches the healthy microbiome, including *Lactobacillus* and *Bifidobacterium*, in the colon. An optimally functioning gut microbial community is vital in the regulation of immunity since it directly impacts the immune cell formation through the control of inflammation (Donadio et al., 2024). This, in turn, supports the maintenance of conditions like inflammatory bowel diseases and digestion in general. Emerging evidence proposes that pectin may have anti-inflammatory characteristics that are significant to maintaining immunological equilibrium in vulnerable groups like newborns and preterm infants (Donadio et al., 2024).

Cellulose and hemicellulose are insoluble fibers in apples and pears. Galactooligosaccharides and fructo-oligosaccharides are also found in small quantities. Fiber affords volume and bulk to ingested food, which affects gastric emptying, nutrient absorption, and satiety. Fiber becomes a substrate for microbial fermentation. Alongside pectin, polyphenols in apples and pears also enhance digestive wellness through anti-inflammatory, antioxidant, and microbiota-modulating effects. Chronic administration of nonsteroidal anti-inflammatory drugs causes damage to the gastric mucosa. It has been reported that indomethacin promotes oxidative stress and lipid peroxidation in gastric and intestinal cells (Jakobek and Matić, 2024). However, it is worth mentioning that polyphenols derived from apples and pears are very helpful in alleviating the risk of the side effects caused by consuming nonsteroidal anti-inflammatory drugs in the digestive tract. Prevention of increased malondialdehyde concentrations leads to the protection of the gastric mucosa against oxidative stress, while prevention of neutrophil infiltration into the mucosa is done through their anti-inflammatory activity. Further, phenolic compounds in apples and pears could stimulate the healing of gastric ulcers and strengthen the gut barrier by enhancing the expression of tight junction proteins. The administration of pome fruits also adds beneficial effects supporting gastrointestinal health.

14.5.2.3 Berries

Berries are of particular interest and are identified as a superfood for gastrointestinal health among fruits due to their high availability of bioactive compounds. The genera *Vaccinium* (blueberries, *Vaccinium vitis-idaea* L. (lingonberries), and *Vaccinium macrocarpon* Aiton (cranberries)), *Rubus* (*Rubus fruticosus* agg. (blackberries), raspberries), *Ribes* (*Ribes nigrum* L. (black currant)), *Fragaria* (strawberries), and *Lycium* (*Lycium chinense* Mill. (goji berries)) are commonly considered under berries, which support gastrointestinal health (Lavefve et al., 2020). Berries are considered to be an important health-promoting food source due to their rich content of phenolic compounds, flavanols (quercetin and kaempferol), vitamins, carotenoids, and minerals. Hydroxycinnamic acids and hydroxybenzoic acids are common phenolic acids in berries. p-hydroxybenzoic, ellagic, salicylic, and vanillic acids are benzoic derivatives, while caffeic acid and ferulic acid are the main representatives of hydroxycinnamic acids. Berries, particularly cranberries, contain tannins. The key anthocyanins available in berries are pelargonidin, cyanidin, delphinidin, petunidin, malvidin anthocyanidins and peonidin (Lavefve et al., 2020). Current literature positions these bioactive compounds as central to the regulation of gut microbiota, and consequently to the maintenance of gastrointestinal health and immunological resilience.

Almost all the administered polyphenols from berries can move down to the colon and be digested at that site, leading to the production of metabolites. The modulation of bacterial populations in the gut could be facilitated by polyphenols and their metabolites. Multiple substrates, such as carbohydrate polymers in the form of insoluble fibers that can be utilized by the microbiome, are present in berries (Feng et al., 2025). Phenolic compounds in berries directly influence bacteria, supporting gut homeostasis. Besides, in view of the poor bioavailability of berry phenolic compounds, these compounds are unable to be absorbed by the small intestine but are metabolized by the gut microbiome or enzymes in the colon. Phenolic compounds further undergo alterations through methylation, glucuronidation, and sulfation reactions once absorbed in the liver and kidneys. Several research studies documented the inhibitory potential of polyphenols and fibers in berries against bacterial (*Vibrio cholera*, *Streptococcus mutans*), viral (herpes simplex virus type 1 and influenza A virus), and fungal pathogens (*Candida albicans* and *Trichophyton rubrum*). The chelation of essential minerals, including iron, copper, and zinc, for the metabolism of pathogens is accelerated by polyphenols in berries (Borowska et al., 2020; Maqsood et al., 2025). Furthermore, polyphenols in berries have the capacity to reduce virulence determinants through the inhibition of quorum sensing and disruption of biofilm formation. This was supported by some in vitro studies on *Pseudomonas aeruginosa*. It is well reported that the ingestion of black and red berries leads to enhanced *Lactobacillus*, *Clostridia*, and *Bifidobacterium* levels (Borowska et al., 2020; Maqsood et al., 2025).

Berries are a great source of antioxidants with the presence of anthocyanins, ellagic acid, resveratrol, etc. Incorporation of berries into the diet, particularly strawberries, could reduce tissue redox imbalance.

Administration of raspberries has the potential to reduce colonic shortening and ulcer formation. In the same way, mulberry and blueberry help to maintain the normal structure of intestinal crypts and enhance the number of goblet cells, resulting in improved barrier function (Perri et al., 2021). Blueberries can be considered a powerhouse of nutrients. The availability of high amounts of fiber in berries facilitates maintaining regular bowel movements and prevents constipation. Insoluble fibers could add bulk to the fecal content, stimulating gastrointestinal peristalsis through physical contact, while fibers with high solubility could modulate the microbiota composition in the intestine. Both soluble and insoluble fibers facilitate the retention of water in stool. Lipids in the diet and bile acids are also adsorbed, making the stool more emollient and voluminous, which stimulates peristalsis and consequently increases the frequency of bowel movements (Perri et al., 2021). The nutrients and bioactive compounds in berries can help to reduce intestinal permeability directly or indirectly. Further, those nutrients and phytochemicals in berries promote the homeostasis of the generation of anorexigenic hormones, digestion, and absorption. Cholecystokinin and serotonin are directly involved in controlling digestion, satiety, and gastrointestinal motility, emphasizing the participation of the enteric nervous system in the pathophysiology of constipation. Thus, the regular use of fruits, including berries and their consequent products, may improve an individual's clinical condition by reducing constipation. In addition, berries, including blueberries and strawberries, have the potential to improve favorable gut bacteria owing to the high value of prebiotics, which serve as precursors to probiotics. In accordance with this, berries are an excellent medicinal fruit for supporting gastrointestinal health.

14.5.2.4 Pomegranate

Pomegranate is widely recognized as a robust medicinal fruit for supporting gastrointestinal health owing to its numerous bioactive compounds, including flavonoids, phenolic acids, condensed tannins, and hydrolysable tannins. Pomegranate could impact the immune system and the integrity of the intestinal system on account of the potent effects of its bioactive compounds on the microbiota and oxidative stress. Polyphenols in pomegranate have been widely studied for their potential to reduce oxidative stress and intestinal inflammation (Buccato et al., 2025). Scientific studies show that polyphenols derived from pomegranate have the potency to inhibit systemic inflammation and gastrointestinal diseases by reducing pro-inflammatory cytokines and modulating the integrity of the intestinal lining (Santacroce et al., 2024; Yin et al., 2024; Buccato et al., 2025). Further, modulation of the intact immune system by reducing intestinal inflammation can be triggered by pomegranate polyphenols. It has been proven that pomegranate polyphenols derived from the peel have the ability to exert immune-enhancing effects through the immunomodulation of T cells, B cells, and macrophages. The body amplifies its immunological defenses, and dysregulation contributes to autoimmune diseases or long-term inflammation (Santacroce et al., 2024). All these properties have sparked scientific interest in pomegranate and invigorated the ingestion of diverse pomegranate-enriched formulations for the treatment and prevention of inflammatory conditions, such as inflammatory bowel disease, and for supporting gastrointestinal health.

14.5.2.5 Banana

Bananas are of particular interest for their production of secondary metabolites such as carotenoids, biogenic amines, phenolic compounds, and flavonoids due to their health benefits. Bananas are rich in fructo-oligosaccharides and resistant starch, which are types of prebiotics that encourage the multiplication and health-promoting functions of the intestinal microbiota (Muñoz et al., 2025). The fermentation of resistant starch by the gut microbiota leads to the production of SCFAs and butyrate, thereby decreasing inflammation, maintaining intestinal motility, and preserving the integrity of the gut. Bananas have recently garnered attention as a traditional dietary intervention for alleviating gastritis. Several studies have identified flavonoids such as leucocyanidin among the key bioactive compounds responsible for the gastroprotective effects of bananas. Compounds derived from bananas, including protocatechuic acid, catechol, catechin, epicatechin, gallocatechin, naringenin, and rutin, have been shown to promote mucosal healing, reduce inflammation, and prevent ulcer formation. Some studies have suggested that the beneficial effects of bananas on gastric ulcers may vary based on the type of banana and its specific bioactive compounds (Muñoz et al., 2025). For instance, a pronounced gastroprotective effect was demonstrated for *Musa* × *paradisiaca* L. (Family: Musaceae). This effect is highly attributed to the ability of banana extracts to enhance gastric pH, decrease ulcer size, and increase gastric wall mucositis.

Apart from the mentioned fruits and berries, *Carica papaya* L. (papaya, Family; Caricaceae), *Ananas comosus* L. (pineapple, Family; Bromeliaceae), *Mangifera indica* L. (mango, Family; Anacardiaceae), *Actinidia chinensis* var. deliciosa (kiwi fruits, Family; Actinidiaceae), grapes, avocado, and *Citrullus lanatus* (Thunb.) Matsum (watermelon, Family: Cucurbitaceae) also confer significant gastrointestinal benefits attributable to their fermentable fiber, carotenoids, and bioactive enzymes that support microbiota balance and mucosal integrity.

14.5.3 Preclinical and Clinical Evidence Supporting Gastrointestinal Benefits

Gastrointestinal health is increasingly reinforced by preclinical and clinical evidence generated through complementary bench-top assays, in vivo, and human intervention studies. The cornerstone of in vitro intestinal models is the epithelial cell layer, which constitutes the intestinal mucosa lining. Researchers select epithelial cells from a variety of options, including immortalized human intestinal cell lines such as Caco-2 and HT-29, primary cells isolated from intestinal tissues of human or animal donors, and stem cells. Intestinal permeability and absorption studies, colorectal cancer, and intestinal permeability and differentiation are assessed mainly using Caco-2 and HT-29 immortalized cell lines. Intestinal development, function, colonic physiology, and responses to inflammation and infection, epithelial transport, and ion secretion are commonly evaluated using primary cells such as colon epithelial cells (HCEC-1CT), intestinal tissue (HIECs), and embryonic intestine (Int-407) (Adesina et al., 2025). In vitro human intestinal models span a spectrum of complexity, including the basic two-dimensional monolayers and Transwell systems to more complex systems like cocultures, spheroids, organoids, flip-well, gut-on-a-chip, and bioengineered models. These models vary in architecture, cellular composition, and how well they replicate the dynamic intestinal environment.

It is important to have effective digestion for potent absorption of nutrients and to maintain overall health. Malabsorption of nutrients is caused by a disruption in digestion. It directly affects the immunological, neurological, and cardiovascular systems. Simulated in vitro digestion systems have come forward as a current alternative to in vivo experiments on food and drug release and absorption in the gastrointestinal tract. (Wu et al., 2025). Static models, dynamic models including the TNO gastrointestinal model, engineered stomach and small intestinal systems, and an artificial gastric digestive system are several in vitro models used in food digestion (Kiran et al., 2025). These models mimic the real physiological settings of the human digestive tract, offering an ethically sound and cost-effective approach to assess the digestion process for applications in gastrointestinal health. Rodent models of diet-induced dysbiosis and chemically induced colitis are employed in vivo to evaluate the effects of foods such as fruits and berries on intestinal permeability, mucus layer thickness, immune response, microbiota structure, often alongside systemic markers such as inflammatory markers and oxidative stress. Promising results obtained through preclinical studies are further extended into clinical trials, where gastrointestinal health is evaluated through validated symptom scores, fecal biomarkers, stool frequency and consistency, intestinal permeability markers, systemic inflammatory indices, and high-throughput analyses of gut microbiota and the metabolome. Together, converging evidence from these preclinical and clinical settings provides a multidimensional understanding of how dietary components, including medicinal fruits and berries, and their bioactive compounds support and restore gastrointestinal health.

An expanding pool of data from both experimental and human studies convincingly demonstrates that medicinal fruits and berries exert multifaceted benefits on gastrointestinal health. The effect of apples (*Malus domestica* Mill) and pears (*Pyrus* spp.) on in vitro fermentation using human fecal microbiota was assessed in numerous scientific studies, and the results highlighted that the availability of polyphenols and fibers could significantly affect the gut mycobiome and SCFA production (Zhao et al., 2022; Ferroni et al., 2024; Hoogeveen et al., 2025; Zhang et al., 2025). In addition, the ability to reduce inflammation and affect the gut microbiome was assessed in a flavonoid-rich fraction derived from *Passiflora foetida* L. (Family: Passifloraceae) fruits by Han et al. (2023). The study found that flavonoids in *P. foetida* fruit have the potential to decrease nitric oxide and interleukin-6 (IL-6) concentrations in immune cells (RAW 264.7), indicating their anti-inflammatory effect. In a dynamic human gut model, *P. foetida* could favorably reshape the microbiota by encouraging the growth of *Bifidobacterium* and *Lactobacillus* and by enhancing the generation of health-supporting SCFAs (Han et al., 2023).

Polyphenols, particularly gallic acid and gallotannins, are present in mango. They are well-known for their pharmacological impact, such as reducing oxidative stress and inflammation and improving resilience against long-term inflammation and related diseases in the intestine (Kim et al., 2021). The protection and improvement of intestinal barrier function were assessed with the introduction of mango polyphenols in human intestinal cell models (Caco2/HT29-MTX) exposed to an inflammatory condition (Pereira-Caro et al., 2024). The impact of fruit and fruit-based product consumption has also been investigated in clinical trials. Asuncion et al. (2023) examined the impact of fresh mango consumption on the gut microbiome, proteins that control gut permeability, and bowel habits in people who are overweight or obese. This study followed a crossover design conducted over a 12-week intervention, with 27 participants consuming 100 kcal/day of either mangos or low-fat cookies. According to the results, the abundance of *Prevotella maculosa*, *Corynebacterium pyruviciproducens*, and *Mogibacterium timidum* increased as potentially beneficial microbes, while *Prevotella copri* decreased with the administration of mango, highlighting the positive effects on gastrointestinal health (Asuncion et al., 2023).

The effects of pineapple and green banana fiber powder ingestion on the host gut microbiome were evaluated by Chong et al. (2024). Stool-derived DNA was prepared for sequencing with the administration of green banana (10.75 g) and pineapple fiber powder (7.41 g) for 60 adults with a mean age of 40 years. The study claims that dietary fiber and resistant starch in green banana and pineapple supported a gradual increase in beneficial bacteria. Green banana powder remarkably elevated seven species, including *Bifidobacterium*, while pineapple fiber powder enriched six species, including *Bacteroides ovatus* and *Bacteroides cellulosilyticus*. Further, based on the data from questionnaire responses, it is highlighted that both green banana and pineapple fiber powder have the potential to improve bowel regularity as well as the metabolism of histidine, biotin, branched-chain amino acids, and SCFA production (Chong et al., 2024).

Several types of berries have been investigated considering their pharmacological impact on gastrointestinal health. Multiple lines of research have highlighted the beneficial properties of blueberries, cranberries, raspberries, and strawberries as modulators that support gastrointestinal health. Six-week-old C57BL/6J male mice were developed into an animal model of metabolic syndrome-induced vascular inflammation to investigate the way in which strawberry consumption diminishes vascular complications and changes in the composition of gut microbes by Miller et al. (2022). Mice fed with a 2.35% freeze-dried strawberry-supplemented diet for 12 weeks exhibited significantly suppressed vascular inflammation by decreasing the elevated serum chemokines, mRNA expression of inflammatory molecules, and monocyte binding to vasculature (Miller et al., 2022). Forty-three patients with functional gastrointestinal diseases were recruited to conduct a double-blind, randomized, crossover trial to evaluate the effect of two doses of 15-g powder on gastrointestinal health. Ingestion was done after breakfast and dinner meals for six weeks. Based on the results, blueberry treatment resulted in relevant abdominal symptom relief in patients compared to the placebo group, highlighting the beneficial effect of consuming blueberries (Wilder-Smith et al., 2023).

Beyond the scientific studies already described for medicinal fruits, including berries, extensive research has emerged from both preclinical and clinical settings. Accordingly, scientific investigations carried out in 2024 and 2025 toward gastrointestinal health, involving specifically medicinal berries, are summarized in Table 14.1, while studies focusing on medicinal fruits (excluding berries) are presented in Table 14.2.

TABLE 14.1
Preclinical and Clinical Studies Carried Out in 2024 and 2025 on Berries for Gastrointestinal Health

Type of Berry	Nature of Study	Biochemical Assessment	Main Results	References
Fragaria × ananassa (Strawberry) *Rubus chamaemorus* (cloudberry) *Rubus idaeus* (Red raspberry) *Vaccinium myrtillus* (bilberry) *Vaccinium vitis idaea* (Lingonberry)	43 healthy adults Dose—200 g/day of mixed berry supplement for four weeks	Analysis of fecal phenolic acids Fecal microbiota profiling Cell viability in human colonic adenocarcinoma (HCA-7), colorectal adenocarcinoma (Caco-2), and the kidney of African green monkey (CV1-P)	Increased fecal levels of p-coumaric and protocatechuic acids Reduced unfavorable gut microbiota and enhanced beneficial microbiota Decreased viability	Onali et al. (2025)
Berry	23 healthy adults Dose—capsule containing 36% polyphenol rich berry extract for 18 days	Metabolomic analysis Microbiota analysis	Reduced intestinal gas and gas-related discomfort Exhibited positive correlation among some taxa, bile acids, and digestive sensation Decreased steroid corresponding metabolite cortolone and bile acid 7-ketodeoxycholic in the urine Induced as an adaptation of gut microbiota, increased favorable microbiota	Barber et al. (2024)
Rubus chingii Hu. (Raspberry)	High-fat diet–induced obesity in C57BL/6 mice Dose—400 mg/kg of raspberry polysaccharides for four weeks	Caco-2 cells culture and transepithelial resistance assay 16S rRNA-sequencing analysis Inflammation indicators	Ameliorated inflammatory infiltration, intestinal villi shedding, and intestinal epithelium permeability Increased generation of SCFAs Profusion of *Ruminococcaceae, Bifidobacterium pseudolongum*, and *Lactobacillus taiwanensis* Decreased expressions of TNF-α, IL-6, and IL-1β	Huang et al. (2024)
Lycium barbarum L. (Goji berry)	Male C57BL/6J mice induced by a high-fat diet Dose—200 mg/kg of lyophilized powder for six weeks	Microbiota DNA sequencing	Increased *Faecalibacterium prausnitzii* and *Akkermansia muciniphila*	Jeong et al. (2024)
Goji berry	In vitro	Digestion followed by fermentation Analysis of SCFAs Analysis of microbiota	Enhanced the bioavailability of phenolic compounds, and the release of phenolics reached their highest in the first 15 min of the gastric phase Improved production of SCFAs while reducing the colon pH Enhanced the gut microbiota profile by increasing useful bacteria	Wang et al. (2024)

Note: TNF-α; tumor necrosis factor-alpha, IL-1β; interleukin-1β, IL-6; interleukin-6, SCFAs; short chain fatty acids.

TABLE 14.2
Preclinical and Clinical Studies Carried Out in 2024 and 2025 on Medicinal Fruits (Except Berries) on Gastrointestinal Health

Type of Fruit	Nature of the Study	Biochemical Assessment/Assessment	Main Results	References
Actinidia deliciosa var Hayward (Green kiwifruit)	A randomized, double-blind, placebo-controlled trial (n=188) with constipation-predominant irritable bowel syndrome Dose—575 mg twice daily for 28 days	Bowel movement frequency using Bristol stool scores Abdominal cramp	Improved bowel habits with improved Bristol stool score (87%—test group vs. 73%—placebo group) Reduced abdominal cramps	Holtmann et al. (2025)
Choerospondias axillaris L. (Nepali hog plum fruit)	In vitro study using polysaccharides isolated from ethanol extract	Gastrointestinal digestion, fecal fermentation, followed by 16S rRNA sequencing	Increased abundance of helpful bacteria—*Megasphaera*, *Bifidobacterium*, and *Megamonas*, Decreased infective bacteria—*Collinsella*, *Gemmiger*, *Citrobacter*, and *Klebsiella*	Dong et al. (2024)
Citrus × *sinensis* (Orange)	Obese patients (n=46) Dose—200 mL/day flavonoid-enriched juice for six weeks	Oxidative stress markers Inflammatory parameters (IFNϒ and TNF-α)	Increased glutathione peroxidase-1 protein expression by 20% Decreased by ~20% of inflammatory markers	Navajas-Porras et al. (2025)
Malus domestica Borkh (Apple)	Dextran sulfate sodium–induced ulcerative colitis model in C57BL/6 male mice. Dose—500 mg/kg body weight/day for 36 days	Assessment of disease activity index Hematoxylin and eosin and alcian blue/periodic acid Schiff staining Protein expressions of colonic MUC-2 by immunohistochemistry Inflammatory markers (IL-1β and TNF-α, and cortical content of 5-HT) Composition of gut microbiota using 16S rDNA sequencing	Reduced the disease activity index score and increased the colon length with the treatment. Ameliorated ulcerative colitis, reduced tissue damage Increased expression Suppressed microglia and astrocytes activation, reduced concentrations of inflammatory markers Restored dominant organism with amplified *Firmicutes*/*Bacteroidetes* ratio and enhanced *Akkermansia* abundance Reduced the *Escherichia–Shigella*	Qian et al. (2025)
Malus domestica Borkh (Apple)	Dextran sulfate sodium–induced colitis model in C57BL/6NRj male mice. Dose—10% lyophilized and milled apple pomace for 24 days	Histological analysis 16S rRNA gene sequencing SCFA analysis Inflammatory marker analysis	Reduced the severity of the intestinal inflammation Enhanced modulation of key taxa *Lachnospiraceae* genera and *Ruminococcus* species in the colonic microbiota Improved the production Reduced significantly IL-10 and TNF-α by ~10% with the treatment compared to the control group	Calvete-Torre et al. (2024)

Malus domestica Borkh (Apple)	Acute ethanol-induced ALI mouse model in C57BL/6J mice Dose—200 mg/kg of lyophilized fruit powder for three weeks	Intestinal microecology	Increased advantageous bacteria—*Dubosiella, Bifidobacterium, Lactobacillus*	Chen et al. (2024)
Musa spp. AAA (Green Dwarf Banana)	Male Wistar rat with trinitrobenzenesulfonic acid model of intestinal inflammation Dose—10% banana flour for days	Inflammatory markers (myeloperoxidase activity and cytokine (IL-1α, IL-1β, IL-6, IL-17, TNF-α)) SCFA determination Histological analysis	Decreased myeloperoxidase activity by 64%, decreased in IL-1β, IL-6, and TNF-α by 53%, 24%, and 59%, respectively Enhanced the production of propionic, acetic, and SCFAs Decreased extension of lesions The recovery process, including membrane-bound mucin and cell cytoarchitecture, was like that of healthy animals	Batista et al. (2025)
Passiflora edulis Sims (Passion Fruit)	In vitro study using pectin polysaccharides isolated from peels	Gastrointestinal digestion and fecal fermentation characteristics SCFA production	Increased entire comparative richness of *Firmicutes* and *Bacteroidota* in the intestinal flora Improved the generation of helpful bacteria—*Prevotella, Megasphaera*, and *Dialister* Suppressed the growth of harmful genera—*Escherichia–Shigella* Increased creation of SCFAs	Yu et al. (2024)
Punica granatum L. (Pomegranate)	In vitro study using ethanol (7): water (3) extract	Gastrointestinal digestion and fecal fermentation are followed by assessing antioxidant activity (TPC, DPPH, ABTS, FRAP) Analysis of SCFAs	Increased TPC, DPPH, and FRAP activity after digestion by 85.99, 20.28, and 22.76%, respectively Only TPC was increased by 73.0% after the fermentation. Increased generation of SCFAs	Buccato et al. (2025)

Note: IFN-γ; interferon-gamma, TNF-α; tumor necrosis factor-alpha, IL-1β; interleukin-1β, IL-6; interleukin-6, TPC; total polyphenol content, DPPH; 2,2-diphenyl-1-picrylhydrazyl, ABTS; 2,2′-azino-bis(3-ethylbenzothiazoline-6-sulfonic acid), FRAP; ferric reducing antioxidant power, SCFAs; short chain fatty acids.

14.6 CONCLUSION AND FUTURE REMARKS

This chapter highlights the profound impact of medicinal fruits and berries on gastrointestinal health. Several preclinical experiments and well-designed clinical trials described the role of medicinal fruits and berries in gastrointestinal health using methods such as in vitro digestion and fermentation, followed by microbiota DNA sequencing and SCFA profiling, histological assessment, and the use of inflammatory markers and patient-reported symptoms to demonstrate meaningful improvement in gastrointestinal function and comfort. Accordingly, phenolic acids, flavonoids, anthocyanins, antioxidants, and fiber in citrus fruits, pomegranate, banana, pineapple, apple, pears, mango, and kiwi fruit support gastrointestinal health prominently, increasing the beneficial microbes in the *Lactobacillus* and *Bifidobacterium* genera and decreasing pathogenic bacteria of the *Escherichia–Shigella* group. Further, these medicinal fruits and berries have the capacity to support the production of SCFAs, strengthen the intestinal barrier, moderate oxidative stress and inflammation, and improve motility and symptom burden in a range of gastrointestinal disturbances.

However, despite the increasing volume of preclinical and clinical evidence on the consumption of fruits and berries, there has been a notable gap in understanding the interaction between diverse types of fruit mixtures and their interference effects in supporting gastrointestinal health and the underlying mechanisms. There is also scope to explore the synergistic combination of medicinal fruits and berries integrated with probiotics or prebiotics, and to extend research avenues to explore clinical outcomes. Many trials use short intervention periods and short follow-up periods with a small sample size. Therefore, further long-term, comprehensive investigations are needed, focusing on well-defined clinical studies to draw potential conclusions. The proposed research would have the potential to translate robust findings on gastrointestinal health into clear, actionable guidance for the public, while laying the groundwork for the development of evidence-based fruit- and berry-derived supplements designed to support individuals with specific nutritional needs.

REFERENCES

Adesina, P. A., Ooka, M., TeKrony, C., & Xia, M. 2025. Emerging advances in intestinal models for in vitro preclinical research. *American Journal of Physiology-Gastrointestinal and Liver Physiology* 329(3): G403–G416.

Ansari, P., Reberio, A. D., Ansari, N. J., Kumar, S., Khan, J. T., Chowdhury, S., Abd El-Mordy, F. M., Hannan, J., Flatt, P. R., & Abdel-Wahab, Y. H. 2025. Therapeutic potential of medicinal plants and their phytoconstituents in diabetes, cancer, infections, cardiovascular diseases, inflammation and gastrointestinal disorders. *Biomedicines* 13(2): 454.

Asuncion, P., Liu, C., Castro, R., Yon, V., Rosas Jr, M., Hooshmand, S., Kern, M., & Hong, M. Y. 2023. The effects of fresh mango consumption on gut health and microbiome–randomized controlled trial. *Food Science & Nutrition* 11(4): 2069–2078.

Barber, C., Sabater, C., Frutos, M. D., Vallejo, F., Guyonnet, D., Daniel, N., Guarner, F., Espín, J. C., Margolles, A., & Azpiroz, F. 2024. Effects of a (poly)phenol-rich berry mix on gas production in healthy individuals: an integrated clinical, metagenomic, and metabolomic proof-of-concept study. *Journal of Functional Foods* 113: 106032.

Batista, G. R., Quaglio, A. E., Almeida Junior, L. D., Sassaki, L. Y., & Di Stasi, L. C. 2025. Dietary supplementation with green fruits of Musa spp. AAA ameliorates acute, subchronic, and chronic intestinal inflammation with relapse induced by trinitrobenzenesulfonic acid in rats. *ACS Food Science & Technology* 5: 2776–2786.

Borowska, S., Tomczyk, M., Strawa, J. W., & Brzóska, M. M. 2020. Estimation of the chelating ability of an extract from Aronia melanocarpa L. berries and its main polyphenolic ingredients towards ions of zinc and copper. *Molecules* 25(7): 1507.

Buccato, D. G., Delgado-Osorio, A., De Lellis, L. F., Morone, M. V., Ullah, H., Izzo, L., Lombardi, S., Di Minno, A., Riccioni, C. V., & Moriki, D. 2025. Exploring the influence of a pomegranate extract on the functionality of healthy and diseased human gut microbiota: an in vitro study. *Molecules* 30(7): 1634.

Calvete-Torre, I., Sabater, C., Cantabrana, B., Margolles, A., Sánchez, M., & Ruiz, L. 2024. Gut microbiota modulation and effects of a diet enriched in apple pomace on inflammation in a DSS-induced colitis mouse model. *Food & Function* 15(4): 2314–2326.

Chen, Y., Ma, H., Liang, J., Sun, C., Wang, D., Chen, K., Zhao, J., Ji, S., Ma, C., & Ye, X. 2024. Hepatoprotective potential of four fruit extracts rich in different structural flavonoids against alcohol-induced liver injury via gut microbiota-liver axis. *Food Chemistry* 460: 140460.

Chong, C. W., Liew, M. S., Ooi, W., Jamil, H., Lim, A., Hooi, S. L., Tay, C. S., & Tan, G. 2024. Effect of green banana and pineapple fiber powder consumption on host gut microbiome. *Frontiers in Nutrition* 11: 1437645.

Donadio, J. L., Fabi, J. P., Sztein, M. B., & Salerno-Gonçalves, R. 2024. Dietary fiber pectin: challenges and potential anti-inflammatory benefits for preterms and newborns. *Frontiers in Nutrition* 10: 1286138.

Dong, J., Wang, W., Zheng, G., Wu, N., Xie, J., Xiong, S., Tian, P., & Li, J. 2024. In vitro digestion and fermentation behaviors of polysaccharides from Choerospondias axillaris fruit and its effect on human gut microbiota. *Current Research in Food Science* 8: 100760.

Feng, Y., Jin, Q., Liu, X., Lin, T., Johnson, A., & Huang, H. 2025. Advances in understanding dietary fiber: classification, structural characterization, modification, and gut microbiome interactions. *Comprehensive Reviews in Food Science and Food Safety* 24(1): e70092.

Ferroni, L., Rubini, A., Bargellini, P., Tremoli, E., Cappucci, I. P., D'Amora, U., Ronca, A., Calogero, G., Panini, P. C., & Bettini, G. 2024. Apple vesicles: revolutionary gut microbiota treatment for inflammatory bowel disease. *Food Bioscience* 62: 105052.

Greenwood-Van Meerveld, B., Johnson, A. C., & Grundy, D. 2017. Gastrointestinal physiology and function. *Gastrointestinal Pharmacology* 2017: 1–16.

Han, X., Song, Y., Huang, R., Zhu, M., Li, M., Requena, T., & Wang, H. 2023. Anti-inflammatory and gut microbiota modulation potentials of flavonoids extracted from Passiflora foetida fruits. *Foods* 12(15): 2889.

He, W., Li, Y., Liu, M., Yu, H., Chen, Q., Chen, Y., Ruan, J., Ding, Z., Zhang, Y., & Wang, T. 2018. Citrus aurantium L. and its flavonoids regulate TNBS-induced inflammatory bowel disease through anti-inflammation and suppressing isolated jejunum contraction. *International Journal of Molecular Sciences* 19(10): 3057.

Holtmann, G., Talley, N. J., Nandurkar, S., & Gibson, P. R. 2025. Randomized, placebo-controlled, double-blind 8-week trial on the efficacy of a proprietary kiwifruit extract on constipation-predominant irritable bowel syndrome. *JGH Open* 9(8): e70250.

Hoogeveen, A. M., Butts, C. A., Kim, C. C., Jobsis, C. M., Parkar, S. G., Stoklosinski, H. M., Sutton, K. H., Davis, P., Hedderley, D. I., & Johnston, J. 2025. The effect of apple and pear cultivars on in vitro fermentation with human faecal microbiota. *Microorganisms* 13(8): 1870.

Huang, Y., Hu, J., Xia, Q., Tang, M., Wang, Y., Wang, G., Shao, X., Yuan, H., Li, S., & Huang, P. 2024. Amelioration of obesity and inflammation by polysaccharide from unripe fruits of raspberry via gut microbiota regulation. *International Journal of Biological Macromolecules* 261: 129825.

Hyży, A., Rozenek, H., Gondek, E., & Jaworski, M. 2025. Effect of antioxidants on the gut microbiome profile and brain functions: a review of randomized controlled trial studies. *Foods* 14(2): 176.

Jakobek, L., & Matić, P. 2024. Phenolic compounds from apples: from natural fruits to the beneficial effects in the digestive system. *Molecules* 29(3): 568.

Jeong, E., Eun, S., Chae, S., & Lee, S. 2024. Prebiotic potential of goji berry (Lycium barbarum) in improving intestinal integrity and inflammatory profiles via modification of the gut microbiota in high-fat diet-fed rats. *Journal of Medicinal Food* 27(8): 704–712.

Kanmani, P., Suganya, K., & Kim, H. 2020. The gut microbiota: how does it influence the development and progression of liver diseases. *Biomedicines* 8(11): 501.

Kim, H., Castellon-Chicas, M. J., Arbizu, S., Talcott, S. T., Drury, N. L., Smith, S., & Mertens-Talcott, S. U. 2021. Mango (Mangifera indica L.) polyphenols: anti-inflammatory intestinal microbial health benefits, and associated mechanisms of actions. *Molecules* 26(9): 2732.

Kiran, B. S., Bashir, O., Pawase, P. A., Dash, K. K., Amin, T., Shams, R., Shaikh, A. M., & Béla, K. 2025. A review of in vitro digestibility models on diverse foods in various segments of human digestive tract. *Discover Food* 5(1): 1–24.

Lavefve, L., Howard, L. R., & Carbonero, F. 2020. Berry polyphenols metabolism and impact on human gut microbiota and health. *Food & Function* 11(1): 45–65.

Manske, S. 2024. Gastrointestinal health: a foundation for whole-body wellness. *Alternative Therapies in Health & Medicine* 30(7): 6–10.

Maqsood, S., Arshad, M. T., Ikram, A., & Gnedeka, K. T. 2025. Fruit-based diet and gut health: a review. *Food Science & Nutrition* 13(5): e70159.
Miller, J. C., Satheesh Babu, A. K., Petersen, C., Wankhade, U. D., Robeson, M. S., Putich, M. N., Mueller, J. E., O'Farrell, A. S., Cho, J. M., & Chintapalli, S. V. 2022. Gut microbes are associated with the vascular beneficial effects of dietary strawberry on metabolic syndrome-induced vascular inflammation. *Molecular Nutrition & Food Research* 66(22): 2200112.
Muñoz, R. L. P., Mora, C. P., Parra-Perdomo, L. V., & Rojas, G. 2025. Healing from the peel: exploring the bioactive potential of bananas for gastric ulcer management. *Journal of Agricultural and Food Chemistry* 73(24): 14689–14706.
Navajas-Porras, B., Bosch-Sierra, N., Grau-del Valle, C., Salazar, J. D., Marqués-Cardete, R., Sáez, G., Morillas, C., & Bañuls, C. 2025. Effects of a flavonoid-enriched orange juice on antioxidant capacity, lipid profile, and inflammation in obese patients: a randomized placebo-controlled trial. *Food Research International* 2025: 116759.
Onali, T., Slabá, H., Jian, C., Koivumäki, T., Päivärinta, E., Marttinen, M., Määttänen, M., Salonen, A., & Pajari, A. M. 2025. Berry supplementation in healthy volunteers modulates gut microbiota, increases fecal polyphenol metabolites and reduces viability of colon cancer cells exposed to fecal water—a randomized controlled trial. *The Journal of Nutritional Biochemistry* 141: 109906.
Pereira-Caro, G., Cáceres-Jiménez, S., Moreno-Ortega, A., Dobani, S., Pourshahidi, K., Gill, C. I., Mena, P., Del Rio, D., Moreno-Rojas, J. M., & Taurino, G. 2024. Colon-available mango (poly)phenols exhibit mitigating effects on the intestinal barrier function in human intestinal cell monolayers under inflammatory conditions. *Food & Function* 15(9): 5118–5131.
Perri, M. R., Romano, C., Marrelli, M., Zicarelli, L., Toma, C.-C., Basta, D., Conforti, F., & Statti, G. 2021. Beneficial role of fruits, their juices, and freeze-dried powders on inflammatory bowel disease and related dysbiosis. *Plants* 11(1): 4.
Petersen, C., Wankhade, U. D., Bharat, D., Wong, K., Mueller, J. E., Chintapalli, S. V., Piccolo, B. D., Jalili, T., Jia, Z., & Symons, J. D. 2019. Dietary supplementation with strawberry induces marked changes in the composition and functional potential of the gut microbiome in diabetic mice. *The Journal of Nutritional Biochemistry* 66: 63–69.
Pinheiro, D. F., Maciel, G. M., Lima, N. P., Lima, N. F., Ribeiro, I. S., & Haminiuk, C. W. I. 2024. Impact of fruit consumption on gut microbiota: benefits, contaminants, and implications for human health. *Trends in Food Science & Technology* 154: 104785.
Qian, Q., Luan, W., Ma, J., Chen, S., Xie, Y., Wu, Z., Yang, H., Ma, J., Shang, S., & Li, J. 2025. Beneficial effects of apple polyphenol extract and time-restricted feeding on ulcerative colitis and accompanying depression-like behaviors are correlated with gut–brain axis modulation. *Journal of Agricultural and Food Chemistry* 73(32): 20219–20234.
Santacroce, L., Bottalico, L., Charitos, I. A., Castellaneta, F., Gaxhja, E., Topi, S., Palmirotta, R., & Jirillo, E. 2024. Exploitation of natural by-products for the promotion of healthy outcomes in humans: special focus on antioxidant and anti-inflammatory mechanisms and modulation of the gut microbiota. *Antioxidants* 13(7): 796.
Shang, Z., Pai, L., & Patil, S. 2024. Unveiling the dynamics of gut microbial interactions: a review of dietary impact and precision nutrition in gastrointestinal health. *Frontiers in Nutrition* 11: 1395664.
Suri, C., Pande, B., Sahu, T., Sahithi, L. S., & Verma, H. K. 2024. Revolutionizing gastrointestinal disorder management: cutting-edge advances and future prospects. *Journal of Clinical Medicine* 13(13): 3977.
Van Hul, M., Cani, P. D., Petitfils, C., De Vos, W. M., Tilg, H., & El-Omar, E. M. 2024. What defines a healthy gut microbiome. *Gut* 73(11): 1893–1908.
Wang, J., Ren, Y., Ye, X., Zhang, H., & Tian, J. 2024. In vitro digestion and fermentation of the whole goji berry: bioactive ingredients change and impacts on human gut microbiota. *Journal of Food Science* 89(10): 6465–6480.
Wilder-Smith, C. H., Materna, A., & Olesen, S. S. 2023. Blueberries improve abdominal symptoms, well-being and functioning in patients with functional gastrointestinal disorders. *Nutrients* 15(10): 2396.
Wu, P., Zhang, P., & Chen, X. D. 2025. Assessing food digestion in the elderly using in vitro gastrointestinal models. *Advances in Food and Nutrition Research* 114: 273–300.
Yin, Y., Martínez, R., Zhang, W., & Estévez, M. 2024. Crosstalk between dietary pomegranate and gut microbiota: evidence of health benefits. *Critical Reviews in Food Science and Nutrition* 64(27): 10009–10035.
Yu, Y. H., Wu, L. B., Liu, X., Zhao, L. C., Li, L. Q., Jin, M. Y., Yu, X., Liu, F., Li, Y., & Li, L. 2024. In vitro simulated digestion and fermentation characteristics of pectic polysaccharides from fresh passion fruit (Passiflora edulis f. flavicarpa L.) peel. *Food Chemistry* 452: 139606.

Zayed, A., Adly, G. M., Khaled, L., Abd El-Al, M. S., & Farag, M. A. 2025. Citrus-derived prebiotics: influence on gut microbiota composition, fermentation metabolites, and functional health effects. *Food Bioscience* 2025: 107535.

Zhang, T., Chang, M., Hou, X., Yan, M., Zhang, S., Song, W., Sheng, Q., Yuan, Y., & Yue, T. 2025. Apple polyphenols prevent patulin-induced intestinal damage by modulating the gut microbiota and metabolism of the gut-liver axis. *Food Chemistry* 463: 141049.

Zhao, Y., Bi, J., Yi, J., Peng, J., & Ma, Q. 2022. Dose-dependent effects of apple pectin on alleviating high fat-induced obesity modulated by gut microbiota and SCFAs. *Food Science and Human Wellness* 11(1): 143–154.

15 Medicinal Fruits and Berries as Nutraceuticals

Uswatun Khasanah, Ari Satia Nugraha, Catherine M. Miller, and Phurpa Wangchuk

15.1 INTRODUCTION

The concept that "food is medicine" was stated by Hippocrates (460 BCE), forming the philosophical foundation of functional foods (Li 2025). This concept was supported by the ancient Roman scholar, Galen, who also documented the medicinal properties of various foods (Hajar 2012). There were two famous historical stories that used food as medicine. First: consuming unpolished rice prevented beriberi, a disease caused by thiamine (vitamin B1) deficiency; second, consuming citrus could prevent and cure scurvy, a disease caused by vitamin C deficiency (Arnold 2010; Nguyen 2020). Both incidents proved that foods could act as medicine to prevent and cure diseases (Mozaffarian et al. 2018).

In 1989, Stephen DeFelice, founder of the "Foundation of Innovation Medicine", combined the words "nutrition" and "pharmaceutical", which initiated the development of nutraceutical terminology (Aronson 2017; Kalra 2003). In the twenty-first century, the nutraceutical industry expanded and integrated into preventive healthcare and functional foods, prompting developed countries to separate regulatory categories for medicines. After 2010, the number of nutraceutical products claimed to have antioxidant and probiotic properties rose significantly, followed by an increase in immune-enhancing bioactives after the COVID-19 pandemic in 2019 (Lordan 2021; Paudel et al. 2022). Currently, the nutraceutical industry develops nutraceuticals from sustainable sources, such as marine, microbial, plant waste, and uncommon fruits and berries (Wan et al. 2023).

Nutraceuticals, whose properties can overlap with supplements and drugs, had no universal legal recognition until 2024. However, according to historical definitions, nutraceuticals are foods or components of food that have medical and health benefits, including disease prevention and treatment, and could be easily integrated into a patient's diet (Husen and Iqbal 2021; Kalra 2003; Silpi 2025). Based on existing definitions, a clear distinction between nutraceuticals, dietary supplements, and functional foods is necessary, as outlined in Table 15.1. From this comparison, an expanded definition is proposed whereby nutraceuticals are defined as foods or food-derived components formulated in a dosage form that exhibit scientific disease-prevention and therapeutic benefits (Kalra 2003; Shahidi 2009).

According to Table 15.1, functional foods do not require scientific proof. However, preclinical and clinical evaluations of nutraceutical products are required to support their claims for disease prevention and treatment. Therefore, depending on the jurisdiction, nutraceuticals are generally required to be registered or notified with regulatory authorities prior to commercialization.

Several fruits and berries, such as *Vaccinium spp.*, *Lycium spp.*, and *Vitis spp.*, have been extensively investigated as sources of commercial nutraceutical products (Martau et al. 2023; Wang et al. 2022a). These fruits possess well-documented bioactivity, supported by in vitro, in vivo, and clinical studies. As a consequence, high demand coupled with insufficient supply has created pressure for cultivation expansion and increased raw material costs. In contrast, certain underutilized fruits and berries, known to contain bioactive compounds used in traditional medicine, remain underexplored.

 DOI: 10.1201/9781003688259-15

TABLE 15.1
Differences Between Functional Foods, Nutraceuticals, and Dietary Supplements

Parameters	Functional Food	Nutraceutical	Dietary Supplement
Definition	Foods consumed as part of a normal diet that provide health benefits (Cencic and Chingwaru 2010; Damian et al. 2022; Shahidi 2020).	Foods/part of foods with combined nutritional and pharmaceutical effects, including disease prevention and/or treatment (Damian et al. 2022; Puri et al. 2022; Shahidi 2020).	Substances added to the diet to treat or prevent a nutritional deficiency (Aronson 2017; Shahidi 2020).
Primary Purpose	Health maintenance and reduction of chronic disease risk.	Therapeutic support, disease prevention, and adjunct treatment.	Correction or prevention of nutrient deficiencies.
Form of Consumption	Eaten as normal foods (natural or processed).	Often in dosage forms, for instance, tablets, capsules, and powders.	Pharmaceutical dosage forms such as tablets, syrups, powders, and capsules.

Research on their bioactivity and phytoconstituents is limited. This chapter examines lesser-known fruits and berries with the potential to develop into nutraceutical raw materials. Expanding research on these unexplored sources may diversify nutraceutical options, enhance local value, and enable advanced formulations, particularly for bitter-tasting varieties unsuitable for direct consumption.

15.2 GLOBAL MARKET OF NUTRACEUTICALS

The global nutraceutical industry has expanded from the niche "health supplement" segment into the mainstream consumer health market (Chopra et al. 2022). Consumers increasingly prioritize health prevention rather than symptom-only treatment, supporting the demand for vitamins, botanicals, and condition-specific formulas (Lordan 2021). In addition, the use of pharmaceutical dosage forms in many nutraceutical delivery systems facilitates consumption and increases the bioavailability of some secondary metabolites, such as polyphenols (Asghar et al. 2018). Based on their bioactivities, immune-related indications dominated nutraceutical consumption during 2020–2021, while metabolic, cardiovascular, gut, and mental health indications became increasingly prominent during 2022–2025 (Djaoudene et al. 2023).

The period from 2020 to 2025 represents one of the most accelerated growth phases in the history of the global nutraceutical market, driven primarily by the COVID-19 pandemic and its lasting effect on consumer health behaviors (Lordan 2021). According to Grand View Research, in 2024, the nutraceutical market reached USD 591.1 billion and is estimated to grow to USD 919.1 billion by 2030 with a compound annual growth rate (CAGR) of 7.6% (MarketResearch.com 2025). The trend in nutraceutical revenue over the last five years is shown in Figure 15.1.

15.3 BIOACTIVE COMPOUNDS OF FRUITS AND BERRIES AS NUTRACEUTICALS

The botanical and herbal ingredients used as raw materials for nutraceuticals are derived from fruits, berries, seeds, and leaves (Rosa et al. 2023). Beyond their role as providers of micronutrients and vitamins, these plant-based foods are associated with a high content of secondary metabolites that exert health-promoting effects (Cisneros-Zevallos 2021). This chemical diversity enables multitarget biological activity and contributes to both disease prevention and overall health support (Dillard and German 2000).

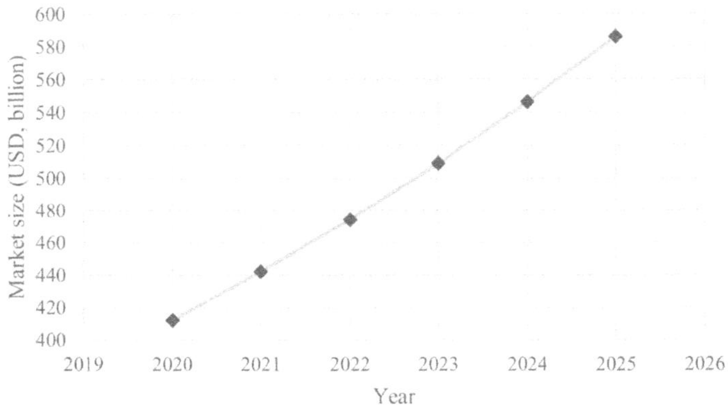

FIGURE 15.1 Global nutraceuticals market from 2020 to 2025 with CAGR 7.3–7.8%.

The major groups of secondary metabolites contributing to the nutraceutical value of fruits and berries include flavonoids, anthocyanins, polyphenols, terpenoids, fatty acids, and polysaccharides, as illustrated in Figure 15.2 (Dillard and German 2000; Rosa et al. 2023). Among these, polyphenols represent the most abundant bioactive compounds in fruits and berries and are central to their nutraceutical potential. This group includes flavonoids, anthocyanins, phenolic acids, and tannins. Anthocyanins, shown in Figure 15.2, such as the aglycones malvidin, cyanidin, delphinidin, and their glycosides, are widely recognized for their antioxidant and anti-inflammatory properties (Khoo et al. 2017). Their chemical structures, characterized by multiple hydroxyl groups, enable effective scavenging of reactive oxygen species (ROS), thereby preventing cellular free radical reactions (Dudek et al. 2022; Kahkonen and Heinonen 2003). In nutraceutical products, berries from the genus *Vaccinium* spp., including blueberries, bilberries, and cranberries, are commonly standardized for anthocyanin content and marketed for cardiovascular health, antioxidant properties as an adjuvant cancer therapy, and immune system enhancement, as summarized in Table 15.2 (Martau et al. 2023).

In addition to anthocyanins, flavonols illustrated in Figure 15.2, such as quercetin, kaempferol, and isorhamnetin, are also abundant in many fruits and berries (Panche et al. 2016). These compounds contribute to vascular protection, metabolic regulation, and antioxidant activity. Moreover, phenolic acids, including gallic acid, ellagic acid, chlorogenic acid, and caffeic acid, complement these effects through synergistic antioxidant and anti-inflammatory mechanisms.

Terpenoid-based secondary metabolites, presented in Figure 15.2, such as tocopherols, lycopene, and carotenoids, are lipid-soluble bioactive compounds that protect against oxidative damage in lipid-rich tissues (Kaulmann and Bohn 2014; Sahin et al. 2018). Carotenoid-rich fruits are frequently incorporated into nutraceutical products targeting eye health, skin health, and immune function (Woodall et al. 1996). Unsaturated fatty acids, including α-linolenic acid, linoleic acid, and oleic acid, support epithelial integrity, skin health, and metabolic regulation (Gutierres et al. 2025).

Fruits and berries also contain polysaccharides, which are commonly employed in nutraceuticals for gut health and microbiota modulation, with important implications for immune function, metabolic regulation, and intestinal inflammation (Slavin 2013). Fruit-derived polysaccharides, including pectins and soluble fibers, can act as prebiotics, stimulating beneficial gut microorganisms and promoting the formation of short-chain fatty acids.

The phytochemical complexity (phytocomplex) of fruits and berries is a key factor distinguishing nutraceuticals from pharmaceutical agents (Natarajan et al. 2019). Rather than acting individually, these compounds function synergistically within the human body, enhancing overall bioactivity while potentially reducing adverse effects, as summarized in Table 15.2 (Darko et al. 2025).

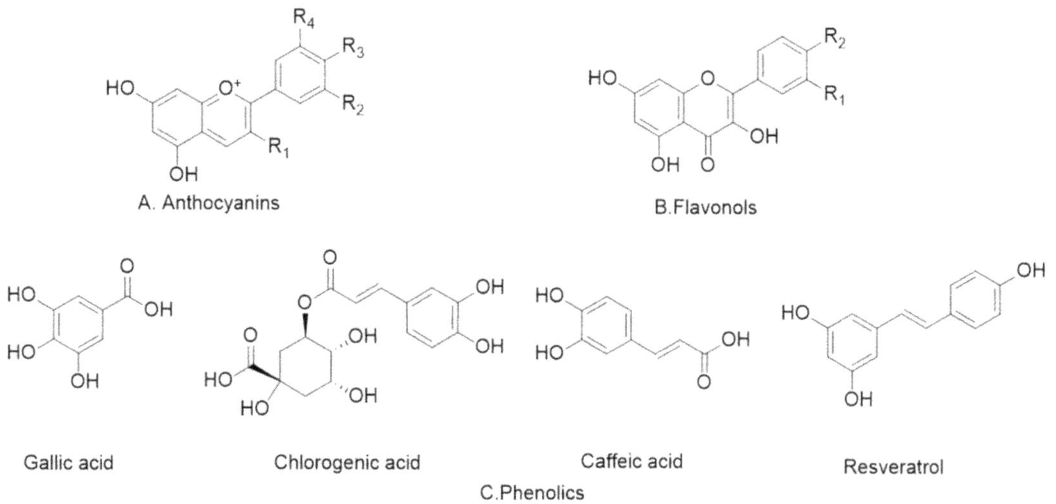

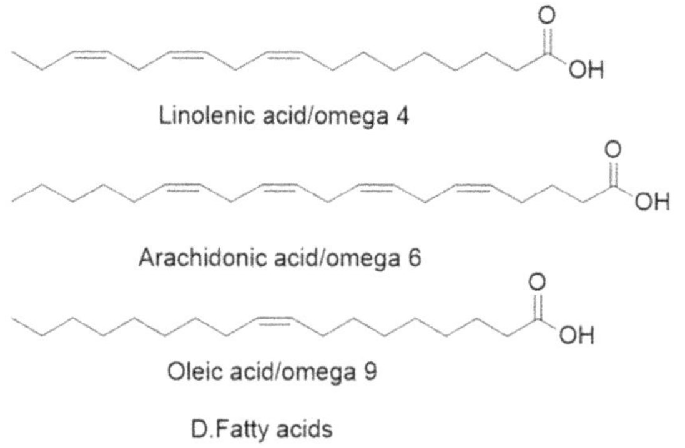

A. Anthocyanins

Compounds	**R1**	**R2**	**R3**	**R4**
Delphinidin	OH	OH	OH	OH
Malvidin	OH	OCH_3	OH	OCH_3
Cyanidin	OH	OH	OH	H

B. Flavonols

Compounds	**R1**	**R2**
Quercetin	OH	OH
Kaempferol	OH	H
Isorhamnetin	OH	OCH_3

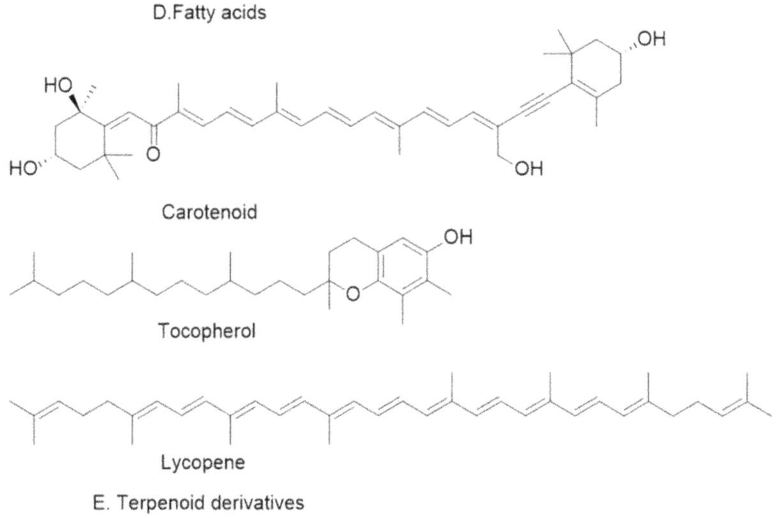

FIGURE 15.2 Major classes of bioactive compounds in fruits and berries used as nutraceuticals.

TABLE 15.2
Common Fruits and Berries that are Commercially Available as Nutraceutical Products

Species	Key Bioactive Compounds	Pharmacological Activity	Level of Evidence
Aronia melanocarpa (Michx.) Elliot	Anthocyanins, procyanidins	Antioxidant and cardiometabolic support	In vivo
Citrus bergamia (Risso) Risso & Poit.	Flavanones	Lipid and cholesterol regulation	Clinical
Euterpe oleracea Mart.	Anthocyanins, phytosterols	Antioxidant and metabolic effects	In vivo
Lycium barbarum L.	Polysaccharides, carotenoids	Metabolic and immune modulation	In vivo/In vitro
Malus domestica (Suckow) Borkh.	Quercetin, pectin	Cardiometabolic and gut health	Epidemiological/In vivo
Punica granatum L.	Punicalagins, ellagic acid, tannins	Antioxidant and cardiometabolic effects	Clinical
Ribes nigrum L.	Anthocyanins, vitamin C	Exercise recovery, vascular health	Clinical
Sambucus nigra L.	Anthocyanins, flavonols	Immune and antiviral activity	Clinical
Vaccinium corymbosum L.	Anthocyanins	Cognitive and cardiovascular health	Clinical
Vaccinium macrocarpon Aiton	A-type proanthocyanidins	Urinary tract health	Clinical
Vaccinium myrtillus L.	Anthocyanins	Eye and vascular health	In vivo/Clinical
Vitis vinifera L.	Oligomeric proanthocyanidins	Blood pressure regulation	Clinical

15.4 THE MECHANISM OF NUTRACEUTICALS IN COMMUNICABLE DISEASES (CDs) PREVENTION AND TREATMENT

Nutraceuticals derived from plants exhibit therapeutic potential against infectious diseases through antioxidant, immunomodulatory, antiviral, and antiparasitic mechanisms (Di Sotto et al. 2020). During infection, pathogenic microorganisms produce toxins and virulence factors that induce cytokine production and activate pro-inflammatory signaling pathways. In addition, ROS are generated during infection and may initiate or amplify inflammatory responses; when excessive, this process can contribute to tissue damage and multiple organ failure. Nevertheless, activation of the immune system during infectious disease is essential for eliminating microbial burden in the human host (Alsaikhan and Farhood 2025).

Several plants exhibit strong immunostimulatory activity and are used as adjuvant treatments for infectious diseases, such as *Phyllanthus niruri* L., Echinacea spp., *Vaccinium* spp., and *Vitis vinifera* L. Those fruits and berries contain vitamin E, Vitamin C, vitamin B6, quercetin, anthocyanin, ellagic acid, gallic acid, catechin, chlorogenic acid, hesperidin, naringin, and beta-carotene, which modulate immune system response (Maheshwari et al. 2022). Vitamins A, E, and C act as antioxidant scavengers that prevent oxidative damage and oxidative stress in DNA, while vitamin B6 is essential for lymphocyte proliferation and antibody generation (Costagliola et al. 2021; Maheshwari et al. 2022). Moreover, flavonoids help maintain the balance of Th1/Th2 cells and impede IgE production, tannins stimulate phagocytic cells, and hesperidin inhibits enzymes that regulate pro-inflammatory mediator biosynthesis (Maheshwari et al. 2022).

Azadirachta indica A.Juss., a fruit commonly used in the treatment of infectious diseases, significantly reduces parasitemia levels in a mouse malaria model compared with artesunate (Igwenyi et al. 2023). GC-MS analysis identified dodecanoic acid, oleic acid, 13-octadecyl, 15-tetracosanoic acid methyl ester, and alkaloids as major bioactive constituents (Igwenyi et al. 2023). Another potent nutraceutical in malaria is rutin, a flavonoid glycoside abundant in buckwheat, berries, and citrus fruits. Oral rutin administration reduced parasitemia percentages, prolonged survival time, restored hemoglobin levels, and showed synergy with chloroquine (Bhatt et al. 2022). Importantly, rutin was effective against both chloroquine-sensitive and resistant *Plasmodium falciparum*.

Beyond malaria, polyphenols also inhibit other parasitic and viral infections (Zothantluanga et al. 2022). Gallic acid enhanced anti-*Leishmania major* activity by stimulating macrophage phagocytosis and intracellular Ca^{2+} signaling. Apigenin blocked chikungunya virus (CHIKV) entry or replication, while other polyphenols damaged parasite lipid bilayers, altered membrane integrity, inhibited essential enzymes, induced oxidative stress, and triggered apoptosis in *Schistosoma mansoni* and *Fasciola gigantica* (Zothantluanga et al. 2022). Further investigation identified phenolic acids (syringic, taxifolin, t-cinnamic, and gallic acid) and gluconic acid as key active compounds conferring bacteriostatic and bactericidal activity against *Escherichia coli* and *Staphylococcus aureus* (Mackin et al. 2023). Collectively, these studies show that nutraceuticals have significant promise as adjunct and alternative therapies for infectious diseases through well-defined bioactive compounds and molecular pathways.

15.5 THE ROLE OF NUTRACEUTICALS IN THE PREVENTION AND TREATMENT OF NONCOMMUNICABLE DISEASES (NCDs)

A growing body of evidence indicates that nutraceuticals exert diverse molecular effects on key pathological drivers of metabolic syndrome (MetS) and noncommunicable diseases (NCDs), such as oxidative stress, chronic inflammation, insulin resistance, adipokine dysregulation, endothelial dysfunction, and mitochondrial impairment. These interrelated mechanisms collectively facilitate the development of type 2 diabetes, obesity, dyslipidemia, hypertension, and cardiovascular diseases (Nelson 2013).

One of the metabolic syndromes leading to atherosclerosis complications is hyperlipidemia, characterized by elevated triglycerides and low-density lipoprotein (LDL) and a decrease in high-density lipoprotein (HDL), in which ROS play an essential role in the pathogenic pathway (Furukawa et al. 2004). Therefore, the administration of antioxidants could contribute to normalizing the lipid profile and preventing further complications. Excess circulating triglycerides exacerbate ROS production, leading to mitochondrial dysfunction and impairment of insulin signaling. Nutraceuticals rich in polyphenols, flavonoids, omega-3 fatty acids, and antioxidant compounds effectively counteract these processes by scavenging ROS, stabilizing mitochondrial redox homeostasis, and restoring pancreatic β-cell integrity (Calder 2015; Shahidi and Ambigaipalan 2015). This redox modulation improves insulin sensitivity and lipid metabolism, which are pivotal for preventing cardiometabolic complications (Olalekan et al. 2024).

Furthermore, elevated lipid profiles and blood glucose levels activate pro-inflammatory cytokines such as tumor necrosis factor-α (TNF-α), leptin, and interleukin-6 (IL-6), driving endothelial dysfunction and accelerating atherosclerotic progression (Kaushal 2024). Nutraceuticals, including berberine, cinnamon, resveratrol, and quercetin, exhibit strong anti-inflammatory and lipid-lowering activity. These compounds suppress cytokine release, downregulate vascular inflammation, improve endothelial nitric oxide bioavailability, and reduce triglyceride and LDL levels. Through these coordinated actions, nutraceuticals mitigate hypertension-induced vascular injury and attenuate cardiovascular risk associated with metabolic syndrome (Dama et al. 2024).

Disruption of insulin receptor signaling remains a central hallmark of metabolic syndrome and induces impairment of the PI3K/Akt cascade, coupled with ROS-induced mitochondrial damage, thereby perpetuating insulin resistance and hyperglycemia (Xu et al. 2018). Nutraceutical antioxidants re-establish insulin receptor substrate activity, restore the downstream of Akt phosphorylation, and improve skeletal muscle glucose transport. This restoration of insulin signaling represents a key molecular foundation for the metabolic benefits consistently observed with bioactive nutraceutical interventions (Dama et al. 2024).

Collectively, these findings suggest that nutraceuticals act as systems-level modulators, influencing oxidative stress, inflammatory signaling, endocrine regulation, and insulin receptor activity. These integrated mechanisms address the metabolic disturbances underlying obesity, type 2 diabetes, dyslipidemia, hypertension, and atherosclerosis. Notably, nutraceuticals demonstrate both preventive and disease-modifying properties, highlighting their potential as adjunctive therapies in the management of metabolic syndrome and related NCDs (Chopra et al. 2022; Shahidi 2020).

15.6 UNDERUTILIZED FRUITS AND BERRIES AS NUTRACEUTICALS

15.6.1 *Picralima nitida* (Stapf) T.Durand & H.Durand

The taxonomy and distribution of *Picralima nitida* are clearly established, with the species belonging to the family Apocynaceae, with a natural occurrence throughout West and Central Africa, particularly in Nigeria (Ige et al. 2025). From an ethnobotanical standpoint, the fruit and seeds of *P. nitida* are empirically used in West African medicine to treat malaria, fever, hypertension, jaundice, gastrointestinal disorders, and diabetes mellitus (Adeola et al. 2023). In diabetic therapy, the powdered fruit seeds are traditionally sprinkled on food before ingestion, reflecting a dietary–medicinal integration (Shittu et al. 2010).

The bioactive compounds of *P. nitida* fruit are dominated by indole alkaloids, including akuammicine, akuammine, akuammidine, burnamine, 10-deoxyakuammine, and picraline, which are responsible for its pharmacological activities (Shittu et al. 2010). Nonalkaloidal bioactives such as erucic acid, 9,12-octadecadienoic acid, n-hexadecanoic acid, phenols, flavonoids, and saponins contribute to its antioxidant and antimicrobial properties (Adeola et al. 2023).

For in vitro activity, aqueous fruit seed extracts exhibited strong antioxidant potential against DPPH with an IC_{50} of 104.30±3.17 μg/mL and showed antimicrobial effects against Gram-positive bacteria, including *Bacillus subtilis*, *Staphylococcus aureus*, and *Streptococcus pneumoniae*, and Gram-negative bacteria, including *Salmonella typhi, Escherichia coli*, and *Pseudomonas aeruginosa* (Adeola et al. 2023). In a metabolic model, the alkaloid akuammicine significantly stimulated glucose uptake in fully differentiated 3T3-L1 adipocytes without cytotoxicity, indicating direct antidiabetic potential (Shittu et al. 2010).

In terms of in vivo pharmacology, oral administration of *P. nitida* extracts at 100 mg/kg BW improved estradiol levels, follicle-stimulating hormone, luteinizing hormone, lipid profile, and glutathione status in female Wistar rats with MSG-induced dysmenorrhea, confirming antioxidant–hormonal modulation (Ige et al. 2025). Moreover, antimalarial efficacy against *Plasmodium berghei* was observed in a mouse malaria model at doses up to 800 mg/kg (Adeyoju et al. 2022).

Currently, no controlled clinical trials have been reported for *P. nitida* fruit, indicating a critical translational research gap. Regarding safety, methanolic extracts showed LD_{50} > 2000 mg/kg BW with no mortality or behavioral toxicity in mice, while only mild uterine vascular congestion was observed histologically in reproductive studies (Adeyoju et al. 2022; Ige et al. 2025).

15.6.2 *Aegle marmelos* (L.) Corrêa

Aegle marmelos (bael or "eagle" fruit) is part of the Rutaceae family, native to dry forests of Southeast Asia and now cultivated widely across India, Nepal, Bangladesh, Sri Lanka, and Myanmar (De Silva et al. 2023; Venthodika et al. 2021). The fruit exhibits a round to oval morphology, typically measuring 5–20 cm in diameter, and possesses a thin, hard, woody, or occasionally soft rind, as illustrated in Figure 15.3 (Sharma and Dubey 2013). In Ayurvedic traditional medicine, the fruits of *A. marmelos* are used to treat diarrhea and dysentery. In addition, the leaves and other plant parts are used to manage inflammation, asthma, hypoglycemia, hepatitis, fever, infections, and various metabolic disorders (Joshi et al. 2024; Venthodika et al. 2021).

FIGURE 15.3 Underutilized fruits and berries potentially developed as nutraceutical products. A. *Aegle marmelos* (L.) Corrêa; B. *Hippophae rhamnoides* L.; C. *Rhus chinensis* Mill.; D. *Syzygium cumini* (L.) Skeels.

Photographs from Phurpa Wangchuk's private collection.

Chemically, bael is rich in coumarins such as marmelosin, auraptene, epoxyauraptene, imperatorin, alloimperatorin, scopoletin, umbelliferone, and marmin; alkaloids including aegeline, marmeline, and phenylethyl-cinnamides; phenylpropanoids (marmenol); flavonoids (rutin, quercetin, kaempferol, flavone glycosides, flavan-3-ols) and phenolic acids (ellagic, gallic, chlorogenic, ferulic, ferulic); tannins; terpenoids (p-cymene, α-phellandrene, limonene, lupeol, cineol, citral), polysaccharides, and pectic gums (Sarkar et al. 2020; Venthodika et al. 2021).

In vitro studies demonstrate that fruit extracts of *A. marmelos* possess broad bioactivity, including inhibition of *Vibrio cholerae*, *Shigella* spp., and *E. coli*, which are causative agents of diarrheal diseases. Seed extracts display antimalarial activity against *Plasmodium berghei* and inhibit *Helicobacter pylori* peptide deformylase (Kamaraj et al. 2012; Venthodika et al. 2021). Bael extracts also exhibit cytotoxicity in various cancer cell lines, as well as anti-HIV and broad-spectrum antiviral activities (Somu et al. 2019). Encapsulation of bael polyphenol fraction inhibits α-amylase and α-glucosidase, demonstrates strong DPPH, ABTS, and FRAP antioxidant activities, and provides membrane-stabilizing anti-inflammatory effects (De Silva et al. 2023).

In vivo studies indicate that *A. marmelos* exhibits antidiarrheal effects and reduces microbial load in rats, while demonstrating anticancer and radioprotective activity in mice. Additional findings include antifertility effects in male rats, antihyperlipidemic and antioxidant effects in obese or high-fat-diet animal models, and antiviral protection in silkworm larvae infected with BmNPV (Sarkar et al. 2020; Somu et al. 2019; Venthodika et al. 2021). Moreover, oral administration of the water extract at doses of 125 and 250 mg per kg body weight, twice daily for four weeks, resulted in significant reductions in blood glucose in diabetic rats (Kamalakkannan and Prince 2003).

Safety studies indicate that *A. Marmelos* is generally nontoxic, as repeated administration of 250 mg/kg BW extract for 14 days in rodents produced no major biochemical or histopathological organ damage (Jagetia et al. 2004; Sarkar et al. 2020).

15.6.3 *Syagrus romanzoffiana* (Cham.) Glassman

Syagrus romanzoffiana, commonly referred to as jerivá or mbocayá, is a member of the Arecaceae family (palms). This species is widely distributed throughout Brazil, Uruguay, Argentina, and Paraguay (Danielski and Shahidi 2024; Mandon and Campagna 2023). Ethnobotanical studies indicate that *S. romanzoffiana* fruit has historically served as a seasonal food source for wildlife and rural communities, especially during periods of winter food scarcity (Rodrigues et al. 2022). The fruit is consumed fresh and is also processed into juice, fermented beverages, ice creams, jams, and sweets. Despite these uses, it remains underutilized in large-scale commercial food systems (Danielski and Shahidi 2024).

Phytochemical analyses reveal that *S. romanzoffiana* fruit is rich in polyphenols and fatty acids. The pulp and seed contain substantial amounts of phenolic acids, flavonoids, tannins, and gallic acid derivatives, which are the primary contributors to its free radical–scavenging capacity (Danielski and Shahidi 2024). The lipid fraction is dominated by polyunsaturated fatty acids, particularly linoleic and linolenic acids, which are associated with cardiometabolic benefits (Mandon and Campagna 2023). Metabolomic profiling has also identified flavonols, lignans, stilbenoids (resveratrol-type compounds), oleic acid, linoleic acid, and linolenic acid, supporting the fruit's antioxidant and neuroactive potential (El-Hawary et al. 2021; Mello et al. 2024).

In vitro studies demonstrate that phenolic-rich extracts from *S. romanzoffiana* fruit exhibit potent antioxidant activity, including suppression of LDL-cholesterol oxidation, inhibition of α-glucosidase, and protection against oxidative DNA damage. These effects support potential antidiabetic, antiobesity, and cardioprotective mechanisms (Danielski and Shahidi 2024). An in vivo study showed that hydroalcoholic extracts administered to rats at a dose of 50–100 mg/kg BW significantly reduced acetylcholinesterase activity, indicating neuroprotective potential (de Lacerda Coriolano et al. 2021). Additionally, *S. romanzoffiana* extracts reduced acetylcholinesterase activity and improved histopathological outcomes in the cerebral cortex and cerebellum in an Alzheimer's disease rat model induced by aluminum chloride, suggesting disease-modifying neuroprotective effects (El-Hawary et al. 2021). Currently, no controlled human clinical trials have evaluated *S. romanzoffiana* fruit, and its therapeutic potential remains at the preclinical stage.

15.6.4 *Myrciaria dubia* (Kunth) Mc Vaugh

Myrciaria dubia (camu-camu) is a Myrtaceae tree native to the Amazon River region of Brazil, Peru, Colombia, and Venezuela (Noguera et al. 2024). Its fruits, particularly the peel, are exceptionally rich in vitamin C, carotenoids, anthocyanins (cyanidin-3-O-glucoside), flavonoids (quercetin, myricetin-O-pentoside, and kaempferol derivatives), tannins, and phenolics (ellagitannins, stilbenes, gallic acid, coumaric acid, 2,3-dihydroxybenzoate, 2',4-dihydroxychalcone, and dihydroxymandelic acid) (Flores Trindade et al. 2025; Oliveira et al. 2025). Camu-camu is traditionally consumed as fresh or processed juice by the Amazonians to treat infectious diseases. However, its high acidity and bitterness limit raw consumption and make it suitable for processed products and nutraceuticals.

In vitro studies of extracts from seed or peel fractions demonstrate strong antioxidant activity through the scavenging of free radicals in DPPH, ABTS, FRAP, ORAC, and Folin assays (Flores Trindade et al. 2025; Manica-Cattani et al. 2022). Seed and peel extracts exhibit cytotoxicity against lung cancer (A549) and colon cancer (HCT-8) cell lines. Moreover, the extract also decreases ROS generation in normal human cells, inhibits peroxide-induced DNA damage, and exhibits antimicrobial activity against *Streptococcus mutans*, *S. sanguinis*, and other Gram-positive bacteria. In addition, phenolic extracts from pulp inhibit the activity of enzymes α-glucosidase and α-amylase, supporting antidiabetic potential (da Silva et al. 2021; Noguera et al. 2024; Oliveira et al. 2025).

The in vivo studies showed that camu-camu juice protects mice from alcohol-induced hepatic damage and bone marrow toxicity, improves lipid profiles, fecal cholesterol excretion, obesity, and insulin resistance in dyslipidemic rats and high-fat-diet mice (Noguera et al. 2024; Schwertz et al. 2012). Additional animal studies suggest improved glycemic control through enhanced insulin signaling and glucose uptake (Dias et al. 2025).

Human trials report that the administration of 70 mL/day of 100% *M. dubia* juice, which is equivalent to 1050 mg vitamin C for seven days in habitual male smokers, reduces the concentration of serum ROS, inflammatory cytokines, and C-reactive protein more effectively than equivalent vitamin C tablets, and that clarified pulp juice in healthy volunteers lowers post-prandial glucose by about one-third (Manica-Cattani et al. 2022; Noguera et al. 2024). The in vitro toxicity study shows no cytotoxicity effects in MDCK (Madin–Darby Canine Kidney cell line) or other normal cell lines up to 800 μg/mL. In addition, oral administration of 25–100% *M. dubia* juice in mice showed no toxicity or genotoxicity effects. In conclusion, camu-camu and its by-products are considered safe, highly bioactive ingredients for food, nutraceutical, and pharmaceutical applications when appropriately formulated.

15.6.5 *Syzygium cumini* (L.) Skeels

Syzygium cumini (syn. *Eugenia jambolana*), commonly known as jamun/jambolan/black plum/juwet, is a member of the family Myrtaceae, growing across South and Southeast Asia, Australia, Africa, and tropical Latin America (Rizvi et al. 2022). The bark, leaves, seeds, fruit, peel, and pulp are used to treat diabetes, dysentery, diarrhea, ulcers, bronchitis, asthma, liver and stomach disorders, hypertension, obesity, and urinary problems in Ayurveda, Unani, Chinese, and various folk medicine systems (Chaudhary and Mukhopadhyay 2012; Prabhakar 2022). Jamun contains a diverse array of bioactive compounds, including anthocyanins (delphinidin, petunidin, and cyanidin), which contribute to its dark purple color (as shown in Figure 15.3), as well as flavonoids, polyphenols, tannins, and phenolic acids (gallic, ellagic, and caffeic). Additional constituents include terpenoids (such as monoterpenes and sesquiterpenes), phytosterols (β-sitosterol), saponins, as well as vitamins (C, A, and E) and minerals (iron, potassium, and zinc) (do Nascimento-Silva et al. 2022; Prabhakar 2022).

In vitro studies consistently show strong antioxidant, antiglycation, and enzyme-modulating activity. Jamun fruit and leaves exhibit DPPH and other radical-scavenging activities and inhibit α-glucosidase, α-amylase, and lipase, supporting their use in the treatment of diabetes and hyperlipidemia (Kumar et al. 2022; Rizvi et al. 2022). Fruit extracts also show anti-inflammatory activity via membrane stabilization and protein denaturation assays, and anthocyanins modulate Nrf2 signaling and reduce ROS in cell models (do Nascimento-Silva et al. 2022; Sousa et al. 2021). Cytotoxic and anticancer effects have been documented against colon, breast, and lung cancer cell lines (Kaur et al. 2024).

An extensive in vivo study, primarily in alloxan- or streptozotocin-induced diabetic rats, demonstrates that jamun seed and pulp extracts reduce fasting blood glucose and HbA1c and improve insulin and lipid profiles, antioxidant enzyme activity, and inflammatory markers (Kumar et al. 2022; Rizvi et al. 2022). The mechanisms include enhanced insulin secretion, improved insulin sensitivity, inhibition of intestinal glucose absorption, and protection of pancreatic β-cell DNA. Additional animal models show antihypertensive effects via reduced vascular tone and Ca^{2+} influx, hepatoprotective effects in CCl_4 and paracetamol liver injury, radioprotective and neuroprotective actions, and anti-inflammatory activity in carrageenan-induced edema and granuloma models (Kumari et al. 2023; Prabhakar 2022; Rizvi et al. 2022).

Administration of jamun seed and fruit powder (4.5–10 g/day) in hyperglycemic, dyslipidemic, or prediabetic patients over 84–90 days improved glycemic control and lipid profiles without major changes in anthropometry, supporting antidiabetic and antihyperlipidemic claims (Kumar et al. 2022; Rizvi et al. 2022). Polyherbal Ayurvedic formulations, including jamun, have produced substantial reductions in HbA1c in single-patient or small-series reports (Prabhakar 2022).

Jamun seed and fruit extract exhibit acute and subacute doses up to 2000 mg/kg BW, while the oral administration of 3000 mg/kg BW *S. cumini* extract induces moderate, reversible renal parameter changes without histological damage (Kumar et al. 2022).

15.6.6 *Hippophae rhamnoides* L.

Sea buckthorn, primarily *Hippophae rhamnoides* L. of the Elaeagnaceae family, is widely distributed across cold-temperate Eurasia, including the Tibetan Plateau, Central Asia, the Himalayas, China, Mongolia, and Russia (Nybom et al. 2023; Wang et al. 2022c). Ethnopharmacologically, sea buckthorn is a traditional medicine used to treat cough, respiratory, and digestive disorders in Tibet and Mongolia. In addition, it is also documented across ancient cultures (Greeks, Hindus/Ayurveda,

Chinese, Mongols, Celts, and Slavs), where it served as a "food medicine" plant and was empirically used as an herbal immunostimulant against the common cold (Gatlan and Gutt 2021). Most *Hippophae rhamnoides* genotypes produce oblong berries exhibiting a range of epicarp colors, from dark orange to light orange and yellow, as illustrated in Figure 15.3 (Ilhan et al. 2021).

Previous investigations have identified that sea buckthorn contains approximately 100 to nearly 200 bioactive compounds, including carotenoids, polyphenols, and flavonoids (isorhamnetin, quercetin, kaempferol, catechin, epicatechin, proanthocyanidins), phenolic acids (gallic, caffeic, p-coumaric acids), tocopherols, and polysaccharides (Dong et al. 2021; Gatlan and Gutt 2021; Mihal et al. 2023). In vitro studies of polyphenol-rich fractions demonstrate suppression of inflammatory mediators such as COX-2, TNF-α, and IL-6, primarily through inhibition of NF-κB and MAPK signaling pathways, thereby reducing oxidative stress and cytokine production (Wang et al. 2022b, 2022c). Anticancer activity has been observed, particularly from isorhamnetin, quercetin, apigenin, luteolin, and myricetin, which induce ROS-mediated apoptosis, mitochondrial membrane depolarization, caspase-3 activation, and cell-cycle arrest, with IC_{50} values generally ranging from 10 to 50 µg/mL (Janceva et al. 2022; Mihal et al. 2023).

Moreover, in vivo studies have demonstrated that sea buckthorn supplementation significantly ameliorates metabolic syndrome, including hyperlipidemia, obesity, and type 2 diabetes (Gatlan and Gutt 2021; Shah et al. 2021; Wang et al. 2022c). These effects are characterized by reductions in total cholesterol, LDL, triglycerides, fasting glucose, and insulin resistance, as well as enhanced activity of antioxidant enzymes, including catalase, superoxide dismutase, and glutathione peroxidase. Hepatoprotective effects have been observed in chemically induced liver injury models, as indicated by reduced aspartate aminotransferase (AST) and alanine aminotransferase (ALT) levels, decreased lipid peroxidation, and diminished inflammatory infiltration. Cardioprotective outcomes include improved endothelial function, increased nitric oxide bioavailability, and reduced atherosclerotic plaque development (Dong et al. 2021).

Although clinical investigations are limited, they provide important translational support. Studies have shown that sea buckthorn oil, juice, and berry-based formulations improve cardiovascular risk profiles, dyslipidemia, skin hydration, mucosal integrity, and postmenopausal vulvovaginal atrophy. Reported benefits include reductions in LDL-cholesterol, C-reactive protein, and oxidative stress markers, as well as enhanced epithelial repair and wound healing (Dong et al. 2021; Gatlan and Gutt 2021; Wang et al. 2022c).

15.6.7 *Rhus chinensis* Mill.

Rhus chinensis Mill. (family Anacardiaceae), presented in Figure 15.3, is a deciduous tree distributed throughout Japan, Korea, China, Nepal, and North-Eastern India (Chhetri et al. 2020). Taxonomically classified within the genus Rhus, this species is recognized for its dual role as a food source and medicinal plant in various Asian cultures (Heirangkhongjam and Ngaseppam 2018; Nemkul et al. 2021).

Phytochemical analyses demonstrate that *R. chinensis* fruits are abundant in phenolic compounds, gallotannin derivatives, flavonoids (myricetin-3-O-rhamnoside, quercetin-3-O-rhamnoside), organic acids, and triterpenoids (Ali et al. 2017; Heirangkhongjam and Ngaseppam 2019; Hwang et al. 2019). Nutritional evaluations further indicate high concentrations of vitamin C, dietary fiber, carbohydrates, and lipids, which support both nutritional and therapeutic applications (Heirangkhongjam and Ngaseppam 2019).

Ethnopharmacological studies indicate that *R. chinensis* fruits are traditionally used to treat diarrhea and dysentery, as supported by antibacterial assays (Nemkul et al. 2021). Furthermore, in China, India, and Japan, both fruits and seeds are used to treat cough, fever, jaundice, stomach ache, indigestion, intestinal worms, skin diseases, and vomiting, and are also considered as antitoxins (Ali et al. 2017).

Further in vitro studies provide substantial pharmacological evidence supporting traditional uses. Fruit extracts demonstrate broad-spectrum antibacterial activity against Gram-positive strains (*Staphylococcus aureus, Bacillus subtilis*, and *Enterococcus faecalis*) with minimum bactericidal concentration (MBC) between 3.12 and 50 mg/mL, and four Gram-negative strains (*Escherichia coli, Klebsiella pneumoniae, Pseudomonas aeruginosa*, and *Shigella dysenteriae)* with MBC ranging from 12.5 to 25 mg/mL (Nemkul et al. 2021; Zhang et al. 2017). Notably, phenolic fractions of the fruit exhibit potent antidiabetic effects, including inhibition of α-glucosidase, DPP-IV, and advanced glycation end-product (AGE) formation, with IC_{50} values of 0.56±0.04; 66.08±1.36; 0.43±0.03 mg/mL, respectively (Liu et al. 2021). Furthermore, fruit extracts inhibit pancreatic lipase, suggesting potential antiobesity applications (Zhang et al. 2017).

Expanding on these pharmacological properties, *R. chinensis* extracts inhibit adipocyte differentiation in multipotent bone marrow stromal cells and reduce fat accumulation in gonadal fat, bone tissues, and liver in ovariectomized mice fed a high-fat diet (Hwang et al. 2019). Furthermore, administration of 800 mg/kg BW of *R. chinensis* fruit extract significantly decreases ALT, AST, ALP, and MDA levels. The extract also lowers TNF-α, IL-6, and IL-1β levels and restores GSH and SOD levels in a rifampicin- and isoniazid-induced liver damage model (Sun et al. 2023).

Safety considerations have also been evaluated in experimental studies. Acute toxicity studies indicate that the LD_{50} of *R. chinensis* fruit extract exceeds 5000 mg/kg BW. No significant adverse effects are observed at subchronic doses of 312 and 625 mg/kg BW. However, higher daily oral doses of 1250 and 2000 mg/kg BW induce weight gain and toxic effects on the liver, spleen, kidney, and heart, as evaluated by hematological, serum biochemical, and histological analyses (Wu et al. 2018).

15.7 CONCLUSION AND FUTURE DIRECTION

The evidence presented in this chapter demonstrates that nutraceuticals derived from fruits and berries possess a wide range of biological activities relevant to both communicable and NCDs. The increasing global demand for nutraceutical products, especially in the past decade, underscores the importance of fruits and berries as sustainable and generally safe raw materials. Despite notable advancements, several challenges persist, including limited access to raw materials, which are often sourced from regions with suitable climatic conditions, constraining large-scale production. Additionally, many fruits and berries lack comprehensive preclinical and clinical evaluation, delaying the development of standardized nutraceutical products. Expanding the diversity of fruit- and berry-based nutraceuticals requires systematic exploration of local and underutilized species, many of which have traditional applications in the prevention or treatment of infectious diseases. However, insufficient in vivo toxicity and clinical data continue to impede their translational development.

This chapter identified several underexplored fruits and berries with promising bioactivity, making them potential nutraceutical candidates. *A. marmelos*, *R. chinensis*, *S. romanzoffiana*, and *P. nitida* contain polyphenols and alkaloids that are active against infectious diseases, although clinical studies remain unavailable. In addition, *Hippophae rhamnoides, S. cumini*, and *M. dubia* have demonstrated favorable in vivo safety profiles, supportive clinical studies, and high levels of anthocyanins and phenolics. Future research should prioritize standardized extracts, rigorous clinical trials, and systematic investigation of underutilized fruits and berries to clarify their preventive and therapeutic potential and to reinforce their integration into evidence-based healthcare systems.

REFERENCES

Adeola, O.E., Akpor, O.A., Adamolekun, M.M., Adewale, O.B., Akpor, O.B. (2023). In vitro antioxidant and antimicrobial potentials of aqueous extract of *Picralima nitida* seeds. *Vegetos*, 37: 486–493.

Adeyoju, E.O., Ajayi, C.O., Adepiti, A.O., Elujoba, A.A. (2022). Comparative in vivo antimalarial activities of aqueous and methanol extracts of MAMA powder – a herbal antimalarial preparation. *Journal of Ethnopharmacology*, 283: 114686.

Ali, M., Sultana, S., Mir, S.R. (2017). Aliphatic alcohols, alkylated aromatic and triterpenic constituents from the aerial parts of *Rhus chinensis* Mill. *Journal of Pharmaceutical and Biological Sciences*, 5: 1–7.

Alsaikhan, F., Farhood, B. (2025). Phytochemical-based immunomodulation: a promising therapeutic approach for viral infections. *Phytotherapy Research*, 40: 398–419.

Arnold, D. (2010). British India and the "beriberi problem", 1798–1942. *Medical History*, 54: 295–314.

Aronson, J.K. (2017). Defining nutraceuticals: neither nutritious nor pharmaceutical. *British Journal of Clinical Pharmacology*, 83: 8–19.

Asghar, A., Randhawa, M.A., Masood, M., Abdullah, M., Irshad, M.A. (2018). Nutraceutical formulation strategies to enhance the bioavailability and efficiency: an overview. In: Grumezescu, A.M., Holban, A.M. (Eds.), *Role of Materials Science in Food Bioengineering*. Academic Press, pp. 329–352.

Bhatt, D., Kumar, S., Kumar, P., Bisht, S., Kumar, A., Maurya, A.K., Pal, A., Bawankule, D.U. (2022). Rutin ameliorates malaria pathogenesis by modulating inflammatory mechanism: an in vitro and in vivo study. *Inflammopharmacology*, 30: 159–171.

Calder, P.C. (2015). Marine omega-3 fatty acids and inflammatory processes: effects, mechanisms and clinical relevance. *Biochimica et Biophysica Acta*, 1851: 469–484.

Cencic, A., Chingwaru, W. (2010). The role of functional foods, nutraceuticals, and food supplements in intestinal health. *Nutrients*, 2: 611–625.

Chaudhary, B., Mukhopadhyay, K. (2012). *Syzygium cumini* (L.) Skeels: a potential source of nutraceuticals. *International Journal of Pharmacy and Biological Sciences*, 2: 46–53.

Chhetri, D.R., Mahanta, J., Chettri, A., Pradhan, A. (2020). Phytochemical and antioxidant attributes of *Rhus chinensis* Mill., an edible wild fruit from Sikkim Himalaya. *Journal of Plant Science Research*, 36: 157–166.

Chopra, A.S., Lordan, R., Horbanczuk, O.K., Atanasov, A.G., Chopra, I., Horbanczuk, J.O., Jozwik, A., Huang, L., Pirgozliev, V., Banach, M., Battino, M., Arkells, N. (2022). The current use and evolving landscape of nutraceuticals. *Pharmacological Research*, 175: 106001.

Cisneros-Zevallos, L. (2021). The power of plants: how fruit and vegetables work as source of nutraceuticals and supplements. *International Journal of Food Sciences and Nutrition*, 72: 660–664.

Costagliola, G., Nuzzi, G., Spada, E., Comberiati, P., Verduci, E., Peroni, D.G. (2021). Nutraceuticals in viral infections: an overview of the immunomodulating properties. *Nutrients*, 13: 1–16.

Dama, A., Shpati, K., Daliu, P., Dumur, S., Gorica, E., Santini, A. (2024). Targeting metabolic diseases: the role of nutraceuticals in modulating oxidative stress and inflammation. *Nutrients*, 16: 1–18.

Damian, M.R., Cortes-Perez, N.G., Quintana, E.T., Ortiz-Moreno, A., Garfias Noguez, C., Cruceno-Casarrubias, C.E., Sanchez Pardo, M.E., Bermudez-Humaran, L.G. (2022). Functional foods, nutraceuticals and probiotics: a focus on human health. *Microorganisms*, 10: 1–13.

Danielski, R., Shahidi, F. (2024). Nutraceutical potential of underutilized tropical fruits and their byproducts. *Journal of Agricultural and Food Chemistry*, 72: 4035–4048.

Darko, D., Kwekutsu, E., Idoko, P.I. (2025). Synergistic effects of phytochemicals in combating chronic diseases with insights into molecular mechanisms and nutraceutical development. *International Journal of Innovative Science and Research Technology*, 10: 1–18.

De Silva, N.D., Attanayake, A.P., Karunaratne, D.N., Arawwawala, L., Pamunuwa, G.K. (2023). Bael (*Aegle marmelos*) fruit extracts encapsulated alginate nanoparticles as a potential dietary supplement with improved bioactivities. *Journal of Food Science*, 88: 4942–4961.

Di Sotto, A., Vitalone, A., Di Giacomo, S. (2020). Plant-derived nutraceuticals and immune system modulation: an evidence-based overview. *Vaccines*, 8: 1–34.

Dias, B.P., Dantas, H.V.P., de Barros Neto, J., Teixeira, T., da Silva, H.R., Aytar, E.C., Carvalho, J.C.T. (2025). From the Amazon to modern pharmacology: medicinal plants as metabolic and cardiovascular modulators in cardiodiabesity. *Inflammopharmacology*, 33: 4355–4368.

Dillard, C.J., German, J.B. (2000). Phytochemicals: nutraceuticals and human health. *Journal of the Science of Food and Agriculture*, 80: 1744–1756.

Djaoudene, O., Romano, A., Bradai, Y.D., Zebiri, F., Ouchene, A., Yousfi, Y., Amrane-Abider, M., Sahraoui-Remini, Y., Madani, K. (2023). A global overview of dietary supplements. *Nutrients*, 15: 1–24.

Dong, K., Binosha Fernando, W.M.A.D., Durham, R., Stockmann, R., Jayasena, V. (2021). Nutritional value, health-promoting benefits and food application of sea buckthorn. *Food Reviews International*, 39: 2122–2137.

Dudek, A., Spiegel, M., Strugala-Danak, P., Gabrielska, J. (2022). Analytical and theoretical studies of antioxidant properties of chosen anthocyanins. *International Journal of Molecular Sciences*, 23.

El-Hawary, S.S., Fathy, F.I., Sleem, A.A., Morsy, F.A., Khadar, M.S., Mansour, M.K. (2021). Anticholinesterase activity and metabolite profiling of *Syagrus romanzoffiana*. *Natural Product Research*, 35: 1671–1675.

Flores Trindade, M.E., Cardoso, C.E.F., Coimbra, P.P.S., Antonio, A.S., Pereira, H.M.G., Veiga Junior, V.F., Elias, M.B., Teodoro, A.J. (2025). Chemical composition and antioxidant activity of Amazonian *Myrciaria dubia* seed. *Natural Product Research*, 1–6.

Furukawa, S., Fujita, T., Shimabukuro, M., Iwaki, M., Yamada, Y., Nakajima, Y., Nakayama, O., Makishima, M., Matsuda, M., Shimomura, I. (2004). Increased oxidative stress in obesity and its impact on metabolic syndrome. *Journal of Clinical Investigation*, 114: 1752–1761.

Gatlan, A.M., Gutt, G. (2021). Sea buckthorn in plant-based diets. *International Journal of Environmental Research and Public Health*, 18: 1–23.

Gutierres, D., Pacheco, R., Reis, C.P. (2025). The role of omega-3 and omega-6 polyunsaturated fatty acid supplementation in human health. *Foods*, 14: 1–26.

Hajar, R. (2012). The air of history (part II): medicine in the Middle Ages. *Heart Views*, 13: 158–162.

Heirangkhongjam, M.D., Ngaseppam, I.S. (2018). Traditional medicinal uses and pharmacological properties of *Rhus chinensis*: a systematic review. *European Journal of Integrative Medicine*, 21: 43–49.

Heirangkhongjam, M.D., Ngaseppam, I.S. (2019). Nutritional evaluation of *Rhus chinensis* and development of value-added products. *Indian Journal of Traditional Knowledge*, 18: 16–24.

Husen, A., Iqbal, M. (2021). Plant-based potential nutraceuticals for improving the human immune system. In: Husen, A. (Ed.), *Traditional Herbal Therapy for the Human Immune System*. CRC Press, pp. 1–11.

Hwang, Y.-H., Jang, S.-A., Kim, T., Ha, H. (2019). Anti-osteoporotic and anti-adipogenic effects of *Rhus chinensis* nutgalls. *Planta Medica*, 85: 1128–1135.

Ige, A.B., Ogunlakin, A.D., Sonibare, M.A. (2025). Antioxidant and ameliorative effects of selected Nigerian plants. *Frontiers in Nutrition*, 12: 1635080.

Igwenyi, I., Onodugo, C.A., Aja, P.M., Elom, S.O., Awoke, J.N., Ibhadode, O.S., Uraku, A.J., Edwin, N., Obasi, A.N., David, E.E., Igwenyi, I.P., Ogbu, P.N., Egwu, C., Atoki, A.V. (2023). *Azadirachta indica* fruit juice clears malaria parasites and replenishes blood levels in *Plasmodium* berghei-infected mice, 4: 1–9.

Ilhan, G., Gundogdu, M., Karlović, K., Židovec, V., Vokurka, A., Ercişli, S. (2021). Agro-morphological and biochemical berry characteristics of sea buckthorn. *Sustainability*, 13: 1–14.

Jagetia, G.C., Venkatesh, P., Baliga, M.S. (2004). Fruit extract of *Aegle marmelos* protects mice against radiation-induced lethality. *Integrative Cancer Therapies*, 3: 323–332.

Janceva, S., Andersone, A., Lauberte, L., Bikovens, O., Nikolajeva, V., Jashina, L., Zaharova, N., Telysheva, G., Senkovs, M., Rieksts, G., Ramata-Stunda, A., Krasilnikova, J. (2022). Sea buckthorn waste biomass as a source of bioactive compounds. *Plants*, 11: 1–20.

Joshi, S., Bhattarai, K., Subedi, A.R., Bhattarai, J., Amatya, S., Baral, B. (2024). Validation of ethnopharmacological findings of *Aegle marmelos*. *Pharmacological Research – Natural Products*, 5: 1–10.

Kahkonen, M.P., Heinonen, M. (2003). Antioxidant activity of anthocyanins and their aglycons. *Journal of Agricultural and Food Chemistry*, 51: 628–633.

Kalra, E.K. (2003). Nutraceutical: definition and introduction. *AAPS PharmSci*, 5: 1–2.

Kamalakkannan, N., Prince, P.S. (2003). Hypoglycaemic effect of *Aegle marmelos* in diabetic rats. *Journal of Ethnopharmacology*, 87: 207–210.

Kamaraj, C., Kaushik, N.K., Rahuman, A.A., Mohanakrishnan, D., Bagavan, A., Elango, G., Zahir, A.A., Santhoshkumar, T., Marimuthu, S., Jayaseelan, C., Kirthi, A.V., Rajakumar, G., Velayutham, K., Sahal, D. (2012). Antimalarial activities of medicinal plants. *Journal of Ethnopharmacology*, 141: 796–802.

Kaulmann, A., Bohn, T. (2014). Carotenoids, inflammation, and oxidative stress—implications of cellular signaling pathways and relation to chronic disease prevention. *Nutrition Research*, 34: 907–929.

Kaur, D., Yousuf, B., Qadri, O.S. (2024). *Syzygium cumini* anthocyanins: recent advances in biological activities, extraction, stability, characterisation and utilisation in food systems. *Food Production, Processing and Nutrition*, 6: 1–17.

Kaushal, A. (2024). Nutraceuticals and pharmacological approaches to balance the transitional microbiome to extend immunity during COVID-19 and other viral infections. *Journal of Translational Medicine*, 22: 847.

Khoo, H.E., Azlan, A., Tang, S.T., Lim, S.M. (2017). Anthocyanidins and anthocyanins: colored pigments as food and pharmaceutical ingredients and their potential health benefits. *Food and Nutrition Research*, 61: 1361779.

Kumar, M., Zhang, B., Nishad, J., Verma, A., Sheri, V., Dhumal, S., Radha, Sharma, N., Chandran, D., Senapathy, M., Dey, A., Rajalingam, S., Muthukumar, M., Mohankumar, P., Amarowicz, R., Pateiro, M.,

Lorenzo, J.M. (2022). Jamun (*Syzygium cumini*) seed: a review on nutritional profile, functional food properties, health-promoting applications, and safety aspects. *Processes*, 10: 1–15.
Kumari, N., Kumar, M., Chaudhary, N., Zhang, B., Radha, Chandran, D., Joshi, S., Singh, D., Dey, A., Rajalingam, S., Natarajan, K., Muthukumar, M., Mohankumar, P., Sheri, V., Dhumal, S., Lorenzo, J.M. (2023). Exploring the chemical and biological potential of jamun (*Syzygium cumini*) leaves: a comprehensive review. *Chemistry and Biodiversity*, 20: e202300479.
de Lacerda Coriolano, D., Menezes Estevam Alves, M.H., Macario Ferro Cavalcanti, I. (2021). Biological potential of products obtained from palm trees of the genus *Syagrus*. *Evidence-Based Complementary and Alternative Medicine*, 2021: 5580126.
Li, D. (2025). The history of nutritional sciences. *Asia Pacific Journal of Clinical Nutrition*, 34: 265–270.
Liu, X., Fu, Y., Ma, Q., Yi, J., Cai, S. (2021). Anti-diabetic effects of different phenolic-rich fractions from *Rhus chinensis* fruits in vitro. *eFood*, 2: 37–46.
Lordan, R. (2021). Dietary supplements and nutraceuticals market growth during the coronavirus pandemic—implications for consumers and regulatory oversight. *PharmaNutrition*, 18: 100282.
Mackin, C., Dahiya, D., Nigam, P.S. (2023). Honey as a natural nutraceutical: combinational therapeutic strategies applicable to blood infections. *Pharmaceuticals*, 16: 1-15.
Maheshwari, S., Kumar, V., Bhadauria, G., Mishra, A. (2022). Immunomodulatory potential of phytochemicals and other bioactive compounds of fruits: a review. *Food Frontiers*, 3: 221–238.
Mandon, E., Campagna, M.N. (2023). Estudio de los frutos de *Butia yatay* y *Syagrus romanzoffiana* como fuente de nutrientes y compuestos bioactivos. *Boletín de la Sociedad Argentina de Botánica*, 58: 533–546.
Manica-Cattani, M.F., Hoefel, A.L., Azzolin, V.F., Montano, M.A.E., da Cruz Jung, I.E., Ribeiro, E.E., Azzolin, V.F., da Cruz, I.B.M. (2022). Amazonian fruits with potential effects on COVID-19 by inflammaging modulation: a narrative review. *Journal of Food Biochemistry*, 46: e14472.
MarketResearch.com (2025). Nutraceuticals: global industry analysis.
Martau, G.A., Bernadette-Emoke, T., Odocheanu, R., Soporan, D.A., Bochis, M., Simon, E., Vodnar, D.C. (2023). *Vaccinium* species: phytochemistry and biological properties of medicinal plants. *Molecules*, 28: 1–30.
Mello, B., Malarski, A., Bohm, V. (2024). Bioactive compounds and antioxidant capacity of pulp, peel and seeds from jeriva (*Syagrus romanzoffiana*). *Antioxidants*, 13: 1–11.
Mihal, M., Roychoudhury, S., Sirotkin, A.V., Kolesarova, A. (2023). Sea buckthorn, its bioactive constituents, and mechanism of action. *Frontiers in Endocrinology*, 14: 1244300.
Mozaffarian, D., Rosenberg, I., Uauy, R. (2018). History of modern nutrition science—implications for current research, dietary guidelines, and food policy. *BMJ*, 361: k2392.
do Nascimento-Silva, N.R.R., Bastos, R.P., da Silva, F.A. (2022). Jambolan (*Syzygium cumini*): a review on its nutrients, bioactive compounds and health benefits. *Journal of Food Composition and Analysis*, 109: 1–9.
Natarajan, T.D., Ramasamy, J.R., Palanisamy, K. (2019). Nutraceutical potentials of synergic foods: a systematic review. *Journal of Ethnic Foods*, 6: 1–7.
Nelson, R.H. (2013). Hyperlipidemia as a risk factor for cardiovascular disease. *Primary Care: Clinics in Office Practice*, 40: 195–211.
Nemkul, C.M., Bajracharya, G.B., Maeda, H., Shrestha, I. (2021). Ethnomedicinal knowledge verification for the antidiarrheal and antioxidant effects of *Rhus chinensis*. *Pharmacognosy Journal*, 13: 37–43.
Nguyen, M. (2020). The forgotten story of vitamin C deficiency. *Blood*, 136: 35.
Noguera, N.H., Noguera, D., Machado, A., Reguengo, L.M., Nascimento, R.P.D. (2024). Emerging berries from the Brazilian Amazon and Atlantic Forest biomes. *Food and Function*, 15: 5752–5784.
Nybom, H., Ruan, C., Rumpunen, K. (2023). The systematics, reproductive biology, biochemistry, and breeding of sea buckthorn—a review. *Genes*, 14: 1–34.
Olalekan, S.O., Bakare, O.O., Okwute, P.G., Osonuga, I.O., Adeyanju, M.M., Olalekan, R.O. (2024). The role of nutraceuticals in managing metabolic syndrome. *Egyptian Journal of Internal Medicine*, 36: 1–15.
Oliveira, B.C.R., Martins, C.P.C., Soutelino, M.E.M., Rocha, R.S., Cruz, A.G., Marsico, E.T., Silva, A.C.O., Esmerino, E.A. (2025). Potential of select edible Amazonian fruits and their applications in dairy products. *Critical Reviews in Food Science and Nutrition*, 65: 5572–5586.
Panche, A.N., Diwan, A.D., Chandra, S.R. (2016). Flavonoids: an overview. *Journal of Nutritional Science*, 5: e47.
Paudel, K.R., Patel, V., Vishwas, S., Gupta, S., Sharma, S., Chan, Y., Jha, N.K., Shrestha, J., Imran, M., Panth, N., Shukla, S.D., Jha, S.K., Devkota, H.P., Warkiani, M.E., Singh, S.K., Ali, M.K., Gupta, G., Chellappan, D.K., Hansbro, P.M., Dua, K. (2022). Nutraceuticals and COVID-19: a mechanistic approach toward attenuating disease complications. *Journal of Food Biochemistry*, 46: e14445.

Prabhakar, P. (2022). Role of Indian fruits in the prevention and management of hypertension. *Journal of the Practice of Cardiovascular Sciences*, 8: 135–143.

Puri, V., Nagpal, M., Singh, I., Singh, M., Dhingra, G.A., Huanbutta, K., Dheer, D., Sharma, A., Sangnim, T. (2022). A comprehensive review on nutraceuticals: therapy support and formulation challenges. *Nutrients*, 14: 1–28.

Rizvi, M.K., Rabail, R., Munir, S., Inam-Ur-Raheem, M., Qayyum, M.M.N., Kieliszek, M., Hassoun, A., Aadil, R.M. (2022). Health benefits of jamun (*Syzygium cumini*) toward metabolic syndrome. *Molecules*, 27: 1–16.

Rodrigues, C.E., Schafer, L., Gregolon, J.G.N., de Oliveira, J.F., Baez, O.P., Deolindo, C.T.P., de Melo, A.P.Z., Singer, R.B., Ledur Kist, T.B., Hoff, R. (2022). Determination of amino acid and fatty acid profiles in non-conventional edible fruits. *Food Research International*, 162: 111995.

Rosa, I.S., Pires, L.O., Ishida, J.K. (2023). Nutraceutical production from plants. In: Upadhyay, S.K.S., Sudhir, P. (Eds.), *Plants as Bioreactors for Industrial Molecules*. John Wiley & Sons, pp. 367–391.

Sahin, K., Gencoglu, H., Bilir, B., Kucuk, O. (2018). Protective role of lycopene against oxidative stress in liver. In: Patel, V.B., Rajendram, R., Preedy, V.R. (Eds.), *The Liver*. Academic Press, pp. 155–167.

Sarkar, T., Salauddin, M., Chakraborty, R. (2020). Pharmacological and nutritional properties of bael (*Aegle marmelos*): a critical review. *Journal of Agriculture and Food Research*, 2: 1–22.

Schwertz, M.C., Maia, J.R.P., de Sousa, R.F.S., Aguiar, J.P.L., Yuyama, L.K.O., Lima, E.S. (2012). Hypolipidemic effect of camu-camu juice in rats. *Revista de Nutrição*, 25: 35–44.

Shah, R.K., Idate, A., Ugale, V., Poorva (2021). Comprehensive review on sea buckthorn: biological activity and potential uses. *Pharma Innovation*, 10: 942–953.

Shahidi, F. (2009). Nutraceuticals and functional foods: whole versus processed foods. *Trends in Food Science and Technology*, 20: 376–387.

Shahidi, F. (2020). Nutraceuticals, functional foods and dietary supplements in health and disease. *Journal of Food and Drug Analysis*, 20: 226–230.

Shahidi, F., Ambigaipalan, P. (2015). Phenolics and polyphenolics in foods, beverages and spices. *Journal of Functional Foods*, 18: 820–897.

Sharma, N., Dubey, W. (2013). History and taxonomy of *Aegle marmelos*: a review. *International Journal of Pure and Applied Bioscience*, 1: 7–13.

Shittu, H., Gray, A., Furman, B., Young, L. (2010). Glucose uptake stimulatory effect of akuammicine from *Picralima nitida*. *Phytochemistry Letters*, 3: 53–55.

Silpi, C. (2025). Nutraceuticals: traditional practices, regulatory guidelines, opportunities and challenges. *Discover Food*, 5: 1–19.

da Silva, J.L., Cadavez, V., Lorenzo, J.M., Figueiredo, E.E.D.S., Gonzales-Barron, U. (2021). Effects of camu-camu (*Myrciaria dubia*) powder on physicochemical and kinetic parameters of deteriorating microorganisms and *Salmonella enterica* in refrigerated vacuum-packed ground beef. *Agriculture*, 11: 1–14.

Slavin, J. (2013). Fiber and prebiotics: mechanisms and health benefits. *Nutrients*, 5: 1417–1435.

Somu, C., Karuppiah, H., Sundaram, J. (2019). Antiviral activity of seselin from *Aegle marmelos*. *Journal of Ethnopharmacology*, 245: 112155.

Sousa, M.M., Lima, R.M.T., Lima, A., Reis, A.C., Cavalcante, A., Sattler, J.A.G., Almeida-Muradian, L.B., Lima Neto, J.S., Moreira-Araujo, R., Nogueira, N.D.N. (2021). Antioxidant action of polyphenols from jambolan. *Food Chemistry*, 363: 130353.

Sun, Y., Zhang, Y., Ma, N., Cai, S. (2023). *Rhus chinensis* fruits alleviate liver injury induced by drugs. *Journal of Ethnopharmacology*, 310: 116387.

Venthodika, A., Chhikara, N., Mann, S., Garg, M.K., Sofi, S.A., Panghal, A. (2021). Bioactive compounds of *Aegle marmelos* and its food applications. *Phytotherapy Research*, 35: 1887–1907.

Wan, C., Langyan, S., Echeverría, J., Devkota, H.P., Tewari, D., Moosavi, M.A., Ezzat, S.M., Perez-Vazquez, A., Fraga-Corral, M., Cravotto, G., Prieto, M.A., Belwal, T., Li, M. (2023). Edible fruits and berries as a source of functional polyphenols. *Phytochemistry Reviews*, 24: 1197–1225.

Wang, B., Han, L., Liu, J.-M., Zhang, J., Wang, W., Li, B.-G., Dong, C.-X., Bai, C.-C. (2022a). *Lycium* genus polysaccharide: extraction, structures and applications. *Separations*, 9: 1–41.

Wang, K., Xu, Z., Liao, X. (2022b). Bioactive compounds, health benefits and functional food products of sea buckthorn: a review. *Critical Reviews in Food Science and Nutrition*, 62: 6761–6782.

Wang, Z., Zhao, F., Wei, P., Chai, X., Hou, G., Meng, Q. (2022c). Phytochemistry, health benefits, and food applications of sea buckthorn (*Hippophae rhamnoides*): a comprehensive review. *Frontiers in Nutrition*, 9: 1–20.

Woodall, A.A., Britton, G., Jackson, M.J. (1996). Dietary supplementation with carotenoids: effects on alpha-tocopherol levels and susceptibility of tissues to oxidative stress. *British Journal of Nutrition*, 76: 307–317.

Wu, Z., Ma, Y., Zhao, L., Cai, S., Cheng, G. (2018). Acute and subchronic toxicities of ethanol and hot-water extracts from Chinese sumac (*Rhus chinensis*) fruits in rats. *Food and Chemical Toxicology*, 119: 14–23.

Xu, H., Li, X., Adams, H., Kubena, K., Guo, S. (2018). Etiology of metabolic syndrome and dietary intervention. *International Journal of Molecular Sciences*, 20: 1–19.

Zhang, C., Ma, Y., Zhao, Y., Hong, Y., Cai, S., Pang, M. (2017). Phenolic composition, antioxidant and pancreatic lipase inhibitory activities of Chinese sumac (*Rhus chinensis*) fruits. *International Journal of Food Science and Technology*, 53: 1045–1053.

Zothantluanga, J.H., Paul, A., Sailo, N., Zonunmawii, Lalthanzara, H., Chetia, D. (2022). Dietary polyphenols in parasitic diseases and neglected tropical diseases. In: Mithun, R. (Ed.), *Dietary Polyphenols in Human Diseases*. CRC Press, pp. 195–206.

16 Synergistic Effects of Herbal Medicines, Medicinal Fruits, and Berries

Sarfaraz Ali Ghumro, Felix Wambua Muema, and Phurpa Wangchuk

16.1 INTRODUCTION

The combined use of herbal medicines, medicinal fruits, and berries has been a fundamental aspect of traditional medical systems, including Traditional Chinese Medicine, Ayurveda, Unani medicine, and various Indigenous healing practices, where complex botanical mixtures are used to boost therapeutic effectiveness (Husen 2021; Chen et al. 2025). In these systems, fruits and berries are not just nutritional elements but are often seen as medicinal agents that support herbs by providing bioactive compounds with antioxidant, anti-inflammatory, and other pharmacological effects (Ferlemi and Lamari 2016). They can also influence the bioavailability, toxicity, and safety of herbal formulations, for example, through known bioenhancer fruits such as *Emblica officinalis* (amla) and other fruit components that modulate the absorption and toxicity of co-administered herbs (Goyal and Chauhan 2024). These complex plant-based mixtures exemplify a holistic strategy in which herbs, fruits, and berries are combined to optimise therapeutic responses in traditional medicine (Rizvi et al. 2022). Contemporary integrative health practices have increasingly adopted traditional multicomponent botanical formulations. This change reflects a growing awareness of the limitations of reductionist, single-compound pharmacotherapy in treating complex, multifactorial diseases, as highlighted in pharmacological synergy research (Wagner and Ulrich-Merzenich 2009).

The prevailing pharmaceutical paradigm has historically emphasised single bioactive molecules with defined molecular targets; however, accumulating evidence indicates that this reductionist strategy has limited effectiveness for complex chronic diseases, which arise from perturbations in interconnected signalling networks rather than isolated pathways (Hopkins 2008 and Medina-Franco et al. 2013). Single-compound approaches often fail to account for compensatory mechanisms, pathway redundancy, and adaptive resistance, which can undermine target-specific interventions, particularly in cancer and chronic inflammatory conditions (Jia et al. 2009). Consequently, therapeutic efficacy may decline. At the same time, adverse effects may increase when purified single bioactive compounds are administered in isolation at higher doses, a limitation that contrasts with the more buffered pharmacological profile of complex phytochemical mixtures (Williamson 2001). Synergistic phytochemical interactions within complex botanical matrices therefore offer an alternative to single-compound approaches, as multiple constituents can modulate diverse biological targets and produce emergent effects that differ from those of isolated compounds (Caesar and Cech 2019).

Medicinal fruits and berries are rich sources of polyphenols, anthocyanins, flavonoids, and organic acids, and these bioactive compounds are associated with reinforced antioxidant defences, modulation of inflammatory mediators, and potential influences on cellular signalling pathways

DOI: 10.1201/9781003688259-16

relevant to chronic disease processes, including carcinogenesis (Del Rio et al. 2013 and Skrovankova et al. 2015). These compounds provide diverse phytochemicals that can act in concert with other botanical constituents to enhance overall therapeutic effects, particularly by strengthening antioxidant capacity and modulating inflammatory responses in a complementary manner (Latif and Nawaz 2025). When combined within complex formulations, their polyphenols, vitamins, and organic acids may support synergistic outcomes by targeting multiple pathways simultaneously, potentially improving efficacy and tolerability compared with single-compound interventions (Caesar and Cech 2019). This chapter highlights the growing scientific rationale for integrated herbal–fruit–berry strategies, positioning synergy as a central concept in the development of next-generation evidence-based therapeutics for complex diseases.

16.2 SYNERGY IN NATURAL PRODUCTS: CONCEPT AND PHYTOCHEMICAL BASIS

Synergy in natural products refers to the phenomenon whereby the combined biological effect of multiple constituents exceeds the sum of their individual effects, resulting in enhanced efficacy, improved safety, or both (Wagner and Ulrich-Merzenich 2009). Traditional medical systems, including Indian Ayurvedic medicine and Bhutanese traditional medicines, have long relied on polyherbal formulations, with the preconceived idea and concept of mitigating toxicity while enhancing potency, broadening therapeutic coverage, and addressing complex symptom patterns through the complementary synergistic actions of diverse constituents mixed in the products (Vaou et al. 2022, Wagner and Ulrich-Merzenich 2009, and Yeshi and Wangchuk 2018). In multicomponent botanical systems, synergy may arise through pharmacodynamic interactions, such as simultaneous modulation of multiple molecular targets, or pharmacokinetic interactions, including improved solubility, absorption, metabolism, and tissue distribution of bioactive compounds (Caesar and Cech 2019).

Herbal medicines, medicinal fruits, and berries contain diverse classes of bioactive constituents that contribute to synergistic effects. Polyphenols, including phenolic acids, flavonoids, and tannins, are among the most extensively studied compounds due to their antioxidant, anti-inflammatory, and anticancer properties (Del Rio et al. 2013). Flavonoids such as quercetin, catechins, and anthocyanins are particularly abundant in berries and medicinal fruits and are known to interact synergistically with terpenoids and alkaloids present in herbal extracts (Caesar and Cech 2019 and Panche et al. 2016). Terpenoids, including monoterpenes and triterpenes, contribute to anti-inflammatory and anticancer activity by modulating signalling pathways and interacting with biological membranes in ways that can influence permeability and cellular uptake of co-existing phytochemicals (Andre et al. 2016 and Tsuchiya 2015). Polysaccharides, especially those derived from medicinal mushrooms, herbs, and certain fruits, represent another important class of synergistic agents. These macromolecules exert immunomodulatory effects and can influence biological activity through gut-associated mechanisms, including interactions with the gut microbiota and immune signalling (Zhao et al. 2023). The concept of matrix effects further emphasises that minor constituents, often present at low concentrations, can significantly influence the stability, solubility, and biological activity of major compounds within complex mixtures (Caesar and Cech 2019 and Williamson 2001). Such interactions challenge the traditional notion of "active principles" and support the holistic evaluation of botanical combinations rather than isolated constituents. Thus, synergy in herbal medicines, fruits, and berries is rooted in chemical diversity, structural complementarity, and dynamic interactions within complex phytochemical matrices, providing a mechanistic foundation for their enhanced therapeutic potential. Recent research suggests that dietary phytochemicals can modulate key regulatory networks involved in inflammation and cellular stress responses, including crosstalk between NF-κB and Nrf2 signalling, consistent with a multitarget rationale for complex plant-based matrices (Gado et al. 2023 and Krajka-Kuźniak and Baer-Dubowska 2021).

16.3 EVIDENCE OF SYNERGISTIC EFFECTS IN HERBAL MEDICINES, FRUITS, AND BERRIES

Traditional medical systems have long relied on polyherbal formulations, in which multiple botanicals are combined to broaden therapeutic coverage, mitigate adverse effects, and address complex symptom patterns through the complementary actions of diverse constituents (Vaou et al. 2022 and Wagner and Ulrich-Merzenich 2009). Table 16.1 presents a comprehensive overview of traditional herb, fruit, and berry formulations that have been used across various medicinal systems for centuries. This table demonstrates that the concept of synergistic combinations is not a modern invention but rather a principle deeply embedded in ancient healing practices. The most prominent example is Triphala, a classical Ayurvedic formulation composed of three medicinal fruits (*Emblica officinalis*, *Terminalia chebula*, and *Terminalia bellirica*) (Figure 16.1 A–C), which is consistently described as a multicomponent preparation with antioxidant and anti-inflammatory relevance in contemporary biomedical analyses of traditional use (Bairwa et al. 2025). These fruits are also extensively used in Bhutanese traditional medicine (locally known as Sowa Rigpa) for the manufacture of Triphala and other medicinal products (Yeshi et al. 2021). A combination of pomegranate (*Punica granatum*) peel (Figure 16.1D) and hawthorn (*Crataegus* spp.) has demonstrated synergistic anti-inflammatory activity, with experimental evidence implicating cytokine regulation and NF-κB/MAPK-associated signalling, illustrating how fruit-derived polyphenol matrices can yield measurable synergy when combined rationally (Table 16.1) (Zhang et al. 2025).

Traditional Chinese medicine similarly incorporates medicinal fruits as harmonising and supportive agents within formulas; for instance, jujube (*Ziziphus jujuba*) fruit is widely included in herbal mixtures to enhance decoction balance and mitigate toxicity, reflecting a traditional synergy-oriented rationale that extends beyond single-ingredient pharmacology (Table 16.1) (Lam et al. 2016 and Zhu et al. 2024). In this traditional context, synergy is treated as an emergent property of the whole prescription built from major actives, supportive constituents, and the broader phytochemical matrix rather than a single isolated compound effect (Wagner and Ulrich-Merzenich 2009).

Modern studies increasingly provide direct experimental evidence that medicinal fruits and berries can strengthen combination-based outcomes in antioxidant, anti-inflammatory, and anticancer domains (Pyo et al. 2024 and Vahapoglu et al. 2021). In anticancer research, Aronia (*Aronia melanocarpa*) berry extract combined with oligomeric proanthocyanidins has been reported to exhibit synergistic anticancer efficacy in colorectal cancer cells, linked in part to apoptosis regulation and LMNB1/AKT-associated signalling, supporting the premise that berry matrices can amplify multi-target anticancer effects in combination designs (Table 16.1) (Li et al. 2024). A well-documented fruit–herb example is *Vaccinium macrocarpon* (cranberry), where A-type proanthocyanidins exhibit

FIGURE 16.1 Photos of the selected fruits that are used in herbal medicines with substantiated synergistic effects. (A) *Terminalia chebula*. (B) *Terminalia bellirica*. (C) *Emblica officinalis*, and (D) *Punica granatum*.

Original photos provided by P. Wangchuk.

TABLE 16.1
Synergistic Effects of Selective Medicinal Herbs, Fruits, and Berries on Inflammatory, Oxidative, Immune, and Cancer-Related Pathways

Medicinal Herb/Fruit/Berry	Combination/Formulation Context	Reported Synergistic Effects	Key Mechanisms/Signalling Pathways	Evidence Type	Key References
Emblica officinalis, Terminalia chebula, Terminalia bellirica (Triphala)	Classical Ayurvedic polyherbal fruit formulation	Broad antioxidant, digestive, detoxifying, and anti-inflammatory synergy; improved therapeutic balance	Complementary phytochemical interactions within whole-formula matrix	Traditional use supported by modern biomedical analyses	Bairwa et al. (2025)
Ziziphus jujuba (Jujube fruit)	Included in Traditional Chinese Medicine prescriptions	Harmonisation of formulations; mitigation of toxicity	Supportive modulation within multiherb systems rather than single-target effects	Traditional medical rationale with contemporary evaluation	Lam et al. (2016) and Zhu et al. (2024)
Indigo naturalis–derived formulation (*Isatis indigotica, Baphicacanthus cusia, Polygonum tinctorium*; indirubin and isatin)	Multiherb Chinese medicinal preparation	Strong anti-inflammatory and immunoregulatory effects in colitis	Inhibition of NF-κB and MAPK; reduced MPO activity; enhanced Foxp3; reduced $CD4^+$ T-cell infiltration	In vivo (DSS-induced UC mouse model)	Gao et al. (2018)
Blueberries, cranberries, blackcurrants	Polyphenol-rich berry mixtures	Enhanced suppression of inflammatory mediators	Inhibition of IκB kinase; reduced NF-κB nuclear translocation; downregulation of COX-2, TNF-α, IL-6, iNOS, IL-1β	Experimental and mechanistic studies	Kannan et al. (2025)
Ampelopsis grossedentata	Polyphenol-rich herbal extract	Synergistic anti-inflammatory activity	Inhibition of IRAK1/TRAF6/NF-κB signalling; suppression of IL-6, IL-8, IL-1β, TNF-α	In vitro inflammatory models	Chen et al. (2018)
Punica granatum (Pomegranate peel) + *Crataegus* spp. (Hawthorn)	Fruit–fruit combination	Synergistic anti-inflammatory effects	Cytokine regulation; NF-κB and MAPK pathway modulation	Experimental inflammatory disease models	Zhang et al. (2025)
Vaccinium macrocarpon (Cranberry A-type proanthocyanidins) + *Camellia sinensis* (Green tea EGCG)	Fruit–herb polyphenol combination	Enhanced reduction of inflammatory mediator secretion	Attenuation of LPS-induced inflammatory signalling	3D oral mucosal co-culture model	Bedran et al. (2015)
Herbal Flavonoid/diterpene/triterpenoid extracts + berry anthocyanins	Herbal–fruit/berry combinations	Sustained and broader anti-inflammatory responses	Concurrent suppression of NF-κB and MAPK (p38, JNK, ERK); redox-mediated kinase regulation	Mechanistic synergy models	Merecz-Sadowska et al. (2020)

(*Continued*)

TABLE 16.1 (Continued)

Medicinal Herb/Fruit/ Berry	Combination/ Formulation Context	Reported Synergistic Effects	Key Mechanisms/ Signalling Pathways	Evidence Type	Key References
Morinda citrifolia (Noni fruit juice)	Fruit-derived antioxidant system	Enhanced antioxidant defence and cytoprotection	Activation of Nrf2; induction of HO-1, GST, SOD, NQO1; Nrf2 nuclear accumulation	Mechanistic and in vitro studies	Manavalan et al. (2021)
Aronia melanocarpa (Aronia berry) + oligomeric proanthocyanidins	Berry–polyphenol combination	Synergistic anticancer activity in colorectal cancer	Apoptosis regulation; LMNB1/AKT-associated signalling	In vitro cancer studies	Li et al. (2024)
Zhuyu Pill (ZYP) – berberine, palmatine, evodiamine, rutaecarpine	Multicomponent Chinese herbal medicine	Immunomodulatory synergy; attenuation of atherosclerosis	Suppression of TLR4/ MyD88/TRAF6–NF-κB; activation of STAT6 and PPAR-γ; reduced M1 and enhanced M2 macrophage markers	In vivo atherosclerosis models	Pan et al. (2023) and Zhao et al. (2025)
Citrus aurantium (Bitter orange fruit) + *Tasmannia lanceolata* (Mountain pepper berry) + herbs	Native Australian botanical blends	Synergistic anti-inflammatory and antioxidant effects	Multitarget redox and inflammatory pathway modulation	Cell-based formulation studies	Kapini et al. (2025)
Citrus reticulata peel + black tea (CRPBT)	Herb–fruit formulation	Synergistic anticancer and anti-metastatic activity	Inhibition of PI3K/ AKT; increased Bax/ Bcl-2 ratio; suppression of MMP-2/3, N-cadherin, vimentin	In vitro and in vivo cancer studies	Wang et al. (2023) and Wen et al. (2020)
Curcuma zedoaria + kelp	Herbal–marine combination	Enhanced inhibition of liver cancer proliferation and metastasis	Reduced endogenous H_2S; downregulation of STAT3/BCL-2 and VEGF signalling	In vitro and in vivo cancer models	Han et al. (2019)
Rubus spp. (Blackberry) + *Anethum graveolens* (Dill)	Fruit–herb extract combination	Improved skin elasticity beyond individual components	Complementary bioactivities affecting skin structure and function	Controlled clinical study	Meza et al. (2020)
Apple products (juice and smoothies; high procyanidins)	Fruit-derived phytochemical mixtures	Amplified antioxidant and chemopreventive responses in colon	Product-dependent activation of Nrf2 signalling	In vivo (Sprague–Dawley rat studies)	Gado et al. (2023)

cooperative anti-inflammatory activity with epigallocatechin gallate (EGCG) from *Camellia sinensis* (green tea) (Table 16.1). This synergy more effectively reduced lipopolysaccharide (LPS)-induced inflammatory mediator secretion than either component alone in a three-dimensional oral mucosal co-culture model (Bedran et al. 2015).

Additional formulation studies further reinforce that fruit and berry components can function as synergy-driving ingredients in mixtures, as demonstrated by investigations of native Australian plant combinations. These blends containing bitter orange fruit (*Citrus aurantium*), mountain pepper berry (*Tasmannia lanceolata*), and complementary herbs showed synergistic suppression of inflammatory markers and enhanced antioxidant responses in cell-based models (Table 16.1) (Kapini et al. 2025). Beyond inflammation and cancer, synergy has also been reported in other bioactivity contexts relevant to botanical product design; for example, a blackberry (*Rubus* spp.)–dill (*Anethum graveolens*) extract combination was shown to synergistically improve skin elasticity relative to each ingredient alone in a controlled clinical study, underscoring that fruit-based extracts can contribute functionally distinct actions when paired with complementary botanicals (Table 16.1) (Meza et al. 2020).

Collectively, these findings support the conclusion that medicinal fruits and berries are not merely supportive nutritional components, but active synergistic partners that expand target coverage, strengthen antioxidant and anti-inflammatory capacity, and contribute mechanistic depth to combination formulations when appropriately selected and validated (Caesar et al. 2019 and Vaou et al. 2022).

16.4 MECHANISTIC INSIGHTS INTO ANTI-INFLAMMATORY AND ANTICANCER SYNERGY

Pharmacodynamic interactions arise when two or more agents, including drugs or bioactive natural compounds, influence the same biological targets or interconnected physiological processes, thereby modifying the overall therapeutic response (Briguglio et al. 2018). These interactions may be synergistic, enhancing therapeutic efficacy and reducing toxicity, or antagonistic, diminishing drug effectiveness or increasing the likelihood of adverse effects. The combined use of plant-derived biomolecules with conventional drugs offers a therapeutic approach that can be highly effective, as the complementary and cooperative actions of both agents enhance overall treatment outcomes (Chaachouay et al. 2022). Figure 16.2 schematically summarises representative anti-inflammatory and anticancer mechanisms discussed in this chapter.

16.4.1 Modulation of Inflammatory Signalling Pathways

Synergistic mixtures of herbal medicines, medicinal fruits, and berries produce strong anti-inflammatory effects by jointly regulating multiple signalling pathways, rather than acting through a single molecular target (Chen et al. 2022). Chronic inflammation is governed by intricate signalling networks that include pattern recognition receptors, intracellular kinases, transcription factors, and downstream effector enzymes. Multicomponent botanical formulations are particularly well-suited to target multiple regulatory points within these pathways at the same time (Yuan et al. 2022). A frequently reported mechanism underlying synergistic anti-inflammatory activity is suppression of the nuclear factor kappa B (NF-κB) signalling axis, which governs the transcription of numerous pro-inflammatory genes (Kannan et al. 2025). Indirubin and isatin, constituents of a Chinese herbal medicine comprising *Isatis indigotica* Fort., *Baphicacanthus cusia* (Nees) Bremek., and *Polygonum tinctorium* Ait, were tested in dextran sulphate sodium (DSS)–induced ulcerative colitis (UC) mice (Table 16.1). The study demonstrated that the herbal medicine suppressed pro-inflammatory mediators and myeloperoxidase (MPO) activity. Furthermore, the drug enhanced anti-inflammatory cytokines and Foxp3 expression, reduced $CD4^+$ T-cell infiltration, attenuated oxidative stress and epithelial cell apoptosis, and inhibited the NF-κB and mitogen-activated protein kinase (MAPK) signalling pathways (Gao et al. 2018).

Berries rich in polyphenols, such as blueberries, cranberries, and blackcurrants, further contribute to anti-inflammatory synergy by inhibiting IκB kinase activity and restricting nuclear translocation

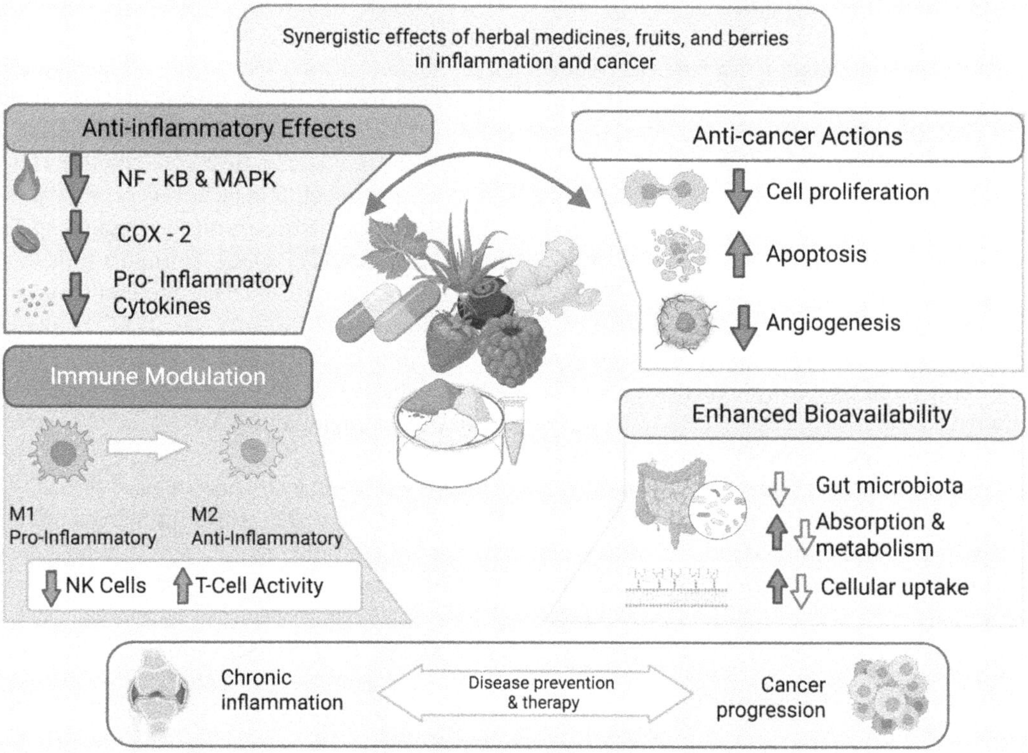

FIGURE 16.2 The figure summarises the multitarget mechanisms through which integrated herbal–fruit–berry formulations modulate chronic inflammation and cancer-related processes. Anti-inflammatory effects involve coordinated inhibition of NF-κB/MAPK signalling, reduced COX-2 activity, and suppression of pro-inflammatory cytokines, together with immune modulation via macrophage polarisation towards an anti-inflammatory (M2) phenotype and regulation of NK and T-cell responses. Anticancer actions include inhibition of tumour cell proliferation, induction of apoptosis, and suppression of angiogenesis. Enhanced bioavailability, mediated by gut microbiota interactions and improved absorption and cellular uptake, further amplifies these effects, highlighting the role of synergistic botanical combinations in disrupting the inflammation–cancer continuum.

of NF-κB (Table 16.1). This leads to reduced expression of inflammatory mediators, including COX-2, TNF-α, IL-6, iNOS, and IL-1β (Kannan et al. 2025). Ampelopsis grossedentata, a Chinese herbal medicine rich in polyphenols, inhibited IRAK1/TRAF6/NF-κB-mediated inflammatory signalling pathway and suppressed the increased expression of IL-6, IL-8, IL-1β, and TNF-α (Table 16.1) (Chen et al. 2018). When combined with herbal extracts containing high levels of flavonoids, diterpenes, or triterpenoids, these actions are strengthened through simultaneous suppression of upstream kinases and key inflammatory enzymes (Table 16.1). MAPK cascades such as p38, JNK, and ERK serve as further nodes of synergistic interaction. Compounds from herbal sources can reduce MAPK phosphorylation, while anthocyanins and phenolic acids from berries help maintain cellular redox balance, thereby indirectly limiting kinase activation (Merecz-Sadowska et al. 2020). Together, the simultaneous inhibition of NF-κB and MAPK pathways produces anti-inflammatory effects that are more extensive and longer lasting than those achieved by single isolated agents.

16.4.2 Regulation of Oxidative Stress and Nrf2 Signalling

Oxidative stress is a key driver in the onset and progression of chronic inflammatory diseases and cancer (Neganova et al. 2021). Overproduction of reactive oxygen species (ROS) damages cellular macromolecules and amplifies both inflammatory signalling and genomic instability (Juan et al. 2021). Botanical combinations frequently display enhanced antioxidant capacity due to complementary redox-modulating mechanisms (Allegra 2021). Medicinal fruits and berries are particularly rich in phenolic acids, flavonoids, anthocyanins, and vitamin-associated antioxidants, which not only neutralise ROS directly but also regulate redox-sensitive signalling pathways involved in inflammation and cellular stress responses (Åhlberg 2021 and Rafaqat et al. 2025). Nuclear factor erythroid 2-related factor 2 (Nrf2) is a master cellular defence transcription factor. Its primary function is to protect cells against oxidative stress, inflammation, and toxic insults by regulating the expression of cytoprotective genes (He et al. 2020). Ripe *Morinda citrifolia* L. (noni) fruit juice (Table 16.1) strongly activates the Nrf2 pathway by enhancing Nrf2 nuclear accumulation and inducing downstream antioxidant genes such as HO-1, glutathione S-transferases (GST), superoxide dismutase (SOD), and reduced nicotinamide adenine dinucleotide phosphate (NAD(P)H) quinone oxidoreductase 1 without cytotoxic effects (Manavalan et al. 2021). This Nrf2 regulatory activity is attributed, at least in part, to bioactive fruit-derived compounds such as scopoletin, which was identified as a potent Nrf2 activator (Qader et al. 2020).

The dual strategy of direct ROS neutralisation coupled with sustained enhancement of endogenous antioxidant defences underlies the superior efficacy of combined herbal–fruit–berry systems. Crucially, restoration of redox homeostasis indirectly suppresses inflammation and carcinogenesis by limiting oxidative activation of NF-κB and other stress-responsive signalling pathways (Manavalan et al. 2021, Singh et al. 2021, and Singh et al. 2021).

16.4.3 Immunomodulation and Macrophage Polarisation

Macrophages are central effectors of innate immunity and represent a key mechanistic link between chronic inflammation and cancer development (Christofides et al. 2022). In response to microenvironmental cues, macrophages polarise towards a classically activated M1 phenotype, characterised by pro-inflammatory and cytotoxic functions, or an alternatively activated M2 phenotype, which is associated with immune regulation, tissue repair, and tumour progression (Cerdeira and Brigagão 2024). Synergistic combinations of herbal medicines, fruits, and berries have been shown to modulate macrophage polarisation, promoting a shift away from excessive M1 activation towards a balanced or M2-skewed phenotype. For example, Zhuyu Pill (ZYP) is a Chinese herbal medicine whose main bioactive metabolites are berberine, evodiamine, and rutaecarpine (Pan et al. 2023). The formulation promotes macrophage polarisation towards an anti-inflammatory M2 phenotype, thereby attenuating atherosclerosis progression (Table 16.1). Mechanistically, ZYP suppresses the pro-inflammatory TLR4/MyD88/TRAF6–NF–κB pathway, leading to reduced expression of M1-associated mediators, including TNF-α, IL-6, iNOS, and NF-κB, and diminished foam cell accumulation (Zhao et al. 2025). Concurrently, ZYP activates STAT6 phosphorylation and the PPAR-γ signalling axis, thereby enhancing the expression of M2 markers such as Arg-1 and increasing levels of the M2-associated cytokines IL-4 and IL-13. Through the coordinated suppression of M1 polarisation and promotion of M2 differentiation, ZYP attenuates inflammatory cell infiltration, improves lipid metabolism, and limits atherosclerotic plaque formation. (Liu et al. 2025 and Pan et al. 2023).

This immunomodulatory synergy is particularly relevant in chronic inflammatory disorders and within the tumour microenvironment, where persistent M1-driven inflammation contributes to tumour initiation, progression, and therapy resistance (Guo and Qi 2025). By reprogramming macrophage responses, combined botanical systems may attenuate pathological inflammation while preserving essential immune surveillance.

16.4.4 Synergistic Actions on Cancer-Related Pathways

Cancer is a multifactorial disease driven by dysregulated cell proliferation, resistance to apoptosis, chronic inflammation, oxidative stress, and immune evasion (Figure 16.2) (Black 2024 and Jomova et al. 2025 and Kapoor et al. 2025). Synergistic botanical combinations exert multitargeted anticancer effects by concurrently modulating these interconnected processes (Jha et al. 2025). Although many medicinal plants have been traditionally used as standalone cancer therapies and offer notable advantages, their clinical effectiveness remains limited. Enhanced anticancer outcomes are frequently achieved when these plants are combined with other natural products (Reyaz et al. 2025). Plant-derived secondary metabolites, including flavonoids, alkaloids, terpenoids, saponins, and related compounds, have been widely recognised as rich sources of bioactive anticancer agents (Muema et al. 2025). Combination strategies improve therapeutic efficacy by lowering the effective dose required, achieved through complementary or synergistic actions on shared molecular targets or signalling pathways (Wang et al. 2023).

Citrus reticulata peel black tea (CRPBT), formulated from *Citrus reticulata* peel and black tea, markedly suppresses PI3K/AKT signalling in liver cancer cells while simultaneously promoting apoptosis through an increased Bax/Bcl-2 ratio (Table 16.1). In addition, CRPBT inhibits tumour invasion and metastasis by downregulating MMP-2/3, N-cadherin, and vimentin, demonstrating how herb–fruit combinations exert complementary anti-proliferative, pro-apoptotic, and anti-metastatic effects in cancer models (Wang et al. 2023 and Wen et al. 2020). Combined use of *Curcuma zedoaria* and kelp has been shown to inhibit liver cancer cell proliferation and metastasis both in vitro and in vivo by reducing endogenous H_2S production and downregulating the STAT3/BCL-2 and VEGF signalling pathways (Table 16.1) (Han et al. 2019). This multitarget interference underscores how complementary marine–herbal formulations can enhance anticancer efficacy by simultaneously impairing survival, angiogenesis, and metastatic signalling networks.

Fruits rich in diverse polyphenols can exert synergistic regulatory effects on cancer-related signalling pathways by collectively enhancing cytoprotective and redox-sensitive mechanisms. In Sprague–Dawley rat studies, consumption of apple juice and smoothies produced a product-dependent activation of colonic Nrf2, with the strongest induction observed for apple products containing the highest levels of procyanidins (Gado et al. 2023). This finding suggests that the combined action of fruit-derived phytochemicals amplifies Nrf2-mediated antioxidant and detoxification signalling, which is closely linked to suppression of oxidative stress-driven carcinogenic processes in the colon. Collectively, these findings highlight that synergistic combinations of medicinal herbs, marine resources, and fruits exert enhanced anticancer effects by convergently modulating multiple cancer-related signalling pathways, reinforcing the therapeutic potential of multicomponent botanical strategies in cancer prevention and treatment.

16.5 THERAPEUTIC IMPLICATIONS AND TRANSLATIONAL POTENTIAL

Integrated modulation of pro-inflammatory and oxidative stress–responsive signalling networks provides a mechanistic basis for therapeutic intervention in chronic inflammatory pathologies, including arthritis, cardiovascular disease, neuroinflammatory disorders, and metabolic syndrome, by attenuating sustained immune activation and redox imbalance that drive disease progression. By targeting multiple inflammatory mediators simultaneously, combined botanical systems may achieve therapeutic efficacy at lower doses compared with single-compound interventions. Such dose-sparing effects are particularly relevant for long-term management of chronic diseases, where safety and tolerability are critical. Moreover, the immunomodulatory properties of herbal–fruit–berry combinations support restoration of immune homeostasis rather than indiscriminate immunosuppression.

Synergistic botanical combinations are increasingly recognised for their role in cancer chemoprevention and as adjuncts to conventional therapies (Desai et al. 2025). Their capacity to suppress inflammation, reduce oxidative DNA damage, modulate immune responses, and sensitise cancer cells to apoptosis supports a holistic and integrative approach to cancer management. Importantly, several studies suggest that such combinations may mitigate chemotherapy or radiotherapy-induced toxicity, thereby improving patient quality of life and treatment adherence (Puranik and Song 2025).

While not intended to replace standard therapies, synergistic botanical systems may enhance therapeutic outcomes when used responsibly.

A major translational advantage of combining herbs with fruits and berries lies in enhanced bioavailability. Organic acids, flavonoids, sugars, and dietary fibres present in fruits and berries improve the solubility, stability, and intestinal absorption of herbal phytochemicals (Davodabadi et al. 2025, Palai et al. 2025, and Rudrapal et al. 2024). Gut microbiota-mediated metabolism of complex polyphenols generates bioactive metabolites that modulate systemic immune responses, suppress chronic inflammation, and interfere with tumorigenic signalling pathways (Cano et al. 2024). By influencing redox balance, cytokine production, and apoptotic regulators, these microbiota-derived compounds extend the protective effects of dietary phytochemicals beyond the gut. This interplay between the food matrix, microbial metabolism, and host signalling highlights the superior efficacy of whole-food or whole-extract interventions over isolated compounds in anticancer and anti-inflammatory strategies.

The demonstrated synergy between herbal medicines and edible fruits and berries provides a strong scientific basis for their incorporation into functional foods and nutraceutical formulations (Mudondo et al. 2025). Such products offer culturally acceptable and accessible strategies for disease prevention and health promotion, aligning traditional knowledge with modern nutritional science.

16.6 CHALLENGES, SAFETY CONSIDERATIONS, AND FUTURE PERSPECTIVES

One of the major challenges in synergy research is the inherent variability in phytochemical composition, which can arise from differences in plant species, cultivar, geographic origin, soil composition, climate, cultivation practices, harvest timing, and post-harvest processing methods. Adulteration with similar, closely related species or look-alike plants has also been linked to quality, safety, and efficacy (including synergistic effects) of polyingredient formulations (Wangchuk and Jamtsho 2023). Such variability can substantially affect the concentration and bioavailability of bioactive compounds, including polyphenols, flavonoids, and other secondary metabolites, and consequently alter their biological activity. Without rigorous standardisation, experimental results are difficult to reproduce, and the translation of preclinical findings to clinical applications remains limited (Paul and Kumar 2023). To address these challenges, comprehensive phytochemical profiling using advanced analytical techniques such as high-performance liquid chromatography (HPLC), liquid chromatography–mass spectrometry (LC–MS), and nuclear magnetic resonance (NMR) is essential to accurately characterise the active constituents (Paul and Kumar 2023). In addition, validated extraction protocols, standardised dosage formulations, and stringent quality control measures are critical for ensuring consistency, enabling meaningful comparisons across studies, and ultimately facilitating the development of safe and effective herbal–fruit–berry combinations for therapeutic applications.

Although synergistic interactions can enhance therapeutic efficacy, they may also increase the risk of adverse effects or herb–drug interactions. This is particularly relevant for patients receiving long-term pharmacotherapy or cancer treatments. Rigorous toxicological evaluation, pharmacokinetic studies, and post-marketing surveillance are required to ensure safety and support evidence-based use of complex botanical combinations (Banerjee et al. 2025). Most synergy studies remain confined to in vitro and preclinical models. Well-designed clinical trials that account for the complexity of multicomponent systems are urgently needed. Emerging approaches such as systems biology, network pharmacology, and multiomics technologies offer powerful tools to elucidate mechanisms of action and guide rational formulation design. Future research should integrate ethnopharmacological knowledge with modern analytical, computational, and clinical methodologies. Precision phytotherapy, personalised nutrition, and microbiome-informed approaches may further enhance the therapeutic potential of synergistic herbal–fruit–berry systems.

16.7 CONCLUSION

The growing body of evidence reviewed in this chapter underscores that the therapeutic efficacy of herbal medicines, particularly medicinal fruits and berries, is largely driven by synergistic interactions among their diverse phytochemical constituents rather than by single isolated compounds. The intrinsic chemical complexity of these plant matrices enables coordinated modulation of multiple biological targets, offering mechanistic explanations for the broad-spectrum and often superior bioactivities observed with whole-plant or whole-fruit preparations. Such synergy manifests through complementary antioxidant and anti-inflammatory actions, enhanced bioavailability, metabolic modulation, and reduced toxicity, collectively contributing to improved therapeutic outcomes.

Medicinal fruits and berries occupy a unique position at the interface of nutrition and medicine, providing a compelling model for studying phytochemical synergy in both preventive and therapeutic contexts. Their demonstrated effects against chronic inflammation, oxidative stress, metabolic disorders, neurodegeneration, and cancer highlight the relevance of multitarget, systems-level interventions in addressing complex diseases. In particular, synergistic modulation of interconnected signalling pathways involved in redox homeostasis, immune regulation, and cell survival reinforces the limitations of reductionist drug discovery approaches when applied to multifactorial pathologies.

Advances in analytical techniques, omics technologies, and computational tools such as network pharmacology have significantly enhanced the capacity to characterise and predict synergistic interactions within complex herbal systems. These methodologies provide a critical bridge between traditional empirical knowledge and modern biomedical science, enabling more rational design, standardisation, and validation of synergistic herbal formulations. Nevertheless, challenges remain, including variability in phytochemical composition, lack of standardised extraction and dosing protocols, and limited clinical validation.

Future research should prioritise integrative experimental designs that combine phytochemical profiling, mechanistic studies, and well-controlled clinical investigations to fully harness the therapeutic potential of herbal synergy. Embracing whole-extract and whole-food paradigms, particularly those based on medicinal fruits and berries, may facilitate the development of safer, more effective functional foods, nutraceuticals, and complementary therapeutic strategies. Overall, recognising and scientifically validating synergistic interactions represents a critical step towards advancing evidence-based herbal medicine within modern healthcare systems.

REFERENCES

Åhlberg, M. K. (2021). A profound explanation of why eating green (wild) edible plants promote health and longevity. *Food Frontiers*, 2, 240–267.

Allegra, M. (2021). Redox systems, oxidative stress, and antioxidant defences in health and disease. In *Antioxidants*, 10, 1955. https://doi.org/10.3390/antiox10121955.

Andre, C. M., Hausman, J.-F., & Guerriero, G. (2016). Cannabis sativa: The plant of the thousand and one molecules. *Frontiers in Plant Science*, 7, 19.

Bairwa, V. K., Kashyap, A. K., Meena, P., & Jain, B. P. (2025). Triphala's characteristics and potential therapeutic uses in modern health. *International Journal of Physiology, Pathophysiology and Pharmacology*, 17, 19.

Banerjee, D., Banerjee, S., Shah, P., Singh, A., & Saharan, V. A. (2025). Evidence-based clinical assessment of herbal formulations: Traditional and modern insights. *Formulating Pharma-, Nutra-, and Cosmeceutical Products from Herbal Substances: Dosage Forms and Delivery Systems*, 703–742.

Bedran, T. B. L., Spolidorio, D. P., & Grenier, D. (2015). Green tea polyphenol epigallocatechin-3-gallate and cranberry proanthocyanidins act in synergy with cathelicidin (LL-37) to reduce the LPS-induced inflammatory response in a three-dimensional co-culture model of gingival epithelial cells and fibroblasts. *Archives of Oral Biology*, 60, 845–853.

Black, H. S. (2024). Oxidative stress and ROS link diabetes and cancer. *Journal of Molecular Pathology*, 5, 96–119.

Briguglio, M., Hrelia, S., Malaguti, M., Serpe, L., Canaparo, R., Dell'Osso, B., et al. (2018). Food bioactive compounds and their interference in drug pharmacokinetic/pharmacodynamic profiles. *Pharmaceutics*, 10, 277.

Caesar, L. K., & Cech, N. B. (2019). Synergy and antagonism in natural product extracts: When 1 + 1 does not equal 2. *Natural Product Reports*, 36, 869–888.
Cano, R., Bermúdez, V., Galban, N., Garrido, B., Santeliz, R., Gotera, M. P., et al. (2024). Dietary polyphenols and gut microbiota cross-talk: Molecular and therapeutic perspectives for cardiometabolic disease: A narrative review. *International Journal of Molecular Sciences*, 25(16), 9118.
Cerdeira, C. D., & Brigagão, M. R. (2024). Targeting macrophage polarization in infectious diseases: M1/M2 functional profiles, immune signaling and microbial virulence factors. *Immunological Investigations*, 53, 1030–1091.
Chaachouay, N., Douira, A., & Zidane, L. (2022). Herbal medicine used in the treatment of human diseases in the Rif, Northern Morocco. *Arabian Journal for Science and Engineering*, 47, 131–153.
Chen, H., Liu, Z., Du, Y., Meng, X., Xue, L., Sun, X., et al. (2025). Comparative analysis of Chinese classical prescriptions and global traditional polyherbal formulations: Insights from the database of global polyherbal formulation (GPFD). *Chinese Medicine*, 20, 1–15.
Chen, X., Li, H., Zhang, B., & Deng, Z. (2022). The synergistic and antagonistic antioxidant interactions of dietary phytochemical combinations. *Critical Reviews in Food Science and Nutrition*, 62, 5658–5677. https://doi.org/10.1080/10408398.2021.1888693
Chen, Y.-L., Zhang, Y.-L., Dai, Y.-C., & Tang, Z.-P. (2018). Systems pharmacology approach reveals the anti-inflammatory effects of Ampelopsis grossedentata on dextran sodium sulfate-induced colitis. *World Journal of Gastroenterology*, 24, 1398.
Christofides, A., Strauss, L., Yeo, A., Cao, C., Charest, A., & Boussiotis, V. A. (2022). The complex role of tumor-infiltrating macrophages. *Nature Immunology*, 23, 1148–1156. https://doi.org/10.1038/s41590-022-01267-2
Davodabadi, F., Nasri, N., Valizadeh, N., Haji Ali, B., Ghotekar, S., Sargazi, S., et al. (2025). Nanotechnology-enhanced delivery systems for bioactive citrus compounds: A comprehensive review. *Critical Reviews in Food Science and Nutrition*, 1–45.
Del Rio, D., Rodriguez-Mateos, A., Spencer, J. P., Tognolini, M., Borges, G., & Crozier, A. (2013). Dietary (poly)phenolics in human health: Structures, bioavailability, and evidence of protective effects against chronic diseases. *Antioxidants & Redox Signaling*, 18, 1818–1892.
Desai, U., Poojara, L., Goswami, D., Pandya, M., Modi, S., Shukla, R., et al. (2025). Medicinal plants in combination therapies for cancer. In *Medicinal plants and their bioactives in human diseases* (pp. 259–291). Springer.
Ferlemi, A.-V., & Lamari, F. N. (2016). Berry leaves: An alternative source of bioactive natural products of nutritional and medicinal value. *Antioxidants*, 5, 17.
Gado, F., Ferrario, G., Della Vedova, L., Zoanni, B., Altomare, A., Carini, M., et al. (2023). Targeting Nrf2 and NF-κB signaling pathways in cancer prevention: The role of apple phytochemicals. *Molecules*, 28(3), 1356.
Gao, W., Zhang, L., Wang, X., Yu, L., Wang, C., & Gong, Y. (2018). The combination of indirubin and isatin attenuates dextran sodium sulfate induced ulcerative colitis in mice. *Biochemistry and Cell Biology*, 96, 636–645.
Goyal, M. R., & Chauhan, A. (2024). Holistic approach of nutrients and traditional natural medicines for human health: A review. *Future Integrative Medicine*, 3, 197–208.
Guo, R., & Qi, F. (2025). The multifaceted role of post-translational modifications in macrophage polarization: From mechanisms to therapeutic targets. *Frontiers in Immunology*, 16, 1749857.
Han, H., Wang, L., Liu, Y., Shi, X., Zhang, X., Li, M., et al. (2019). Combination of curcuma zedoary and kelp inhibits growth and metastasis of liver cancer in vivo and in vitro via reducing endogenous H_2S levels. *Food & Function*, 10, 224–234.
He, F., Ru, X., & Wen, T. (2020). NRF2, a transcription factor for stress response and beyond. *International Journal of Molecular Sciences*, 21, 4777.
Hopkins, A. L. (2008). Network pharmacology: The next paradigm in drug discovery. *Nature Chemical Biology*, 4, 682–690.
Husen, A. (2021). *Traditional herbal therapy for the human immune system*. CRC Press. https://doi.org/10.1201/9781003137955
Jha, S. K., Singh, N., Shanker, O. R., Antil, I., Baghel, J. S., Huddar, V., et al. (2025). A review on integrative approaches in oncology: Bridging ayurvedic medicine and modern cancer therapeutics. *Frontiers in Natural Products*, 4, 1635197.
Jia, J., Zhu, F., Ma, X., Cao, Z. W., Li, Y. X., & Chen, Y. Z. (2009). Mechanisms of drug combinations: Interaction and network perspectives. *Nature Reviews Drug Discovery*, 8, 111–128.

Jomova, K., Alomar, S. Y., Valko, R., Fresser, L., Nepovimova, E., Kuca, K., et al. (2025). Interplay of oxidative stress and antioxidant mechanisms in cancer development and progression. *Archives of Toxicology*, 100, 1–47.

Juan, C. A., Pérez de la Lastra, J. M., Plou, F. J., & Pérez-Lebeña, E. (2021). The chemistry of reactive oxygen species (ROS) revisited: Outlining their role in biological macromolecules (DNA, lipids and proteins) and induced pathologies. *International Journal of Molecular Sciences*, 22, 4642.

Kannan, G., Paul, B. M., & Thangaraj, P. (2025). Stimulation, regulation, and inflammaging interventions of natural compounds on nuclear factor kappa B (NF-κB) pathway: A comprehensive review. *Inflammopharmacology*, 33, 145–162.

Kapini, R., Chang, D., Münch, G., Carroll, L., & Zhou, X. (2025). Synergistic combinations of native Australian plants for skin inflammation and wound healing. *Biomedicines*, 13, 1754.

Kapoor, G., Prakash, S., Jaiswal, V., & Singh, A. K. (2025). Chronic inflammation and cancer: Key pathways and targeted therapies. *Cancer Investigation*, 43, 1–23.

Krajka-Kuźniak, V., & Baer-Dubowska, W. (2021). Modulation of Nrf2 and NF-κB signaling pathways by naturally occurring compounds in relation to cancer prevention and therapy: Are combinations better than single compounds? *International Journal of Molecular Sciences*, 22, 8223.

Lam, C. T., Chan, P. H., Lee, P. S., Lau, K. M., Kong, A. Y., Gong, A. G., et al. (2016). Chemical and biological assessment of jujube (Ziziphus jujuba)-containing herbal decoctions: Induction of erythropoietin expression in cultures. *Journal of Chromatography B*, 1026, 254–262.

Latif, R., & Nawaz, T. (2025). Medicinal plants and human health: A comprehensive review of bioactive compounds, therapeutic effects, and applications. *Phytochemistry Reviews*, 1–44.

Li, Y., Xu, C., Weng, W., & Goel, A. (2024). Combined treatment with aronia berry extract and oligomeric proanthocyanidins exhibit a synergistic anticancer efficacy through LMNB1–AKT signaling pathways in colorectal cancer. *Molecular Carcinogenesis*, 63, 2145–2157.

Liu, S., Jing, J., Zhang, Y., Sun, J., & Zhu, C. (2025). Potential of herbal formulas and bioactive metabolites in treating atherosclerosis: Targeted modulation of macrophage polarization. *Frontiers in Pharmacology*, 16, 1631274.

Manavalan, A., Qader, M., Wu, X., & Cao, S. (2021). Nrf2 activation by Morinda citrifolia L. (noni) fruit juices. *World Journal of Traditional Chinese Medicine*, 7, 184–191. https://doi.org/10.4103/wjtcm.wjtcm_24_21

Medina-Franco, J. L., Giulianotti, M. A., Welmaker, G. S., & Houghten, R. A. (2013). Shifting from the single to the multitarget paradigm in drug discovery. *Drug Discovery Today*, 18, 495–501.

Merecz-Sadowska, A., Sitarek, P., Sliwiński, T., & Zajdel, R. (2020). Anti-inflammatory activity of extracts and pure compounds derived from plants via modulation of signaling pathways, especially PI3K/AKT in macrophages. *International Journal of Molecular Sciences*, 21, 9605.

Meza, D., Li, W. H., Seo, I., Parsa, R., Kaur, S., Kizoulis, M., et al. (2020). A blackberry–dill extract combination synergistically increases skin elasticity. *International Journal of Cosmetic Science*, 42, 444–451.

Mudondo, J., Happy, K., Gang, R., Ban, Y., & Kang, Y. (2025). From nature to nutrition: Exploring the synergistic benefits of functional foods and herbal medicines for holistic health. *Applied Biological Chemistry*, 68, 17.

Muema, F. W., Miller, C. M., Hebbard, L., & Wangchuk, P. (2025). Phytophenolic compounds and their anticancer properties. In *Plant secondary metabolites* (pp. 40–55). CRC Press. https://doi.org/10.1201/9781003518358-3

Neganova, M., Liu, J., Aleksandrova, Y., Klochkov, S., & Fan, R. (2021). Therapeutic influence on important targets associated with chronic inflammation and oxidative stress in cancer treatment. *Cancers*, 13, 6062.

Palai, S., Priyadarshini, A., & Sadangi, S. (2025). Challenges in effective delivery and utilization of bioactive compounds. In S. A. Wani & H. R. Naik (Eds.), *Harnessing nanoencapsulation: Valorization of bioactive compounds for health and beyond* (pp. 53–72). Springer Nature Switzerland. https://doi.org/10.1007/978-3-032-03329-1_3

Pan, Y., Feng, X., Song, W., Zhou, X., Zhou, Z., Chen, G., et al. (2023). Effects and potential mechanism of zhuyu pill against atherosclerosis: Network pharmacology and experimental validation. *Drug Design, Development and Therapy*, 17, 597–612.

Panche, A. N., Diwan, A. D., & Chandra, S. R. (2016). Flavonoids: An overview. *Journal of Nutritional Science*, 5, e47.

Paul, T., & Kumar, K. J. (2023). Standardization of herbal medicines for lifestyle diseases. In A. K. Dhara & S. C. Mandal (Eds.), *Role of herbal medicines: Management of lifestyle diseases* (pp. 545–557). Springer Nature Singapore. https://doi.org/10.1007/978-981-99-7703-1_27

Puranik, N., & Song, M. (2025). Phytochemical-based strategies in enhancing chemotherapy and radiotherapy efficacy in cancer treatment. In Nidhi Pura (Ed.), *Nano-formulation of dietary phytochemicals for cancer management* (pp. 177–209). Springer.

Pyo, Y., Kwon, K. H., & Jung, Y. J. (2024). Anticancer potential of flavonoids: Their role in cancer prevention and health benefits. *Foods*, 13, 2253.

Qader, M., Xu, J., Yang, Y., Liu, Y., & Cao, S. (2020). Natural Nrf2 activators from juices, wines, coffee, and cocoa. *Beverages*, 6, 68.

Rafaqat, S., Saqib, H. M. J., Yasir, M. H., Rasheed, M. M., Fatima, M. N., & Malik, F. (2025). A comprehensive review on the sources, stability, and health implications of bioactive compounds present in foods. *Indus Journal of Bioscience Research*, 3, 112–118.

Reyaz, A., Qadri, S. S., Javaid, D., Ganie, S. Y., & Reshi, M. S. (2025). Exploring the role of traditional medicinal plants in cancer therapy: Present efficacy and future directions. *Phytomedicine Plus*, 5, 100845. https://doi.org/10.1016/j.phyplu.2025.100845

Rizvi, S. A., Einstein, G. P., Tulp, O. L., Sainvil, F., & Branly, R. (2022). Introduction to traditional medicine and their role in prevention and treatment of emerging and re-emerging diseases. *Biomolecules*, 12, 1442.

Rudrapal, M., Rakshit, G., Singh, R. P., Garse, S., Khan, J., & Chakraborty, S. (2024). Dietary polyphenols: Review on chemistry/sources, bioavailability/metabolism, antioxidant effects, and their role in disease management. *Antioxidants*, 13, 429.

Singh, S., Nagalakshmi, D., Sharma, K., & Ravichandiran, V. (2021). Natural antioxidants for neuroinflammatory disorders and possible involvement of Nrf2 pathway: A review. *Heliyon*, 7. DOI: 10.1016/j.heliyon.2021.e06216

Skrovankova, S., Sumczynski, D., Mlcek, J., Jurikova, T., & Sochor, J. (2015). Bioactive compounds and antioxidant activity in different types of berries. *International Journal of Molecular Sciences*, 16, 24673–24706.

Tsuchiya, H. (2015). Membrane interactions of phytochemicals as their molecular mechanism applicable to the discovery of drug leads from plants. *Molecules*, 20, 18923–18966.

Vahapoglu, B., Erskine, E., Gultekin Subasi, B., & Capanoglu, E. (2021). Recent studies on berry bioactives and their health-promoting roles. *Molecules*, 27, 108.

Vaou, N., Stavropoulou, E., Voidarou, C., Tsakris, Z., Rozos, G., Tsigalou, C., et al. (2022). Interactions between medical plant-derived bioactive compounds: Focus on antimicrobial combination effects. *Antibiotics*, 11, 1014.

Wagner, H., & Ulrich-Merzenich, G. (2009). Synergy research: Approaching a new generation of phytopharmaceuticals. *Phytomedicine*, 16, 97–110.

Wang, Y., Li, J., & Xia, L. (2023). Plant-derived natural products and combination therapy in liver cancer [Review]. *Frontiers in Oncology*, 13. https://doi.org/10.3389/fonc.2023.1116532

Wangchuk, P., & Jamtsho, T. (2023). Phytopharmaceutical properties and quality assessment of two Himalayan medicinal plants, *Meconopsis horridula* and *Meconopsis simplicifolia*. *Journal of Herbal Medicine*, 38, 100628.

Wen, S., Sun, L., An, R., Zhang, W., Xiang, L., Li, Q., et al. (2020). A combination of *Citrus reticulata* peel and black tea inhibits migration and invasion of liver cancer via PI3K/AKT and MMPs signaling pathway. *Molecular Biology Reports*, 47, 507–519.

Williamson, E. M. (2001). Synergy and other interactions in phytomedicines. *Phytomedicine*, 8, 401–409.

Yeshi, K., Aagaard-Hansen, J., & Wangchuk, P. (2021). Medicinal, nutritional, and spiritual significance of plants in Bhutan: Their biodiscovery potential and conservation status. In Arshad Mehmood Abbasi, & Rainer W. Bussmann (Eds.), *Ethnobiology of mountain communities in Asia* (pp. 1–25). Springer.

Yeshi, K., & Wangchuk, P. (2018). Geopharmaceuticals of Himalayan Sowa Rigpa medicine: Pharmacopoeial description, geochemical identification, their diversity and current utilization in Bhutan. *Journal of Ethnopharmacology*, 223, 99–112.

Yuan, S., Wang, Q., Li, J., Xue, J.-C., Li, Y., Meng, H., et al. (2022). Inflammatory bowel disease: An overview of Chinese herbal medicine formula-based treatment. *Chinese Medicine*, 17, 74.

Zhang, S., Qiu, Q., Yuan, M., Yu, J., Gao, W., Wang, X., et al. (2025). Synergistic anti-inflammatory effects of pomegranate peel–hawthorn combinations in ulcerative colitis: Network pharmacology prediction and experimental validation. *Current Issues in Molecular Biology*, 47, 243.

Zhao, J., Hu, Y., Qian, C., Hussain, M., Liu, S., Zhang, A., et al. (2023). The interaction between mushroom polysaccharides and gut microbiota and their effect on human health: A review. *Biology*, 12, 122.

Zhao, M., He, Q., Shu, X., Xu, R., Zhang, Z., Mou, Y., et al. (2025). Zhuyu pill attenuates metabolic-associated fatty liver disease by regulating macrophage polarization through TLR4 signaling pathway. *Phytomedicine*, 138, 156439.

Zhu, D., Jiang, N., Wang, N., Zhao, Y., & Liu, X. (2024). A literature review of the pharmacological effects of jujube. *Foods*, 13, 193.

17 Lesser-Known Medicinal Fruits and Berries

Tanfidz Alishlah, Firdha Aprillia Wardhani, Hery Diar Febryanto, Phurpa Wangchuk, and Ari Satia Nugraha

17.1 INTRODUCTION

Fruits and berries develop from flowers. A fruit is the fully developed ovary of a flower, and it usually contains seeds. Berries include both true and false berries. While true berries with a fleshy interior develop from an ovary (e.g., bananas), false berries (aggregate fruits) form from multiple ovaries (e.g., strawberries), where the "seed" is actually the fruit. The vast array of fruit and berry species has evolved over millions of years through adaptation to diverse ecological niches and selective breeding (Giampieri et al., 2012). Both have been valuable commodities throughout human history, serving as notable sources of medicine and food. For example, commonly consumed berries are blueberries (*Vaccinium corymbosum* L.), raspberries (*Rubus idaeus* L.), and blackberries (*Rubus armeniacus* Focke).

Berries are rich in dietary fibers, antioxidants, vitamins, minerals, and other interesting secondary metabolites (SMs), especially polyphenols (anthocyanins and flavonoids), and carotenoids (Kong et al., 2003; Panche et al., 2016). These SMs possess strong antioxidant properties, endow diverse tastes and characteristic colors to the fruits and berries (Giampieri et al., 2012), and support animal and human health.

They can reduce cholesterol, cardiovascular disease, cancer, and other chronic diseases (Aguilera, 2024). For example, apple (*Malus domestica* Borkh.), rose hip (*Rosa canina* L.), and hawthorn (*Crataegus* spp.) are high in nutrients and antioxidant properties. Similarly, lingonberries (*Vaccinium vitis-idaea* L.), bilberries (*Vaccinium myrtillus* L.), and cranberries (*Vaccinium macrocarpon* Aiton) are good for vascular and urinary health (Khoo et al., 2017). These health benefits have made berries one of the world's superfoods, with Mexico becoming a prominent berry supplier. Its market size reached $26.97 billion in 2025 and is expected to grow to $34.67 billion in 2030 (The Business Research Company, 2026).

This chapter focuses on lesser-known berries, namely, *Aglaia edulis, Ardisia sanguinolenta, Catunaregam spinosa, Merope angulata, Morinda citrifolia*, and *Trapa natans* (Figure 17.1). It discusses their taxonomy, ethnopharmacology, phytochemistry, and pharmacology.

17.2 *AGLAIA EDULIS (ROXBURGH) WALLICH*

17.2.1 Taxonomy, Biology, and Distribution

Aglaia is the largest genus, comprising 150 species, within the Meliaceae family (Awang et al., 2012; Su et al., 2006). Eight species of this genus have been identified in southern China, and it has been found in the tropical and subtropical forests of southern Asia, northern Australia, and the Pacific region (Xie et al., 2007; Yodsaoue et al., 2012). Within Asia, it is widely distributed in India, Thailand, Vietnam, Malaysia, and Indonesia (Joycharat et al., 2008; Zhang et al., 2010). In Indonesia, known as Mahogani, there are 65 species endemic to the country (Heyne, 1987; Mabberley et al., 1995).

DOI: 10.1201/9781003688259-17

FIGURE 17.1 Lesser-known fruits of *Morinda Citrifolia* grown in the archipelago of Indonesia.

Source: from the author's personal documentation.

Several plants in the Aglaia genus have been used to treat various illnesses, including skin conditions, wounds, fevers, influenza, and cough (Mabberley et al., 1995; Schulz et al., 2021).

Aglaia edulis (Roxburgh) Wallich is classified within the Kingdom Plantae, Phylum Streptophyta, Class Equisetopsida, Subclass Magnoliidae, Order Sapindales, Family Meliaceae, and Genus Aglaia, and it has been historically known under 37 synonymous names, including *Milnea edulis* Roxb. (1824) as the basionym, along with numerous heterotypic synonyms such as *Aglaia acida*, *A. barberi*, *A. cambodiana*, *A. curranii*, *A. diffusa*, A. indica, *A. khasiana*, *A. latifolia*, *A. magnifoliola*, *A. minahassae*, *A. montrouzieri*, *A. motleyana*, *A. mucronulata*, *A. oblonga*, *A. pirifera*, *A. rugosa*, *A. samarensis*, *A. sulingi*, *A. testicularis*, *A. undulata*, *A. verrucosa*, *Amoora verrucosa*, *Beddomea indica*, *B. sarmentosa*, the not validly published *Camunium bengalense*, *Lepiaglaia montrouzieri*, several Milnea species (*M. cambodiana*, *M. pirifera*, *M. racemosa*, *M. rugosa*, *M. sulingi*, *M. verrucosa*), the not validly published Milnea undulata, Nyalelia racemosa, and the illegitimate name *Achras retusa* (POWO, 2026).

Aglaia edulis is a small- to medium-sized evergreen tree. It has red and hard timber. Trees are with rufous bark that are 5–9 m tall and up to 30 cm. Its young branches are glabrous, pale brown, and have faint lenticels. The leaves are approximately 25–30 cm long. Its fruits are indehiscent, brown, elliptic, about 5.5 × 3–3.5 cm, yellowish lenticellate, thick brown squamate, basally condensed into a 1–1.5 cm stipe, and with a plump, turbinate apex and an inconspicuous persisting calyx. Each fruit has one to three ellipsoid seeds, each measuring around 4 cm, with a hilum of up to 3 cm (Wu et al., 2008).

This species is one of the *Aglaia* species that is predominantly found in Yunnan, China. It is indigenous to evergreen broad-leaved forests, primarily on limestone slopes and hillsides at elevations up to 5,000 feet, but it can also be found near rocky coastlines(Missouri Botanical Garden, 2026). *A. edulis* is native to Assam, Bangladesh, Borneo, Cambodia, China South-Central, East Himalaya, India, Java, Laos, Lesser Sunda Islands, Malaya, Maluku, Myanmar, Nicobar Islands, Philippines, Sulawesi, Sumatra, Thailand, and Vietnam (POWO, 2026).

17.2.2 Ethnopharmacology

For many years, the ethnopharmacological use of the *Aglaia* species has been widespread in Southeast Asia for the treatment of various illnesses. For instance, in Southeast Asia, the Aglaia genus has been widely used as a medicinal treatment for bruises, heart disease, and fever, and as an insect repellent (Proksch et al., 2001; Yodsaoue et al., 2012). Furthermore, according to Chinese folklore, the *Aglaia* species, sometimes called Mi-Zi-Lan or Shu-Lan, is used as a febrifuge or heart stimulant and as a herb to cure toxin-induced coughs, inflammations, injuries, and vomiting (Liu et al., 2014; Proksch et al., 2001). Another species, from the Aglaia genus, is used in Indonesia to treat bruises, ulcers, menstrual problems, body odor, and diarrhea, while the flowers are frequently used as tea, a scent for clothing, and medication for flatulence, coughing, and dizziness (Hayashi et al., 1982; Heyne, 1987)

Specific to *A. edulis*, the tree was used for furniture, house posts, boats, and firewood for a long time in Mizoram, Northeast India (Laha et al., 2018). The fruit of *A. edulis*, known locally as Sanulahsune, has been used by the indigenous communities of the Sikkim Himalayas as an edible fruit with a slightly acidic flavor and to prevent diarrhea (Dutta et al., 2025; Harneti & Supratman, 2021; Lemmens & Bunyapraphatsara, 2003). In Cambodia, the ripe fruits of the trees *Aglaia edulis*, known as *mchu:l*, are eaten, while the wood is used in temporary constructions and as firewood (Phon, 2000).

17.2.3 Phytochemistry

A. edulis has attracted considerable scientific interest due to its rich and structurally diverse phytochemical profile. Phytochemical investigations over the past decades have demonstrated that *A. edulis* is a prolific source of biologically active secondary metabolites, encompassing multiple compound classes such as flavaglines (cyclopenta[b]benzofurans and cyclopenta[bc]benzopyrans), bisamides, limonoids, terpenoids, and steroids (Brader et al., 1998; Harneti & Supratman, 2021; Kim et al., 2006).

Flavaglines are a distinctive class of bioactive secondary metabolites predominantly found in the genus Aglaia (Meliaceae) and serve as chemotaxonomic markers of this genus. Structurally, flavaglines are characterized by a unique cyclopenta[b]benzofuran or cyclopenta[bc]benzopyran core skeleton, biosynthesized from flavonoid precursors via complex oxidative cyclization. Since their initial discovery, flavaglines have attracted significant attention for their remarkable structural diversity and potent biological activities, including cytotoxic, anti-inflammatory, antiviral, and insecticidal effects. Numerous phytochemical studies have demonstrated that *Aglaia* species, particularly *A. edulis*, are prolific sources of flavaglines, such as rocaglamide and aglaroxin, and their analogues, many of which exhibit strong activity against cancer cell lines and viral pathogens. The restricted occurrence of flavaglines almost exclusively within the genus *Aglaia* highlights their importance not only as bioactive natural products but also as valuable chemical markers for taxonomic and evolutionary studies within the Meliaceae family (Harneti & Supratman, 2021; Kim et al., 2006).

Early research revealed the nine new flavaglines, cyclopenta[bc]benzopyrans (thapsakins) and benzo[b]oxepines (thapoxepines) from the lipophilic root extract of *A. edulis*, along with two known cyclopenta[b]benzofurans, aglaroxin A and pannellin (Bacher et al., 1999). In 2005, two benzo[b]oxepine derivatives, edulisones A and B, bearing a 2-aminopyrrolidine side chain, were isolated from the bark of *A. edulis* collected in Indonesia. The cytotoxic activities of edulisones A and B were evaluated in a small panel of human cancer cell lines, but neither compound exhibited discernible activity (Kim et al., 2005). In further study, Kim et al. (2006) reported a phytochemical investigation of *A. edulis*, revealing a structurally diverse profile of secondary metabolites, particularly flavaglines and bisamide derivatives, isolated from different plant parts and characterized by spectroscopic techniques. From the bark, the study identified a known cyclopenta[b]benzofuran compound, aglaroxin A, along with two new cyclopenta[b]benzofurans, aglaroxin A 1-O-acetate and 3′-methoxyaglaroxin A 1-O-acetate, a new benzo[b]oxepine (19,20-dehydroedulisone A), and five new cyclopenta[bc]benzopyrans, including edulirin A, edulirin A 10-O-acetate, 19,20-dehydroedulirin A, isoedulirin A, and edulirin B. Another study also reported 18 isolated cyclopenta[b]benzopyran derivatives with cytotoxic activity (Yi et al., 2025).

Bisamides are a group of compounds found in some species of Aglaia. Researchers have detected bisamide, derived from putrescine, known as edulimide, in the leaf extract of *A. edulis*. Edulimide is the equivalent of cinnamic/dihydrocinnamic acid bisamide, which has an extra acetic acid moiety that forms an imide group (Brader et al., 1998). The leaves of *A. edulis* were also identified as unique bisamide constituents—aglaiduline, aglaithioduline, and aglaidithioduline—whose structures were elucidated using detailed spectroscopic analyses, including nuclear magnetic resonance (NMR) spectroscopy and mass spectrometry. These compounds are characterized by sulfur-containing amide functionalities, representing a relatively uncommon class of natural products within the genus *Aglaia* and the Meliaceae family (Saifah et al., 1999). Additionally, from the leaves and twigs, four new bisamide compounds (aglamides A–D) were isolated. Aglamide A and aglamide B were identified as sulfur-containing bisamides, with aglamide B being the inaugural sulfoxide-containing molecule from the genus Aglaia. The other known phytochemicals identified in this research are agalactone, scopoletin, and 5-hydroxy-3,6,7,4′-tetramethoxyflavone (Kim et al., 2006). From a phytochemical perspective, the discovery of these bisamides highlights the metabolic diversity of *A. edulis* beyond its more widely reported terpenoid and flavagline constituents.

Recently, terpenoids and steroids were identified in an expanded chemical landscape of *A. edulis*. The phytochemical investigation of *A. edulis* led to the isolation and characterization of a chemically diverse array of terpenoids from the plant's twigs and leaves. A total of 29 terpenoids were identified, comprising 19 triterpenoids, 6 diterpenoids, and 4 sesquiterpenoids, of which 4 compounds were reported as new natural products. The dominance of highly oxygenated triterpenoids underscores the biosynthetic capacity of *A. edulis* to produce structurally elaborate terpenoid scaffolds characteristic of the Meliaceae family (Li et al., 2025a). From the twigs and leaves of *A. edulis*, 4 steroids and 15 terpenoids, including 2 new steroids and 3 new terpenoids alongside known analogues, which were revealed to be a chemically diverse assembly of secondary metabolites, have been isolated. The coexistence of structurally complex steroids alongside multiple classes of terpenoids underscores the metabolic versatility of *A. edulis* and reflects biosynthetic features characteristic of the Meliaceae family. The isolated compounds displayed notable structural variation in oxygenation patterns and ring systems, contributing to their chemical distinctiveness (Chen et al., 2025).

Limonoids, a class of tetranortriterpenoids formed by the loss of four terminal carbon of the side chain in the apotirucallane or apoeuphane skeleton and cyclized to form the 17-β-furan rings, belong to the Meliaceae and Rutaceae families. Due to their diverse skeletons and wide range of biological properties, particularly their cytotoxic and anti-inflammatory effects, limonoids have attracted significant attention in chemistry and biology (Tan & Luo, 2011). In an extensive phytochemical investigation of the fruits of *A. edulis*, four biogenic A/D-seco limonoid analogues with unusual β-substituents at C-6 were found, as well as Aglatestine A, an unprecedented 3/6/6 tricarbocyclic limonoid framework. High-resolution electrospray ionization mass spectrometry (HRESIMS), NMR spectroscopy, electronic circular dichroism (ECD), X-ray crystallography, and quantum-chemical calculations were used to determine their structures and absolute configurations (Sun et al., 2021). Further research isolated a structurally remarkable series of limonoids, significantly expanding the known secondary metabolites of this species beyond previously characterized flavaglines and terpenoids. The study resulted in the discovery of eleven previously undescribed limonoids, collectively named agledulines A–K, which include eight biogenic A/D-rings-seco limonoid analogs (agledulines A-H), one D-ring seco limonoid (agleduline I), and two A-ring seco limonoids featuring a rare $\Delta^{4,28}$ moiety (agledulines J-K), alongside twelve known limonoid compounds. These complex triterpenoid derivatives were structurally elucidated through a combination of one- and two-dimensional NMR spectroscopy, HRESIMS, single-crystal X-ray diffraction, ECD, and exciton chirality methods, providing unambiguous stereochemical assignments for several key structures (Sun et al., 2022). Highly oxidized three new limonoids (agledulines L-N) and ten limonoids have been successfully elucidated from *A. edulis* fruits. These limonoids demonstrated cytotoxic, anti-dengue virus (DENV), and anti-inflammatory activity (Lv et al., 2025; Yi et al., 2025).

Collectively, accumulated phytochemical investigations demonstrate that *A. edulis* represents a chemically rich and biosynthetically versatile species within the Meliaceae family. Studies have revealed a wide spectrum of structurally diverse secondary metabolites, including flavaglines, terpenoids, limonoids, steroids, and bisamides, many of which exhibit complex molecular architectures and significant biological activities (Kim et al., 2006; Li et al., 2025a; Saifah et al., 1999). The prominence of unique flavagline scaffolds, alongside highly oxygenated terpenoids and rearranged limonoids, highlights *A. edulis* as an important source of chemotaxonomically significant and pharmacologically promising natural products (Sun et al., 2022). These multiclass secondary metabolites with structural diversity underpin a range of biological activities, align with broader phytochemical patterns noted across the *Aglaia* genus, and highlight its value for continued natural product research (Harneti & Supratman, 2021).

17.2.4 Pharmacology

A. edulis has emerged as a pharmacologically significant plant owing to the broad spectrum of biological activities exhibited by its secondary metabolites. Pharmacological investigations have demonstrated that extracts and isolated compounds from *A. edulis* possess notable anti-cancer, anti-inflammatory, antiviral, cytoprotective, and immunomodulatory activities.

In anticancer studies, early investigations revealed that flavaglines and related compounds isolated from *A. edulis* exhibit pronounced cytotoxicity against multiple human cancer cell lines. Rinner et al. (2004) evaluated the antiproliferative and pro-apoptotic effects of extracts from several medicinal plants, including species of the genus Aglaia (Meliaceae), on medullary thyroid carcinoma (MTC) cell lines. MTC, a neuroendocrine malignancy known for its relative insensitivity to standard chemotherapy, poses significant therapeutic challenges; thus, alternative phototherapeutic approaches have been explored. Using in vitro assays, including viability and proliferation analyses, the study reported a strong antiproliferative effect exerted by each *Aglaia* extract on MTC cell growth, indicating that *Aglaia*-derived compounds can significantly inhibit tumor cell proliferation. The LC_{50} of the *A. edulis* extract was 0.5 mg/mL. Cell numbers decreased in a dose-dependent manner. Cell accumulation was unaffected. Bcl-2 studies and DAPI labeling revealed no apoptotic cells. After 48 hours, the strongest caspase-3 staining was observed, and it did not exceed 15% (Rinner et al., 2004).

In particular, 12 new compounds were tested for in vitro cytotoxicity in several human cancer cell lines (Lu1, LNCaP, and MCF-7) and a nontumorigenic cell line (HUVEC). The cyclopenta[b]benzofuran derivatives, such as aglaroxins, showed strong growth-inhibitory effects on Lu1, LNCaP, and MCF-7 cancer cells, with ED_{50} values ranging from 0.001 to 0.8 μg/mL across these isolates. Aglaroxin A 1-O-acetate was subsequently evaluated in the in vivo P388 lymphocytic leukemia model via intraperitoneal administration but was found to be inert. However, other classes, such as the cyclopenta[bc]benzopyrans and the new benzo[b]oxepine, showed limited cytotoxicity (Kim et al., 2006). Another study reported that out of the isolated 18 cyclopenta[b]benzopyran derivatives and 10 limonoids from *A. edulis*, nine compounds exhibited notable cytotoxicity against three cancer cell lines, with IC_{50} values ranging from 3 to 900 nM (Yi et al., 2025). These findings highlight the potential of *A. edulis*–derived flavaglines as lead compounds for anticancer drug development. The unique cyclopenta[b]benzofuran framework of these metabolites contributes to their cell proliferation inhibitory properties, which are often superior to those of other secondary metabolite classes isolated from this species and are considered characteristic chemotaxonomic markers of the *Aglaia* genus linked to antitumor potency. Although not limited to *A. edulis*, broader phytochemical reviews of the genus have affirmed that flavagline derivatives, such as silvestrol and rocaglamide, exhibit exceptionally strong cytotoxicity across multiple cancer models and are thus central to antineoplastic research within this plant group (Harneti & Supratman, 2021).

Beyond anticancer activity, *A. edulis* has also been shown to possess anti-inflammatory properties, as demonstrated by more recent studies focusing on its terpenoid and steroidal constituents. Chen et al. (2025) reported that several steroids and terpenoids isolated from the twigs and leaves of *A. edulis* significantly inhibited nitric oxide (NO) production in lipopolysaccharide (LPS)-stimulated macrophages, a widely accepted in vitro model of inflammation. The result showed IC_{50} values ranging from 3.41 ± 0.05 to 26.7 ± 0.22 μM. In particular, aglatestates A exhibited an equivalent potential inhibitory effect on NO production (IC_{50}: 3.41 ± 0.05 μM) to the positive control, pyrrolidine dithiocarbomate ammonium (IC_{50}: 3.26 ± 0.26 μM). According to another biological assessment, an A/D-seco limonoid analogue from *A. edulis* fruits demonstrated anti-inflammatory activity by inhibiting NO production in LPS-activated RAW 264.7 macrophages (IC_{50}: 35.72 ± 1.96 μM) (Sun et al., 2021). These results suggest that the anti-inflammatory effects of *A. edulis* are mediated, at least in part, through the modulation of inflammatory signaling pathways, further expanding its pharmacological relevance beyond cytotoxicity.

Related to anti-inflammatory activity, a comprehensive in silico investigation aimed at identifying phytocompounds capable of modulating epigenetic targets relevant to rheumatoid arthritis (RA). RA is an autoimmune disease characterized by chronic inflammation that affects about 1% of the population. Deshpande et al. (2023) employed similarity indexing and molecular docking to explore compounds structurally analogous to the histone deacetylase (HDAC) inhibitor SAHA, a drug used clinically for certain inflammatory conditions, including RA. Using the Tanimoto similarity coefficient to screen a large natural compound database, the study identified aglaithioduline, a sulfur-containing phytocompound found abundantly in *Aglaia* species such as *A. leptantha* and *A. edulis*, exhibiting approximately 70% structural similarity to SAHA. Molecular docking against the HDAC8 protein, an epigenetic enzyme implicated in RA pathogenesis, revealed a favorable binding affinity (–8.4 kcal/mol) for aglaithioduline, and subsequent molecular dynamics simulations confirmed the stability of the HDAC8–aglaithioduline complex, with stable root-mean-square deviation (RMSD) and sustained protein–ligand interactions throughout simulation. These computational findings suggest that aglaithioduline may act as an effective HDAC8 inhibitor, paralleling the mechanism of HDAC inhibitors known to ameliorate inflammatory responses in RA, thereby supporting the traditional empirical use of *A. edulis*-related extracts in treating pain and joint disorders and justifying further in vitro and in vivo pharmacological validation (Deshpande et al., 2023)

Antiviral activities have also been documented for compounds derived from *A. edulis*. Saifah et al. reported that the isolated bisamides from *A. edulis* exhibited mild antiviral activity against herpes simplex virus types 1 and 2, suggesting a potential functional relevance of this compound class (Saifah et al., 1999). More recently, Yi et al. (2025) reported that cyclopenta[bc]benzopyran derivatives and limonoids from *Aglaia edulis* displayed cytotoxicity and inhibitory effects against dengue virus (DENV), with selectivity indices ranging from 13.0 to 532.6, highlighting the species' potential as a source of antiviral agents. An examination of the mechanism of action indicated that the compound may act as an eIF4E activator, potentially inhibiting E protein expression and thereby exhibiting substantial efficacy against dengue virus.

In addition, as cytoprotective agents, Li et al. (2025a) demonstrated that several terpenoids from *A. edulis* twigs and leaves protected human lung epithelial cells against cigarette smoke extract–induced injury. The protective impact of these compounds on lung epithelial cells was evaluated using NNK-induced MLE-12 cells. Biological assays demonstrated that compounds 1–2, 4–6, 8, 10, and 12–15 may mitigate NNK-induced damage in MLE-12 cells, suggesting a possible role in the prevention or management of respiratory disease. Collectively, these findings highlight *A. edulis* as a rich source of structurally diverse terpenoids with notable biological relevance, reinforcing its value in natural product chemistry and providing a phytochemical basis for further pharmacological exploration.

Further biological evaluation underscored the functional relevance of compounds from *A. edulis*. Among the isolates, agleduline C exhibited significant cytotoxicity (IC_{50} value 10.05 μM), while 11α-acetoxygedunin demonstrated notable anti-inflammatory activity (IC_{50} value 4.70 μM), and both agleduline I and 11α-acetoxygedunin reversed multidrug resistance in MCF-7/Dox cells. These findings not only enrich the phytochemical profile of *Aglaia edulis* but also highlight the potential pharmacological importance of its limonoid constituents as bioactive leads for anticancer, anti-inflammatory, and multidrug-resistance–modulating agents (Sun et al., 2022).

Moreover, antifungal activity was reported by Engelmeier et al. (2000), who found that six cyclopenta[b]benzofuran derivatives, including several isolated from *A. edulis*, were evaluated against plant pathogens, including *Pyricularia grisea*. The results demonstrated strong antifungal activity of the benzofuran derivatives, especially rocaglaol, against rice blast fungus. In addition, the anti-amoebic activity of the stem bark and leaves of *Aglaia edulis* was also investigated. The *A. edulis* extract shows weaker anti-amoebic activity against *Entamoeba histolytica* than *Aglaia elaeagnoidea* and *Aglaia odorata* (Tasanor et al., 2007).

In conclusion, these pharmacological studies indicate that *A. edulis* exhibits multifaceted biological activities, including anticancer, anti-inflammatory, antiviral, and cytoprotective effects, which are closely linked to its diverse phytochemical composition. The convergence of strong bioactivities across multiple compound classes underscores the therapeutic potential of *A. edulis* and supports continued pharmacological and mechanistic investigations aimed at drug discovery and development.

17.3 *CATUNAREGAM SPINOSA* (THUNB.) TIRVENG.

17.3.1 Taxonomy, Biology, and Distribution

Catunaregam spinosa (Thunb.) Tirveng. is a species belonging to the Kingdom Plantae, Phylum Streptophyta, Class Equisetopsida, Subclass Magnoliidae, and Order Gentianales. It is part of the family Rubiaceae and the genus Catunaregam. This species has a complex synonym, with 32 known synonyms. Homotypic synonyms include *Canthium spinosum* (Thunb.) J.St.-Hil., *Gardenia spinosa* Thunb., *Randia spinosa* (Thunb.) Poir., and *Xeromphis spinosa* (Thunb.) Keay. Heterotypic synonyms include names such as *Canthium chinense* Pers., Catunaregam *brandisii* Kottaim., *Ceriscus malabaricus* Gaertn., *Gardenia dumetorum* Retz., and many others, reflecting the species' diverse taxonomic history. Commonly, it is referred to as the "spiny gardenia" or "emetic nut" (POWO, 2026).

C. spinosa is a thorny shrub or small tree in the Rubiaceae family. Its leaves are opposite and elliptical, and its spines are thick and sharp. The species has white, tubular flowers with a pleasant scent, and its fruit is a round berry with a hard shell that turns yellowish when ripe and contains many seeds in a pulpy, bitter matrix. It can live in dry and semidry environments due to xerophytic adaptations, including thorns and hard fruit (Timalsina et al., 2021).

This species is widely distributed across tropical and subtropical regions of South and Southeast Asia, including India, Nepal, Sri Lanka, Myanmar, Thailand, and parts of Indonesia. It typically thrives in dry deciduous forests, rocky slopes, and scrublands, adapting well to arid environments (Crous et al., 2021).

17.3.2 Ethnopharmacology

The ethnopharmacological significance of *C. spinosa* is deeply embedded in the traditional medical systems of South and Southeast Asia. In folk medicine across Asia, its fruits and roots are utilized for treating gastrointestinal, hepatic, and inflammatory ailments (Alsharif & Boylan, 2025; Khanavi et al., 2020). Moreover, across the Indian subcontinent, this species is widely recognized in Ayurveda under the Sanskrit name Madanaphala, where it is categorized among the Vamaka dravyas, substances that induce therapeutic emesis (Vendrapati et al., 2024). The fruit has been an essential component of Panchakarma detoxification therapy, specifically in the Vamana karma procedure, which aims to eliminate accumulated toxins and excess kapha dosha from the body (Singh, 2012). In this practice, powdered or decocted fruits are administered with honey or warm water under the supervision of a physician to cleanse the upper gastrointestinal tract and restore doshic balance (Timalsina et al., 2021).

Beyond classical Ayurveda, the ethnomedicinal use of *C. spinosa* is widespread among rural and tribal communities across India, Nepal, Sri Lanka, Myanmar, Thailand, and Indonesia. In the hilly regions of Nepal, traditional healers prepare fruit decoctions to treat fever, dysentery, and diarrhea, often combined with other digestive herbs (Timalsina et al., 2021). In Thai folk medicine, boiled fruits are consumed to reduce body heat and cleanse the digestive system, reflecting their perceived cooling and detoxifying nature. Similarly, in Indonesian traditional medicine (Jamu), infusions made from the fruit are used as natural purgatives and tonics for intestinal cleansing, while traditional systems of medicine in Sri Lanka have documented the use of numerous plant species for diabetes management, highlighting the potential for discovering novel antidiabetic compounds from such ethnobotanical knowledge (Ajaib et al., 2022; Sathasivampillai et al., 2017).

Different parts of the plant are valued for distinct therapeutic purposes. The fruits, being the most potent part, are used internally as emetics, purgatives, or febrifuges, whereas the bark and roots are externally applied for treating skin diseases, ulcers, and swelling. Decoctions and infusions are the most common forms of preparation, but powders and macerations are also used depending on regional practices. The bitter and astringent taste of the fruit, owing to its iridoid and triterpenoid constituents, has contributed to its use as a digestive stimulant and liver tonic, often consumed in small doses to enhance appetite and metabolism (Dhatwalia et al., 2021).

The ethnomedical applications of *C. spinosa* are remarkably diverse, spanning gastrointestinal, hepatic, respiratory, and inflammatory disorders. Traditional healers prescribe its fruit pulp for liver dysfunction, jaundice, chronic cough, and parasitic infections. In certain Indian tribes, it is even used as an antidote for snakebite and as a general remedy for fever and malaise (Basini et al., 2019). These wide-ranging applications indicate a deep empirical understanding of the plant's purgative, detoxifying, and protective properties, which likely correspond to its bioactive phytochemical profile.

Despite its broad traditional use, *C. spinosa* remains relatively lesser-known in modern pharmacognosy and phototherapeutic research compared to more prominent medicinal fruits (Timalsina et al., 2021). Its rediscovery within ethnopharmacological and phytochemical studies could offer new insights into its potential as a source of bioactive compounds. Altogether, the ethnopharmacological evidence indicates that *C. spinosa* is a multipurpose medicinal fruit, with rich traditional knowledge supporting its therapeutic versatility. Its persistent role in diverse healing systems, from Ayurvedic detoxification rituals to rural remedies for fever and infection, demonstrates not only its medicinal potential but also its deep cultural and symbolic resonance as a plant of purification and healing.

17.3.3 Phytochemistry

The phytochemical profile of *C. spinosa* reflects a remarkable diversity of bioactive compounds, consistent with its wide range of ethnomedicinal uses. Studies conducted on different parts of the plant, particularly the fruits, bark, and roots, have revealed the presence of several secondary metabolite classes, including iridoids, triterpenoids, flavonoids, saponins, tannins, phenolics, and alkaloids (Abdulrahman et al., 2022; Timalsina et al., 2021). These phytoconstituents are believed to collectively contribute to the plant's emetic, hepatoprotective, antioxidant, and anti-inflammatory properties, as observed in traditional medicine and experimental models.

Among the compounds identified, iridoid glycosides represent one of the dominant groups in *C. spinosa*. Notable iridoids such as geniposidic acid, gardenoside, and scandoside have been isolated from the fruit pulp and bark. These compounds are known to exert hepatoprotective and anti-inflammatory effects by modulating oxidative stress pathways and stabilizing cellular membranes. The presence of these iridoids also aligns the chemical composition of *C. spinosa* with other

Rubiaceae members, such as *Gardenia jasminoides* and *Morinda citrifolia*, which are similarly rich in iridoid glycosides (Chedraoui et al., 2017; Sun et al., 2023).

Triterpenoids constitute another major group of bioactive compounds in *C. spinosa*, particularly lupeol, oleanolic acid, and betulinic acid. These pentacyclic triterpenes have been well documented for their anti-inflammatory, hepatoprotective, and antimicrobial activities. Lupeol, for instance, inhibits key enzymes involved in the inflammatory cascade, such as cyclooxygenase (COX) and lipoxygenase (LOX), while oleanolic acid has shown protective effects against oxidative liver injury. The combined presence of these triterpenoids strengthens the pharmacological rationale behind the plant's use in treating hepatic and gastrointestinal disorders (Basini et al., 2019; Singh, 2012).

Flavonoids and phenolic compounds further enhance the plant's therapeutic potential. Compounds such as quercetin, kaempferol, and their glycosides have been reported in the methanolic extracts of the fruit, exhibiting strong antioxidant and free radical–scavenging activity. Total phenolic and flavonoid content analyses have confirmed that *C. spinosa* extracts possess significant reducing power, correlating with their ability to mitigate oxidative stress in biological systems (Bacchetti et al., 2022; Naffati et al., 2025). These phenolic constituents are also responsible for the fruit's bitter taste, which has traditionally been associated with its detoxifying and digestive properties.

Preliminary nutritional analyses indicate that the fruit pulp contains modest amounts of carbohydrates and essential minerals such as calcium and iron, along with trace levels of vitamins and antioxidant compounds that may contribute to its tonic effects. Although it is not typically consumed as a dietary fruit due to its bitterness, these nutritional and biochemical components add to its therapeutic value in small medicinal doses (Bacchetti et al., 2022).

Analytical investigations employing chromatographic and spectroscopic techniques, such as high-performance liquid chromatography (HPLC), liquid chromatography–mass spectrometry (LC-MS), and NMR spectroscopy, have been instrumental in identifying and characterizing the major metabolites of *C. spinosa*. LC-MS profiling has revealed a complex mixture of iridoids and triterpenoid glycosides, while NMR analyses have elucidated their structural frameworks, providing chemotaxonomic evidence linking *C. spinosa* to other medicinally important genera within Rubiaceae (Naffati et al., 2025; Sun et al., 2023; Timalsina et al., 2021; Verma et al., 2025). Such findings underscore the chemical richness of this species and justify its continued exploration for potential pharmaceutical and nutraceutical applications.

In summary, the phytochemistry of *C. spinosa* reveals a rich interplay of bioactive compounds, particularly iridoids, triterpenoids, and flavonoids, that underpin its wide range of pharmacological effects. The combination of these molecules imparts antioxidant, anti-inflammatory, and hepatoprotective properties, supporting the ethnopharmacological uses of the plant in traditional medicine. However, despite considerable progress in compound identification, comprehensive metabolomic profiling and bioassay-guided isolation remain limited. Future studies should focus on establishing standardized extraction protocols and correlating specific chemical constituents with biological activities to fully harness the therapeutic potential of this lesser-known medicinal fruit.

17.3.4 Pharmacology

The pharmacological activities of *C. spinosa* have attracted growing scientific attention due to its longstanding ethnomedicinal reputation and its rich profile of bioactive phytochemicals. A wide range of preclinical studies, both in vitro and in vivo, have provided evidence supporting many of the traditional claims associated with this plant, particularly its antioxidant, anti-inflammatory, antimicrobial, hepatoprotective, and analgesic effects. These findings demonstrate that *C. spinosa* is not merely a folk remedy but a promising source of pharmacologically active compounds with diverse therapeutic applications.

Antioxidant and free radical–scavenging activities represent one of the most extensively documented pharmacological properties of *C. spinosa* (Basini et al., 2019; Naffati et al., 2025). Methanolic and aqueous extracts of the fruit and bark have been shown to neutralize reactive oxygen species

(ROS) in standard assays, including DPPH, ABTS, and NO radical scavenging assays. These antioxidant effects are largely attributed to the presence of phenolic and flavonoid compounds, including quercetin, kaempferol, and their glycosides, which donate hydrogen atoms or electrons to stabilize free radicals. This antioxidative potential underlies many of the plant's protective effects against oxidative stress–related diseases, including liver and cardiovascular disorders (Ajaib et al., 2022; Basini et al., 2019).

The anti-inflammatory and analgesic activities of *C. spinosa* have been demonstrated in several *in vivo* studies using animal models. Ethanolic extracts of the fruit have been reported to significantly reduce carrageenan-induced paw edema and acetic acid-induced writhing in mice, comparable to standard anti-inflammatory agents such as indomethacin. The mechanism is thought to involve inhibition of pro-inflammatory mediators, including prostaglandins and cytokines, by suppressing the COX and LOX pathways. The presence of triterpenoids such as lupeol and oleanolic acid likely contributes to these effects, given their known ability to modulate inflammatory signaling cascades and oxidative stress (Alqahtani et al., 2023; El Azhary et al., 2017; Hajimehdipoor et al., 2024; Li et al., 2015).

Antimicrobial activity has also been widely observed, supporting the plant's traditional use in treating infections and skin disorders. Extracts of *C. spinosa* have exhibited broad-spectrum antibacterial activity against Gram-positive and Gram-negative bacteria, including *Staphylococcus aureus, Escherichia coli*, and *Pseudomonas aeruginosa*. Antifungal effects have been noted against *Candida albicans* and *Aspergillus niger*, with methanolic fruit extracts showing the strongest inhibition zones in agar diffusion assays. The antimicrobial potency is believed to arise from the synergistic action of saponins, tannins, and alkaloids, which disrupt microbial membranes and inhibit enzymatic metabolism (Annaz et al., 2022; Raish et al., 2025; Silva et al., 2016).

In addition to these effects, *C. spinosa* has demonstrated significant hepatoprotective properties. In experimental models of chemically induced liver injury, such as carbon tetrachloride (CCl_4) and paracetamol toxicity, administration of fruit or bark extracts significantly reduced serum levels of hepatic enzymes (SGOT, SGPT, and ALP) and improved histopathological architecture. The hepatoprotective mechanism is linked to the antioxidant and membrane-stabilizing activities of iridoids and triterpenoids, which protect hepatocytes from lipid peroxidation and cellular damage. These findings validate the plant's traditional use in jaundice and liver ailments across South and Southeast Asia (El-Hawary et al., 2018; Kalantari et al., 2018).

Toxicological evaluations indicate that *C. spinosa* extracts are generally safe at therapeutic doses. Acute oral toxicity tests in rodents have shown no mortality or significant behavioral alterations at doses up to 2000 mg/kg body weight, suggesting a wide safety margin. However, given the plant's known emetic properties, higher doses or improper preparations may cause nausea or vomiting, a characteristic also noted in its traditional use as an emetic and purgative (Al-Anazi et al., 2021; Salirrosas et al., 2021). Therefore, standardization and dose optimization are essential for safe pharmacological applications.

Overall, pharmacological evidence supports the traditional ethnomedicinal claims of *C. spinosa* and highlights its potential as a source of bioactive compounds for modern drug development. The diverse pharmacological profile, spanning antioxidant and hepatoprotective to antimicrobial and anti-inflammatory effects, reflects the synergistic interplay among iridoids, triterpenoids, flavonoids, and phenolics within the plant matrix. Nevertheless, most existing data are preclinical, and rigorous clinical investigations remain lacking. Future research should focus on isolating and characterizing specific active principles, elucidating molecular mechanisms, and assessing pharmacokinetic and safety profiles to fully explore the therapeutic promise of this underutilized medicinal fruit.

17.4 *MEROPE ANGULATA* (WILLD.) SWINGLE

The exploration of the botanical world for therapeutic agents has long been a cornerstone of both traditional healing systems and modern drug discovery. Among the diverse ecosystems that harbor unique biodiversity, mangrove and tidal forests stand out as particularly harsh yet chemically rich environments. One such plant species, *Merope angulata* (Willd.) Swingle, commonly known as Mangrove Lime or Limau Lelang, is a remarkable example of an understudied medicinal plant that occupies a specialized ecological niche. Found in the transition zones between land and sea across South and Southeast Asia, this monotypic genus exhibits morphological and biochemical adaptations that have captured the attention of both ethnobotanists and pharmacologists. The Rutaceae family, to which *M. angulata* belongs, is globally recognized for its significant contributions to the citrus industry, as well as its wealth of secondary metabolites such as alkaloids, coumarins, and essential oils. While the genus *Citrus* is the most well-known member of this family, many of its relatives, including *Merope*, have evolved to survive in extreme environmental conditions (Buenavista & Purnobasuki, 2023).

The significance of studying medicinal plants like *M. angulata* cannot be overstated. In an era marked by rising antimicrobial resistance and an urgent need for novel anticancer agents, the specialized metabolites produced by halophytic plants, those adapted to high-salinity environments, represent a promising frontier in pharmaceutical research. Mangrove species and their associated plants are known to synthesize complex secondary metabolites to mitigate oxidative stress caused by hypersalinity, high UV radiation, and tidal inundation. These compounds, including various phenolics, flavonoids, and acridone alkaloids, often exhibit potent biological activities that hold great potential for human health applications(Buenavista & Purnobasuki, 2023).

17.4.1 Taxonomy, Biology, and Distribution

Taxonomically, *M. angulata* belongs to the Kingdom Plantae, Phylum Streptophyta, Class Equisetopsida, Subclass Magnoliidae, Order Sapindales, Family Rutaceae, and Genus *Merope*. In addition, it also belongs to the subfamily Aurantioideae and the tribe Citreae, placing it in close phylogenetic relation to the genus *Citrus*. However, its monotypic status within the genus *Merope* highlights its unique evolutionary path. It has synonyms that are classified as homotypic—*Atalantia angulata* (Willd.) Engl., *Citrus angulata* Willd., *Gonocitrus angulatus* (Willd.) Kurz, and *Limonia angulata* (Willd.) Wight & Arn. ex Voigt—and heterotypic—*Atalantia longispina* Kurz, *Atalantia spinosa* (Blume) Hook. ex Koord., *Glycosmis spinosa* (Blume) D. Dietr., *Merope spinosa* (Blume) M.Roem., *Paramignya angulata* Kurz, *Paramignya longispina* Hook.f., and *Sclerostylis spinosa* Blume (POWO, 2026).

M. angulata is an erect, thorny shrub or small tree that typically grows up to 3 m in height. It is characterized by its sparse stems that emerge from a robust root crown, an adaptation that provides stability in the shifting, waterlogged sediments of brackish wetlands. This species is characterized by stout, sharp spines measuring 1.5–3.5 cm, found in the axils of juvenile stems. Its leaves are simple, alternate, and leathery, measuring 4.5–16 cm. A distinctive feature of *M. angulata* is the presence of numerous small, translucent glands, oil-filled dots, scattered across the leaf surface. When crushed, these glands release a resinous, lime-like aroma (Buenavista & Purnobasuki, 2023).

The native distribution of *M. angulata* is broad but fragmented, extending from the mouth of the Ganges in India and Bangladesh to the Indonesian archipelago and New Guinea (POWO, 2026). Despite this wide range, the species faces increasing threats from habitat loss due to coastal development, aquaculture, and agriculture. Its conservation status varies across regions, with the plant being critically endangered in Singapore and endangered in Peninsular Malaysia. These concerns highlight the urgent need for both ecological protection and the scientific documentation of its traditional uses (Das & Supriya Devi, 2020; Sarkar et al., 2023).

17.4.2 Ethnopharmacology

The ethnopharmacological profile of *M. angulata* is closely tied to the maritime cultures of Southeast Asia and the Indian subcontinent. Known as “Limau Lelang” in Malay and “Bano Nimbu” in Oria, this plant has been a staple in the traditional pharmacies of coastal communities for generations. Its integration into these healing systems reflects not only its practical uses but also a profound understanding of the plant's seasonal availability and the specific preparation methods needed to unlock its therapeutic benefits. In traditional Malay medicine, *M. angulata* has long been recognized for its role in maternal care. It is particularly noted for its ability to treat abdominal issues and assist in womb contraction after childbirth. This application suggests the presence of bioactive compounds with potential oxytocic or smooth-muscle-stimulant properties, which are valued in postpartum recovery to prevent hemorrhage and promote systemic healing. The use of *M. angulata* spans a wide range of communities, from the villagers of the Ayeyarwaddy Delta in Myanmar to the Munda ethnic group near the Sundarbans in Bangladesh. For the Munda people, the mangrove forest serves as a "universal resource supply system," and *M. angulata* is among many species used for primary healthcare needs. Ethnobotanical studies have also shown that the plant is used to alleviate body pain, a property corroborated by research on the medicinal value of mangrove-associated species. The therapeutic benefits attributed to *M. angulata* in traditional health systems are often linked to its strong aromatic properties. In Malaysia, the juice extracted from the leaves is applied topically to treat rheumatism and neuralgia, as well as to prevent hair loss (alopecia). The resinous, lime-like aroma, especially when the leaves are bruised, is traditionally believed to indicate the presence of "heat-producing" elements, which are thought to expel dampness or treat cold-related ailments like body aches and respiratory congestion. Additionally, the practice of chewing the leaves to ward off snake bites has been documented in coastal regions, suggesting a belief in the plant's prophylactic or neutralizing properties. This belief may be linked to its bitter alkaloids and volatile oils, which could act as mild stimulants or repellents. The juice of the leaves is also used to treat wounds inflicted by poison arrows, suggesting an ancient understanding of its antiseptic or antitoxic potential (Islam et al., 2021; Ono & Suzuki, 2013).

The integration of *M. angulata* into traditional health systems is particularly notable for its role in complex polyherbal formulations. In the North Maluku Province of Indonesia, for example, mangrove plants are seldom used in isolation; rather, they are combined with other species to treat endemic diseases such as malaria. While *M. angulata* may not always be the primary ingredient in these malaria remedies, its presence within the "biocultural refugia" of the mangrove ecosystem ensures its continued relevance in the local pharmacopeia. Public perception and acceptance of *M. angulata* as a medicinal plant remain strong in rural coastal areas, where access to formal healthcare is often limited. The plant is regarded as a "botanical amphibian," able to thrive in conditions that would be lethal to ordinary plants, thus endowing it with a cultural status of resilience and potency. However, as urban development increasingly encroaches upon back-mangrove habitats, the traditional knowledge surrounding *M. angulata* is at risk of fading. This biocultural erosion is particularly evident in regions like Singapore and Peninsular Malaysia, where the species has become rare, and its medicinal applications are increasingly relegated to historical texts rather than being actively practiced (Buenavista & Purnobasuki, 2023; Tamalene et al., 2021).

17.4.3 Phytochemistry

The chemical diversity of *M. angulata* is a direct result of its adaptation to the harsh intertidal environment. As a member of the Rutaceae family, this plant is expected to produce a wide range of secondary metabolites. However, its halophytic nature has led to the development of a unique array of compounds specifically designed to manage oxidative stress and protect against specialized pests and pathogens. Phytochemical investigations have revealed a complex mixture of nitrogen-containing compounds, phenolics, and volatile oils. Research into the chemical constituents of *M. angulata* has identified several major classes of bioactive molecules, including alkaloids, flavonoids, terpenoids, tannins, and steroids. The presence of these metabolites is characteristic of the Rutaceae family, which is renowned for its chemical complexity. One of the most significant phytochemical findings in *M. angulata* is the presence of acridone alkaloids. Acridones are a group of alkaloids with a limited distribution in the plant kingdom, primarily occurring in the Rutaceae family. Specific compounds isolated from *M. angulata* include acrifoline and 5-hydroxy-1-demethyl acronycine. These compounds are of great interest to the pharmacological community due to their DNA-intercalating properties and their potential as scaffolds for the development of anticancer drugs. The isolation of these alkaloids typically involves the use of chloroform or methanol extracts, followed by chromatographic separation techniques. In addition to alkaloids, the leaves and fruit of *M. angulata* are rich in flavonoids and other phenolics, which act as primary antioxidants. Preliminary screenings have confirmed the presence of these compounds in various solvent extracts, with methanol proving particularly effective in their recovery. These phenolics play a crucial role in the plant's defense against UV radiation and salinity-induced ROS. The essential oil of *M. angulata* is predominantly composed of monoterpenoids. These analyses have revealed that these volatile compounds make up a significant portion of the leaf's aromatic profile. Major compounds identified include beta-pinene and cis-verbenol, which contribute to the resinous, citrus-like scent characteristic of the species. Beyond their sensory properties, these terpenoids also contribute to the plant's antimicrobial and analgesic effects (Kuspradini et al., 2020; Ramakrishna Pillai et al., 2022).

The successful identification of bioactive compounds in *M. angulata* depends on the application of appropriate extraction and analytical methodologies. Due to the variability in the chemical nature of its metabolites, ranging from highly volatile essential oils to polar phenolic glycosides, a multifaceted approach is essential. Solvent extraction remains the most widely used method for isolating secondary metabolites from *M. angulata*. Maceration with methanol or ethanol is commonly employed to recover a broad range of compounds, including alkaloids and flavonoids. For the isolation of essential oils, steam distillation is the preferred technique, as it enables the collection of volatile monoterpenes without the risk of thermal degradation associated with more aggressive extraction methods (Malik et al., 2022).

Several advanced analytical techniques are employed to identify and quantify the bioactive compounds in *M. angulata*, each suited to different types of metabolites. GC-MS is considered the gold standard for analyzing volatile components, such as the monoterpene β-pinene, in the plant's essential oils. This technique enables the identification of compounds based on their retention times and mass spectral fragmentation patterns, providing valuable insights into the plant's aromatic profile. HPLC is used for the quantification of nonvolatile metabolites, including acridone alkaloids and phenolic acids. HPLC offers high sensitivity and precision, making it ideal for assessing the chemical profile of different plant parts. Additionally, TLC serves as a vital tool for preliminary screening and for monitoring the progress of chromatographic separations. It is particularly useful in detecting the presence of alkaloids, often utilizing reagents such as Mayer's or Hager's reagent to highlight the compounds of interest (Isomiddin & Hamzayeva, 2024).

The biological activity of the phytochemicals found in *M. angulata* is intricately linked to the specific chemical structures of its metabolites. The plant's antioxidant properties, for example, are primarily attributed to its rich content of phenolics and flavonoids. These compounds are effective free radical scavengers and play a significant role in preventing lipid peroxidation, which is a key process in cellular damage. In DPPH assays, ethanolic extracts from the fruits and leaves of various rutaceous plants, including *M. angulata*, frequently demonstrate remarkable radical scavenging activity. In many cases, the inhibition of free radicals can exceed 80%, a performance often comparable to, or even surpassing, well-known standards such as ascorbic acid, which is commonly used as a reference antioxidant. The antimicrobial effects of *M. angulata* are primarily driven by its terpenoid and alkaloid content. Specific compounds, such as β-pinene, have been shown to disrupt the structural integrity of bacterial cell membranes, leading to the leakage of intracellular contents and ultimately causing cell death. This antimicrobial potential makes *M. angulata* a promising candidate for the development of natural antimicrobial agents. Additionally, the plant's anti-inflammatory activity, which is well documented in traditional medicine, is likely mediated by the presence of flavonoids. These bioactive molecules have been shown to inhibit the activity of proinflammatory enzymes, such as COX, and to modulate the production of proinflammatory cytokines, contributing to a reduction in inflammation. Furthermore, the synergy between the various chemical classes present in *M. angulata*, including alkaloids, terpenoids, flavonoids, and phenolics, ensures that the plant offers a broad therapeutic window. The combination of antioxidant, antimicrobial, and anti-inflammatory properties allows for a versatile therapeutic profile, making *M. angulata* a valuable resource for addressing a variety of health concerns. Its multifaceted biological activity underscores its potential as a source of novel therapeutic agents and justifies the continued exploration of its phytochemical composition (Islam et al., 2021; Kuspradini et al., 2020; Ramakrishna Pillai et al., 2022).

17.4.4 Pharmacology

The pharmacological evaluation of *M. angulata* has evolved from basic observations of its traditional uses to more sophisticated in vitro and in vivo studies. These modern scientific investigations are beginning to map both the systemic and local effects of the plant's extracts, providing a solid scientific foundation for its inclusion in folk medicine as well as its potential applications in contemporary pharmacotherapies. The unique environmental stressors that *M. angulata* faces in its intertidal habitat have contributed to a "bioactive density" within the plant, making it particularly effective in combating various human disease models. The chemical profile of *M. angulata*, shaped by the plant's adaptation to extreme conditions, appears to enhance its therapeutic potential. Pharmacological testing has evaluated the effects of *M. angulata* on various biological systems, revealing a wide range of systemic and local actions. In animal models, extracts from the plant have been used to assess its influence on critical physiological parameters, including blood glucose levels, cholesterol concentrations, and hematological markers. These studies have begun to identify the plant's broad spectrum of bioactivities, suggesting that it could play a multifaceted role in human health. Subchronic toxicity studies conducted in albino rats have demonstrated that oral administration of *M. angulata* extracts can lead to significant changes in physiological markers. Notably, these studies have shown a reduction in the total cholesterol levels and an increase in red blood cell counts. These findings are particularly noteworthy, as they suggest that the plant may offer beneficial effects on the cardiovascular system and the hematopoietic system. This supports its traditional use in managing systemic issues such as "body aches" and general fatigue, which are often linked to imbalances in these systems. In addition, some rutaceous extracts, including those from *M. angulata*, have shown α-glucosidase inhibitory activity, indicating a potential role for the plant in blood glucose regulation. This mechanism is of significant interest for its possible applications in the management of diabetes, a disease characterized by impaired blood glucose control. The ability to regulate glucose levels, coupled with the plant's effects on cardiovascular and hematological systems, further underscores *M. angulata*'s potential as a therapeutic agent, both in traditional and modern medicine (Basak et al., 2022; Sunday & Ilesanmi, 2023).

Locally, the essential oil and leaf extracts of *M. angulata* are known for their potent antimicrobial and anti-inflammatory properties. In traditional medicine, topical applications of these extracts are commonly used to treat conditions such as neuralgia and rheumatism. This suggests that the bioactive compounds within the plant can penetrate the skin barrier, delivering an analgesic effect that aligns with its traditional use in alleviating pain and inflammation. In addition to these topical applications, in vitro studies have demonstrated that the essential oil is highly effective against oral pathogens, particularly *Streptococcus mutans*, the bacteria responsible for dental caries. This provides pharmacological validation for the traditional practice of chewing the leaves, a common remedy for oral health. The biological activities of *M. angulata* are diverse, reflecting the complexity of its phytochemical composition. Its primary areas of pharmacological interest include antimicrobial, anticancer, and anti-inflammatory properties. Studies conducted by Ravikumar et al. (2012) and other researchers have established that *M. angulata* possesses broad-spectrum antimicrobial activity. The essential oil has been shown to produce inhibition zones ranging from 11.44 mm to 50.00 mm against a variety of pathogens, including *S. aureus* and *Candida albicans*. This antimicrobial effect is attributed to the presence of monoterpenoids, such as β-pinene, which are known to destabilize the lipid bilayer of microbial cell walls, leading to cell disruption and death. These findings underscore the potential of *M. angulata* as a source of novel bioactive compounds that can be developed for medical and therapeutic applications. Its ability to combat both bacterial and fungal pathogens, combined with its anti-inflammatory and analgesic effects, positions *M. angulata* as a promising candidate for further pharmacological exploration and validation in modern medicine (Islam et al., 2021; Kuspradini et al., 2020; Sen & Bhakat, 2019).

One of the most promising areas of pharmacological research for *M. angulata* is its potential for cancer treatment. Extracts from the plant have demonstrated cytotoxic effects against several cancer cell lines, suggesting its role as a source of potential anticancer agents. In the case of A549 (lung carcinoma), Rutaceous extracts have been shown to inhibit the growth of these cancer cells. This effect is attributed to the induction of apoptosis through the generation of ROS and the sensitization of the mitochondrial membrane, which disrupts the integrity of cancer cell survival mechanisms. This highlights the plant's ability to target key cellular processes involved in tumor progression. For MIA PACA-2 (pancreatic cancer) cells, certain analogues isolated from *M. angulata* have exhibited remarkable potency, with IC_{50} values as low as 0.65 μM, indicating a strong inhibitory potential. The anticancer activity of these extracts is largely attributed to the presence of acridone alkaloids, which interfere with essential cellular processes such as DNA synthesis and topoisomerase activity. These findings suggest that *M. angulata* may have a significant role in modulating molecular targets critical for tumor cell proliferation and survival. In addition to its anticancer properties, the anti-inflammatory effects of *M. angulata* were first highlighted by Mani Senthil Kumar et al. (2008). The plant's ability to alleviate pain and swelling is thought to be mediated through the inhibition of pro-inflammatory mediators, such as NO and prostaglandins. This activity is particularly relevant in the treatment of chronic inflammatory conditions, such as rheumatism, for which the plant has been traditionally used. The dual action of *M. angulata*—both as an anticancer and anti-inflammatory agent—further demonstrates its therapeutic versatility and potential in modern pharmacological applications (Islam et al., 2021; Karker et al., 2016; Sf et al., 2013).

Ensuring the safe therapeutic use of *M. angulata* requires a thorough understanding of its toxicological profile. Although the plant has a long history of use in traditional medicine, scientific validation of its safety is essential for its potential incorporation into modern pharmacological treatments. As with any medicinal plant, safety assessments are a critical step in determining the appropriate dosage and therapeutic window for human use. In experimental models using albino rats, ethanolic extracts from the leaves of *M. angulata* were administered at doses of 50, 100, and 200 mg/kg body weight. At these doses, no toxic effects were observed. The extracts were shown to alter biochemical parameters, such as cholesterol levels, without inducing any signs of behavioral distress or damage to vital organs. This suggests that, at these doses, the plant's extracts exhibit a favorable safety profile. Regarding acute toxicity, the median lethal dose (LD_{50}) for *M. angulata* extracts was determined to be greater than 5,000 mg/kg. This places the plant extract in the "practically non-toxic" category when administered orally, according to standard toxicological classification. This finding supports the plant's safe use within the specified dosage ranges, although further studies are needed to evaluate its long-term safety and the potential for cumulative toxic effects (Sunday & Ilesanmi, 2023).

Despite the high LD_{50} value, certain precautions must be taken when using *M. angulata* therapeutically. In vitro brine shrimp lethality tests (BSLTs) have demonstrated that although the extract is nontoxic at moderate concentrations (up to 500 mg/mL), it can cause significant mortality at much higher concentrations, such as 1,000 mg/mL. This dose-dependent toxicity underscores the importance of standardized preparation and administration, ensuring that the appropriate dosage is maintained to avoid harmful effects. Furthermore, the traditional use of *M. angulata* to assist in womb contraction suggests that the plant contains compounds with potent effects on reproductive tissues. While this action may be beneficial in certain contexts, its use during pregnancy, particularly before labor, could potentially lead to adverse outcomes. Therefore, the plant should be strictly regulated in such instances, and its use during pregnancy should be approached with caution, if at all.

Chronic use or administration of very high doses of *M. angulata* could also result in mild toxicity, particularly in the liver and kidneys. This is consistent with findings from sub-chronic studies on other Rutaceous essential oils, where prolonged exposure led to elevated liver enzyme levels and changes in lymphocyte counts, indicating possible hepatic and renal stress. These findings highlight the need for further investigation into the long-term safety and potential organ-specific effects of *M. angulata* to ensure safe therapeutic applications (Adokoh et al., 2019; Kuspradini et al., 2020.

17.5 *MORINDA CITRIFOLIA* L.

17.5.1 Taxonomy, Biology, and Distribution

Morinda citrifolia L., widely known as noni, is classified under the Kingdom Plantae, Phylum Streptophyta, Class Equisetopsida, Subclass Magnoliidae, Order Gentianales, Family Rubiaceae, and the genus Morinda. This species has 16 recognized synonyms, including the homotypic synonym *Samama citrifolia* (L.) Kuntze, as well as heterotypic synonyms such as *Morinda citrifolia* f. *potteri* (O.Deg.) H.St.John, *M. citrifolia* var. *potteri* O.Deg., *M. ligulata* Blanco, *M. litoralis* Blanco, *M. macrophylla* Desf., *M. mudia* Buch.-Ham., *M. multiflora* Roxb., *M. nodosa* Buch.-Ham., *M. quadrangularis* G.Don, *M. teysmanniana* Miq., *M. tinctoria* Noronha, *M. tinctoria* var. *multiflora* (Roxb.) Hook.f., *M. zollingeriana* Miq., *Platanocephalus orientalis* Crantz, and *Sarcocephalus leichhardtii* F.Muell.(POWO, 2026).

M. citrifolia is a small evergreen tree most recognizable by its unique, composite, ovoid fruit, which transitions in color from green to yellowish-white as it matures. The fruit features a soft, translucent pulp and emits a strong, pungent aroma (Potterat & Hamburger, 2007). Inside, numerous small, hard, reddish-brown seeds are embedded within the flesh. Notably, *M. citrifolia* is capable of producing flowers and fruits throughout the year in favorable conditions (Almeida et al., 2019). An image of *M. citrifolia* is shown in Figure 17.1.

The native range of this species extends from tropical and subtropical Asia to northern Australia, where it typically grows as a shrub or tree within wet tropical environments (POWO, 2026). Today, it is extensively cultivated across tropical and subtropical regions, including the Pacific Islands, Puerto Rico, the Dominican Republic, and Brazil, especially in the Central-West and Northeast regions (Nparks, 2026).

17.5.2 Ethnopharmacology

The ethnomedicinal use of *M. citrifolia* has been widely practiced by numerous ethnic groups around the world as a component of traditional healing systems. The Tahitians are believed to have recognized the medicinal benefits of *M. citrifolia* as early as 400 A.D. In their traditional medicinal practices, the bark, leaves, and unripe (green) fruit of the plant are mainly used for the treatment of various local ailments (Caramel et al., 2015). Bark is frequently applied in cases of bacterial infections or to induce abortion. Leaves are commonly used to address bacterial infections, inflammation, abdominal pain, and diarrhea(Chander et al., 2014). The green fruit is used for a wide range of purposes, including relieving bad breath, treating bacterial and fungal infections, and alleviating menstrual cramps, arthritis, gastric ulcers, mouth ulcers, toothache, and indigestion. Meanwhile, ripe fruit is sometimes used for combatting parasitic and bacterial infections, as a means to induce menstruation (emmenagogue), or for intestinal cleansing (de Farias et al., 2020). In Indonesia, noni is also widely used as a folk remedy for ailments like diabetes (Rahmawati, 2020). Traditionally, the plant is prepared as a decoction, made into juice, or chewed fresh according to customary practice. Additionally, there exists a long-established herbal formula called Pikad Tri-phol-sa-mut-than (also known as Phikud Tri-Phon or PTP) that combines noni, coriander, and dried mukul fruits in equal proportions (1:1:1 w/w) (Kongpuckdee et al., 2020). This ancient recipe is used as a tonic to treat gastrointestinal disturbances, fever, and trichotomies, and is also relied upon for its anti-emetic and laxative effects. In Asia, fruits, roots, barks, and leaves are used for the treatment of diabetes, arthritis, and high blood pressure (Souza et al., 2018). In Malaysia, roots, barks, fruits, flowers, and stems are widely used in the treatment of diabetes, high blood pressure, and many other illnesses. The fruit juice is highly effective in preventing diabetes, hypertension, and arteriosclerosis(Abdul Majid et al., 2020). In South-East Asia, fruits are used to treat diabetes, swollen spleen, liver diseases, and cough. In India, leaves, roots, and seeds are curative agents for diarrhea, fevers, urinary tract infections, constipation, ulcers, nausea, and asthma (Upadhyay et al., 2025). In the Benin Republic, different parts of the plant are used in the treatment of tuberculosis, cough, and skin infections. In Africa and Asia, the plant is used to treat blood pressure, anxiety, diabetes, colds, flu, and depression (Sina et al., 2021)

17.5.3 Phytochemistry

The chemical composition of medicinal plants is fundamental to their pharmacological activities. Research on the phytochemistry of *M. citrifolia* has primarily focused on its roots, leaves, and fruits, where the key compounds include anthraquinones, iridoids, flavonoids, and coumarins. Among these, the main constituents of *M. citrifolia* are damnacanthal, scopoletin, rutin, ursolic acid, and asperuloside (Pawlus & Kinghorn, 2007; Potterat & Hamburger, 2007; Wang et al., 2002). The compounds extracted from the roots predominantly consist of anthraquinones (Lv et al., 2011), such as damnacanthal and sterols (Ee et al., 2009). The leaves of *M. citrifolia* are rich in various iridoids, flavonoids, and triterpenoids (Deng et al., 2008; Sang et al., 2003; Siddiqui et al., 2003; Takashima et al., 2007).

Given that noni juice is widely used as a dietary supplement, extensive analysis has been conducted on its fruit composition, including phenolic compounds, polysaccharides, coumarins, fatty acids, cyclic enol ether terpenes, flavonoids, carotenoids, and essential oils (Bui et al., 2006; Samoylenko et al., 2006; Wang et al., 2021). Furthermore, the presence of vitamins, amino acids, and trace minerals has also been documented. This diverse chemical profile underpins the biological activities attributed to *M. citrifolia*, supporting its traditional and experimental applications in health and medicine (Almeida et al., 2019).

Iridoids are an intriguing class of compounds found across a wide variety of plants and some insects, including families such as Aspidistra, Gentianae, Lamiaceae, Lauraceae, Rubiaceae, Sedum, and Verbenaceae, as well as in butterflies. Chemically, iridoids belong to the monoterpenoid subclass and are acetal derivatives of iridodial. They share a common structural feature: a cyclopentadiene-fused pyran ring. Iridoids are categorized into four main types: iridoid glycosides, secoiridoid glycosides, bis-iridoids, and nonglycosidic iridoids. These compounds have been recognized for a broad spectrum of biological activities, including anti-inflammatory, antibacterial, antiviral, neuroprotective, hepatoprotective, hypoglycemic, antitumor, and antioxidant effects (Wang, Wang, Wu, et al., 2022; Wang, Wang, Zhang, et al., 2022)

Anthraquinones are a group of chemical compounds characterized by a structure in which three benzene rings are fused together in a specific arrangement. These compounds are widely present in nature, occurring in various plants such as those from the Polygonum and Rubiaceae families, especially in the rhizomes and bark, as well as in certain fungi. In *M. citrifolia*, anthraquinones represent one of the main bioactive groups, predominantly found in the roots (García-Vilas et al., 2017). The biological activities of anthraquinones are influenced by the presence of hydroxyl groups on their rings, which are linked to a variety of effects, including anti-cancer, antimicrobial, antioxidant, and anti-osteoporosis activities. The antimicrobial potency of these compounds generally increases with the polarity of the substituent groups attached to the anthraquinone ring. However, the attachment of sugar chains to anthraquinones tends to decrease their antioxidant effectiveness (Li et al., 2025b). Among the anthraquinones present in *M. citrifolia*, damnacanthal is the most noteworthy due to its ability to target multiple tyrosine kinases, exhibiting promising anticancer properties. Other anthraquinones isolated from *M. citrifolia* include alizarin, morindadiol, nordamnacanthal, rubiadin, ibericin, tectoquinone, and lucidin, as well as several methyl and ether derivatives of these compounds (Garcia-Vilas et al., 2015).

Coumarins are compounds characterized by a distinctive bicyclic structure composed of a benzene ring fused with a pyran ring. Variations in the substituent groups attached to this bicyclic framework contribute to the wide-ranging biological activities exhibited by coumarins, including antioxidants, antibacterial, and anti-inflammatory effects. These compounds are frequently found in plants belonging to the Leguminosae, Rosaceae, and Cuscuta families (Wu et al., 2015). For example, scopoletin is the predominant coumarin found in *M. citrifolia*, present at a concentration of about 6 mg per gram in the aqueous extract of its leaves (Mohamad Shalan et al., 2017).

Flavonoids are specialized secondary metabolites found in plants, consisting of small natural molecules that exhibit a wide range of pharmacological activities such as anticancer, antioxidant, anti-inflammatory effects, and the ability to decrease vascular fragility. Structurally, flavonoids are composed of two aromatic rings linked by a three-carbon bridge, forming a diphenylpropane backbone with phenolic or polyphenolic groups positioned variably. The evaluation of noni extract quality includes measuring its total flavonoid content (Singh & Sharma, 2020), commonly using rutin as a reference standard in assays (Chen et al., 2024b). Interestingly, the flavonoid content in noni extract shows slight variations correlated with the fruit's ripeness, generally increasing as the fruit matures. Research conducted by Zhang et al. (2020) and Chen et al. (2024b) found that the total flavonoid content in noni juice was about 234.42 mg/L, while in noni fruits, it ranged between 12.56 and 14.48 g/kg.

Polysaccharides are large biomolecules composed of over 10 monosaccharide units linked by glycosidic bonds. They play vital roles in nature and exhibit various biological activities, including immune system modulation, antioxidant effects, anti-inflammatory, anti-cancer, and blood sugar–lowering properties. Several studies on different noni extracts have demonstrated notable anti-inflammatory and immunomodulatory effects attributed to their polysaccharide components. Typically, noni polysaccharides are extracted by processing the juice or puree with distilled water, followed by centrifugation and precipitation with alcohol (Li et al., 2020; Sousa et al., 2018; Yang et al., 2020; Zou et al., 2019). During fruit maturation, the crude polysaccharide content in noni decreases significantly—from 1.72 g per 100 g in unripe fruit to 1.07 g per 100 g in ripe fruit. This reduction is thought to result from the enzymatic breakdown of cell wall polysaccharides as the fruit softens (Batista et al., 2020).

Minerals make up about 8.4% of the dry weight of noni fruit, and their levels can vary depending on the fruit's stage of ripeness. Research by West et al. found potassium to be the most abundant mineral in noni puree, with a concentration of roughly 214.34 mg per 100 grams. Other minerals identified in noni fruit include calcium, iron, sodium, and selenium. Regarding vitamins, noni fruit is particularly rich in vitamin C, with content ranging from 24 to 158 mg per 100 grams of dry matter. The vitamin C concentration in noni puree is approximately 1.13 mg per gram (West et al., 2018; West et al., 2007). For a summary of the phytochemicals in *M. citrifolia* and their associated benefits, please refer to Table 17.1.

17.5.4 Pharmacology

17.5.4.1 Anticancer

A growing body of evidence highlights the potential of *M. citrifolia* (noni) as a natural anticancer agent. In vitro studies have demonstrated that noni juice exerts significant inhibitory effects on A549 lung cancer cells by suppressing cell proliferation, migration, and invasion, while simultaneously promoting apoptosis. Mechanistically, these effects are mediated through the downregulation of AKT, p50, and STAT3 phosphorylation. Complementary in vivo experiments further confirmed its efficacy, showing reduced tumor growth, decreased expression of proliferation markers such as Ki67 and PCNA, as well as the antiapoptotic Bcl-2, and increased activation of caspase-3.

TABLE 17.1
Phytochemicals in *M. citrifolia* Fruits

Phytochemical Group	Types of Compounds	Main Benefits	Key References
Iridoids	Iridoid glycosides, secoiridoid glycosides, bis-iridoids, nonglycosidic iridoids	Anti-inflammatory, antibacterial, antiviral, neuroprotective, hepatoprotective, hypoglycemic, antitumor, antioxidant	Wang, Wang, Wu, et al. (2022); Wang, Wang, Zhang, et al. (2022)
Anthraquinones	Damnacanthal, alizarin, morindadiol, nordamnacanthal, rubiadin, ibericin, tectoquinone, lucidin, derivatives	Anticancer, antimicrobial, antioxidant, anti-osteoporosis	Garcia-Vilas et al. (2015); García-Vilas et al. (2017)
Coumarins	Scopoletin	Antioxidant, antibacterial, anti-inflammatory	Mohamad Shalan et al. (2017); Wu et al. (2015)
Flavonoids	Kaempferol, rutin, quercetin, narcissoside, acacetin-7-*O*-β-D-glucopyranoside, quercetin-3-*O*-β-D-glucopyranoside, others	Antioxidant, anticancer, anti-inflammatory, reduces vascular fragility	Singh and Sharma, (2020); Chen et al. (2024b); Zhang et al. (2020); Chen et al. (2024b)
Polysaccharides	Complex polysaccharides	Immunomodulation, anti-inflammatory, antitumor, antioxidant, hypoglycemic	Li et al. (2020); Sousa et al. (2018); Yang et al. (2020); Zou et al. (2019); Batista et al. (2020)
Minerals and Vitamins	Potassium, calcium, iron, sodium, selenium, vitamin C	Support cardiovascular health, immune system, antioxidant	West et al. (2018); West et al. (2007)

Collectively, these findings indicate that noni juice may exert its antitumor effects through modulation of the AKT/NF-κB signaling pathway (Ma et al., 2021).

Beyond the whole juice, specific phytochemicals from *M. citrifolia* roots have also shown remarkable anticancer activities. Damnacanthal, an anthraquinone derivative, has been particularly well studied in colorectal cancer (CRC). This compound demonstrated superior growth-inhibitory effects compared to the standard chemotherapeutic agent 5-fluorouracil (5-FU), primarily through the downregulation of cyclin D1 expression. In animal models, damnacanthal was able to suppress tumor growth by two- to three-fold while exhibiting low oral toxicity, with an LD_{50} of 2500 mg/kg. These results highlight damnacanthal as a promising candidate for further development in CRC therapy (Woradulayapinij et al., 2022). CRC, one of the most prevalent malignancies worldwide, is frequently associated with mutations in tumor suppressor and oncogenic pathways, particularly TP53 and KRAS. This has spurred investigations into phytochemicals capable of modulating these molecular targets. In this context, two anthraquinones from *M. citrifolia*, damnacanthal and morindone, were evaluated against HCT116 and HT29 CRC cell lines. Both compounds exhibited potent dose- and time-dependent antiproliferative activity, induced G1 phase cell cycle arrest, and triggered apoptosis. Importantly, morindone was found to downregulate mutant TP53 and KRAS gene expression. This effect was further enhanced by siRNA-mediated knockdown, which sensitized cells to anthraquinone treatment and significantly reduced IC_{50} values. Although increased protein expression of TP53 and KRAS was observed, the overall outcome favored apoptosis induction and cell cycle inhibition. Taken together, these results underscore morindone's potential as a multitargeted therapeutic candidate for CRC (Chee et al., 2024).

In addition to its effects on colorectal and lung cancers, noni juice has also been studied in the context of breast cancer, both as a standalone treatment and in combination with conventional chemotherapy.

When tested alongside cyclophosphamide (CYL), noni demonstrated synergistic antitumor activity in both in vitro and in vivo models. While CYL alone induced notable myelotoxicity and hepatotoxicity, evidenced by alterations in hematological and biochemical parameters, coadministration with noni not only enhanced the antitumor efficacy but also ameliorated the associated toxicities. These findings suggest that noni juice may serve as an effective adjuvant therapy by potentiating the anticancer effects of chemotherapeutic agents while mitigating their adverse effects (Ali et al., 2024).

Overall, cancer continues to be one of the leading causes of morbidity and mortality worldwide, creating an urgent need for novel therapeutic strategies. Medicinal plants such as *M. citrifolia* have attracted increasing scientific attention due to their broad pharmacological activities. A systematic review encompassing 51 studies 41 focused on efficacy and 10 on safety—concluded that noni exhibits a wide spectrum of anticancer mechanisms, including antitumor, antiproliferative, pro-apoptotic, antiangiogenic, antimigratory, anti-inflammatory, and immunomodulatory properties. Although these preclinical and clinical findings provide strong support for the therapeutic potential of noni, more rigorous and large-scale studies are required to fully establish its efficacy and safety profile in cancer management (Chanthira Kumar et al., 2022).

17.5.4.2 Antidiabetic

M. citrifolia has emerged as a promising candidate for antidiabetic therapy. Previous *in vivo* studies demonstrated that its extract inhibited α-amylase activity and lowered blood glucose levels in streptozotocin-induced diabetic mice. Molecular docking analyses further confirmed that ursolic acid and sterol constituents interacted with key residues at the active sites of α-amylase and α-glucosidase. In this study, molecular dynamics simulations (100 ns, GROMACS 2016.3, AMBER99SB-ILDN force field, 310 K) were employed to assess the stability of stigmasterol—isolated for the first time from *M. citrifolia* fruit—and β-sitosterol in complex with both enzymes. The results indicated that stigmasterol formed more stable complexes with α-amylase and α-glucosidase compared to acarbose, a standard inhibitor, while β-sitosterol exhibited superior stability with α-glucosidase. Binding energy analysis further demonstrated stronger affinities of stigmasterol and β-sitosterol toward both enzymes relative to acarbose. Collectively, these findings suggest that stigmasterol and β-sitosterol from *M. citrifolia* fruit possess significant potential as natural α-amylase and α-glucosidase inhibitors for the development of novel antidiabetic agents (Lolok et al., 2024).

In another study, the polysaccharide NJSPd-1 was successfully isolated and characterized from fermented noni (Morinda citrifolia) fruit juice. NJSPd-1 has a molecular weight of 18,545 Da and is composed of galacturonic acid, galactose, rhamnose, and arabinose in specific molar ratios. Structural analysis revealed the presence of distinct glycosidic linkages, such as →4)-α-D-GalAp-(1→ and →3)-α-L-Araf-(1→. In vitro assays on HepG2 cells exposed to high-glucose conditions demonstrated that NJSPd-1 could attenuate oxidative stress by modulating the expression of Nrf2, Keap1, HO-1, and NQO1 genes and proteins. These findings indicate that NJSPd-1 exerts protective effects against oxidative stress through activation of the Nrf2/HO-1/NQO1 signaling pathway, suggesting its potential as a therapeutic candidate for type 2 diabetes mellitus (Zhang et al., 2025).

17.5.4.3 Anti-Inflammatory

Inflammation plays a key role in the onset and progression of various chronic and severe diseases. Evidence indicates that *M. citrifolia* exerts anti-inflammatory effects through several mechanisms involving bioactive constituents, such as phenolic compounds, polysaccharides, and specific proteins. These effects have been confirmed through diverse experimental models, including those mimicking LPS-induced inflammation, inflammatory bowel disease, esophagitis, rheumatoid arthritis, pneumonia, edema, and steatohepatitis in animals (Tanikawa et al., 2021).

In cell culture studies, noni seed extract was shown to suppress tumor necrosis factor-α (TNF-α) and inducible nitric oxide synthase (iNOS) expression in RAW 264.7 cells stimulated with LPS.

In addition, a Thai herbal formulation known as PTP, which contains *M. citrifolia*, exhibited strong anti-inflammatory effects in rat models of ear and paw edema induced by ethyl phenylpropionate, carrageenan, or arachidonic acid. Fermented noni juice was also reported to downregulate NLRP3 and TNF-α expression while alleviating ankle swelling caused by monosodium urate, supporting its potential benefit for acute gouty arthritis (Tan et al., 2022).

Phenolic extracts from ripe *M. citrifolia* fruits were found to decrease intestinal inflammatory markers in animals fed a high-fat diet, consistent with the observations made in Dyah Aninta Kustiarini's study. Moreover, polysaccharides isolated from the plant inhibited activation of the NF-κB pathway and helped modulate gut microbiota, protecting intestinal tissue and improving inflammatory bowel disease symptoms (Kustiarini et al., 2019).

Deacetylasperulosidic acid (DAA), a major cyclic terpene compound present in *M. citrifolia* fruit, was shown to regulate mitogen-activated protein kinase (MAPK) phosphorylation in human HaCaT cells exposed to TNF-α and interferon-γ. This suggests DAA's potential as a therapeutic agent for atopic dermatitis by restoring the Th1/Th2 immune balance and strengthening the skin barrier (Kim & Jeong, 2014).

The noni stem and seeds, usually considered nonmedicinal parts, also demonstrate anti-inflammatory activities. The stem bark inhibits NO production, and the seeds contain a heat-stable lipid transfer protein, McLTP1, that alleviates inflammation and mitigates the side effects of irinotecan-induced intestinal mucositis (do Carmo et al., 2022).

17.5.4.4 Hypoglycemic Activity

For centuries, *noni* has been traditionally utilized in Indonesia and Polynesia as a natural remedy for diabetes. To scientifically validate this traditional use, diabetic models are generally established through a combination of a high-fat and high-sugar diet, administration of streptozotocin (50 mg/kg body weight), and tetracycline treatment. Findings from experimental studies indicate that the bioactive constituents of *M. citrifolia* exhibit therapeutic potential against type 2 diabetes by inhibiting the activities of α-amylase and α-glucosidase while exerting insulin-like effects (Lolok et al., 2022).

Human pancreatic α-amylase serves as a critical target in the management of type 2 diabetes mellitus. Through molecular docking and molecular dynamics simulation analyses, ursolic acid demonstrated optimal binding energy with α-amylase, suggesting its potential as a promising therapeutic compound for diabetes treatment (Lolok et al., 2024). In a separate investigation, Xu identified 15 compounds in *M. citrifolia* seeds that showed inhibitory effects on α-glucosidase activity (Xu et al., 2022).

Another study evaluated the combined extract of noni, mango, and pineapple for α-glucosidase and α-amylase inhibitory activities using acarbose as a reference compound. The results revealed that the mixture exhibited remarkable antidiabetic activity in vitro, which was further supported by in vivo validation using streptozotocin-induced diabetic rat models, confirming its potential as a natural agent for diabetes management (Vejayan et al., 2022).

17.5.4.5 Antimicrobial Activity

Antimicrobial activity is commonly evaluated using the disk diffusion method. In one study, dried *M. citrifolia* root powder was sequentially extracted with hexane, dichloromethane, ethyl acetate, and ethanol. The results showed that the hexane extract effectively inhibited *S. aureus*, *S. epidermidis*, and *Bacillus cereus*. The dichloromethane extract demonstrated inhibitory activity against *E. coli*, producing a 5.5 mm inhibition zone, while the ethyl acetate extract was active against *P. aeruginosa* and *S. epidermidis* (Bhavsar et al., 2025).

Additionally, the antitubercular potential of *M. citrifolia* was assessed using the microplate Alamar Blue assay, where the ethanol extract of the leaves achieved 89% inhibition of *Mycobacterium tuberculosis* at a concentration of 100 μg/mL. In dental applications, water extracts of *M. citrifolia* proved more effective in disinfecting metal–ceramic crowns than both conventional disinfectants and photodynamic therapy. Furthermore, cerium oxide nanoparticles derived from noni displayed notable bacteriostatic activity against both Gram-positive and Gram-negative bacteria, surpassing the antimicrobial effectiveness of amoxicillin (Elderdery et al., 2022).

17.5.4.6 Antioxidant

ROS are naturally generated during mitochondrial energy production and play a critical role in immune defense through phagocytosis. The body's antioxidant defense system works to maintain low basal ROS levels by scavenging these reactive molecules and converting them into harmless substances via various mechanisms. When ROS accumulates excessively, oxidative stress occurs, contributing to inflammation and tissue damage, which are key factors in the progression of many diseases and injuries. To evaluate antioxidant potential, in vitro assays such as DPPH and ABTS radical scavenging tests and measurements of antioxidant enzyme levels (like superoxide dismutase [SOD] and glutathione peroxidase [GPx]) are commonly performed. In vivo, animal models with liver damage induced by a high-fat diet are used for evaluation.

In *M. citrifolia*, the primary antioxidant compounds include polyphenols, iridoids, scopoletin, and polysaccharides. Phenolic extracts from noni fruits notably boost antioxidant enzyme activities, such as glutathione (GSH) and catalase (CAT) (Wang, Wang, Wu, et al., 2022), thereby effectively reducing oxidative stress in mouse models of nonalcoholic fatty liver disease induced by high-fat diets. Separate research demonstrated antioxidant properties of silver nanoparticles synthesized from Gerbera leaves in vitro (Datkhile et al., 2024). A traditional Thai herbal formula named PTP, comprising *Aegle marmelos*, *M. citrifolia*, and *Coriandrum sativum*, exhibited a DPPH free radical scavenging IC_{50} value of 92.4 ± 6.6 μg/mL (Ngamdokmai et al., 2022). Additionally, noni wine was found to regulate the Nrf2 signaling pathway and enhance expression of antioxidant enzyme genes in mice fed a high-fat diet (Zhang et al., 2020).

The molecular weight of noni polysaccharides significantly influences their antioxidant effectiveness (Cai et al., 2022). Research has shown that noni polysaccharides exert strong antioxidant action by scavenging ROS (with DPPH radical inhibition rates exceeding 70%) and activating the Nrf2/Keap1 pathway (Yang et al., 2020). In high-fat diet animal models, these polysaccharides substantially raised hepatic SOD and CAT levels while decreasing malondialdehyde (Pointinger et al., 2008), indicating their dual capacity to mitigate oxidative stress and inflammation, partly through modulation of the gut microbiota. Traditional fermentation methods not only reduce noni's unpleasant odor but also enhance its immune-stimulating and antioxidant benefits. Studies have consistently reported that over 70% DPPH radical scavenging activity for noni polysaccharides and their derivatives in vitro, highlighting their potent antioxidant potential (Li et al., 2020).

One study found that ethanolic leaf extracts of *M. citrifolia* improved antioxidant enzyme levels in rats with liver damage. Furthermore, lignin isolated from noni leaves showed strong free radical scavenging activity, likely due to its phenolic constituents. This body of evidence supports the significant antioxidant role of various bioactive components in *M. citrifolia* (da Silva et al., 2021).

17.5.4.7 Gastroprotective Activity

The gastroprotective effects of polyphenols present in *M. citrifolia* fruit were explored in a study conducted by Chen. In this research, dried noni fruit powder was used to better control and standardize conditions that simulate human digestion while preserving sensitive bioactive compounds. The polyphenols underwent oral, gastric, and intestinal digestion phases before being fermented in the colon for up to 24 hours. During this fermentation, the release of polyphenols, their antioxidant activities, and changes in gut microbiota composition were monitored. Findings revealed that most phenolic compounds linked to dietary fibers were released during colon fermentation, which also positively influenced the intestinal microbiota by promoting the growth of beneficial bacterial genera, implying a strong prebiotic effect on gastrointestinal health (Chen et al., 2024a).

In another study, the therapeutic potential of *M. citrifolia* fruit extract was evaluated for treating acute pancreatitis induced by L-arginine in male rats. Rats treated with the extract at doses of 200 and 400 mg/kg for six days prior to disease induction showed a significant reduction in blood amylase and lipase levels, markers of pancreatic damage. Histological examination indicated reduced pancreatic tissue injury, while DNA analysis demonstrated the extract's ability to prevent DNA fragmentation, suggesting protection against cell death via apoptosis (Campos et al., 2017).

Further, Chong et al. (2019) investigated the properties of naturally fermented noni juice in rodent models of liver fibrosis induced by thioacetamide. Animals treated with various doses of fermented noni juice exhibited an increase in antioxidant capacity, which correlated with protection against liver fibrosis.

17.6 *TRAPA NATANS* L.

17.6.1 Taxonomy, Biology, and Distribution

Trapa natans L. is called water caltrop, water chestnut, or singhara. This species is taxonomically classified within the Kingdom Plantae, Phylum Streptophyta, Class Equisetopsida, Subclass Magnoliidae, Order Myrtales, Family Lythraceae, and Genus Trapa. There are eight accepted infra-specific varieties of *Trapa natans*: *T. natans* var. *africana*, *T. natans* var. *bispinosa*, *T. natans* var. *japonica*, *T. natans* var. *magnicorona*, *T. natans* var. *natans*, *T. natans* var. *pseudincisa*, *T. natans* var. *quadricaudata*, and *T. natans* var. *quadrispinosa* (POWO, 2026).

T. natans is a free-floating, annual aquatic plant that grows in shallow, slow-moving, or still freshwater bodies (Banu et al., 2023). It is characterized by a rosette of rhomboid or triangular floating leaves with inflated petioles, feathery underwater structures, and a distinctive hard, nut-like fruit with two to four sharp, horn-like spines that encloses a large starchy seed(Adkar et al., 2014).

The native range of *T. natans* is Eurasia, Europe, Asia, and Africa. It is a hydroannual plant that primarily grows in temperate biomes and thrives in shallow water bodies, swampy areas, or ponds. This species is now considered extinct in Spain and Sweden, but also as an invasive plant in some regions (POWO, 2026).

17.6.2 Ethnopharmacology

Trapa natans plays a significant role in traditional medicine, especially within Ayurvedic, Unani, and Chinese herbal systems, where its fruits, seeds, leaves, and pericarp are employed to treat various ailments (Banu et al., 2023; Matovic et al., 2025; Shalabh et al., 2012). People eat its fruits as a tonic for illness, addressing fatigue, and sexual debility when combined with honey and ghee. Leaves are used to treat burning sensations, diarrhea, indigestion, inflammation, bronchitis, skin diseases such as leprosy, and fractures, while the stem bark is used to treat sores and ulcers. In Thai and Southeast Asian traditions, pericarp decoctions ease sore throats, urethral discharge, bleeding, and irregular menstruation; the roots manage wounds and dysmenorrhea (Adkar et al., 2014; Mondal et al., 2024).

17.6.3 Phytochemistry

Phytochemical studies on *T. natans* revealed a broad spectrum of primary and secondary metabolites that contribute to its nutritional benefits and therapeutic properties. Bioactive components in fruit pericarps, stems, leaves, and seeds showed diverse pharmacological effects. For example, fruits and pericarp are rich in triterpenoids such as ursolic acid, olenolic acid, cycloeucalenol, and 2β,3α,23-trihydroxyurs-12-en-28-oic acid. Flavonoids (quercetin and kaempferol) and phenolic compounds (gallic acid, ellagic acid, chlorogenic acid, syringic acid, and ferulic acid) (Banu et al., 2023; Devi & Gupta, 2025). A methanolic leaf extract yielded several phenolic acids and their derivatives (gallic acid, protocatechuic acid, *p*-hydroxybenzoic acid, *p*-coumaric acid, ellagic acid, *p*-coumaroyl-digalloyl-*O*-glucose, and ferulic acid), flavonoids (quercetin, pinobanksin, naringenin, and rhamnetin), as well as flavonoid glycosides (naringenin-7-*O*-hexoside, kaempferol, rutin, kaempferol-3-*O*-glucoside (astragalin), quercetin-3-*O*-rhamnoside, and quercetin-3-*O*-galactoside) (Corovic et al., 2021; Devi & Gupta, 2025).

17.6.4 Pharmacology

T. natans or *T. bispinosa* showed significant pharmacological potential. Recent studies confirmed its antibiotic, antidiabetic, analgesic, and immunomodulatory activities, with the methanol extracts from the roots exhibiting dose-dependent nootropic enhancement, pain relief, neuroprotection, and antiulcer properties in rat models. These pharmacological effects correspond to the plant's abundant phytochemicals, including elevated levels of saponins (36.92%), alkaloids (0.775%), flavonoids, and phenols, which underpin its anticholinergic, spasmolytic, and antioxidant properties (Adkar et al., 2014). Research studying four *T. natans* L. leaf extracts assessed in vitro antimicrobial and antibiofilm activities via microdilution, tissue culture plate, colorimetric methods, GC-FID, and GC-MS showed the strongest antibacterial activity, especially against *P. aeruginosa* (MIC 313 μg/mL; superior to controls and matched antibiofilm potency). Effects were better on Gram-positive bacteria (MICs 78–625 μg/mL); ethanol and ethyl acetate extracts had BIC < MIC. The extracts potently inhibited *Aspergillus restrictus* (MICs 78–156 μg/mL). Ethyl acetate GC-MS revealed 22 compounds, led by (all-*E*)-squalene (20.2%), n-alkanes, and hinokiresinol (Radojevic et al., 2016). *T. natans* showed strong antioxidant activity by scavenging DPPH/ABTS radicals and elevating SOD, CAT, and GSH levels in oxidative stress models. It exerts anti-inflammatory activity by downregulating TNF-α and IL-6, blocking NF-κB, and promoting M2 macrophage polarization, particularly in cisplatin-induced cardiotoxicity. Antidiabetic properties reduced blood glucose levels in STZ-induced diabetic rats by enhancing GLUT4 expression, preserving β-cell function, and improving insulin sensitivity. Hepatoprotection actions counter drug-induced damage by reducing ROS. Antimicrobial efficacy inhibits Gram-positive/-negative bacteria and fungi (MIC 50–200 mg/mL). Additional properties include nanoparticle-enhanced anticancer apoptosis, analgesia, antiulcer effects, immunomodulation, and nootropic activity. Clinical potential remains underexplored, warranting human trials (Majee et al., 2021).

17.7 CONCLUSION AND FUTURE REMARKS

Fruits and berries are an essential part of the diet. The lesser-known medicinal fruits and berries *Aglaia edulis, Ardisia sanguinolenta, Catunaregam spinosa, Merope angulata, Morinda citrifolia,* and *Trapa natans* have been used to treat several infectious and noninfectious ailments, although they are not economically important in the global market. Phytochemical evaluation revealed that berries contain several compound classes, including flavonoids, saponins, tannins, phenolics, and alkaloids, which are responsible for a range of activities, including antimicrobial, antiviral, antifungal, anti-inflammatory, antidiabetic, and antihyperlipidemic.

Nevertheless, further studies evaluating their efficacy in animal models are necessary, as well as studies on their general and specific toxicity to warrant their long-term health benefits. In addition, an agronomic approach is necessary to ensure the supply of these superfood berries to consumers worldwide.

REFERENCES

Abdul Majid, N., Abdul Hamid, A., Salleh, S. Z., Saari, N., Abas, F., Pak Dek, M. S., Ramli, N. S., & Jaafar, A. H. (2020). Metabolomics approach to investigate the ergogenic effect of *Morinda citrifolia* L. leaf extract on obese Sprague Dawley rats. *Phytochemical Analysis, 31*(2), 191–203. https://doi.org/10.1002/pca.2880

Abdulrahman, M. D., Zakariya, A. M., Hama, H. A., Hamad, S. W., Al-Rawi, S. S., Bradosty, S. W., & Ibrahim, A. H. (2022). Ethnopharmacology, biological evaluation, and chemical composition of *Ziziphus spina-christi* (L.) Desf.: A review. *Advances in Pharmacological and Pharmaceutical Sciences, 2022*, 4495688. https://doi.org/10.1155/2022/4495688

Adkar, P., Dongare, A., Ambavade, S., & Bhaskar, V. (2014). *Trapa bispinosa* Roxb.: A review on nutritional and pharmacological aspects. *Advances in Pharmacological and Pharmaceutical Sciences, 2014*, 959830.

Adokoh, C. K., Asante, D. B., Acheampong, D. O., Kotsuchibashi, Y., Armah, F. A., Sirikyi, I. H., Kimura, K., Gmakame, E., & Abdul-Rauf, S. (2019). Chemical profile and in vivo toxicity evaluation of unripe *Citrus aurantifolia* essential oil. *Toxicology Reports*, *6*, 692–702. https://doi.org/10.1016/j.toxrep.2019.06.020

Aguilera, J. M. (2024). Berries as foods: Processing, products, and health implications. *Annual Review of Food Science and Technology*, *15*, 1–26. https://doi.org/10.1146/annurev-food-072023-034248

Ajaib, M., Ishtiaq, S., Ishtiaq, M., Maqbool, M., Bhatti, K. H., Khan, A., Afreen, A., Hussain, T., Sardar, T., Gul, A., & Azeem, M. (2022). Analysis of antidiabetic, antiulcer, and analgesic potential of traditional ethnomedicinal plant *Emex spinosa* (L.) Campd. from Azad Jammu and Kashmir. *PLoS One*, *17*(10), e0274706. https://doi.org/10.1371/journal.pone.0274706

Al-Anazi, K. M., Al-Mareed, A. A., Farah, M. A., Ali, M. A., Hailan, W. A. Q., & Al-Hemaid, F. M. (2021). Protective effect of *Capparis spinosa* extract against potassium bromate-induced oxidative stress and genotoxicity in mice. *Evidence-Based Complementary and Alternative Medicine*, *2021*, 8875238. https://doi.org/10.1155/2021/8875238

Ali, M., Manjula, S. N., Mohiuddin, I., Mruthunjaya, K., Shakeel, F., Mir, S. A., & Wani, S. U. D. (2024). Noni enhances the anticancer activity of cyclophosphamide and suppresses myelotoxicity and hepatotoxicity in tumor-bearing mice. *Journal of Cancer Research and Clinical Oncology*, *150*(4), 212. https://doi.org/10.1007/s00432-024-05734-1

Almeida, É. S., de Oliveira, D., & Hotza, D. (2019). Properties and applications of *Morinda citrifolia* (noni): A review. *Comprehensive Reviews in Food Science and Food Safety*, *18*(4), 883–909. https://doi.org/10.1111/1541-4337.12456

Alqahtani, A. S., Nasr, F. A., Ahmed, M. Z., Mansour, M. Y. B., Biksmawi, A. A., Noman, O. M., Herqash, R. N., Al-Zahrani, M., Qurtam, A. A., & Rudayni, H. A. (2023). In vitro protective and anti-inflammatory effects of *Capparis spinosa* and its flavonoids profile. *Open Chemistry*, *21*(1). https://doi.org/10.1515/chem-2023-0186

Alsharif, B., & Boylan, F. (2025). *Capparis* L. (Capparaceae): A scoping review of phytochemistry, ethnopharmacology, and pharmacological activities. *Molecules*, *30*(18), 3705. https://doi.org/10.3390/molecules30183705

Annaz, H., Sane, Y., Bitchagno, G. T. M., Ben Bakrim, W., Drissi, B., Mahdi, I., El Bouhssini, M., & Sobeh, M. (2022). Caper (*Capparis spinosa* L.): An updated review on its phytochemistry, nutritional value, traditional uses, and therapeutic potential. *Frontiers in Pharmacology*, *13*. https://doi.org/10.3389/fphar.2022.878749

Awang, K., Loong, X.-M., Leong, K. H., Supratman, U., Litaudon, M., Mukhtar, M. R., & Mohamad, K. (2012). Triterpenes and steroids from the leaves of *Aglaia exima* (Meliaceae). *Fitoterapia*, *83*(8), 1391–1395. https://doi.org/10.1016/j.fitote.2012.10.004

Bacchetti, T., Campagna, R., Sartini, D., Cecati, M., Morresi, C., Bellachioma, L., Martinelli, E., Rocchetti, G., Lucini, L., Ferretti, G., & Emanuelli, M. (2022). *C. spinosa* L. subsp. *rupestris* phytochemical profile and effect on oxidative stress in normal and cancer cells. *Molecules*, *27*(19), 6488. https://doi.org/10.3390/molecules27196488

Bacher, M., Hofer, O., Brader, G., Vajrodaya, S., & Greger, H. (1999). Thapsakins: Possible biogenetic intermediates towards insecticidal cyclopenta[b]benzofurans from *Aglaia edulis*. *Phytochemistry*, *52*(2), 253–263.

Banu, Z. W., Dasgupta, D., Hazarika, I., Laloo, D., & Kalita, J. M. (2023). *Trapa natans* L.: A journey from traditional to contemporary therapies—A review. *The Natural Products Journal*, *13*(8), 122–134.

Basak, S. K., Ray, K., Giri, C. K., Saha, S., Kotal, H. N., Mandal, A., Chowdhury, R., & Begam, M. M. (2022). Effect of degradation stressors on loss of original habitat and biodiversity of "rare and threatened" mangrove species and its associates in fringe mangroves of Indian Sundarbans. *Bilateral International Conference on Ecotoxicology & Environmental Sciences (ICEES)*, Khulna University, Bangladesh.

Basini, J., Swetha, D., & Mallikarjuna, G. (2019). Anti-hyperglycaemic and antioxidant activity of *Catunaregam spinosa* (Thunb.) against dexamethasone-induced diabetes in rats. *International Journal of Pharmacy and Pharmaceutical Sciences*, 56–61. https://doi.org/10.22159/ijpps.2019v11i6.32563

Batista, J. A., Magalhães, D. A., Sousa, S. G., Ferreira, J. D. S., Pereira, C. M. C., Lima, J., de Albuquerque, I. F., Bezerra, N., de Brito, T. V., Monteiro, C., Franco, A. X., Di Lenardo, D., Oliveira, L. A., Feitosa, J. P. A., de Paula, R. C. M., Barros, F. C. N., Freitas, A. L. P., de Oliveira, J. S., Vasconcelos, D. F. P., … Barbosa, A. (2020). Polysaccharides derived from *Morinda citrifolia* Linn reduce inflammatory markers during experimental colitis. *Journal of Ethnopharmacology*, *248*, 112303. https://doi.org/10.1016/j.jep.2019.112303

Bhavsar, B. A., Jain, V., Kumari, A., Shukla, P., Pandya, R., Kumar, M. R., & Singh, S. (2025). Evaluate the antimicrobial properties of double antibiotic paste (DAP) and *Morinda citrifolia* and propolis paste

(MCP) used in regenerative endodontics: An in vitro study. *Journal of Pharmacy & Bioallied Sciences*, *17*(Suppl 1), S703–S705. https://doi.org/10.4103/jpbs.jpbs_1988_24

Brader, G., Vajrodaya, S., Greger, H., Bacher, M., Kalchhauser, H., & Hofer, O. (1998). Bisamides, lignans, triterpenes, and insecticidal cyclopenta[b]benzofurans from *Aglaia* species. *Journal of Natural Products*, *61*(12), 1482–1490.

Buenavista, D., & Purnobasuki, H. (2023). People and mangroves: Biocultural utilization of mangrove forest ecosystem in Southeast Asia. *Journal of Marine and Island Cultures*, *12*(2). https://doi.org/10.21463/jmic.2023.12.2.07

Bui, A. K., Bacic, A., & Pettolino, F. (2006). Polysaccharide composition of the fruit juice of *Morinda citrifolia* (noni). *Phytochemistry*, *67*(12), 1271–1275. https://doi.org/10.1016/j.phytochem.2006.04.023

Cai, J., Liang, Z., Li, J., Manzoor, M. F., Liu, H., Han, Z., & Zeng, X. (2022). Variation in physicochemical properties and bioactivities of *Morinda citrifolia* L. (noni) polysaccharides at different stages of maturity. *Frontiers in Nutrition*, *9*, 1094906. https://doi.org/10.3389/fnut.2022.1094906

Campos, D. C. O., Costa, A. S., Luz, P. B., Soares, P. M. G., Alencar, N. M. N., & Oliveira, H. D. (2017). *Morinda citrifolia* lipid transfer protein 1 exhibits anti-inflammatory activity by modulation of pro- and anti-inflammatory cytokines. *International Journal of Biological Macromolecules*, *103*, 1121–1129. https://doi.org/10.1016/j.ijbiomac.2017.05.148

Caramel, S., Marchionni, M., & Stagnaro, S. (2015). *Morinda citrifolia* plays a central role in the primary prevention of mitochondrial-dependent degenerative disorders. *Asian Pacific Journal of Cancer Prevention*, *16*(4), 1675.

Chander, M. P., Kartick, C., Gangadhar, J., & Vijayachari, P. (2014). Ethnomedicine and healthcare practices among Nicobarese of Car Nicobar—An indigenous tribe of Andaman and Nicobar Islands. *Journal of Ethnopharmacology*, *158*(Pt A), 18–24. https://doi.org/10.1016/j.jep.2014.09.046

Chanthira Kumar, H., Lim, X. Y., Mohkiar, F. H., Suhaimi, S. N., Mohammad Shafie, N., & Chin Tan, T. Y. (2022). Efficacy and safety of *Morinda citrifolia* L. (noni) as a potential anticancer agent. *Integrative Cancer Therapies*, *21*, 15347354221132848. https://doi.org/10.1177/15347354221132848

Chedraoui, S., Abi-Rizk, A., El-Beyrouthy, M., Chalak, L., Ouaini, N., & Rajjou, L. (2017). *Capparis spinosa* L.: A systematic review—A xerophilous species of multi values and promising potentialities for agrosystems under the threat of global warming. *Frontiers in Plant Science*, *8*, 1845. https://doi.org/10.3389/fpls.2017.01845

Chee, C. W., Mohd Hashim, N., & Nor Rashid, N. (2024). Morindone as a potential therapeutic compound targeting TP53 and KRAS mutations in colorectal cancer cells. *Chemico-Biological Interactions*, *392*, 110928. https://doi.org/10.1016/j.cbi.2024.110928

Chen, J., Chen, X., Zhang, Y., Feng, Z., Zhu, K., Xu, F., & Gu, C. (2024a). Bioactivity and influence on colonic microbiota of polyphenols from noni (*Morinda citrifolia* L.) fruit under simulated gastrointestinal digestion. *Food Chemistry X*, *21*, 101076. https://doi.org/10.1016/j.fochx.2023.101076

Chen, X., Gu, C., Zhu, K., Xu, F., Feng, Z., & Zhang, Y. (2024b). Insight into the effects of different ripeness levels on the quality and flavor chemistry of noni fruit (*Morinda citrifolia* L.). *Food Chemistry*, *434*, 137408. https://doi.org/10.1016/j.foodchem.2023.137408

Chen, X., Li, Z., Yang, X., Li, Y., Jin, J., Liu, J., Hao, X., Yuan, C., & Yi, P. (2025). Steroids and terpenoids from *Aglaia edulis* and their anti-inflammatory activities. *Journal of Molecular Structure*, *1328*, 141195.

Chong, C. L. G., Hussan, F., & Othman, F. (2019). Hepatoprotective effects of *Morinda citrifolia* leaf extract on ovariectomized rats fed with thermoxidized palm oil diet: Evidence at histological and ultrastructural level. *Oxidative Medicine and Cellular Longevity*, *2019*, 9714302. https://doi.org/10.1155/2019/9714302

Corovic, R. C., Bradic, J., Tomovic, M., Dabanovic, V., Jakovljevic, V., Zarkovic, G., & Rogac, Z. (2021). Chemical composition and biological activity of *Trapa natans* L. *Serbian Journal of Experimental and Clinical Research*, *2*, 123–134.

Crous, P. W., Osieck, E. R., Jurjević, Ž., Boers, J., Iperen, A. L., Starink-Willemse, M., Dima, B., Balashov, S., Bulgakov, T. S., Johnston, P. R., Morozova, O. V., Pinruan, U., Sommai, S., Alvarado, P., Decock, C. A., Lebel, T., McMullan-Fisher, S., Moreno, G., Shivas, R. G., … Groenewald, J. Z. (2021). Fungal planet description sheets: 1284–1382. *Persoonia*, *47*, 178–374. https://doi.org/10.3767/persoonia.2021.47.06

da Silva, P. R., do Carmo Alves de Lima, M., Souza, T. P., Sandes, J. M., da Conceição Alves de Lima, A., Neto, P. J. R., Dos Santos, F. A. B., Alves, L. C., da Silva, R. M. F., de Moraes Rocha, G. J., & da Cruz Filho, I. J. (2021). Lignin from *Morinda citrifolia* leaves: Physical and chemical characterization, in vitro evaluation of antioxidant, cytotoxic, antiparasitic, and ultrastructural activities. *International Journal of Biological Macromolecules*, *193*(Pt B), 1799–1812. https://doi.org/10.1016/j.ijbiomac.2021.11.013

Das, A., & Supriya Devi, R. (2020). Mangrove plants of Odisha: An ecological indicators & sources of bioactive compounds.

Datkhile, K. D., Chakraborty, S., Durgawale, P. P., & Patil, S. R. (2024). Green synthesis of silver nanoparticles using *Morinda citrifolia* Linn leaf extract and its antioxidant, antibacterial, and anticancer potential. *Pharmaceutical Nanotechnology*, *12*(4), 340–350. https://doi.org/10.2174/2211738511666230913095001

de Farias, V. A., da Rocha Lima, A. D., Santos Costa, A., de Freitas, C. D. T., da Silva Araújo, I. M., Dos Santos Garruti, D., de Figueiredo, E. A. T., & de Oliveira, H. D. (2020). Noni (*Morinda citrifolia* L.) fruit as a new source of milk-clotting cysteine proteases. *Food Research International*, *127*, 108689. https://doi.org/10.1016/j.foodres.2019.108689

Deng, S., West, B. J., & Jensen, C. J. (2008). Simultaneous characterisation and quantitation of flavonol glycosides and aglycones in noni leaves using a validated HPLC-UV/MS method. *Food Chemistry*, *111*(2), 526–529. https://doi.org/10.1016/j.foodchem.2008.04.021

Deshpande, S. H., Bagewadi, Z. K., Khan, T. Y., Mahnashi, M. H., Shaikh, I. A., Alshehery, S., Khan, A. A., Patil, V. S., & Roy, S. (2023). Exploring the potential of phytocompounds for targeting epigenetic mechanisms in rheumatoid arthritis: An in silico study using similarity indexing. *Molecules*, *28*(6), 2430.

Devi, V., & Gupta, D. (2025). *Trapa natans*: A comprehensive review of its phytochemistry, pharmacological, and therapeutic potential. *European Journal of Pharmaceutical and Medical Research*, *12*(9), 61–66.

Dhatwalia, J., Kumari, A., Verma, R., Upadhyay, N., Guleria, I., Lal, S., Thakur, S., Gudeta, K., Kumar, V., Chao, J. C. J., Sharma, S., Kumar, A., Manicum, A.-L. E., Lorenzo, J. M., & Amarowicz, R. (2021). Phytochemistry, pharmacology, and nutraceutical profile of *Carissa* species: An updated review. *Molecules*, *26*(22). https://doi.org/10.3390/molecules26227010

do Carmo, L. D., de Fátima Pinheiro Rangel, G., Rabelo, L. M. A., de Fátima Goebel de Souza, T., Júnior, R., Wong, D. V. T., de Carvalho Leitão, R. F., da Silva, A. A. V., Magalhães, P. J. C., Costa, A. S., de Carvalho Oliveira Campos, D., de Alencar, N. M. N., & de Oliveira, H. D. (2022). Therapeutic effects of a lipid transfer protein isolated from *Morinda citrifolia* L. (noni) seeds on irinotecan-induced intestinal mucositis in mice. *Naunyn-Schmiedeberg's Archives of Pharmacology*, *395*(9), 1097–1107. https://doi.org/10.1007/s00210-022-02267-7

Dutta, S. K., Chettri, P., & Mishra, V. (2025). Documentation of folk-medicinal uses of fruit tree species by the indigenous communities of Sikkim Himalayas. *Phytomedicine Plus*, *5*(3), 100838.

Ee, G. C., Wen, Y. P., Sukari, M. A., Go, R., & Lee, H. L. (2009). A new anthraquinone from *Morinda citrifolia* roots. *Natural Product Research*, *23*(14), 1322–1329. https://doi.org/10.1080/14786410902753138

El Azhary, K., Tahiri Jouti, N., El Khachibi, M., Moutia, M., Tabyaoui, I., El Hou, A., Achtak, H., Nadifi, S., Habti, N., & Badou, A. (2017). Anti-inflammatory potential of *Capparis spinosa* L. in vivo in mice through inhibition of cell infiltration and cytokine gene expression. *BMC Complementary and Alternative Medicine*, *17*(1), 81. https://doi.org/10.1186/s12906-017-1569-7

Elderdery, A. Y., Alzahrani, B., Alabdulsalam, A. A., Hamza, S. M. A., Elkhalifa, A. M. E., Alhamidi, A. H., Alanazi, F., Mohamedain, A., Subbiah, S. K., & Ling Mok, P. (2022). Structural, optical, antibacterial, and anticancer properties of cerium oxide nanoparticles prepared by green synthesis using *Morinda citrifolia* leaves extract. *Bioinorganic Chemistry and Applications*, *2022*, 6835625. https://doi.org/10.1155/2022/6835625

El-Hawary, S. S., Taha, K. F., Kirillos, F. N., Dahab, A. A., El-Mahis, A. A., & El-Sayed, S. H. (2018). Complementary effect of *Capparis spinosa* L. and silymarin with/without praziquantel on mice experimentally infected with *Schistosoma mansoni*. *Helminthologia*, *55*(1), 21–32. https://doi.org/10.1515/helm-2017-0055

Engelmeier, D., Hadacek, F., Pacher, T., Vajrodaya, S., & Greger, H. (2000). Cyclopenta[b]benzofurans from *Aglaia* species with pronounced antifungal activity against rice blast fungus (*Pyricularia grisea*). *Journal of Agricultural and Food Chemistry*, *48*(4), 1400–1404.

Garcia-Vilas, J. A., Medina, M. A., Melo, F. R., Pejler, G., & Garcia-Faroldi, G. (2015). Damnacanthal inhibits IgE receptor-mediated activation of mast cells. *Molecular Immunology*, *65*(1), 86–93. https://doi.org/10.1016/j.molimm.2015.01.008

García-Vilas, J. A., Pino-Ángeles, A., Martínez-Poveda, B., Quesada, A. R., & Medina, M. (2017). The noni anthraquinone damnacanthal is a multi-kinase inhibitor with potent anti-angiogenic effects. *Cancer Letters*, *385*, 1–11. https://doi.org/10.1016/j.canlet.2016.10.037

Giampieri, F., Tulipani, S., Alvarez-Suarez, J. M., Quiles, J. L., Mezzetti, B., & Battino, M. (2012). The strawberry: Composition, nutritional quality, and impact on human health. *Nutrition*, *28*(1), 9–19. https://doi.org/10.1016/j.nut.2011.08.009

Hajimehdipoor, H., Rajabi, S., Jafari, F., & Keramatian, B. (2024). The effect of *Capparis spinosa* on inflammatory responses in a mouse model of allergic asthma. *Research Journal of Pharmacognosy*, *11*(3). https://doi.org/10.22127/rjp.2024.457051.2454

Harneti, D., & Supratman, U. (2021). Phytochemistry and biological activities of *Aglaia* species. *Phytochemistry*, *181*, 112540.

Hayashi, N., Lee, K.-H., Hall, I. H., McPhail, A. T., & Huang, H.-C. (1982). Structure and stereochemistry of (−)-odorinol, an antileukemic diamide from *Aglaia odorata*. *Phytochemistry*, *21*(9), 2371–2373.

Heyne, K. (1987). *The useful Indonesian plants*. Research and Development Agency, 845–846.

Islam, A. T. M. R., Hasan, M. M., & Islam, T. (2021). Ethnobotanical study of plants used by the Munda ethnic group living around the Sundarbans. *Journal of Ethnopharmacology*, *285*. https://doi.org/10.1016/j.jep.2021.114853

Isomiddin, B., & Hamzayeva, N. (2024). *Physalis angulata* phytochemical bioactive composition (Review). *JournalNX - A Multidisciplinary Peer Reviewed Journal*, *10*(6), 67–69. https://repo.journalnx.com/index.php/nx/article/view/5392

Joycharat, N., Greger, H., Hofer, O., & Saifah, E. (2008). Flavaglines and triterpenoids from the leaves of *Aglaia forbesii*. *Phytochemistry*, *69*(1), 206–211. https://doi.org/10.1016/j.phytochem.2007.06.016

Kalantari, H., Foruozandeh, H., Khodayar, M. J., Siahpoosh, A., Saki, N., & Kheradmand, P. (2018). Antioxidant and hepatoprotective effects of *Capparis spinosa* L. fractions and quercetin. *Journal of Traditional and Complementary Medicine*, *8*(1), 120–127. https://doi.org/10.1016/j.jtcme.2017.04.010

Karker, M., Falleh, H., Msaada, K., Smaoui, A., Abdelly, C., Legault, J., & Ksouri, R. (2016). Antioxidant, anti-inflammatory and anticancer activities of *Reaumuria vermiculata*. *EXCLI Journal*, *15*, 297–307. https://doi.org/10.17179/excli2016-187

Khanavi, M., Ara, L., Khavasi, N., & Hajimehdipoor, H. (2020). *Capparis spinosa*: A comparative study of raw and processed fruits. *Journal of Medicinal Plants*, *19*(73), 91–99. https://doi.org/10.29252/jmp.1.73.91

Khoo, H. E., Azlan, A., Tang, S. T., & Lim, S. M. (2017). Anthocyanidins and anthocyanins. *Food & Nutrition Research*, *61*. https://foodandnutritionresearch.net/index.php/fnr/article/view/1257

Kim, M. H., & Jeong, H. J. (2014). Damnacanthal inhibits the NF-κB/RIP-2/caspase-1 pathway. *Immunopharmacology and Immunotoxicology*, *36*(5), 355–363. https://doi.org/10.3109/08923973.2014.952819

Kim, S., Chin, Y.-W., Su, B.-N., Riswan, S., Kardono, L. B., Afriastini, J. J., Chai, H., Farnsworth, N. R., Cordell, G. A., & Swanson, S. M. (2006). Cytotoxic flavaglines and bisamides from *Aglaia edulis*. *Journal of Natural Products*, *69*(12), 1769–1775.

Kim, S., Su, B.-N., Riswan, S., Kardono, L. B., Afriastini, J. J., Gallucci, J. C., Chai, H., Farnsworth, N. R., Cordell, G. A., & Swanson, S. M. (2005). Edulisones A and B from *Aglaia edulis*. *Tetrahedron Letters*, *46*(52), 9021–9024.

Kong, J.-M., Chia, L.-S., Goh, N.-K., Chia, T.-F., & Brouillard, R. (2003). Analysis and biological activities of anthocyanins. *Phytochemistry*, *64*(5), 923–933. https://doi.org/10.1016/S0031-9422(03)00438-2

Kongpuckdee, S., Ungphaiboon, S., Sungkharak, S., Phadoongsombut, N., & Mahattanadul, S. (2020). Efficacy of aqueous *Morinda citrifolia* extract gel. *Current Pharmaceutical Biotechnology*, *21*(15), 1699–1710. https://doi.org/10.2174/1389201021666200711151925

Kuspradini, H., Putri, A., Sinta, & Diana, R. (2020). Toxicity and antioxidant ability of *Litsea angulata*. *Agriculture and Natural Resources*, *54*, 223–228. https://doi.org/10.34044/j.anres.2020.54.2.15

Kustiarini, D. A., Nishigaki, T., Kanno, H., & To, H. (2019). Effects of *Morinda citrifolia* on rheumatoid arthritis. *Biological & Pharmaceutical Bulletin*, *42*(3), 496–500. https://doi.org/10.1248/bpb.b18-00480

Laha, R., Lalhriatpuia, L. R., Ralte, L., & Lalremruata, P. (2018). Diversity and ethnobotanical uses of wild edible fruits. *International Journal of Pharma and Bio Sciences*, *8*(2), 132–142.

Lemmens, R. H. M. J., & Bunyapraphatsara, N. (2003). *Medicinal and poisonous plants 3*. Backhuys Publishers.

Li, J., Huang, X., Jiang, X.-H., Zhu, Q.-F., Yang, Y., & Gao, G.-C. (2015). Catunarosides I–L, four new triterpenoid saponins from the stem bark of *Catunaregam spinosa*. *Chemical & Pharmaceutical Bulletin*, *63*(5), 388–392. https://doi.org/10.1248/cpb.c15-00055

Li, J., Niu, D., Zhang, Y., & Zeng, X. A. (2020). Physicochemical properties, antioxidant and antiproliferative activities of polysaccharides from *Morinda citrifolia* L. (noni). *International Journal of Biological Macromolecules*, *150*, 114–121. https://doi.org/10.1016/j.ijbiomac.2019.12.157

Li, L., Yin, Q., Zhang, Y., Yang, L., Gao, Y., Hao, J., Lin, B., Zhou, D., & Li, N. (2025a). Terpenoids from *Aglaia edulis* and their lung epithelial cells protective activity. *Phytochemistry*, *239*, 114599.

Li, Q., Chai, J., Deng, S., Xu, J., Feng, Y., Yang, R., & Li, W. (2025b). The comparison of the profile of phenolic compounds in noni (*Morinda citrifolia* L.) fruit by different drying methods. *Foods*, *14*(8). https://doi.org/10.3390/foods14081398

Liu, S., Liu, S.-B., Zuo, W.-J., Guo, Z.-K., Mei, W.-L., & Dai, H.-F. (2014). New sesquiterpenoids from *Aglaia odorata* var. *microphyllina* and their cytotoxic activity. *Fitoterapia*, *92*, 93–99.

Lolok, N., Sumiwi, S. A., Muhtadi, A., Susilawati, Y., Hendriani, R., Ramadhan, D. S. F., Levita, J., & Sahidin, I. (2022). Molecular docking and molecular dynamics studies of bioactive compounds contained in noni fruit (*Morinda citrifolia* L.) against human pancreatic α-amylase. *Journal of Biomolecular Structure and Dynamics*, *40*(15), 7091–7098. https://doi.org/10.1080/07391102.2021.1894981

Lolok, N., Sumiwi, S. A., Ramadhan, D. S. F., Levita, J., & Sahidin, I. (2024). Molecular dynamics study of stigmasterol and beta-sitosterol of *Morinda citrifolia* L. towards α-amylase and α-glucosidase. *Journal of Biomolecular Structure and Dynamics*, *42*(4), 1952–1955. https://doi.org/10.1080/07391102.2023.2243519

Lv, H.-S., Jiang, H.-L., Sun, Y.-J., Ying, P., Zheng, Q., Kong, L.-Y., & Luo, J. (2025). Three new limonoids with high oxidation from the fruits of *Aglaia edulis*. *Journal of Asian Natural Products Research*, 1–7.

Lv, L., Chen, H., Ho, C. T., & Sang, S. (2011). Chemical components of the roots of noni (*Morinda citrifolia*) and their cytotoxic effects. *Fitoterapia*, *82*(4), 704–708. https://doi.org/10.1016/j.fitote.2011.02.008

Ma, L. D., Lin, G. B., Yang, L. B., Cao, J. L., Wang, J., Chen, Q. D., Li, W. Q., & Zhong, W. J. (2021). *Morinda citrifolia* (noni) juice suppresses A549 human lung cancer cells via inhibiting AKT/NF-κB signaling pathway. *Chinese Journal of Integrative Medicine*, *27*(9), 688–695. https://doi.org/10.1007/s11655-020-3421-y

Mabberley, D., Pannell, C. M., & Sing, A. (1995). *Meliaceae. Flora Malesiana-Series 1, Spermatophyta*, *12*(1), 1–407.

Majee, C., Mazumder, R., & Choudhary, A. N. (2021). Pharmacognostical investigation of *Trapa natans* L. leaves. *Research Journal of Pharmacy and Technology*, *14*(9), 4680–4684.

Malik, B., Dar, F. A., Pirzadah, T. B., Zari, A., Zari, T. A., Alharby, H. F., Hakeem, K. R., & Rehman, R. U. (2022). Molecular and phytochemical characterizations of *Cichorium intybus* L. *Applied Sciences*, *12*(23). https://doi.org/10.3390/app122312061

Mani Senthil Kumar, K. T., Gorain, B., Roy, D. K., Zothanpuia, Samanta S. K., Pal, M., Biswas, P., Roy, A., Adhikari, D., Karmakar, S., & Sen, T. (2008). Anti-inflammatory activity of *Acanthus ilicifolius*. *Journal of Ethnopharmacology*, 120(1), 7–12. https://doi.org/10.1016/j.jep.2008.07.024

Matovic, V., Ljujic, B., Radojevic, I., Đelic, G., Kovacevic, M. M., Zivanovic, S., Papic, M., Milivojevic, N., Petrovic, I., & Gazdic Jankovic, M. (2025). *Trapa natans* L. extract attenuates inflammation and oxidative damage in cisplatin-induced cardiotoxicity in rats. *Mediators of Inflammation*, *2025*(1), 6587305.

Missouri Botanical Garden. (2026). *Aglaia edulis* Facilitated by Missouri Botanical Garden. Published on the internet. Retrieved 20 January 2026 from www.missouribotanicalgarden.org/PlantFinder/PlantFinderDetails.aspx?taxonid=292754

Mohamad Shalan, N. A. A., Mustapha, N. M., & Mohamed, S. (2017). Chronic toxicity evaluation of *Morinda citrifolia* fruit and leaf in mice. *Regulatory Toxicology and Pharmacology*, *83*, 46–53. https://doi.org/10.1016/j.yrtph.2016.11.022

Mondal, T., Sarkar, T., Sengupta, S., & Kundu, S. (2024). Nutritional and pharmacological aspects of *Trapa natans*. *International Journal of Agronomy*, *7*(4), 69–72.

Naffati, A., Mišan, A., Kenig, S., Kramberger, K., Jenko Pražnikar, Z., Petelin, A., Kljakić, A. C., Benmerzoug, A., Elfarrah, N., & Stupar, A. (2025). Unlocking the functional food potential of *Capparis spinosa*. *Foods*, *14*(10), 1768. https://doi.org/10.3390/foods14101768

Ngamdokmai, N., Ingkaninan, K., Scholfield, C. N., Insumrong, K., Neungchamnong, N., Minale, G., & Warinhomhoun, S. (2022). A Thai traditional triple-fruit formulation "Phikud Tri-Phon". *Foods*, *11*(19). https://doi.org/10.3390/foods11193067

NParks. (2026). *Morinda citrifolia L.* NParks Flora & Fauna Web. Retrieved January 24, 2026, from https://www.nparks.gov.sg/florafaunaweb/

Ono, K., & Suzuki, K. (2013). Assessment of subsistence plant resource of the mangrove forest in the Ayeyarwady Delta, Myanmar. *Grassland Science*, 17(2), 223–232. https://doi.org/10.57466/ger.17.2_223

Panche, A. N., Diwan, A. D., & Chandra, S. R. (2016). Flavonoids: An overview. *Journal of Nutritional Science*, *5*, e47. https://doi.org/10.1017/jns.2016.41

Pawlus, A. D., & Kinghorn, A. D. (2007). Review of the ethnobotany, chemistry, biological activity and safety of *Morinda citrifolia*. *Journal of Pharmacy and Pharmacology*, *59*(12), 1587–1609. https://doi.org/10.1211/jpp.59.12.0001

Phon, P. D. (2000). *Plantes utilisées au Cambodge*. Dy Phon Pauline.

Pointinger, S., Promdang, S., Vajrodaya, S., Pannell, C. M., Hofer, O., Mereiter, K., & Greger, H. (2008). Phytochemistry study. *Phytochemistry*, 69, 2696.

Potterat, O., & Hamburger, M. (2007). *Morinda citrifolia* (noni) fruit—phytochemistry, pharmacology, safety. *Planta Medica*, *73*(3), 191–199. https://doi.org/10.1055/s-2007-967115

POWO. (2026). *Plants of the World Online*. Royal Botanic Gardens, Kew. Retrieved January 24, 2026, from https://powo.science.kew.org/

Proksch, P., Edrada, R., Ebel, R., Bohnenstengel, F. I., & Nugroho, B. W. (2001). Chemistry and biological activity of rocaglamide derivatives. *Current Organic Chemistry*, *5*(9), 923–938.

Radojevic, I. D., Vasic, S. M., Dekic, M. S., Radulovic, N. S., Delic, G. T., Durdevic, J. S., & Comic, L. R. (2016). Antimicrobial and antibiofilm effects of extracts from *Trapa natans* L., evaluation of total phenolic and flavonoid contents and GC-MS analysis. *Acta Poloniae Pharmaceutica*, *73*(6), 1565–1574.

Rahmawati, R. (2020). Exploring medications for hypertension advertised online: A qualitative study in Indonesia. *Journal of Pharmacy & Bioallied Sciences*, *12*(Suppl 2), S691–S697. https://doi.org/10.4103/jpbs.JPBS_259_19

Raish, S. M. A., Almasri, R. S., Bedir, A. S., & Elkahwagy, A. A. (2025). Phytochemical composition, bioactive compounds, and antidiabetic potential of four medicinal plants native to the UAE: *Capparis spinosa*, *Citrullus colocynthis*, *Morus alba*, and *Rhazya stricta*. *Biology*, *14*(9). https://doi.org/10.3390/biology14091146

Ramakrishna Pillai, J., Wali, A. F., Menezes, G. A., Rehman, M. U., Wani, T. A., Arafah, A., Zargar, S., & Mir, T. M. (2022). Chemical composition analysis, cytotoxic, antimicrobial and antioxidant activities of *Physalis angulata* L.: A comparative study of leaves and fruit. *Molecules*, *27*(5). https://doi.org/10.3390/molecules27051480

Ravikumar, S., Raja, M., & Gnanadesigan, M. (2012). Antibacterial potential of benzoate and phenylethanoid derivatives isolated from *Acanthus ilicifolius* L. leaf extracts. *Natural Product Research*, 26(23), 2270–2273. https://doi.org/10.1080/14786419.2011.652962

Rinner, B., Siegl, V., Pürstner, P., Efferth, T., Brem, B., Greger, H., & Pfragner, R. (2004). Activity of novel plant extracts against medullary thyroid carcinoma cells. *Anticancer Research*, *24*(2A), 495–500.

Saifah, E., Suttisri, R., Shamsub, S., Pengsuparp, T., & Lipipun, V. (1999). Bisamides from *Aglaia edulis*. *Phytochemistry*, *52*(6), 1085–1088.

Salirrosas, D., Reategui-Pinedo, N., Crespo, J. P., Sánchez-Tuesta, L., Arqueros, M., Cabrera, A., Martinez, R. M., Ayala, C., Baby, A. R., & Prieto, Z. A. (2021). Safety profile of *Caesalpinia spinosa* aqueous extract tested in *Oreochromis niloticus*. *Frontiers in Sustainability*, *2*. https://doi.org/10.3389/frsus.2021.696289

Samoylenko, V., Zhao, J., Dunbar, D. C., Khan, I. A., Rushing, J. W., & Muhammad, I. (2006). New constituents from noni (*Morinda citrifolia*) fruit juice. *Journal of Agricultural and Food Chemistry*, *54*(17), 6398–6402. https://doi.org/10.1021/jf060672u

Sang, S., Liu, G., He, K., Zhu, N., Dong, Z., Zheng, Q., Rosen, R. T., & Ho, C. T. (2003). New unusual iridoids from the leaves of noni (*Morinda citrifolia* L.). *Bioorganic & Medicinal Chemistry*, *11*(12), 2499–2502. https://doi.org/10.1016/s0968-0896(03)00180-9

Sarkar, B., Kotal, H. N., Giri, C. K., Mandal, A., Hudait, N., Madhu, N. R., Saha, S., Basak, S. K., Sengupta, J., & Ray, K. (2023). Detection of a bibenzyl core scaffold in mangrove species. *Frontiers in Plant Science*, *14*, 1291805. https://doi.org/10.3389/fpls.2023.1291805

Sathasivampillai, S. V., Rajamanoharan, P. R. S., Munday, M., & Heinrich, M. (2017). Plants used to treat diabetes in Sri Lankan Siddha medicine. *Journal of Ethnopharmacology*, *198*, 531–599. https://doi.org/10.1016/j.jep.2016.07.053

Schulz, G., Victoria, C., Kirschning, A., & Steinmann, E. (2021). Rocaglamide and silvestrol: From anti-tumor to antiviral compounds. *Natural Product Reports*, *38*(1), 18–23.

Sen, U. K., & Bhakat, R. (2019). Ethnobotanical study on sand-dune based medicinal plants. *European Journal of Medicinal Plants*, *26*, 1–19. https://doi.org/10.9734/EJMP/2018/46231

Sf, Y., Rlf, L., Sheue, C.-R., & Yong, J. (2013). The current botanical status of mangrove forests in Singapore.

Shalabh, B., Akash, J., & Jasmine, C. (2012). *Trapa natans* (water chestnut): An overview.

Siddiqui, B. S., Ismail, F. A., Gulzar, T., & Begum, S. (2003). Isolation and structure determination of compounds from *Aspergillus niger*. *Natural Product Research*, *17*(5), 355–360. https://doi.org/10.1080/1057563031000075485

Silva, A. P. S. A., Nascimento da Silva, L. C., Martins da Fonseca, C. S., Araújo, J. M., Correia, M. T. D. S., Cavalcanti, M. D. S., & Lima, V. L. D. M. (2016). Antimicrobial activity and phytochemical analysis of *Cleome spinosa*. *Frontiers in Microbiology*, *7*, 963. https://doi.org/10.3389/fmicb.2016.00963

Sina, H., Dramane, G., Tchekounou, P., Assogba, M. F., Chabi-Sika, K., Boya, B., Socohou, A., Adjanohoun, A., & Baba-Moussa, L. (2021). Phytochemical composition and biological activities of *Morinda citrifolia*. *Saudi Journal of Biological Sciences*, *28*(2), 1331–1335. https://doi.org/10.1016/j.sjbs.2020.11.059

Singh, B., & Sharma, R. A. (2020). Indian *Morinda* species: A review. *Phytotherapy Research*, *34*(5), 924–1007. https://doi.org/10.1002/ptr.6579

Singh, N. (2012). Panchakarma: Cleaning and rejuvenation therapy for curing the diseases. *Journal of Pharmacognosy and Phytochemistry*, 1(2), 1–9.

Sousa, S. G., Oliveira, L. A., de Aguiar Magalhães, D., de Brito, T. V., Batista, J. A., Pereira, C. M. C., de Souza Costa, M., Mazulo, J. C. R., de Carvalho Filgueiras, M., Vasconcelos, D. F. P., da Silva, D. A., Barros, F. C. N., Sombra, V. G., Freitas, A. L. P., de Paula, R. C. M., de Andrade Feitosa, J. P., & Dos Reis Barbosa, A. L. (2018). Chemical structure and anti-inflammatory effect of polysaccharide extracted from *Morinda citrifolia* Linn (noni). *Carbohydrate Polymers*, *197*, 515–523. https://doi.org/10.1016/j.carbpol.2018.06.042

Souza, A. A., Costa, A. S., Campos, D. C. O., Batista, A. H. M., Sales, G. W. P., Nogueira, N. A. P., Alves, K. M. M., Coelho-de-Souza, A. N., & Oliveira, H. D. (2018). Lipid transfer protein isolated from noni seeds displays antibacterial activity. *Biochimie*, *149*, 9–17. https://doi.org/10.1016/j.biochi.2018.03.011

Su, B.-N., Chai, H., Mi, Q., Riswan, S., Kardono, L. B. S., Afriastini, J. J., Santarsiero, B. D., Mesecar, A. D., Farnsworth, N. R., Cordell, G. A., Swanson, S. M., & Kinghorn, A. D. (2006). Activity-guided isolation of cytotoxic constituents from *Aglaia crassinervia*. *Bioorganic & Medicinal Chemistry*, *14*(4), 960–972. https://doi.org/10.1016/j.bmc.2005.09.012

Sun, Y., Cui, L., Sun, Y., Li, Q., Li, Y., Wang, Z., Xu, W., Kong, L., & Luo, J. (2022). A/D-rings-seco limonoids from the fruits of *Aglaia edulis*. *Phytochemistry*, *195*, 113049.

Sun, Y., Yang, T., & Wang, C. (2023). *Capparis spinosa* L. as a potential source of nutrition and its health benefits. *Food Chemistry*, *409*, 135258. https://doi.org/10.1016/j.foodchem.2022.135258

Sun, Y., Yin, Y., Sun, Y., Li, Q., Cui, L., Xu, W., Kong, L., & Luo, J. (2021). Aglatestine A, a rearranged limonoid from *Aglaia edulis*. *The Journal of Organic Chemistry*, *86*(16), 11263–11268.

Sunday, R., & Ilesanmi, O. (2023). Toxicity studies of *Physalis angulata* leaves extract in albino rats. *Journal of Pharmacology and Toxicology*, *18*(2), 83–88. https://doi.org/10.3923/jpt.2023.83.88

Takashima, J., Ikeda, Y., Komiyama, K., Hayashi, M., Kishida, A., & Ohsaki, A. (2007). New constituents from the leaves of *Morinda citrifolia*. *Chemical & Pharmaceutical Bulletin*, *55*(2), 343–345. https://doi.org/10.1248/cpb.55.343

Tamalene, M. N., Sen, U. K., Bhakat, R., Vianti, E., & Bahtiar, B. (2021). Utilization of mangrove plants as a source of malaria medicine. *Asian Journal of Ethnobiology*, *4*. https://doi.org/10.13057/asianjethnobiol/y040203

Tan, B., Chiranthanut, N., Chansakaow, S., Sireeratawong, S., Khonsung, P., Nimlamool, W., Takuathung, M. N., & Lertprasertsuke, N. (2022). Anti-inflammatory effects of Pikad Tri-phol-sa-mut-than remedy. *Journal of Ethnopharmacology*, *298*, 115639. https://doi.org/10.1016/j.jep.2022.115639

Tan, Q.-G., & Luo, X.-D. (2011). Meliaceous limonoids: Chemistry and biological activities. *Chemical Reviews*, *111*(11), 7437–7522.

Tanikawa, T., Kitamura, M., Hayashi, Y., Tomida, N., Uwaya, A., Isami, F., & Inoue, Y. (2021). Anti-inflammatory effects of *Morinda citrifolia* extract. *Medicines*, *8*(8). https://doi.org/10.3390/medicines8080043

Tasanor, O., Brem, B., Leitsch, D., Binder, M., Duchêne, M., Greger, H., & Wernsdorfer, W. H. (2007). Development of a pharmacodynamic screening model with *Entamoeba histolytica*. *Wiener Klinische Wochenschrift*, *119*(Suppl 3), 88–95.

The Business Research Company. (2026). *Berry market report 2026* (Vol. 2026). The Business Research Company. Retrieved April 2 from https://www.thebusinessresearchcompany.com/report/berry-global-market-report

Timalsina, D., Devkota, H. P., Bhusal, D., & Sharma, K. R. (2021). *Catunaregam spinosa*: A review of traditional uses and pharmacological activities. *Evidence-Based Complementary and Alternative Medicine*, *2021*, 3257732. https://doi.org/10.1155/2021/3257732

Upadhyay, S., Rajan, S., Shrivastava, S., & Jeengar, M. K. (2025). Molecular insights into anti-inflammatory activities of selected Indian herbs. *Journal of Ayurveda and Integrative Medicine*, *16*(2), 101081. https://doi.org/10.1016/j.jaim.2024.101081

Vejayan, J., Bathmanathan, R., Tuan Said, S. A., Chakravarthi, S., & Ibrahim, H. (2022). Fruit extract mixture with antidiabetic activities. *Food Technology and Biotechnology*, *60*(3), 375–385 https://doi.org/10.17113/ftb.60.03.22.7456

Vendrapati, R., Bhat, S., Bindu, K., Mahadevaswamy, S., Latheef, S., Kalaivanan, D., Shiddamallayya, N., Narayanam, S., & Acharya, R. (2024). A comprehensive review of traditional medicinal plants. *Pharmacognosy Reviews*, *18*(35), 24–46. https://doi.org/10.5530/phrev.2024.18.4

Verma, A., Chattopadhaya, A., Gupta, P., Tiwari, H., Singh, S., Kumar, L., & Gautam, V. (2025). Integration of hyphenated techniques for natural product profiling. *Chemistry & Biodiversity*, *22*(9), e202500234. https://doi.org/10.1002/cbdv.202500234

Wang, M. Y., West, B. J., Jensen, C. J., Nowicki, D., Su, C., Palu, A. K., & Anderson, G. (2002). *Morinda citrifolia* (noni): A literature review and recent advances. *Acta Pharmacologica Sinica, 23*(12), 1127–1141.
Wang, R., Wang, L., Wu, H., Zhang, L., Hu, X., Li, C., & Liu, S. (2022). Noni fruit phenolic extract supplementation ameliorates NAFLD. *Food Research International, 160*, 111732. https://doi.org/10.1016/j.foodres.2022.111732
Wang, R., Wang, L., Zhang, L., Wan, S., Li, C., & Liu, S. (2022). Solvent effects on phenolics and antioxidant activity of noni. *Food Chemistry, 377*, 131989. https://doi.org/10.1016/j.foodchem.2021.131989
Wang, Z., Dou, R., Yang, R., Cai, K., Li, C., & Li, W. (2021). Changes in phenols and volatile profiles of noni juice. *Molecules, 26*(9). https://doi.org/10.3390/molecules26092604
West, B. J., Deng, S., Isami, F., Uwaya, A., & Jensen, C. J. (2018). Potential health benefits of noni juice. *Foods, 7*(4). https://doi.org/10.3390/foods7040058
West, B. J., Tani, H., Palu, A. K., Tolson, C. B., & Jensen, C. J. (2007). Safety tests of noni leaf. *Journal of the Science of Food and Agriculture, 87*(14), 2583–2588. https://doi.org/10.1002/jsfa.3007
Woradulayapinij, W., Pothiluk, A., Nualsanit, T., Yimsoo, T., Yingmema, W., Rojanapanthu, P., Hong, Y., Baek, S. J., & Treesuppharat, W. (2022). Acute oral toxicity of damnacanthal. *Toxicology Reports, 9*, 1968–1976. https://doi.org/10.1016/j.toxrep.2022.10.015
Wu, Y., Girmay, S., da Silva, V. M., Perry, B., Hu, X., & Tan, G. T. (2015). Role of endophytic fungi in anticancer activity of *Morinda citrifolia. Evidence-Based Complementary and Alternative Medicine, 2015*, 393960. https://doi.org/10.1155/2015/393960
Wu, Z., Raven, P., & Hong, D. (2008). *Flora of China 11: 180–190*. Missouri Botanical Garden Press.
Xie, B.-J., Yang, S.-P., Chen, H.-D., & Yue, J.-M. (2007). Agladupols A–E, triterpenoids from *Aglaia duperreana. Journal of Natural Products, 70*(9), 1532–1535. https://doi.org/10.1021/np0702842
Xu, M., Yu, Z. X., Zhang, B., Li, X. B., Li, M. Y., Xu, W., & Chen, G. Y. (2022). Chemical constituents with α-glucosidase inhibitory activities from seeds of *Morinda citrifolia. Zhongguo Zhong Yao Za Zhi, 47*(13), 3519–3525. https://doi.org/10.19540/j.cnki.cjcmm.20211228.201
Yang, F., Zhu, W., Sun, S., Ai, Q., Edirisuriya, P., & Zhou, K. (2020). Isolation of bacterial β-glucuronidase inhibitors from noni fruits. *Journal of Natural Products, 83*(4), 825–833. https://doi.org/10.1021/acs.jnatprod.9b00279
Yi, P., Qiu, J.-F., Yang, X.-M., Chen, F.-F., Yang, J., Liu, J., Jin, J., Qi, L.-X., Hao, X.-J., & Wu, J.-H. (2025). Cyclopenta[bc]benzopyran derivatives and limonoids from *Aglaia edulis. Journal of Natural Products, 88*(1), 119–132.
Yodsaoue, O., Sonprasit, J., Karalai, C., Ponglimanont, C., Tewtrakul, S., & Chantrapromma, S. (2012). Diterpenoids and triterpenoids from *Aglaia odorata. Phytochemistry, 76*, 83–91. https://doi.org/10.1016/j.phytochem.2012.01.015
Zhang, B., Wei, X., Du, P., Luo, H., Hu, L., Guan, L., & Chen, G. (2025). Structural characterization of polysaccharides from noni juice. *Molecules, 30*(5). https://doi.org/10.3390/molecules30051103
Zhang, F., Wang, J.-S., Gu, Y.-C., & Kong, L.-Y. (2010). Triterpenoids from *Aglaia abbreviata* and their cytotoxic activities. *Journal of Natural Products, 73*(12), 2042–2046. https://doi.org/10.1021/np100599g
Zhang, K., Meng, J., Li, X., Tang, X., Ma, S., Lv, Y., & Yang, S. (2020). Noni wine prevents oxidative stress and obesity in mice. *Journal of Food Biochemistry, 44*(11), e13460. https://doi.org/10.1111/jfbc.13460
Zou, R., Lan, Z., Wu, T., Jia, D., & Yang, Z. (2019). Optimization of noni callus induction. *Sheng Wu Gong Cheng Xue Bao, 35*(2), 298–306. https://doi.org/10.13345/j.cjb.180141

18 Unveiling the Mechanisms Behind Therapeutic Power of Medicinal Fruits and Berries

Gazala Tabassum, Sandeep K. Behera, Nimisha Patel, Ahmad Ali, and Hina Alim

18.1 INTRODUCTION

The history of plants in the form of food spans over 100,000 years, transforming from nature- based sourcing to the highly evolved modern agriculture of 2025 (Mercader, 2009). The period from pre-10,000 to 10,000 BC years is described as the Foraging Era, when early staples like wild grains were eaten, and tropical foraging, where Taro (one of the world's oldest food crops) around 28,700 years ago was introduced. This was followed by the 10,000–3,000 BC Neolithic revolution, when the beginning of agriculture happened independently at different global centres in which founder crops were domesticated, for instance, white millets in China, squash, chilli, maize, and potatoes in America, sorghum, yams, sugarcane, and bananas were grown in Africa and New Guinea. From 1492 to 1800s, the era of globalisation and exchange when the most important phase of plant exchange took place in history between the Western and Eastern hemispheres by introducing potatoes, tomatoes, maize, and cassava to Europe, Asia, and Africa, and in return rice, coffee, wheat, and citrus fruits were introduced to the Americans by Europeans (Scarcelli et al., 2019; Shen et al., 2021). The years starting from 1900 to the present are known as the modern era, which brought the Green Revolution through the development of high-yielding crops as well as fertilisers. Fruits are a specialised gift to human beings. It is defined as the only fleshy growth that develops from a flower's ovary, which is the pulpy, sweet, juicy edible part of a plant or tree that contains seeds and a seed coat. The development of fruits can be characterised by three stages of growth: the growth stage, which consists of cell division and cell enlargement, leading to an increase in fruit size, the maturation stage, which causes the fruits for the development of flavour and also the addition of sugar followed by the senescence stage, when due to ageing tissue degradation takes place under the influence of different biological pathways. A berry is a type of fruit that develops from a single mature ovary, but not all fruits are berries. However, botanically, a true berry is categorised as a fleshy fruit that is characterised as developing from a single flower with one ovary containing multiple seeds, like grapes, bananas, tomatoes, etc. In contrast, fruits commonly called strawberries and raspberries are technically aggregate or false berries because they are formed from multiple ovaries or other parts of the flower in addition to the ovary.

Ancient civilisations utilised fruits and berries for their key role in indigenous medical systems across cultures from Ayurveda and traditional Chinese medicine to European folk practices (Toplak, 2005; Bojadzievski, 1992; Bottcher, 1965; Wiart, 2006). In the initial decades of the 20th century, studies were more inclined towards fruits and vegetables primarily as sources of vital vitamins (especially vitamins A and C) and minerals, based on preventing deficiency conditions like scurvy (vitamin C deficiency). Early modern dietary science, which is also known as the era of vitamins and minerals, focussed mainly on the identification and prevention of acute deficiency diseases (Lukic, 1985).

DOI: 10.1201/9781003688259-18

Fruits were utilised especially as an essential source of vitamins (like vitamin C, the scurvy preventative) and essential minerals. It proved to be a crucial first step, establishing the necessity of their nutritional requirement, but it could only scratch the surface of their full potential to some extent. However, in the late 20th century, also known as the dawn of phytochemical discovery, modern research focus shifted with the discovery and identification of thousands of phytochemicals (bioactive non-nutrient plant compounds), such as polyphenols and carotenoids (Kovacevic, 2000). These foods bring about remarkable changes in measurable disease risks in very large populations. These greatly influence biological pathways related to oxidative stress, inflammation, blood sugar, and blood pressure (Hanahan, 2022; Dave et al., 2020; Kumar and Gupta, 2021). These are safe enough to be consumed every day, throughout their lifetime. The substances that impart the vibrant colours, that is, vivid reds, blues, and purples in berries, with their distinct aromas and flavours, are identified.

18.2 PHYTOCHEMICAL CONSTITUENTS OF MEDICINAL FRUITS AND BERRIES

Phytochemicals are secondary plant products that are nature-borne and are power pockets of medicinal fruits and berries, which protect the human body against various kinds of stress. Research studies have shown that these compounds weren't just decorative means, but they possessed significant potent pharmacological activities, which led to the identification of over 5,000 different phytochemicals. Medicinal fruits and berries are composed of these chemical species, which contribute various colour pigments and have potent effects. Several preeminent examples of them are described below:

***Polyphenols*:** Polyphenols are naturally sourced secondary plant products that are found in bulk populations in medicinal fruits and berries. They show many hydroxyphenyl groups (-OH) in their chemical structure (Singla et al., 2019). These are of two types, flavonoids and non-flavonoids, based on their intricate nature. Polyphenols are found in nature either in sugar-bound, that is, glycoside, or unconjugated form, that is, aglycones. Examples of non-flavonoid polyphenols are phenolic acids: 4-hydroxybenzoic, 3,4-dihydroxybenzoic, 3,5-dimethoxy-4-hydroxybenzoic, vanillic, caffeic, ferulic, para-coumaric, and sinapic acids, stilbenes (e.g., resveratrol), anthocyanidins, lignans, and polyphenol amides, whereas common flavonoids polyphenols are quercetin, catechin, anthocyanin, hesperidin, and kaempferol. Phenolic chemicals also constitute tanniferous and resinous compounds (Bravo, 1998; Molino et al., 2023; Belščak-Cvitanović et al., 2018). These biologically active molecules are found in various forms.

***Flavonoids*:** Flavonoids are chemical substances that have a diphenylpropane carbon backbone and are widely prevalent in medicinal fruits and berries (Toh et al., 2013). These are divided according to the presence of ketone, hydroxy, and phenol groups and position in the basic chemical structure of flavonoids (Kumar and Pandey, 2013). Flavonoids in plants are present in glycosylated form (i.e., flavonoidal glycoside) (Panche et al., 2016) or non-conjugated form, that is, polymorphic. Polymorphic flavonols (myricetin, quercetin, kaempferolcatechin, and epicatechin), flavones (apigenin, luteolin), isoflavones (genistein, daidzein), and flavanones (hesperidin, naringin).

Volatile compounds: These chemical species are of low molecular mass and are usually derivatives of terpenes (isoprene) and terpenoids. The component elements of volatile organic compounds are monoterpenes and diterpenes (Zuzarte and Salgueiro, 2015). The fruits belonging to the genus Citrus (family Rutaceae) display a richness of volatile oils. Sweet orange, lime, lemon, mandarin, and grapefruit are the principal suppliers of volatile oils, especially in their rinds (Espina et al., 2011). These volatile oils have been shown to exert profound positive health effects in humans and have been proven to be gastric protective, antiflatulent, oxidative stress inhibitor, inflammation-suppressing, painkiller, microbicidal, cancer-preventive, antihyperglycaemic, hepatoprotective, and cardiotonic functions (Bakkali et al., 2008; Matera et al., 2023; Seow et al., 2014; Tabassum and Vidyasagar, 2013). Besides these, volatile oils are utilised as food preservation additives and in essential oil therapy (Falleh et al., 2020) and as a flavorant in the scent industry (Sharmeen et al., 2021).

***Glycosides*:** Glycosides are also plant secondary compounds that display many biological actions. Glycosides result in a sugar part, that is a glycone, and a non-sugar constituent, that upon hydrolysis, yields genin (Bartnik and Facey, 1917). These exist in multiple forms of saponins, anthraquinone, flavonol-glycosides, glycol-alkaloids, etc. (Kytidou et al., 2020; Tuygunovna, 2023).

***Lipids*:** Lipids are naturally occurring chemical compounds found in both plants and animals. These compounds are the reduced forms of hydrocarbons or esters of fatty acids with polyols and alcohols (Sargent et al., 2002). Lipids are classified into simple lipids, which are fatty acid esters combined with alcohols, such as triglycerides (including fats, oils, and waxes), and compound lipids like glycolipids and phospholipids. When lipids undergo hydrolysis, they yield fatty acids, which are long-chain carboxylic acids that can be saturated, monounsaturated, or polyunsaturated. Examples of saturated fatty acids include lauric acid, myristic acid, palmitic acid, stearic acid, and arachidic acid. On the other hand, monounsaturated fatty acids include oleic acid and palmitoleic acid, while the polyunsaturated fatty acids consist of linoleic acid, linolenic acid, and arachidonic acid.

***Phytosterols*:** The derivatives of steroidal structures, referred to as sterols when found in plants, are termed phytosterols. This group includes compounds such as stigmasterol, β-sitosterol, and campesterol, which are primarily located in medicinal fruits and berries (Moreau et al., 2002). Phytosterols are widely acknowledged for their role in regulating biological pathways, managing cell sorting, and affecting the fluidity of cellular membranes (Ostlund, 2002). They are known to trigger anti-inflammatory effects (Vilahur et al., 2019), exhibit anticancer properties (Bradford and Awad, 2007), and provide cardiovascular protection (Marangoni and Poli, 2010).

***Sugars and polysaccharides*:** Fruits and berries eaten provide an abundant supply of sugars, which are categorised as healthy ones and have proved to be a substitute for table sugar. They are all sweet in taste because of different varieties of sugars (Kuzma et al., 2017). Polysaccharides also form very crucial compounds.

***Miscellaneous compounds*:** Medicinal fruits and berries are rich in various phytochemical compounds, including proteins, peptides, vitamins, enzymes, and minerals (Sosalagere et al., 2022; Ranjith et al., 2022). Fruits and berries are known to contain a significant amount of proteolytic enzymes, also referred to as proteases (Saleem et al., 2024).

18.3 CORE MECHANISM BEHIND THERAPEUTIC POWERS OF MEDICINAL FRUITS AND BERRIES

18.3.1 The Strength Hidden Behind Fruits and Berries That Makes Them Powerful

It has been clearly proved now that the protective power of fruits and berries lies not only in any single "magic bullet" compound. Instead, it is the sophisticated interplay of this vast array of vitamins (especially vitamins C, E, folate, and carotenoids) (Mohammadzadeh et al., 2017), minerals (potassium, manganese, magnesium), fibres (soluble and insoluble), and phytochemicals, especially polyphenols (anthocyanins, flavonols, phenolic acids, tannins) (Yahia et al., 2018; Salamon et al., 2021). Polyphenols serve as antioxidants and also act to modulate enzymes, receptors, and signalling pathways, which affect inflammation, blood vessel tone, cell growth, and metabolism The pigment anthocyanins make berries blue, red, or purple, and are closely associated with improvement in vascular function, reduced oxidative damage, and betterment of metabolic health.

18.3.2 Protection of Heart and Blood Vessels

Fruits and berries have a strong impact on the health of the heart and blood vessels. The deep-rooted quest for empirical knowledge has established the foundation for a modern scientific exploration to technically unveil the mechanisms that modify fruits and berries into powerful medicinal allies against the ill effects on one's health. Exhaustive studies and meta-analyses have proved that higher fruit consumption is directly proportional to a lower risk of coronary heart disease, stroke, and hypertension. Potassium (for regulation of blood pressure), when combined with fibre (for management of cholesterol), and flavonoids, results in an improvement of endothelial function, for example, affecting the health of blood vessel linings. It was also confirmed that berries, blueberries, strawberries, and other berries cause a lowering of low-density lipoprotein (LDL) or "bad" cholesterol and thus improve the functioning of blood vessels in both healthy people and those at risk (Kovacevic, 2000). Berries have extra influence because they are very rich in antioxidants and anti-inflammatory compounds. Polyphenols prevent overaggregation of platelets and cause a reduction in vascular tone, helping to smooth blood flow and reduce the risk of clotting. This proves that eating fruits and berries daily is one of the correct ways to protect people's future self from diseases of the heart and strokes without feeling like one is "on treatment."

18.3.3 Anti-Inflammatory and Immune-Modulating Actions of Quiet Allies

In the power-packed therapeutic foods, fruits and berries, the real superpower is their capability to act as master conductors of the internal defence systems of the body, namely inflammation and immunity. These fine-tune the complex machinery of the immune system, preparing human bodies to find balance and resilience. The problem that people are facing in modern society is chronic inflammation. This means having the alarm blaring continuously for months or years, without seeing any immediate danger. This continuous, low internal stress harms the tissues and organs with time, thus paving the path for chronic disease (Dennis et al., 2019; Guan et al., 2021). To calm the storm, these allies of fruits and berries are required to turn off chronic alarms without affecting the entire defence system. That is where their "anti-inflammatory" (actively suppressing the damaging, inflammatory responses in excesses) and "immune-modulating" (restoring balance in the immune mechanism, by not suppressing it) activities turn up (Figure 18.1).

18.3.4 The Powerful Players: Phytochemicals With Their Magical Mechanisms

The chief secret agents in this whole process are the polyphenols, which are plant compounds like anthocyanins (the pigments responsible for the blue colour of blueberries and the red colour of raspberries), quercetin, and ellagic acid. These chemical compounds act through a variety of pathways, which are described below.

(a) **By switching off the "alarm bell" (NF-κB pathway)**
The central control panel in cells is called the nuclear factor kappa-light-chain-enhancement of activated B cells (NF-κB), which controls the switch that triggers hundreds of pro-inflammatory genes and messenger molecules (cytokines like TNF-α, IL-6, and IL-1β) (Tuli et al., 2019).
Action of fruits and berries: Polyphenols found in blackberries and blueberries inactivate and block NF-κB translocation physically, thus preventing the "alarm" signal from reaching the nucleus and switching on the expression of inflammatory genes (Rahman et al., 2018). This makes it one of the most crucial processes by which they control inflammation at its origin.

(b) **Modulation of army comprising immune cells**
Fruits and berries prevent inflammation by actually training immune cells to work more effectively; the soldiers of the army are mentioned below:

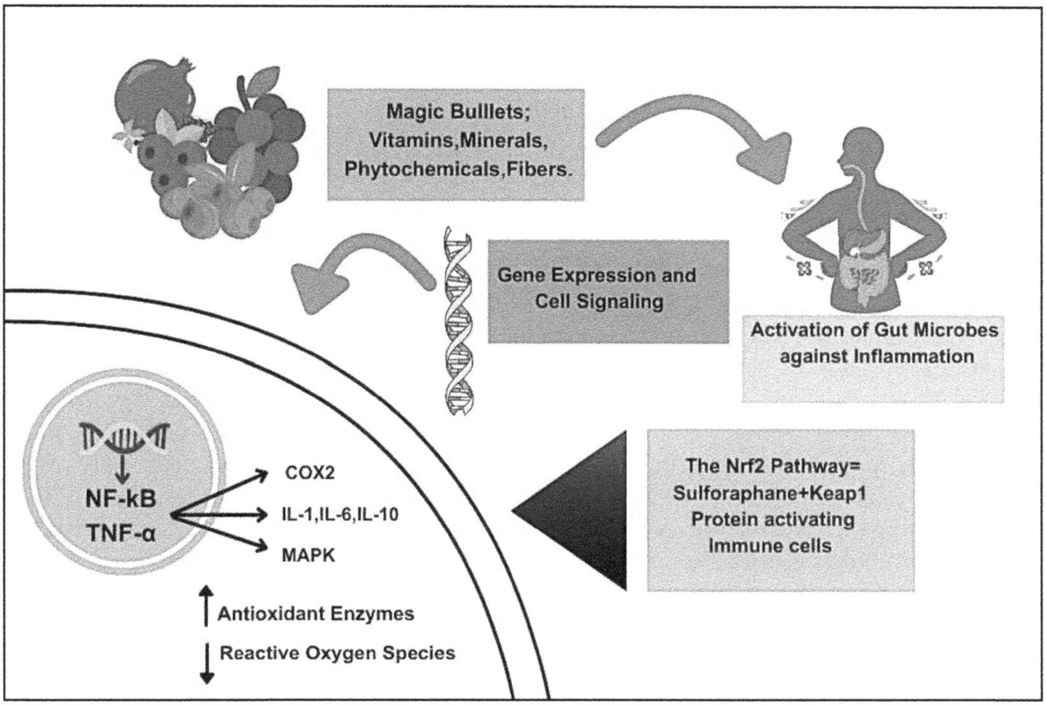

FIGURE 18.1 Powerful impacts of medicinal fruits and berries on the human body.

T-cells: They are the body's adaptive immune system commanders that help in maintaining a balanced response (Kumar and Gupta, 2021; Singh et al., 2022; Vendrame and Klimis-Zacas, 2019).

Natural killer (NK) cells: These are called the "seek and destroy" cells that help in the removal of infected or cancerous cells. Research studies have shown the efficacy of consuming blueberries in doubling the counts of NK cells in marathon runners, and thus enhancing the immune surveillance system (McAnulty et al., 2011).

Macrophages and dendritic cells: These are called "sentinel" cells, which can perceive threats. The chemical compounds of berries modulate their action, stopping them from continuously pumping out inflammatory signals even when there is no danger (Zhang et al., 2024).

(c) **Partnership with the gut bacteria**

Intensive work on bioactive compounds has shown that they carve the design of the microbiome's adaptive resilience. A combination of control of calories and antioxidant-rich plant foods results in the healthy microbiome diversity in the gut, which is followed by switching from inflammatory responses to a larger thematic focus, that is, constituents of fruit and berry behave like an ecological architect that helps in reducing the stressful gut environment. The microbiome of the gut plays a significant role in modulating inflammation throughout the whole body. A large portion of the polyphenols is not absorbed immediately in the body. They rather reach the colon, where they are broken down into even smaller, more bioactive compounds (metabolites) by beneficial gut bacteria (Capuano et al., 2017). The newly formed compounds come to interact with the gut lining and immune cells, thus helping to stop inflammation in the gut and making the gut barrier stronger.

(d) **Effects of antioxidants: a new perspective beyond the old idea of free radical scavenging**
Antioxidants have been proven to be the "power packed heroes" in food (especially in fruits and berries) who move around bodies and zap the "villain," that is, free radicals, resulting in the prevention of damage and disease, finally giving healthy and strong life to people (Wang et al., 2021). These passive clean-up crews play the role of biological messengers, systems architects, and master switch-operators within human cells, making the body defend and repair itself on a fundamental basis. These function like a sensitive thermostat, supporting optimal cellular activity and reinforcing the body's own innate defence mechanisms (Kristo et al., 2016; Olas, 2016, 2017).

(e) **Activation of the body immune system (the Nrf2 pathway)**
Recently, the most remarkable discovery is the way these chemical compounds trigger the body's own built-in factories of antioxidants, the mechanism known as the nuclear factor erythroid 2-related factor 2-Kelch-like ECH-associated protein 1 or Nrf2/Keap1 pathway, and it's a breakthrough in understanding the superpower of medicinal fruits and berries (Nile and Park, 2014). The phytochemicals in berries, that is, sulforaphane (also present in vegetables of the cruciferous family, which show similar pathways) and specific flavonoids, function as mild stressors tricking the cells into strengthening its own defences (Battino et al., 2021). The interaction of these compounds with a protein called Keap1, which normally keeps Nrf2 locked up, activates Nrf2. When Nrf2 gets "moving," it reaches the nucleus of cells and turns on the genes for mighty internally derived antioxidant enzymes. Usually, one dietary molecule of antioxidant neutralises one free radical (in a 1:1 ratio), but activation of the Nrf2 pathway can result in the making of hundreds of the body's own antioxidant enzymes, becoming a much more robust and sustained defence. This is a comprehensive system modification rather than a simple cleanup task. This entire action works against the damage of DNA and causes the regulation of cellular signalling pathways, which are essential for the prevention of diseases like cancer and neurodegenerative disorders (Kabir et al., 2022).

(f) **Role in fine-tuning gene expression and cellular signalling**
The new findings emphasise that compounds of berries are active leaders, regulating one's gene activity and how one's cells communicate with each other. Phytochemicals turn genes "on" or "off." For instance, they activate the switch on mode of genes which are responsible for cell repair and programmed cell death (apoptosis) in impaired or premalignant cells (Mandal, 2020; Ranjan et al., 2019), whereas they cause the switch off mode of genes that may lead to irregular cellular growth or inflammatory responses (Tena et al., 2020; Teresita et al., 2023). This great extent of control supersedes simple free radical scavenging (Uddin et al., 2020). Also, by interfering with or triggering specific cell signalling cascades, called the mitogen-activated protein kinase (MAPK) pathway, helps in cell proliferation and cell viability. They also play a crucial role in maintaining cellular homeostasis.

(g) **Amplified action: naturally integrated nutrient effect**
When high-dose failure occurs, supplements of a single compound bring about the necessity of other crucial aspects of the new understanding: the significance of the whole food matrix. Fruits and berries consist of thousands of bioactive compounds, which include fibres, vitamins, minerals, and various polyphenols, all found in a balanced form (Rauf et al., 2018). This intricate blend forms a synergy where the overall effect is strongest. Inside this natural framework, antioxidants renew each other. The regeneration of vitamin E by vitamin C after neutralising a free radical with the other phytochemicals helps this network. This superfine system disappears when a single compound is used and concentrated into a nutrient capsule (Table 18.1).

TABLE 18.1
Various Activities of Medicinal Fruits and Berries

Job Description	Elucidation	Fruits and Their Phytochemicals as Action Heroes	Key References
Antioxidant as The Cleanup Crew	Scavenges free radicals and causes reduction in oxidative stress	Berries, cherries, citrus fruits, grapes, and pomegranates with their bioactive compounds: flavonols, anthocyanins, phenolic acids, stilbenes, carotenoids, ellagic acid, etc.	Skrovankova et al. (2015)
Anti-inflammatory responses as The Peacekeepers	Helps in inhibition of inflammatory pathways and mediators (e.g., COX enzymes, TNF-α).	All fruits and berries contain anthocyanins, flavonols, tannins, carotenoids, etc.	Liang et al. (2022)
Microbe-fighting agent as The Security guards	Prevention of microbial growth (bacteria, viruses, and fungi)	Cranberries, blueberries, raspberries, gooseberries, apples, and mangoes with their phytochemicals: phenolic compounds, carotenoids, terpenoids, saponins	Farhadi et al. (2019)
Anticancerous as The Bodyguards	Regulates expression of genes and aids in detoxification	Apples, red grapes, watermelon, citrus fruits, black raspberries, strawberries: anthocyanins, quercetin, apigenin, ellagic acid, and phytosterols	De Mattia et al. (2017)

(h) **Metabolic wonders in case of Diabetes**
For diabetics, fruit science brings a change in the power of insulin resistance in the intestinal habitat and vascular complexities. Recent research studies have made it clear that these phytochemicals regulate and help in lowering glucose (fasting) and glycated haemoglobin in patients with type 2 diabetes. The bioactives from berries, polyphenols, cause an increase in the synthesis of insulin, provide protection of pancreatic β-cells from physiological cell death, encouraging their proliferation, by the activation of the adenosine monophosphate-activated protein kinase (AMPK) and glucose transporter (GLUT) signalling regulation (Arunachalam and Brindha, 2023; Malarvizhi et al., 2025). These chemical compounds lead to modification of reactive oxygen species (ROS)-induced stress, inflammatory reactions, and cellular energy-monitoring pathways, whose function is similar to that of some anti-diabetic pills.

18.4 MEDICINAL FRUITS AND BERRIES: SOME COMMON EXAMPLES IN DAILY LIFE

Examples of berries, that is, blueberries, strawberries, blackberries, raspberries, cranberries, and black currants have unexpectedly high contents of anthocyanins and vitamin C which are found linked to health of the heart, brain, and eyes, and benefitting metabolism (Assmann et al., 2022); citrus fruits, that is, oranges, grapefruits, and lemons are found to be rich in vitamin C, flavanones, and soluble fibre that enhance health of the vascular system, boosting immunity, and iron absorption; pomegranates and grapes are enriched with polyphenols, for instance resveratrol and ellagitannins, that assist in proper vascular function and antioxidant defence system (Slika et al., 2022; Ren et al., 2021); and tropical fruits, that is, papaya, mango, and pineapple are packed with carotenoids, vitamin C, and digestive enzymes and which are also found to display cancer-preventive and gut-supportive activities (Limketkai et al., 2020; Li et al., 2019).

Apples and pears offer pectin and phenolic acids, which feed healthy gut microbes and help in lowering cardiovascular (Molan et al., 2014) and diabetes risks (Raimundo et al., 2020). The strong impact of the environment, such as temperature variations, exposure to UV, and hydrological availability, affects the biological milieu by regulating the biochemical synthesis pathways which produce bioactive compounds.

18.5 CONCLUSION AND FUTURE PERSPECTIVE

In 2026, medicinal fruits and berries are not mere superfoods that work like a living pharmacy, which gets chemically translated into the language of our cells, our microbes, and our brain simultaneously. A fruit bowl represents a kind of daily pharmacy, in which each colour and texture has an individual therapeutic profile. Different types of fruits and berries across seasons and colours broaden the phytochemical spectrum. Recently, these phytochemicals have been tracked starting from the plate, passing through the microbiome of the gut, flowing into the blood, and eventually into the brain and vascular cell-signalling cascade with a precision that seemed like science fiction. This silent magic happens, with our daily foods—berries mixed with curd, post-meal fruit servings which involve choreography of a number of communication pathways (Nrf2/Keap1, NF-κB, AMP-regulated kinase, endothelium-derived nitric oxide, GLP-1, microbiota-generated SCFAs, etc). It was discovered that in nutritional psychiatry, fruits and berries act as adjuncts for neurodegenerative disorders such as mood and anxiety disorders because they mitigate oxidative stress levels, neuroinflammatory response, and disrupted gut microbiome by triggering metabolic pathways increasingly implicated in brain health. Their compounds might not be a substitute for therapy or medication when needed, but a daily habit of fruits and berries becomes a gentle support of a clearer, more resilient mind and body. The future of food technology is concentrating on delivering fruit and berry bioactive compounds in such a way that stays close to nature, but their availability will be more practical, suiting the modern lifestyle. There is work going on for green nutrition in which medicinal fruits and berries are grown for maximum content of phytochemicals. Scientists are exploring delivery systems for compounds in the form of an encapsulated or food matrix that consists of delicate compounds which can be ingested by the process of digestion, and thus releasing them in the gut for their maximum impact.

REFERENCES

Arunachalam, K., Brindha, P. 2023. Antidiabetic phytochemicals from medicinal plants: Prospective candidates for drug discovery and development. *Frontiers in Endocrinology* 14: 800714.

Assmann, C.E., Weis, G.C.C., da Rosa, J.R., Bonadiman, B.D.S.R., Alves, A.D.O., Borsoi, F.T., Bagatini, M.D. 2022. Anthocyanins and flavonols: Therapeutic implications of natural compounds cancer. In Kumar, R., ed., *Handbook of Oxidative Stress in Cancer: Mechanistic Aspects*. Springer Nature, Singapore, pp. 1–14.

Bakkali, F., Averbeck, S., Averbeck, D., Idaomar, M. 2008. Biological effects of essential oils—A review. *Food and Chemical Toxicology* 46(2): 446–475.

Bartnik, M., Facey, P. 1917. Fundamentals, Applications and Strategy. *Pharmacognosy* 8: 101–161.

Battino, M., Giampieri, F., Cianciosi, D., Ansary, J., Chen, X., Zhang, D., Gila, E., Forbes-Hernández, T. 2021. The roles of strawberry and honey phytochemical on human health: A possible clue on the molecular mechanisms involved in the prevention of oxidative stress and inflammation. *Phytomedicine* 86: 153170.

Belščak-Cvitanović, A., Durgo, K., Huđek, A., Bačun-Družina, V., Komes, D. 2018. Overview of polyphenols and their properties. In Galanakis, C.M., ed., *Polyphenols: Properties, Recovery, and Applications*, pp. 3–44. Woodhead Publishing.

Bojadzievski, P. 1992. The health services in Bitola through the centuries. In *Bitola Society of Science and Art*, pp. 15–27. Drustvo za nauka i umetnost.

Bottcher, H. 1965. Miracle drugs. In *Zagreb Zora*, pp. 23–139.

Bradford, P.G., Awad, A.B. 2007. Phytosterols as anticancer compounds. *Molecular Nutrition Food Research* 51(2): 161–170.

Bravo, L. 1998. Polyphenols: chemistry, dietary sources, metabolism, and nutritional significance. *Nutrition Reviews* 56(11): 317–333.

Capuano, E., Dekker, M., Verkerk, R., Oliviero, T. 2017. Food as pharma? The case of glucosinolates. *Current Pharmaceutical Design* 23(19): 2697–2721.

Dave, A., Parande, F., Park, E.J., Pezzuto, J.M. 2020. Phytochemicals and cancer chemoprevention. *Journal of Cancer Metastasis Treatment* 6: 1–36.

Dennis, K.K., Go, Y.M., Jones, D.P. 2019. Redox systems biology of nutrition and oxidative stress. *Journal of Nutrition* 149: 553–565.

Di Mattia, C., Martuscelli, M., Sacchetti, G., Beheydt, B., Mastrocola, D., Pittia, P. 2017. Effect of drying techniques on antioxidant activity of fruits and vegetables. *Critical Reviews in Food Science and Nutrition*, 57(2): 187–211.

Espina, L., Somolinos, M., Lorán, S., Conchello, P., García, D., Pagán, R. 2011. Chemical composition of commercial citrus fruit essential oils and evaluation of their antimicrobial activity acting alone or in combined processes. *Food Control* 22(6): 896–902.

Falleh, H., Jemaa, M.B., Saada, M., Ksouri, R. 2020. Essential oils: A promising eco-friendly food preservative. *Food Chemistry* 330: 127268.

Farhadi, F., Khameneh, B., Iranshahi, M., Iranshahy, M. 2019. Antibacterial activity of flavonoids and their structure–activity relationship: An update review. *Phytotherapy Research* 33: 13–40.

Guan, R., Quyet, Van Le, Yang, H., Zhang, D., Gu, H., Yang, Y., Sonne, C., Lam, S.S., Zhong, J., Jianguang, Z., Liu, R., Peng, W. 2021. A review of dietary phytochemicals and their relation to oxidative stress and human diseases. *Chemosphere* 271: 129499.

Hanahan, D. 2022. Hallmarks of cancer: New dimensions. *Cancer Discovery* 12: 31–46.

Kabir, M.T., Rahman, M.H., Shah, M., Jamiruddin, M.R., Basak, D., Al-Harrasi, A., Bhatia, S., Ashraf, G.M., Najda, A., El-Kott, A.F. 2022. Therapeutic promise of carotenoids as antioxidants and anti-inflammatory agents in neurodegenerative disorders. *Biomedicine Pharmacotherapy* 146: 112610.

Kovacevic, N. 2000. Fundamentals of pharmacognosy. In *Beograd Personal Edition*, pp. 170–171.

Kristo, A.S., Klimis-Zacas, D., Sikalidis, A.K. 2016. Protective role of dietary berries in cancer. *Antioxidants* 5: 1–23.

Kumar, S., Gupta, S. 2021. Dietary phytochemicals and their role in cancer chemoprevention. *Journal of Cancer Metastasis Treatment* 7: 51.

Kumar, S., Pandey, A.K. 2013. Chemistry and biological activities of flavonoids: An overview. *The Scientific World Journal* 1: 162750.

Kuzma, J.N., Schmidt, K.A., Kratz, M. 2017. Prevention of metabolic diseases: Fruits including fruit sugars vs. vegetables. *Current Opinion in Clinical Nutrition Metabolic Care* 20(4): 286–293.

Kytidou, K., Artola, M., Overkleeft, H.S., Aerts, J.M. 2020. Plant glycosides and glycosidases: A treasure-trove for therapeutics. *Frontiers in Plant Science* 11: 357.

Li, S., Wu, B., Fu, W., Reddivari, L. 2019. The anti-inflammatory effects of dietary anthocyanins against ulcerative colitis. *International Journal of Molecular Sciences* 20(10): 2588.

Liang, Xue., Chang, Liu., Hang, Ma., Navindra, P. Seeram., Catherine, C. Neto. 2022. Anti-inflammatory activities of cranberry fruit extracts in human THP-1 monocytes are influenced by their phytochemical composition. *ACS Food Science Technology* 2(1): 75–83.

Limketkai, B.N., Gordon, M., Mutlu, E.A., De Silva, P.S., Lewis, J.D. 2020. Diet therapy for inflammatory bowel diseases: A call to the dining table. *Oxford University Press US* 26: 510–514.

Lukic, P. 1985. Pharmacognosy. In *Beograd SSO Faculty of Pharmacy*, pp. 8–22.

Malarvizhi, G., Anusha, G., Manjari, K.A., Tamilselvan, S., Santhakumari, S., Sivamani, S. 2025. An overview on the mechanisms of encapsulation of polyphenols used to help fight diabetes mellitus. *Discover Food* 5: 185.

Mandal, M., Sahoo, S.K., Patra, P., Mallik, S., Zhao, Z. 2020. In silico ranking of phenolics for therapeutic effectiveness on cancer stem cells. *BMC Bioinformatics* 21: 499.

Marangoni, F., Poli, A. 2010. Phytosterols and cardiovascular health. *Pharmacological Research* 61(3): 193–199.

Matera, R., Lucchi, E., Valgimigli, L. 2023. Plant essential oils as healthy functional ingredients of nutraceuticals and diet supplements: A review. *Molecules* 28(2): 901.

McAnulty, L.S., Nieman, D.C., Dumke, C.L., Shoote, L.A., Henson, D.A., Utter, A.C., Milne, G., McAnulty, S.R. 2011. Effect of blueberry ingestion on natural killer cell counts, oxidative stress, and inflammation prior to and after 2.5 h of running. *Applied Physiology Nutrition Metabolism* 36(6): 976–984.

Mercader, J. 2009. Mozambican grass seed consumption during the Middle Stone Age. *Science* 326(5960): 1680–1683.

Mohammadzadeh Honarvar, N., Saedisomeolia, A., Abdolahi, M., Shayeganrad, A., Taheri Sangsari, G., Hassanzadeh Rad, B., Muench, G. 2017. Molecular anti-inflammatory mechanisms of retinoids and carotenoids in Alzheimer's disease: A review of current evidence. *Journal of Molecular Neuroscience* 61: 289–304.

Molan, A.L., Liu, Z., Plimmer, G. 2014. Evaluation of the effect of blackcurrant products on gut microbiota and on markers of risk for colon cancer in humans. *Phytotherapy Research* 28: 416–422.

Molino, S., Pilar Francino, M., Ángel Rufián Henares, J. 2023. Why is it important to understand the nature and chemistry of tannins to exploit their potential as nutraceuticals? *Food Research International* 173: 113329.

Moreau, R.A., Whitaker, B.D., Hicks, B. 2002. Phytosterols, phytostanols, and their conjugates in foods: structural diversity, quantitative analysis, and health-promoting uses. *Progress in Lipid Research* 41(6): 457–500.

Nile, S. H., Park, S. W. 2014. Edible berries: Bioactive compounds and their effect on human health. *Nutrition* 30: 134–144.

Olas, B. 2016. Sea buckthorn as a source of important bioactive compounds in cardiovascular diseases. *Food Chemistry Toxicology* 97: 199–204.

Olas, B. 2017. The multifunctionality of berries toward blood platelets and the role of berry phenolics in cardiovascular disorders. *Platelets* 28: 540–549.

Ostlund Jr, R.E. 2002. Phytosterols in human nutrition. *Annual Review of Nutrition* 22(1): 533–549.

Panche, A.N., Diwan, A.D., Chandra, S.R. 2016. Flavonoids: an overview. *Journal of Nutritional Science* 5: e47.

Rahman, S.U., Li, Y., Huang, Y., Zhu, L., Feng, S., Wu, J., Wang, X. 2018. Treatment of inflammatory bowel disease via green tea polyphenols: Possible application and protective approaches. *Inflammopharmacology* 26: 319–330.

Raimundo, A.F., Félix, F., Andrade, R., García-Conesa, M.T., González-Sarrías, A., Gilsa-Lopes, J., Do, Ó.D., Raimundo, A., Ribeiro, R., Rodriguez-Mateos, A. 2020. Combined effect of interventions with pure or enriched mixtures of polyphenols and anti-diabetic medication in type 2 diabetes management: A meta-analysis of randomized controlled human trials. *European Journal of Nutrition* 59: 1329–1343.

Ranjan, A., Ramachandran, S., Gupta, N., Kaushik, I., Wright, S., Srivastava, S., Das, H., Srivastava, S., Prasad, S., Srivastava, S.K. 2019. Role of phytochemicals in cancer prevention. *International Journal of Molecular Sciences* 20: 4981.

Ranjith, F.H., Adhikari, B., Muhialdin, B.J., Yusof, N.L., Mohammed, N.K., Ariffin, S.H., Hussin, A.S. 2022. Peptide-based edible coatings to control postharvest fungal spoilage of mango Mangifera indica L fruit. *Food Control* 135: 108789.

Rauf, A., Imran, M., Khan, I.A., Ur-Rehman, M., Gilani, S.A., Mehmood, Z., Mubarak, M.S. 2018. Anticancer potential of quercetin: A comprehensive review. *Phytotherapy Research* 32: 2109–2130.

Ren, B., Kwah, M.X.-Y., Liu, C., Ma, Z., Shanmugam, M.K., Ding, L., Xiang, X., Ho, P.C.-L., Wang, L., Ong, P.S., et al. 2021. Resveratrol for cancer therapy: Challenges and future perspectives. *Cancer Letters* 515: 63–72.

Salamon, I., Şimşek Sezer, E.N., Kryvtsova, M., Labun, P. 2021. Antiproliferative and antimicrobial activity of anthocyanins from berry fruits after their isolation and freeze-drying. *Applied Sciences* 11(5): 2096.

Saleem, M., Ali, M., Gulshan, A.B. 2024. Nutritional uses of the family Rutaceae. In *Phytochemical and Pharmacological Investigation of the Family Rutaceae*, pp. 147–160. Apple Academic Press.

Sargent, J.R., Tocher, D.R., Bell, J.G. 2002. The lipids. In Halver, J.E., Hardy, R.W., eds., *Fish Nutrition* 3rd ed., pp. 181–257. San Diego, CA: Elsevier Academic Press.

Scarcelli, N., Cubry, P., Akakpo, R., Thuillet, A.C., Obidiegwu, J., Baco, M.N., Otoo, E., Sonké, B., Dansi, A., Djedatin, G., Mariac, C., Couderc, M., Causse, S., Alix, K., Chaïr, H., François, O., Vigouroux, Y. 2019. Yam genomics supports West Africa as a major cradle of crop domestication. *Science Advances* 5: eaaw1947.

Seow, Y.X., Yeo, C.R., Chung, H.L., Yuk, H.G. 2014. Plant essential oils as active antimicrobial agents. *Critical Reviews in Food Science and Nutrition* 54(5): 625–644.

Sharmeen, J.B., Mahomoodally, F.M., Zengin, G., Maggi, F. 2021. Essential oils as natural sources of fragrance compounds for cosmetics and cosmeceuticals. *Molecules* 26(3): 666.

Shen, Q.K., Peng, M.S., Adeola, A.C., Kui, L., Mansaray, L.R., Nganga, S.I., Atushhumuze, E., Agwanda, B., Isah, S., Wang, M.S., Wu, D.D., Zhang, Y.P. 2021. Genomic analyses unveil helmeted guinea fowl Numida meleagris domestication in West Africa. *Journal of Genetics and Genomics* 48(4): 313–323.

Singh, B., Singh, J.P., Kaur, A., Singh, N. 2022. Immunomodulatory potential of phytochemicals and other bioactive compounds of fruits: A review. *Food Frontiers* 3(1): 15–32.

Singla, R.K., Dubey, A.K., Garg, A., Sharma, R.K., Fiorino, M., Ameen, S.M., Haddad, M.A., Al-Hiary, M. 2019. Natural polyphenols: Chemical classification, definition of classes, subcategories, and structures. *Journal of AOAC International* 102(5): 1397–1400.

Skrovankova, S., Sumczynski, D., Mlcek, J., Jurikova, T., Sochor, J. 2015. Bioactive compounds and antioxidant activity in different types of berries. *International Journal of Molecular Sciences* 16: 24673–24706.

Slika, H., Mansour, H., Wehbe, N., Nasser, S.A., Iratni, R., Nasrallah, G., Shaito, A., Ghaddar, T., Kobeissy, F., Eid, A.H. 2022. Therapeutic potential of flavonoids in cancer: ROS mediated mechanisms. *Biomedicine Pharmacotherapy* 146: 112442.

Sosalagere, C., Kehinde, B.A., Sharma, P. 2022. Isolation and functionalities of bioactive peptides from fruits and vegetables: A review. *Food Chemistry* 366: 130494.

Tabassum, N., Vidyasagar, G.M. 2013. Antifungal investigations on plant essential oils: A review. *International Journal of Pharmacy and Pharmaceutical Sciences* 5(2): 19–28.

Tena, N., Martín, J., Asuero, A.G. 2020. State of the art of anthocyanins: Antioxidant activity, sources, bioavailability, and therapeutic effect in human health. *Antioxidants* 9(5): 451.

Teresita, Y., Lorena, P.P., Estefania, M., Marlon, O. 2023. Mechanisms of action of fruit and vegetable phytochemicals in colorectal cancer prevention. *Molecules* 28(11): 4322.

Toh, J.Y., Tan, V.M., Lim, P.C., Lim, S.T., Chong, M.F. 2013. Flavonoids from fruit and vegetables: a focus on cardiovascular risk factors. *Current Atherosclerosis Reports* 15: 1–7.

Toplak, Galle K. 2005. *Domestic medicinal plants*, pp. 60–61. Mozaic Book.

Tuli, H.S., Tuorkey, M.J., Thakral, F., Sak, K., Kumar, M., Sharma, A.K., Sharma, U., Jain, A., Aggarwal, V., Bishayee, A. 2019. Molecular mechanisms of action of genistein in cancer: Recent advances. *Frontiers in Pharmacology* 10: 1336.

Tuygunovna, S.S. 2023. Medicinal plants containing glycosides. *European Journal of Business Startups and Open Society* 3(12): 24–27.

Uddin, M.S., Hossain, M.F., Al Mamun, A., Shah, M.A., Hasana, S., Bulbul, I.J., Sarwar, M.S., Mansouri, R.A., Ashraf, G.M., Rauf, A. 2020. Exploring the multimodal role of phytochemicals in the modulation of cellular signaling pathways to combat age-related neurodegeneration. *Science of the Total Environment* 725: 138313.

Vendrame, S., Klimis-Zacas, D. 2019. Dietary supplementation with blueberry partially restores T-cell-mediated immune response in mice with diet-induced obesity. *British Journal of Nutrition* 121(11): 1251–1259.

Vilahur, G., Ben-Aicha, S., Diaz-Riera, E., Badimon, L., Padró, T. 2019. Phytosterols and inflammation. *Current Medicinal Chemistry* 26(37): 6724–6734.

Wang, X.Q., Wang, W., Peng, M., Zhang, X.Z. 2021. Free radicals for cancer theranostics. *Biomaterials* 266: 120474.

Wiart, C. 2006. *Ethnopharmacology of medicinal plants*, pp. 1–50. Humana Press, Totowa, NJ.

Yahia, E.M., García-Solís, P., Celis, M.E.M. 2018. Postharvest physiology and biochemistry of fruits and vegetables. In *Contribution of Fruits and Vegetables to Human Nutrition and Health*, pp. 19–45. Elsevier, Amsterdam.

Zhang, Y., Wang, L., Li, J., Gao, H., Hu, W., Zhang, Z., Xu, X. 2024. Evaluation of immunomodulatory potential of Neolamarckia cadamba fruit extract in mice. *3 Biotech* 14: 106.

Zuzarte, M., Salgueiro, L. 2015. Essential oils chemistry. *Bioactive Essential Oils and Cancer* 30: 19–61.

19 Sustainable Harvesting and Cultivation Practices for Medicinal Fruits and Berries

Thushara Indika Sampath, Anoja Priyadarshani Attanayake, and Nayomi Deshani De Silva

19.1 INTRODUCTION

19.1.1 Role of Fruits and Berries in Traditional and Modern Medicines

Fruits and berries have been used as both traditional and modern medicines for thousands of years by many cultures worldwide. Many people have relied on fruits and berries mainly as nutritional supplements and for medicinal purposes throughout history. Ancient people discovered fruits and berries with healing properties, and this knowledge has grown gradually. Today, Ayurveda, Native American herbal practices, Chinese traditional medicine, and Greco-Arab medicine prescribe many natural and nature-based medicines that are obtained from fruits, berries, or their extracts to treat diseases such as diabetes, cardiovascular diseases, infections, cancer, and so on (Husen 2021, 2022, 2024; Latif and Nawaz 2025). Examples of commonly used medicinal fruits and berries, along with their traditional uses and pharmacological activities, are presented in Table 19.1.

Fruits and berries are an important source of a healthy diet. There are four key risk factors in the epidemiology of non-communicable diseases, including physical inactivity, unhealthy diet, chronic use of alcohol, and tobacco, that have a major impact on public health. According to the estimation of the World Health Organization (WHO) report in 2017, around 3.9 million deaths were caused by poor consumption of fruits and vegetables (WHO 2023). The WHO has recommended increasing attention to non-communicable diseases by focusing on prevention and management through health-endorsed lifestyle modifications (WHO 2025). The recommended approach to recovering from this issue is to increase consumption of fruits and berries, which provide important health benefits. The health benefits of consuming fruits and berries are mainly due to their fiber, protein, vitamin, and micronutrient content, as well as their phytochemicals (Ciccoritti et al. 2024).

There are several ways to use fruits and berries in traditional applications. As examples, grapes (*Vitis vinifera* L., Family: Vitaceae), guava (*Psidium guajava* L., Family: Myrtaceae), noni (*Morinda citrifolia* L., Family: Rubiaceae), papaya (*Carica papaya* L., Family: Caricaceae), and garcinia (*Garcinia gummi-gutta* L., Family: Clusiaceae) fruits have been used for wound healing properties and are used in traditional practices in India and other regions (Chaudhury et al. 2024; Joshi and Kharkwal 2025; Swathi et al. 2024). Fruits such as kiwi (*Actinidia chinensis* var. *deliciosa*, Family: Actinidiaceae), amla (*Phyllanthus emblica* L., Family: Phyllanthaceae), and papaya (*Carica papaya* L., Family: Caricaceae) facilitate digestion, while blueberries (*Vaccinium* spp., Family: Ericaceae), blackberries (*Rubus fruticosus* L., Family: Rosaceae), and chokeberries (*Aronia melanocarpa* (Michx.) Elliott, Family: Rosaceae) are used as stimulants. Cranberries (*Vaccinium macrocarpon*

DOI: 10.1201/9781003688259-19

TABLE 19.1
Major Medicinal Fruits and Berries with Traditional Uses and Pharmacological Activities

Medicinal Fruit and Berries	Origin	Disease/Traditional Uses	Pharmacological/Medicinal Activities	Key References
Amla/Indian gooseberry (*Phyllanthus emblica* L.) Family: Phyllanthaceae	Tropical and Southern Asia	Rejuvenator, diarrhea, jaundice, and inflammation	Antioxidant, antidiabetic, cardioprotective, and hepatoprotective	Saini et al. (2022)
Apple (*Malus domestica*) Family: Rosaceae	Non-native	Exhilarant, cardiotonic, stomachic, and astringent	Antioxidant, anti-inflammatory, antimicrobial, cardioprotective, and gut health	Hammad and Malik (2021); Zagórska-Dziok et al. (2025)
Blackberry (*Rubus fruticosus* L.) Family: Rosaceae	Native chiefly to North temperate regions	Tonic and astringent	Antioxidant, anti-inflammatory, and antimicrobial	Aliyu et al. (2025); Eshonkhujaev (2025); Lee et al. (2024)
Blackcurrant (*Ribes nigrum* L.) Family: Grossulariaceae	Northern Europe and Northern Asia	Colds, inflammation, and infections	Antioxidant, antihyperlipidemic, antibacterial, antitumor, antiviral, and neuroprotective	da Costa et al. (2025); Izteleuova et al. (2025)
Blueberry (*Vaccinium* spp.) Family: Ericaceae	North America, Northern and Central Europe, and the Northern part of Asia	Stomachaches, fever, and diarrhea	Antioxidant, anti-inflammatory, cardioprotective, cognitive benefits, and anticancer	Białoszycka et al. (2025); Chatterjee et al. (2025)
Chokeberry/Aronia (*Aronia melanocarpa* (Michx.) Elliott) Family: Rosaceae	Native to eastern North America	Tonic, used in juices, jams, wines, and as dietary supplements	Antioxidant, anti-inflammatory, antidiabetic, antimicrobial, anticancer, and cardioprotective	Sagandyk et al. (2024)
Cranberry (*Vaccinium macrocarpon* Ait. L) Family: Ericaceae	North America	Treat against urinary problems, urinary disorders,	Antioxidant, anti-inflammatory, antidiabetic, antimicrobial, and anticancer	Al-Zubaidi et al. (2025); Xu et al. (2025)
Elderberry (*Sambucus nigra* L.) Family: Adoxaceae	Native plants across Europe, North America, and Western Asia	Colds, influenza, and skin inflammation	Antiviral, immunomodulatory, antioxidant, and antimicrobial	Uhl and Mitchell (2024)
Goji/Wolfberry (*Lycium barbarum* L.) Family: Solanaceae	Native to Asia	Nourishing the liver and kidneys, and a vision tonic	Antioxidant, immunomodulatory, hepatoprotective, antidiabetic, and neuroprotective	Zhang et al. (2025); Shi et al. (2025)
Grape (*Vitis vinifera* L.) Family: Vitaceae	Western Asia	Use as laxatives, carminatives, and as drug therapy for cold, flu, anemia, and wound healing	Resveratrol/ proanthocyanidins: antioxidant, cardioprotective, and anticancer	Devi et al. (2024)
Guava (*Psidium guajava* L.) Family: Myrtaceae	Southern Mexico into or through Central America	Diarrhea, wound healing, and treatment against gastrointestinal disorders	Antioxidant, antimicrobial, antidiarrheal, antidiabetic, and anti-inflammatory	Mahavy et al. (2024); Möwes et al. (2025)

(*Continued*)

TABLE 19.1 (Continued)

Medicinal Fruit and Berries	Origin	Disease/Traditional Uses	Pharmacological/Medicinal Activities	Key References
Hawthorn (*Crataegus rhipidophylla* Gand.) Family: Rosaceae	Northern Europe	Used to strengthen the stomach and promote blood circulation	Antioxidant, anti-inflammatory, antibacterial, and intestinal health, antidiabetic, cardioprotective, anticancer	Ma et al. (2024); Yang et al. (2025)
Kiwi (*Actinidia chinensis* var. *deliciosa*) Family: Actinidiaceae	Central and South China	Digestive tonic	Antidiabetic, anticancer, anti-inflammatory, anti-ulcer, antioxidant activity, and cardioprotective	Li et al. (2023); Moysidou et al. (2024)
Mangosteen (*Garcinia mangostana* L.) Family: Clusiaceae	Southeast Asia	Inflammation and skin disorders	Antioxidant, anti-inflammatory, antibacterial, antiviral, antifibrotic, antimalarial, anticancer, and neuroprotective activities	Arshad et al. (2025); Zhang et al. (2024)
Mulberry (*Morus alba* L.) Family: Moraceae	Asia and North America	Pain-relieving, tonic	Antioxidant, hypoglycemic, lipid-lowering, hepatoprotective, anticancer, and anti-inflammatory	Fatima et al. (2024); Uyak et al. (2024)
Noni (*Morinda citrifolia* L.) Family: Rubiaceae	Southeast Asia and Australia	Hypertension and diabetes	Antibacterial, antiviral, antifungal, anticancer, anthelminthic, analgesic, hypotensive, anti-inflammatory, and immune-boosting effects	Chavda et al. (2025); Ponmurugan et al. (2025)
Papaya (*Carica papaya* L.) Family: Caricaceae	Tropical American fruit tree	Dengue fever, digestive aid, intestinal helminthiasis, and hypertension	digestive aid, antioxidant, anti-inflammatory, antimicrobial, treatment of mouth ulcers, gum disease, and toothaches	Babalola et al. (2024); Fernando et al. (2024)
Pomegranate (*Punica granatum* L.) Family: Lythraceae	Originated from Iran	Stomach diseases, skin diseases, mouth and throat inflammations, and intestinal bleeding	Antioxidant, hypoglycemic, anti-inflammatory, analgesic effect, cardioprotective, antimicrobial, anticancer	Khadivi et al. (2024); Mirzayeva and Jo'raboyev (2025); Wang et al. (2024)
Raspberry (*Rubus idaeus* L.) Family: Rosaceae	Native to Europe and Northern Asia	Digestive aid and tonic	Antioxidant, antimicrobial, anti-inflammatory, antidiabetic, and anticancer	Ispiryan et al. (2024); Vukelić-Nikolić et al. (2025)
Strawberry (*Fragaria* spp.) Family: Rosaceae	Native to the Northern Hemisphere	Diuretic, laxative, carminative, digestive, odontalgic, and cardiotonic	Antioxidant, anti-inflammatory, antimicrobial, and cognitive/skin benefits	Bebek Markovinović et al. (2022); Hussain et al. (2024)

Ait. L., Family: Ericaceae) are mainly used to treat urinary problems in traditional practices. Indeed, most of the fruits are used as stimulants (Nollet and Ahamad 2025). Scientific studies have validated many of the important traditional uses. These fruits and berries are known to be a rich source of bioactive phytochemicals such as alkaloids, flavonoids, anthocyanins, phenolic acids, tannins, terpenes, and steroids (Otero-Guzman and Andrade-Pizarro 2025). The bioactivities of these phytochemicals mainly include antioxidant, anti-inflammatory, antidiabetic, cardioprotective, hepatoprotective, neuroprotective, antimicrobial, cognitive benefits, anticancer, and other bioactivities (Hossain et al. 2025). The phytochemical composition of fruits and/or berries depends on many factors such as plant species, variety, growing conditions, soil type, climate conditions, ripening stage, post-harvesting, handling, etc. (Kaur et al. 2024; Mieszczakowska-Frąc et al. 2025). In modern medicine and nutraceutical research, the extraction and isolation of bioactive compounds from these fruits and berries have led to the development of new drug entities for the management of chronic diseases, modulating immunity, and promoting a healthy life. The transition from traditional medicines to research-based applications demonstrates how these medicinal fruits and berries connect cultural knowledge with modern medicines.

19.1.2 Global Demand and Rising Market

The global demand for fruits and berries has been rising tremendously as consumers pay more attention to human wellbeing, especially in terms of human health. Consumers are tending toward natural products and nature-based supplements as a part of preventive healthcare and a wellness lifestyle. This trend has expanded the market demand for fruits and berries. There are different types of product categories of fruits and berries in the market, with premium berries accounting for 55.3% of the market share, while specialty fruits are estimated to grow at a 6.7% compound annual growth rate in 2024 (Pulidindi and Ahuja 2025). The demand for fruits and berries varies by region, and there is still a huge demand for these products. North America is still the biggest market by a long way for superior fruits and berries (Pulidindi and Ahuja 2025). Europe is well known for its application in organic and greenhouse production. Consumers in the North America and Europe regions are mainly seeking superfoods and functional foods with health benefits. These regions are mainly focused on supplement and nutraceutical types (Shan et al. 2025). The Asia Pacific region is the fastest-growing market, and many countries are fast-growing for the cultivation of fruits and berries (Teng 2025). Africa, Latin America, and the Middle East are also introducing modern cultivation practices and developing the capacity for exporting fruits (Pulidindi and Ahuja 2025). To fulfil this requirement, globalization and the opening up of international markets during this time have increased competition among producers of fruits and berries from different countries globally (Mukhametzyanov et al. 2023). The rising market value also puts extreme harvesting pressure on wild populations and reduces the recovery of the wild population. Moreover, the lack of an efficient supply chain has decreased the market from being fully satisfied.

19.1.3 Risk of Overharvesting

Many medicinal fruits and berries are still obtained from wild populations instead of through cultivation. The world biodiversity disaster is reaching a critical stage, with increasing overharvesting as a dominant risk to the ecological stability of the planet. Overharvesting of natural resources breaks genetic diversity, disrupts the natural regeneration cycle, and risks the survival of the native plant populations (Sana et al. 2025). For example, the wild sea buckthorn (*Hippophae rhamnoides* L., Family: Elaeagnaceae) fruit plant population has been reduced significantly in parts of Central Asia due to overharvesting of the plant, and this is a good example of how overharvesting can affect biodiversity loss and disrupt the natural regeneration cycle, ultimately leading to ecological damage (Fang et al. 2025). Overharvesting of amla/Indian gooseberry (*Phyllanthus emblica* L., Family: Phyllanthaceae) is another example and has led to loss of natural stands and habitat degradation in the Central Western Ghats region in India (Geethika et al. 2024). The increase in monoculture plantations for high-demand fruits may result in deforestation, loss of soil fertility, and reduced local biodiversity. In addition to the medicinal properties of these fruits and berries, they play a critical role in maintaining the food chain by providing nutrition to animals. Unsustainable removal of these resources can also disrupt the food chain, disrupt feeding patterns, and destabilize ecological interactions. Further, this leads to accelerated soil erosion, reduces water retention, and impairs the overall health of the ecosystem. These overharvesting and unsustainable cultural practices represent a significant warning to ecological balance and biodiversity conservation.

19.1.4 Importance of Sustainable Harvesting and Cultivation Practices

Sustainable harvesting of fruits and berries is important to address the increasing global demand, reduce biodiversity loss, and stabilize the ecosystem (Simsek 2025). As shown in Figure 19.1, the sustainable use of medicinal fruits and berries depends on several interconnected factors such as ecological sustainability, community participation, technological support, and strong policy actions. These factors together highlight that sustainable harvesting requires conserving natural habitats and biodiversity, ensuring local engagement, applying modern propagation and post-harvesting techniques, and following national and global conservation guidelines. Sustainable harvesting can mitigate the environmental impact of farming and efficiently manage renewable resources by applying eco-friendly farming practices. Sustainable harvesting methods not only protect natural resources but also play an important role in climate change mitigation. These methods facilitate responsible management, the development of biodiversity conservation, and the stabilization of natural habitats. Moreover, sustainable harvesting of fruits and berries in wild habitats assists in conserving ecosystem services such as better water quality, low soil erosion, and ultimately leads to forestation (Sharma and Birman 2024). It is important to highlight that sustainable harvesting of fruits and berries contributes to the economy while preserving the native plant species for future generations.

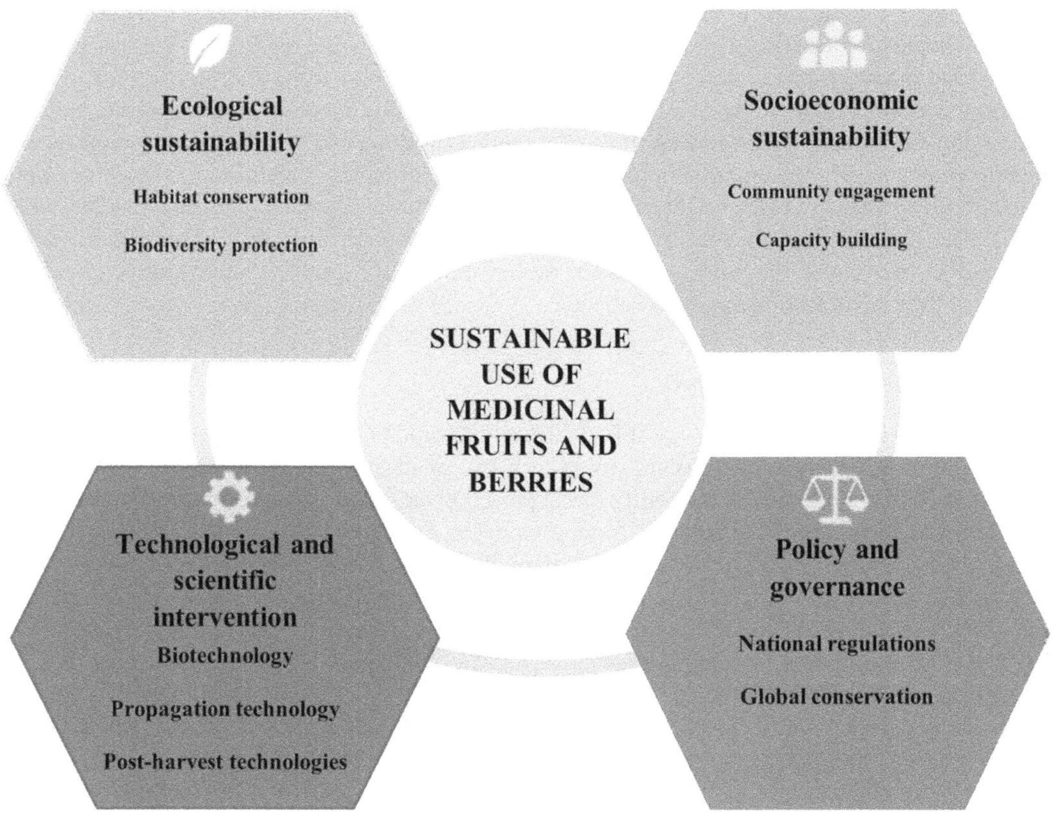

FIGURE 19.1 Conceptual framework for sustainable use of medicinal fruits and berries.

Therefore, allowing for sustainable harvesting is important for managing a balance between resource use and environmental conservation.

19.1.5 Roadmap for Sustainable Harvesting and Cultivation Practices for Medicinal Fruits and Berries

This chapter explains the sustainable harvesting and cultivation practices of medicinal fruits and berries. It outlines the importance of fruits and berries in traditional and modern medicine, along with global demand trends and economic implications. This further aims to discuss the ecological threats connected to overharvesting and unsustainable cultivation and to present sustainable approaches for harvesting and farming, integrating both traditional knowledge and modern science. In addition, this aims to offer a framework for policy, research, and community action to ensure the continued availability of medicinal fruits and berries for future generations. These visions provide the knowledge needed to guide resource management and conservation planning.

19.2 THREATS AND CHALLENGES IN SUSTAINABLE UTILIZATION

19.2.1 Overexploitation and Habitat Destruction

Overexploitation and habitat destruction are considered major risk factors for sustainable utilization (Goswami 2024). Overexploitation, also known as overharvesting or overshoot of an ecosystem, refers to using a renewable resource at a rate that leads to reduced returns (Torrance and Tomlinson 2025). This overuse can decrease the resources as their natural regeneration capacity becomes insufficient. The growing global demand for medicinal fruits and berries has increased harvesting pressures on both wild and cultivated populations. There are some fruit and berry species, such as elderberry (*Sambucus nigra* L., Family: Adoxaceae), sea buckthorn (*Hippophae rhamnoides* L., Family: Elaeagnaceae), and camu-camu (*Myrciaria dubia* (Kunth) McVaugh, Family: Myrtaceae), that remain primarily found in wild habitats (Aguilera and Toledo, 2024). Uncontrolled and unregulated collection of these fruits and berries can reduce fruiting individuals, change reproductive cycles, and limit seed dispersal, ultimately weakening population regeneration. Overharvesting not only reduces the availability of resources but also risks the survival of fruit species that are already rare or slow-growing. Habitat loss of fruit or berry species also increases the pressure on their long-term survival (Guo et al. 2024). According to a study by Guo et al. (2024), it was shown that habitat loss significantly reduces the distribution of apricot (*Prunus armeniaca* L., Family: Rosaceae), wild apple (*Malus sieversii* (Ledeb.) M.Roem., Family: Rosaceae), and dwarf Russian almond (*Prunus ledebouriana* (Schltdl.) Y.Y.Yao, Family: Rosaceae), and these fruit species are shifting toward cooler northern or higher elevation regions. The loss of suitable habitat for these species is due to climatic niche shifts, which increase their vulnerability to population failure and local extinction (Guo et al. 2024). Natural habitats for medicinal fruits are being limited to high-altitude ecosystems due to agricultural practices, urbanization, and other infrastructure developments (Shinwari et al. 2024). When these pressures combine with the overexploitation, they place the medicinal fruit and berry species at serious risk and disrupt the ecological functions that depend on them. Proper cultivation practices and avoiding overharvesting facilitate restoring the survival of the species and maintaining ecosystem stability.

19.2.2 Climate Changes

Climate change also creates a fundamental threat to the sustainable utilization of medicinal fruit and berry species. Temperature changes and precipitation patterns can influence the phenology, growth, and reproductive success of medicinal fruit plants (Szot and Łysiak 2025). These factors are altered because of climate change, which is driven by global warming caused by carbon pollution. The distribution of many berry and tropical fruit plants has limited ecological ranges. Well-known examples include goji berries (*Lycium barbarum* L., Family: Solanaceae) and acai berries (*Euterpe oleracea* L., Family: Arecaceae), which grow best under specific climatic conditions. The quality of berries strongly depends on a cool, dry climate with abundant sunlight conditions (Song et al. 2025; Zhao et al. 2024). Acai berries grow best in warm, humid tropical climates with abundant rainfall and consistent temperatures (Costa et al. 2024; Teixeira et al. 2025). Changes in rainfall patterns, increasing temperatures, and extreme weather events can decrease yields of fruits and berries, and further expose these plants to new pests and pathogens (Deori et al. 2024). Moreover, climate change can increase species vulnerability, which is already exacerbated by overexploitation and habitat loss. These plant species may fail to migrate or adapt quickly to climate change, which ultimately leads to local extinctions. Conservation plans, therefore, must incorporate climate forecasts to identify optimal habitats, implement assisted migration where appropriate, and ensure that harvesting does not increase climate-related pressures on medicinal fruit and berry cultivation, while urgently addressing carbon pollution and preparing for the impacts of global warming. These actions can be implemented by introducing strong policies to address climate change, working with businesses to reduce carbon emissions, and raising awareness of how to address global warming.

19.2.3 Unsustainable Agricultural Practices

Unsustainable agriculture means the implementation of cultivation practices that ultimately lead to a decline in the quality of field soil, land, and other natural resources that are needed for sustainable farming (Mishra 2025). Several factors contribute to unsustainable agriculture, such as monoculturing, technological mess, the excessive use of fertilizers, pesticides, and herbicides, and poor farming techniques (Jocien and Nfor 2023). Large-scale monocultures for popular fruits such as blueberries (*Vaccinium* spp., Family: Ericaceae), grapes (*Vitis vinifera* L., Family: Vitaceae), and pomegranates (*Punica granatum* L., Family: Lythraceae) regularly require extensive pesticide and fertilizer use (Woldegebriel 2025). The addition of extensive chemicals to the soil can degrade soil quality, increase water contamination, and reduce biodiversity around the field. The use of new technology for harvesting and land clearing further damages the local ecosystems and may eliminate native species that play key roles in the ecosystem. Large-scale monoculturing of acai (*Euterpe oleracea* L., Family: Arecaceae) or goji (*Lycium barbarum* L., Family: Solanaceae) berries has replaced the diversity of agroforestry systems in pursuit of maximum yield. Overall, these unsustainable agricultural practices threaten the long-term productivity of the medicinal fruits and berries and may ultimately disrupt the ecological balance.

19.2.4 Socioeconomic Pressure on Local Communities

Local communities always face the challenge of meeting customer requirements while protecting their natural resources. Selling fruits and berries provides a source of income for rural and indigenous populations, especially in developing countries. Local communities in these regions always try to fulfil the market demand for high-demand species without following sustainable harvesting and cultivation practices, which may be due to limited access to proper education on these cultivation practices, credit, or technical support. Another barrier to sustainable harvesting is that large-scale companies in this production system can capture most of the economic value, and most of them do not invest in sustainable practices. As a result, local communities can prioritize short-term harvests and increase the exploitation of natural resources while disregarding conservation strategies. These efforts ultimately build up pressure on local communities and turn the issue into a socioeconomic problem.

19.2.5 Genetic Erosion and Loss of Wild Relatives

Another serious risk to the sustainable utilization is the genetic erosion of medicinal fruits and berries along with their wild relatives. Genetically modified cultivation always focuses on high-yielding and commercially attractive varieties of fruits and berries. This can lead to a reduction in the genetic diversity of the cultivated population (Yadav et al. 2024). The mother plant passes key traits to the next generation, such as disease resistance and climate tolerance, and losing these traits can reduce future improvement of the selective crop. For example, wild populations of blueberries (*Vaccinium* spp., Family: Ericaceae) contain unique genetic traits such as high anthocyanin diversity, exceptional cold hardiness, unique flavor and aroma compounds, etc. (Chu et al. 2025). Overharvesting and genetic modification of these wild populations can result in permanent loss of the selective alleles and reduce the adaptive capacity of the species. Maintaining *in situ* conservation areas and *ex situ* seed banks is therefore essential to safeguard the genetic variations necessary for both ecological resilience and future breeding programs.

19.3 PRINCIPLE OF SUSTAINABLE HARVESTING

Sustainable harvesting refers to harvesting a natural resource within the limits of its capacity for self-regeneration (Wu et al. 2024). The idea of sustainable harvesting of medicinal fruits and berries is to harvest them within the limits that allow the plants to grow and produce in the future. The concept of sustainable harvesting of medicinal fruits and berries starts with an understanding of both ethical and ecological responsibilities by farmers and cultivators. Ethical concerns in sustainable harvesting highlight different factors such as respect for nature, equal resource access, and acknowledgment of indigenous knowledge (Akalibey et al. 2024). Ecological concerns include protecting habitat, maintaining biodiversity, and supporting ecosystem services (Li et al. 2024). Ethical harvesting practices help long-term sustainability by saving resources and promoting social fairness through the careful control of human needs and ecological limits. Harvesting guidelines are crucial to implement sustainable practices. Key principles include selective and seasonal collection and minimizing mechanical damage to harvesting plants (Praveen and Akash 2025). Selective harvesting aims to harvest mature fruits while leaving younger fruits to support regeneration. Seasonal fruit collection does not interfere with flowering or seed dispersal, which maintains population dynamics. Harvesting techniques like handpicking, the use of non-invasive tools, and limiting plant harvest minimize mechanical damage to the plant structure. Avoiding bark stripping or uprooting fruiting shrub practices is also critical to this concept. Local and indigenous community engagement is another key point in sustainable harvesting (Srinivasu et al. 2024). Local communities always have deep traditional knowledge of these practices. However, they need more knowledge of sustainable techniques for future initiatives, and these communities can also be encouraged through profit sharing, which helps them focus on long-term conservation. When local and indigenous knowledge work together with these practices, harvesting programs show that sustainability is readily achievable. Legal certification processes and regulatory guidelines in cultivation and harvesting practices provide additional support for maintaining sustainable harvesting. Organic certification and biodiversity-friendly labels are good examples of initiatives that establish standards for ecological protection (Alcázar Marín 2025). These frameworks guide harvesters to follow best practices in maintaining agreement while providing market identification for sustainably developed products. Altogether, these considerations describe the principle of sustainable harvesting. Several studies have shown the effective use of sustainable harvesting and cultivation practices in the field, and these findings are summarized in Table 19.2.

19.4 CULTIVATION STRATEGIES FOR SUSTAINABILITY

19.4.1 Selection of Fruit and Berry Varieties

Like other plants, the selection of suitable species for the cultivation of medicinal fruits and berries is always a balance between economic requirements and biological knowledge (Dalir et al. 2025). This cultivation begins with the careful selection of species and varieties suited to the target environment.

TABLE 19.2
Sustainable Harvesting and Cultivation Practices of Medicinal Fruits and Berries

Medicinal Fruits and Berries	Sustainable Harvesting Practices	Sustainable Cultivation Practices	References
Amla-/Indian gooseberry (*Phyllanthus emblica* L.) Family: Phyllanthaceae	Selective fruit collection without harming palms	Agroforestry systems with other native crops	Charoenphun et al. (2025); Khedikar and Chaudhari (2024)
Apple (*Malus domestica*) Family: Rosaceae	Hand-picking mature fruits to avoid tree damage	Mixed cropping with medicinal plants, organic farming, and soil conservation practices	Hueppe and Zander (2024)
Blackberry (*Rubus fruticosus* L.) Family: Rosaceae	Seasonal harvesting at maturity to reduce waste	Integrated pest management, use of resistant cultivars, and organic fertilizers	Moreno-Miranda and Dries (2022)
Blackcurrant (*Ribes nigrum* L.) Family: Grossulariaceae	Harvesting ripe fruits to encourage natural regeneration	Low-input farming and hedgerow maintenance for biodiversity	Elqadini et al. (2024); Petrescu and Hoza (2024)
Blueberry (*Vaccinium* spp.) Family: Ericaceae	Mechanical harvesting with minimal plant stress	Crop rotation, reduced pesticide use, and soil organic matter enrichment	Negrușier et al. (2024); Yang and Lin (2025)
Chokeberry/Aronia (*Aronia melanocarpa* (Michx.) Elliott) Family: Rosaceae	Hand or soft mechanical harvesting to reduce plant injury	Sustainable irrigation, mulching, and pollinator-friendly practices	Karakayacı (2024)
Cranberry (*Vaccinium macrocarpon* Ait. L) Family: Ericaceae	Hand-harvesting for local use, and mechanical harvesting for large scale	Low-input farming, hedgerow planting for ecosystem services	Pisani Gareau et al. (2024)
Elderberry (*Sambucus nigra* L.) Family: Adoxaceae	Wet harvesting in flooded bogs with water recycling	Controlled bog management, reduced agrochemicals, and wetland conservation	Brodt et al. (2024); Huchteman et al. (2025)
Goji/Wolfberry (*Lycium barbarum* L.) Family: Solanaceae	Collecting only mature umbels to avoid overharvesting	Intercropping, minimal pesticide use, and preserving wild stands	Rehman et al. (2024)
Grape (*Vitis vinifera* L.) Family: Vitaceae	Hand harvesting to avoid fruit bruising	Drip irrigation, organic farming, reducing chemical inputs, and greenhouse farming	Khadatkar et al. (2025); Ubaidillah et al. (2025)
Guava (*Psidium guajava* L.) Family: Myrtaceae	Selective harvesting to maintain vine health	Organic viticulture, cover crops, and reduced water usage	Jain et al. (2025); Rangare et al. (2025)
Hawthorn (*Crataegus rhipidophylla* Gand.) Family: Rosaceae	Harvesting at physiological maturity and avoiding tree stripping	Agroecological methods, organic fertilizers, and pest-resistant varieties	Xiao et al. (2025)
Kiwi (*Actinidia chinensis* var. *deliciosa*) Family: Actinidiaceae	Limited collection from wild shrubs to preserve biodiversity	Hedgerow cultivation, soil conservation, and organic farming	Supreetha et al. (2025); Tian et al. (2025)
Mangosteen (*Garcinia mangostana* L.) Family: Clusiaceae	Manual harvesting to avoid vine stress	Sustainable trellising, reduced chemical inputs, and water-efficient irrigation	Jaroensutasinee et al. (2024); Tengsetasak et al. (2024)

(*Continued*)

TABLE 19.2 (Continued)

Medicinal Fruits and Berries	Sustainable Harvesting Practices	Sustainable Cultivation Practices	References
Mulberry (*Morus alba* L.) Family: Moraceae	Hand-picking ripe fruits to avoid damaging branches	Shade-grown agroforestry systems and organic manures	Chanotra et al. (2024); Rohela et al. (2025)
Noni (*Morinda citrifolia* L.) Family: Rubiaceae	Regular pruning and fruit collection without bark stripping	Silkworm–mulberry integrated systems and agroforestry	Prakash et al. (2024)
Papaya (*Carica papaya* L.) Family: Caricaceae	Harvesting mature fruits without harming roots or shoots	Mixed cropping in coastal agroecosystems and organic methods	Mendoza-Grimón et al. (2024); Salinas et al. (2024)
Pomegranate (*Punica granatum* L.) Family: Lythraceae	Hand-picking mature fruits and avoiding damaging the latex flow	Crop rotation, organic fertilizers, and biological pest control	Shiwangee et al. (2025); Vadivel et al. (2024)
Raspberry (*Rubus idaeus* L.) Family: Rosaceae	Manual harvesting to avoid skin cracking	Drip irrigation, organic fertilizers, and intercropping	Zejak et al. (2025)
Strawberry (*Fragaria* spp.) Family: Rosaceae	Selective picking to maintain cane health	Mulching, biological pest management, and low-input systems	González-Pedraza et al. (2024); Zeneli et al. (2024)

Plants that are naturally adapted to specific conditions reduce the need for substantial inputs such as irrigation and chemical applications, and these plant species can be easily used for cultivation (Getahun et al. 2024). Native species always show more resistance to pests, diseases, and abiotic stresses (Yasseen and Al-Thani 2025). The use of native species thereby enhances both productivity and ecological compatibility. In addition, the selection of genetically different varieties is important for long-term sustainability. Cultivated plants with high genetic variability can maintain resistance to emerging pathogens and environmental changes (Raza et al. 2025). For example, pomegranate (*Punica granatum* L., Family: Lythraceae) cultivars in the Mediterranean region warrant stable fruit production despite variable rainfall patterns, while multiple blueberry (*Vaccinium* spp., Family: Ericaceae) varieties in North America support resistance against frost and disease pressures (Chu et al. 2025; Raimondo et al. 2025). Therefore, this selection of species and varieties of medicinal fruits and berries to meet market demand forms the basis of sustainable cultivation.

19.4.2 Organic and Agroeconomic Practices

These practices play a key role in reducing environmental impacts while maintaining productivity. Organic farming removes synthetic chemical applications such as fertilizers, instead relying on natural soil alterations, compost, and biological pest control (Panday et al. 2024). Organic practices improve soil fertility, support a healthy microbial population, and reduce chemical runoff into nearby ecosystems (Singh et al. 2024). Agroecological methods apply ecological concepts to farming systems. Methods like intercropping, polycultures, and crop rotations can mimic natural ecosystems, increase resilience, and reduce the vulnerability of medicinal fruits and berries to pests and diseases (Zou et al. 2024). As an example, intercropping amla with nitrogen-fixing legumes improves soil fertility while supporting biodiversity (Mudalagiriyappa et al. 2022). All these practices eventually sustain long-term productivity.

19.4.3 Agroforestry Systems

Trees, shrubs, and crops are part of agroforestry, and this system provides a sustainable way to cultivate medicinal fruits and berries. Figure 19.2 shows how trees, shrubs, and crops grow together in an agroforestry system, and this concept shows a sustainable way to grow medicinal fruits and berries. This system mimics natural forest structure, where climate inputs such as humidity, temperature, precipitation, and solar radiation interact with cultivation practices to affect key physiological processes. The main processes in this system are photosynthesis, transpiration, carbon allocation, and nutrient uptake. Agroforestry improves plant health and makes the environment more resilient when production follows these natural processes. Farmers and cultivators can enhance the ecosystem by combining medicinal plants with other compatible tree species. Acai fruit (*Euterpe oleracea* L., Family: Arecaceae) plants grow well under shade and thrive under mixed canopy systems that mimic their natural rainforest habitat (Young 2024). This system also offers farmers multiple sources of income. Timber, fodder, medicinal herbs, and other forest products can be grown alongside fruits to reduce dependence on a single crop. Agroforestry ensures both productivity and environmental sustainability by aligning cultivation with natural ecological processes.

19.4.4 Soil, Water, and Nutrient Management

Management of soil, water, and nutrients is also crucial for sustainable cultivation. Mulching, cover cropping practices, and organic compost application can improve soil structure, water storage, and be beneficial for microbial populations (Khan et al. 2024). Reduced tillage methods further protect soil integrity and control soil erosion. Water management strategies such as drip irrigation and rainwater storage optimize water-use efficiency while reducing stress on natural resources. Nutrient management relies on organic fertilizers, crop residues, and legume intercropping to supply key nutrients naturally (Sharawat and Shilpa 2025). Cultivation systems can maintain high productivity without compromising long-term ecosystem health by managing these key areas.

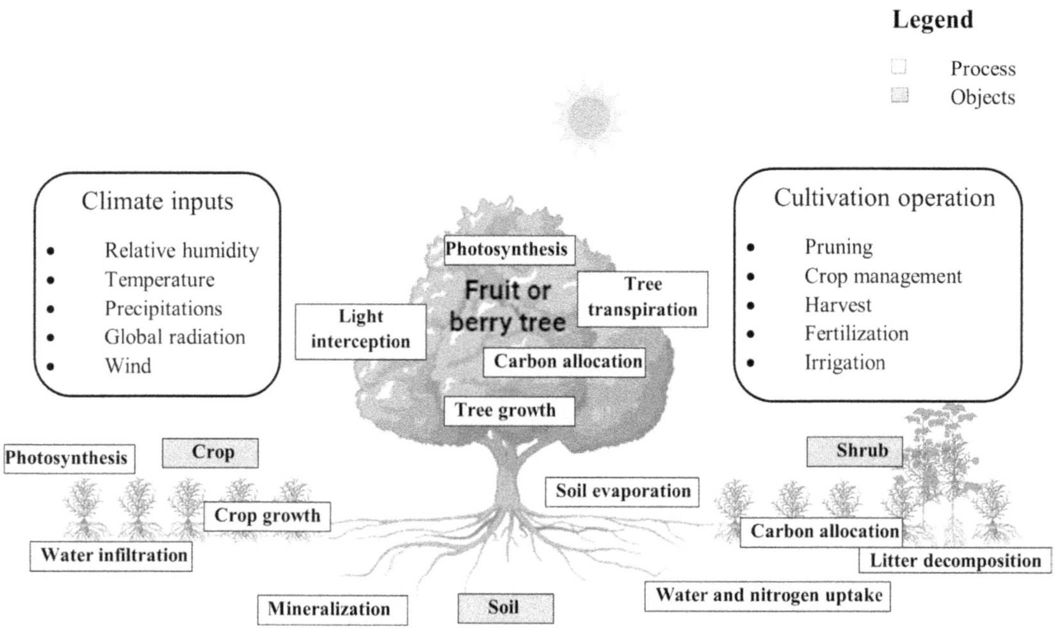

FIGURE 19.2 Agroforestry model for sustainable cultivation of medicinal fruits and berries.

19.4.5 Integrated Pest and Disease Management

Pests and diseases are major concerns in the cultivation of medicinal fruits and berries. Integrated pest management provides an eco-friendly and sustainable basis to address these challenges. Integrated pest management integrates biological control agents, cultivation practices, resistant cultivars, and minimal chemical use to control pests below economic thresholds (Zhou et al. 2024). There are also biological pest control methods, such as introducing beneficial insects or microorganisms, and targeting specific pests without harming other species (Curk and Trdan 2024). Cultivation practices such as sanitation, pruning, crop rotation, and intercropping can also disturb pest life cycles. Integrated pest management methods can reduce chemical usage by integrating multiple control measures.

19.4.6 Post-harvest Handling

Post-harvest handling is a key aspect of sustainable cultivation that directly affects the product quality, shelf life, and market value of medicinal fruits and berries (Lalpekhlua et al. 2024). It is necessary to conduct proper handling of the product during collection to minimize bruising and physical damage. Fruits should be harvested at optimal ripeness to meet market preferences. There are different post-harvest methods, such as sorting, washing, cooling, and storage conditions, and these methods can prevent spoilage and maintain phytochemical quality (Osei-Kwarteng et al. 2024). Drying, freezing, or extraction methods for processed products should be improved to retain bioactive compounds (Gębczyński et al. 2024). Sustainable post-harvest management also includes packaging practices that reduce environmental impact while protecting the product. These factors facilitate the maintenance of the therapeutic and nutritional quality of the products.

19.5 TECHNOLOGICAL AND SCIENTIFIC INNOVATIONS

19.5.1 Biotechnological Tools for Conservation and Cultivation

Biotechnology is a powerful tool to conserve resources and ensure sustainable production for future generations. Tissue culture techniques allow fast growth of rare plant varieties, and this ensures the conservation of genetic resources while reducing pressure on wild populations (Salaria et al. 2025). In vitro propagation provides a consistent method used in biotechnology for producing disease-free plants and is suitable for large-scale cultivation. This technology uses molecular markers to identify and characterize the genetic diversity in wild and cultivated populations (Bunjkar et al. 2024). Genetic studies are useful for developing improved cultivars with disease resistance, thus supporting both ecological and economic sustainability.

19.5.2 Propagation Techniques

Conventional propagation methods are broadly used in the cultivation of fruits and berries; however, the success of these methods always depends on species-specific requirements and local environmental conditions (Manohar et al. 2024). The most common conventional propagation techniques are seed sowing, grafting, layering, and cuttings in the management of fruit and berry cultivation (Khaitov et al. 2024). Among modern techniques, rapid and large-scale production of high-quality plant material is facilitated through micropropagation and somatic embryogenesis (Bogdanović and Ćuković 2025; Sota et al. 2025). Well-known examples include the micropropagation of goji berry (*Lycium barbarum* L., Family: Solanaceae) and sea buckthorn (*Hippophae rhamnoides* L., Family: Elaeagnaceae), which have allowed the establishment of uniform orchards with improved yield and disease resistance (Balasubramaniam et al. 2024; Rehman et al. 2024). Vegetative propagation methods are particularly valuable for controlling necessary traits in medicinal fruits like pomegranate

(*Punica granatum* L., Family: Lythraceae) and blueberry (*Vaccinium* spp., Family: Ericaceae) (Khaitov et al. 2024). The selection of proper propagation techniques can help with sustainable cultivation and reduce pressure on wild populations.

19.5.3 Advances in Post-Harvesting Technology

Advanced technology in post-harvest management aims to preserve product quality, increase shelf life, and reduce waste. The preservation of bioactive compounds and the inhibition of microbial contamination can be controlled through atmosphere storage, refrigeration, vacuum packaging, and modified atmosphere packaging (Habiba et al. 2025). Innovations in drying, freezing, and cold chain logistics increase the stability and potency of blueberries (*Vaccinium* spp., Family: Ericaceae), acai (*Euterpe oleracea* L., Family: Arecaceae), and elderberry fruits (Araujo et al. 2024). Improved processing techniques enabled by new technology also allow the preparation of high-quality extracts and powders while preserving therapeutic properties. These advanced technologies contribute to sustainability by reducing post-harvest losses and minimizing the environmental impacts associated with wasted product.

19.6 POLICY AND INTERNATIONAL GUIDELINES

All the described methods and concepts need to be guided by policy and regulations. The Convention on Biological Diversity highlights the conservation of biodiversity, sustainable use of biological resources, and benefits-sharing from traditional knowledge (Cariño and Zaltzman 2025). This provides a basis for national legislation and community-based resource management, which ensures that the use of medicinal plants aligns with conservation goals. The Nagoya Protocol, which is under the Convention on Biological Diversity, addresses access to genetic resources and the fair sharing of benefits (Lachenmeier and Montagnon 2024). It indicates that benefits are equitably shared with local communities and indigenous populations. These global conventions and protocols provide legal and ethical guidance for sustainable utilization and influence national policies and local practices. Policies are designed at the national and regional levels to balance resource conservation with economic development. Many countries have introduced regulations in the field of wild harvesting, cultivation, trade, and export of medicinal fruits. These regulations usually set limits on harvesting, establish protected areas, and provide guidelines for sustainable collection to protect local ecosystems from overexploitation. As an example, India has developed policies integrating traditional knowledge systems, forest conservation, and community-based management for species such as amla, elderberry, and neem (Sharma et al. 2025). European Union guidelines regulate the commercial trade of medicinal plants and establish quality standards for phytopharmaceutical products (Rausch et al. 2024). These policies provide directions for sustainable production and encourage collaboration among farmers, cooperatives, and government agencies.

19.7 CONCLUSION AND FUTURE REMARKS

The sustainable harvesting and cultivation practices of medicinal fruits and berries focus on several factors, including ecological, social, and economic considerations, to ensure successful long-term sustainability. This chapter describes the role of fruits and berries in traditional and modern medicines, highlighting their nutraceutical benefits, pharmaceutical potential, and economic value for local communities. Moreover, this information indicates that the growing global demand and market expansion have placed pressure on the wild population of the fruits and berries, which has resulted in overharvesting, biodiversity loss, and genetic erosion. Climate change and unsustainable cultivation practices further affect the sustainability of these resources. Balancing resource utilization with conservation strategies is a key factor in achieving sustainable management. In addition, populations of medicinal fruits and berries can be maintained through ethical harvesting practices, rotational collection, and adherence to certification and regulatory guidelines. This chapter highlights that sustainable cultivation strategies can be achieved mainly with species selection, organic and agro-ecological practices, agroforestry systems, soil and water management, integrated pest management, and post-harvest handling. These factors warrant the high productivity of selective medicinal fruits and berries in cultivation without harming ecological balance. New technological inventions and innovations also play a significant role in sustainable harvesting and cultivation practices for the long-term supply of these resources. Altogether, it is important to consider the sustainable harvesting and cultivation practices of medicinal fruits and berries to meet the global requirements while conserving biodiversity.

Future scenarios can focus on the importance of climate-resilient cultivation, sustainable supply chains, conservation of medicinal fruits and berries, value addition, and connecting the traditional knowledge with modern medicines. Long-term sustainability requires a universal vision that addresses the above factors. The future sustainability of fruit and berry production can be strengthened by encouraging collaboration among researchers, policymakers, local communities, and the commercial sectors. These collective approaches protect these resources and knowledge for future generations, supporting human nutrition and health, and provide a plan for sustainable production of medicinal fruits and berries while conserving the resources.

REFERENCES

Aguilera, J.M. and T. Toledo. 2024. Wild berries and related wild small fruits as traditional healthy foods. *Crit. Rev. Food Sci. Nutr.* 64(16): 5603–5617.

Akalibey, S., P. Hlaváčková, J. Schneider, J. Fialová, S. Darkwah and A. Ahenkan. 2024. Integrating indigenous knowledge and culture in sustainable forest management via global environmental policies. *J. For. Sci.* 70: 265–280.

Alcázar Marín, É. 2025. The organic production system: regulation and certification. In: C. Palacios Riocerezo and M. Gonzalez Ronquillo (eds.), *Traditional livestock production*. Springer, Cham.

Aliyu, N.O., M.O. Welcome, N. Medugu, A. Asalu, I. Odo, T.O. Ogundeko et al. 2025. Phytochemical evaluation and pharmacological activities of leaf and fruit extracts of African blackberry nightshade *Solanum nigrum* Linn. (NIPRD H 7357). *Path Sci.* 11(1): 4001–4008.

Al-Zubaidi, A.M., Z.M. Hassan and G.Q. AbdAlkhaliq. 2025. Study the effectiveness of plant extracts against urinary tract infections. *J. Prospect. Res.* 25(4): 22–31.

Araujo, N.M.P., P. Berni, L.R. Zandoná, N.M.V. de Toledo, P.P.M. da Silva, A.A. de Toledo and M.R. Maróstica Junior. 2024. Potential of Brazilian berries in developing innovative, healthy, and sustainable food products. *Sustain. Food Technol.* 2: 506–530.

Arshad, M.T., N.S. Basher, N.A. Ibrahim, A. Ikram, S. Maqsood, A. Rasheed et al. 2025. Nutritional, pharmacological and industrial applications of mangosteen and passion fruit: a review. *Food Sci. Nutr.* 13: e70574.

Babalola, B.A., A.I. Akinwande, A.A. Otunba, G.E. Adebami, O. Babalola and C. Nwufo. 2024. Therapeutic benefits of *Carica papaya*: a review on its pharmacological activities and characterization of papain. *Arab. J. Chem.* 17(1): 105369.

Balasubramaniam, S., S. Ganeshan, S. Natesan and K. Kapildev. 2024. A review on genetic diversity, micropropagation and transformations in the high-value medicinal plant of Himalayas—seabuckthorn (*Hippophae* sp.). *J. Pharmacogn. Phytochem.* 13(1): 360–366.

Bebek Markovinović, A., I. Brčić Karačonji, K. Jurica, D. Lasić, M. Skendrović Babojelić, B. Duralija et al. 2022. Strawberry tree fruits and leaves (*Arbutus unedo* L.) as raw material for sustainable functional food processing: a review. *Horticulturae* 8(10): 881.

Białoszycka, Z., M. Białoszycka, A. Pachevska, V. Istoshyn and A. Biloshytska. 2025. Application of medicinal properties of blueberries (*Vaccinium myrtillus*). *J. Educ. Health Sport* 80: 58369.

Bogdanović, M.D. and K.B. Ćuković. 2025. Secondary somatic embryogenesis in plants: from cellular mechanisms to biotechnological potential. *Plants* 14(22): 3413.

Brodt, S., G. Engelskirchen and K. Fyhrie. 2024. Native blue elderberry in hedgerows bridges revenue and conservation goals. *HortScience* 78(3–4).

Bunjkar, A., P. Walia and S.S. Sandal. 2024. Unlocking genetic diversity and germplasm characterization with molecular markers: strategies for crop improvement. *J. Adv. Biol. Biotechnol.* 27: 160–173.

Cariño, J. and S.H. Zaltzman. 2025. The Indigenous World 2025: Convention on Biological Diversity (CBD). *Indigenous World* 37: 598.

Chanotra, S., M. Attri and G.R. Verma. 2024. Mulberry-based integrated cropping system: an ideal approach for effective utilization of land resources in sericulture. *Int. J. Agric. Ext. Soc. Dev.* 7(SP-Issue 4): 36–39.

Charoenphun, N., Z. Ban, L. Liu, B. Paulraj and K. Venkatachalam. 2025. A review of post-harvest management practices and quality optimization strategies for Indian gooseberry (*Phyllanthus emblica* L.). *Trends Sci.* 22(12): 11546.

Chatterjee, S., S.K. Singh Yadav and V. Singh. 2025. The myths, legends, and folktales surrounding the blueberries. *Agri Express* 2(2): V02I02.05.

Chaudhury, C.M., V.J. Das and N. Das. 2024. Noni fruit: a powerful natural remedy for enhancing wound healing and skin repair. *Int. J. Pharm. Sci. Rev. Res.* 84(11): 57–60.

Chavda, M., L. Manani, C. Chandarana and P. Chavda. 2025. A review on noni: insights into botany, ethnopharmacology, phytochemistry, and commercial prospects. *IJP* 12(2): 100–112.

Chu, Y., J. Clevenger, K. Lee, J. Zhang and C. Li. 2025. Genetic breeding to improve freeze tolerance in blueberries: a review. *Horticulturae* 11: 614.

Ciccoritti, R., R. Ciorba, D. Ceccarelli, M. Amoriello and T. Amoriello. 2024. Phytochemical and functional properties of fruit and vegetable processing by-products. *Appl. Sci.* 14: 9172.

Costa, M.G., I.J.M. Viégas and R.A.M. Cordeiro. 2024. Dynamics, requirements, and use efficiency of magnesium throughout the life cycle of acai palm plants. *Agric. Res.* 13: 498–508.

da Costa, P., M.R.C. Schetinger, J. Baldissarelli, K.P. Reichert, N. Stefanello, N.B. Bottari et al. 2025. Blackcurrant (*Ribes nigrum* L.) and its association with donepezil restore cognitive impairment, suppress oxidative stress and pro-inflammatory responses, and improve purinergic signaling in a scopolamine-induced amnesia model in mice. *Neurochem. Res.* 50: 79.

Curk, M. and S. Trdan. 2024. Benefiting from complexity: exploring enhanced biological control effectiveness via the simultaneous use of various methods for combating pest pressure in agriculture. *Agronomy* 14: 199.

Dalir, M., S. Choobchian and E. Abbasi. 2025. Identifying the relative importance of cultivating medicinal plants in improving rural livelihoods. *Environ. Dev.* 54: 101159.

Deori, M., S. Manasa, P.B. Kakade, D.R.K. Saikanth, G. Ranganna, R.N. Deshmukh et al. 2024. A comprehensive review on the impact of climate change on fruit yield and quality in modern horticultural practices. *Int. J. Plant Soil Sci.* 36: 177–187.

Devi, P., G. Singh, Y. Suneja, H.S. Rattanpal, M. Gupta and R. Lhamu. 2024. Potential value of bioactive and enzymatic antioxidant compounds in grapefruit (*Citrus* × *paradisi* Macf.) varieties. *Food Anal. Methods* 17: 834–846.

Elqadini, M.E.B.R., R.V. Prodanović, V. Bižić, S.D. Vještica, S.M. Paunović et al. 2024. Bioeconomic potential of valorizing blackcurrant (*Ribes nigrum* L.) press residues. In: *Economics: theory and practice*. Novi Sad: Fakultet za ekonomiju i inženjerski menadžment, pp. 141–159.

Eshonkhujaev, O.O. 2025. Obtaining and classifying products for folk medicine from the blackberry (*Rubus*) plant. *Int. J. Sci. Res.* 11(2): 195–196.

Fang, H., Y. Ren and J. Tian. 2025. Origin, distribution, germplasm, and cultivation of sea buckthorn. In: J. Tian et al. (eds.), *Sea buckthorn*. Singapore: Springer.

Fatima, M., M.A. Dar, M.J. Dhanavade, S.Z. Abbas, M.N. Bukhari, A. Arsalan et al. 2024. Biosynthesis and pharmacological activities of the bioactive compounds of white mulberry (*Morus alba*): current paradigms and future challenges. *Biology* 13(7): 506.

Fernando, D.T.K., D.S.H.S. Peiris, S.P.N.N. Senadeera and C.B. Ranaweera. 2024. Therapeutic potential of *Carica papaya* Linn.: a comprehensive review of its pharmacological properties and health benefits, with emphasis on dengue fever healing. *Asian Plant Res. J.* 12(1): 32–48.

Gębczyński, P., M. Tabaszewska, K. Kur, M. Zbylut-Górska and J. Słupski. 2024. Effect of the drying method and storage conditions on the quality and content of selected bioactive compounds of green legume vegetables. *Molecules* 29(8): 1732.

Geethika, E., S. Ramamoorthy and G. Ravikanth. 2024. Does long-term harvesting impact genetic diversity and population genetic structure? A study of Indian gooseberry (*Phyllanthus emblica*) in the central Western Ghats region in India. *Plant Syst. Evol.* 310: 22.

Getahun, S., H. Kefale and Y. Gelaye. 2024. Application of precision agriculture technologies for sustainable crop production and environmental sustainability: a systematic review. *Sci. World J.* 2024: 2126734.

González-Pedraza, A.F., E. González Sarmiento and L. Castellanos González. 2024. Sustainable management practices and soil quality in strawberry cultivation in Norte de Santander, Colombia. *J. Agric. Rural Dev. Trop. Subtrop.* 125(1): 103–113.

Goswami, D. 2024. Addressing the challenge of overexploitation of natural resources in pursuit of development. *Sprin J. Arts Humanit. Soc. Sci.* 3(3): 65–67.
Guo, F., Y. Yang and G. Gao. 2024. Climate change impact on three important species of wild fruit forest ecosystems: assessing habitat loss and climatic niche shift. *Forests* 15(8): 1281.
Habiba, U., A. Bajpai, Z. Shafi, V.K. Pandey and R. Singh. 2025. Advancing sustainability through modified atmosphere packaging (MAP) for fresh food preservation: a critical review. *J. Stored Prod. Res.* 112: 102657.
Hammad, A.R. and R. Malik. 2021. A review of apple and its medicinal uses in Unani medicine. *Int. J. Unani Integr. Med.* 5(2): 165–169.
Hossain, M.S., M.A. Wazed, S. Asha, M.R. Amin and I.M. Shimul. 2025. Dietary phytochemicals in health and disease: mechanisms, clinical evidence, and applications—a comprehensive review. *Food Sci. Nutr.* 13(3): e70101.
Huchteman, M.W., S.D. Moore, J.C. Garcia, R.J. Smeda, T. Skevas, D. Khatiwada et al. 2025. American elderberry performance, yield, and potential profitability are influenced by weed management method. *HortScience* 60(12): 2209–2217.
Hueppe, R. and K. Zander. 2024. Perfect apples or sustainable production? Consumer perspectives from Germany. *J. Consum. Behav.* 23: 698–710.
Husen, A. 2021. *Traditional herbal therapy for the human immune system*. Boca Raton, FL: CRC Press. https://doi.org/10.1201/9781003137955.
Husen, A. 2022. *Herbs, shrubs and trees of potential medicinal benefits*. Boca Raton, FL: CRC Press.
Husen, A. 2024. *Medicinal spices and condiment crops*. Boca Raton, FL: Taylor & Francis Group https://doi.org/10.1201/9781003387046.
Hussain, A., A. Batool, S. Yaqub, A. Iqbal, S. Kauser and M.R. Arif et al. 2024. Effects of spray drying and ultrasonic-assisted extraction on the phytochemicals, antioxidant and antimicrobial activities of strawberry fruit. *Food Chem. Adv.* 5: 100755.
Ispiryan, A., V. Atkociuniene, N. Makstutiene, A. Sarkinas, A. Salaseviciene, D. Urbonaviciene et al. 2024. Correlation between antimicrobial activity values and total phenolic content/antioxidant activity in *Rubus idaeus* L. *Plants* 13(4): 504.
Izteleuova, E.Y., G.T. Zhumashova, T.S. Bekezhanova, Z.B. Allambergenova, K.A. Zhaparkulova, A.A. Karaubayeva et al. 2025. The current state of knowledge on *Ribes* spp. (currant) plants. *Plants* 14(20): 3196.
Jain, S., H. Chawla, R.P. Singh, P. Chakradhar, S. Khanikar and S. Jain et al. 2025. Bahar treatment in guava: a key strategy for year-round fruit production. *Cuest. Fisioter.* 54(3): 3614–3632.
Jaroensutasinee, K., M. Jaroensutasinee, P. Boonsanong and E. Sparrow. 2024. Mangosteen yield and fruit quality under regulated irrigation with sensors and IoT. *J. Hum. Earth Future* 5(2): 205.
Jocien, S.K. and F. Nfor. 2023. Unsustainable agricultural practices as a driver of soil degradation in Santa Sub-Division, North West Region of Cameroon. *Asian Soil Res. J.* 7: 19–33.
Joshi, D.D. and H. Kharkwal. 2025. *Dietary and medicinal significance of wild edible fruiting plants*. Boca Raton, FL: CRC Press.
Karakayacı, Z. 2024. Energy efficiency in organic aronia (chokeberry) production. *Appl. Fruit Sci.* 66: 2365–2370.
Kaur, S., R. Shams, K.K. Dash, V.K. Pandey, A.M. Shaikh, E. Harsányi et al. 2024. Phytochemical and pharmacological characteristics of phalsa (*Grewia asiatica* L.): a comprehensive review. *Heliyon* 10(2): e25046.
Khadatkar, A., C.P. Sawant, D. Thorat, A. Gupta, S. Jadhav, D. Gawande et al. 2025. A comprehensive review on grapes (*Vitis* spp.) cultivation and its crop management. *Discov. Agric.* 3: 9.
Khadivi, A., M. Rezagholi and M. Shams. 2024. Phytochemical properties and bioactive compounds of pomegranate (*Punica granatum* L.). *J. Hortic. Sci. Biotechnol.* 99(6): 639–652.
Khaitov, M., I. Rasulov and J. Karshiyev. 2024. Technology and methods of propagation of fruit seedlings. *Sci. Innov.* 3(D9): 276–279.
Khan, M.T., J. Aleinikovienė and L.-M. Butkevičienė. 2024. Innovative organic fertilizers and cover crops: perspectives for sustainable agriculture in the era of climate change and organic agriculture. *Agronomy* 14(12): 2871.
Khedikar, R.S. and V.M. Chaudhari. 2024. Review on cultivation practices of amla (*Phyllanthus emblica*). *Int. J. Ayurveda Pharma Res.* 12(9): 62–66.
Lachenmeier, D.W. and C. Montagnon. 2024. Convention on Biological Diversity (CBD) and the Nagoya Protocol: implications and compliance strategies for the global coffee community. *Foods* 13(2): 254.

Lalpekhlua, K., A. Tirkey, S. Saranya and P.J. Babu. 2024. Post-harvest management strategies for quality preservation in crops. *Int. J. Veg. Sci.* 30: 587–635.
Latif, R. and T. Nawaz. 2025. Medicinal plants and human health: a comprehensive review of bioactive compounds, therapeutic effects, and applications. *Phytochem. Rev.* (in press). 25: 2299–2342.
Lee, H., Z. Wang, Z. Deng and Y. Wang. 2024. Assessment of six blackberry cultivars using a combination of metabolomics, biological activity, and network pharmacology approaches. *Antioxidants* 13(3): 319.
Li, K., L. Liu, D.J. McClements, Z. Liu, X. Liu and F. Liu. 2023. A review of the bioactive compounds of kiwifruit: bioactivity, extraction, processing and challenges. *Food Rev. Int.* 40(3): 996–1027.
Li, K., X. Yan, Y. Hou, B. Lv, Y. Huang, J. Liu et al. 2024. Can ecological zoning act as an environmental management tool for protecting regional habitat quality: causal evidence from the national key ecological function zone in China. *J. Clean. Prod.* 475: 143623.
Ma, J.-X., W. Yang, C.Y.J. Ng, X.-D. Tang, S. Wong, R.-Y. Gan et al. 2024. The hawthorn (*Crataegus pinnatifida* Bge.) fruit as a new dietary source of bioactive ingredients with multiple beneficial functions. *Food Front.* 5: 1534–1558.
Mahavy, C.E., A.J. Razanatseheno, A. Mol, J. Ngezahayo, P. Duez, M. El Jaziri et al. 2024. Edible medicinal guava fruit (*Psidium guajava* L.) are a source of anti-biofilm compounds against *Pseudomonas aeruginosa*. *Plants* 13(8): 1122.
Manohar, K.A., G. Shukla, N.N. Shahina, K. Sivasankarreddy, S.S. Ravuther and S. Chakravarty. 2024. Conventional versus non-conventional methods of propagation of forest tree species: applications and limitations. In: T.D. Thomas, M.K. Razdan and A. Kumar (eds.), *Biotechnological approaches for sustaining forest trees and their products*. Singapore: Springer.
Mendoza-Grimón, V., R. Amorós, J.R. Fernández-Vera, E. Lopes da Veiga and M.D.P. Palacios-Díaz. 2024. Utilising reclaimed water for papaya (*Carica papaya* L.) cultivation in Cape Verde: a detailed case study. *Agronomy* 14(11): 2726.
Mieszczakowska-Frąc, M., N.J. Dickinson and D. Konopacka. 2025. Effect of postharvest ripening on the phytochemical composition and antioxidant properties of fruits from ten plum (*Prunus domestica* L.) cultivars. *Agronomy* 15(6): 1351.
Mirzayeva, S. and A. Jo'raboyev 2025. Use of pomegranate in folk medicine and its role in modern medicine. *Int. J. Med. Sci.* 1(3): 111–115.
Mishra, H. 2025. Environmental degradation and impacts on agricultural production: a challenge to urban sustainability. In: J.A. Parray, N. Shameem and A.K. Haghi (eds.), *Sustainable urban environment and waste management*. Singapore: Springer.
Moreno-Miranda, C. and L. Dries. 2022. Assessing the sustainability of agricultural production—a cross-sectoral comparison of the blackberry, tomato and tree tomato sectors in Ecuador. *Int. J. Agric. Sustain.* 20(7): 1373–1396.
Möwes, M., G.K. Kandanda, L.N. Nangolo, F.S. Shafodino and L.M. Mwapagha. 2025. Qualitative phytochemical profiling, and in vitro antimicrobial and antioxidant activity of *Psidium guajava* (guava). *PLoS One* 20(4): e0321190.
Moysidou, A.M., K. Cheimpeloglou, S.I. Koutra, M.A. Finos, A. Ofrydopoulou, A. Tsoupras et al. 2024. A comprehensive review on the antioxidant and anti-inflammatory bioactives of kiwi and its by-products for functional foods and cosmetics with health-promoting properties. *Appl. Sci.* 14(14): 5990.
Mudalagiriyappa, T., Ramachandrappa, B.K., Vasanti, B.G., Savitha, M.S. and KM, P. 2022. Growth, yield and economics of amla (*Emblica officinalis* L.) based agri-horticultural system in Alfisols of semi-arid tropic. *Environ. Conserv. J.* 23(3): 439–449.
Mukhametzyanov, R.R., T.M. Dzhancharov, A.M. Khezhev, L.A. Sviridova and D.V. Snegirev et al. 2023. Factors and trends in the development of international trade in fruit and berry products. In: E.G. Popkova et al. (eds.), *Digital agriculture for food security and sustainable development of the agro-industrial complex*. Cham: Springer.
Negrușier, C., A. Colișar, S. Rózsa, M.S. Chiș, S.-M. Sîngeorzan, O. Borsai et al. 2024. Bilberries vs. blueberries: a comprehensive review. *Horticulturae* 10(12): 1343.
Nollet, L. and J. Ahamad. 2025. *Bioactive compounds of edible fruits and berries: health benefits, nutritional importance, and analysis*. Boca Raton, FL: CRC Press.
Osei-Kwarteng, M., M.C. Ogwu, G.K. Mahunu and N.A. Afoakwah. 2024. Post-harvest food quality and safety in the global South: sustainable management perspectives. In: M.C. Ogwu, S.C. Izah and N.R. Ntuli (eds.), *Food safety and quality in the global South*. Singapore: Springer.
Otero-Guzman, N. and R. Andrade-Pizarro. 2025. Bioactive compounds from tropical fruit by-products: extraction, characterization and therapeutic potential. *J. Agric. Food Res.* 21: 101983.

Panday, D., N. Bhusal, S. Das and A. Ghalehgolabbehbahani. 2024. Rooted in nature: the rise, challenges, and potential of organic farming and fertilizers in agroecosystems. *Sustainability* 16: 1530.

Petrescu, A. and D. Hoza. 2024. A review of blackcurrant culture technology. *Sci. Pap. Ser. B, Hortic.* LXVIII(1): 111–121.

Pisani Gareau, T.L., L. Gao and B.J. Gareau. 2024. The enduring nature of cranberry production in a changing climate: the interplay of extreme weather, knowledge networks, and adaptation. *PLoS Clim.* 3(5): e0000350.

Ponmurugan, P., A. Anitha and N. Mathivanan. 2025. Biomedical applications and health benefits of wild noni and its fruits. In: P. Thangaraj (ed.), *Pomotherapeutic insights on wild edible fruits*. Singapore: Springer.

Prakash, R., A.D. Jokhan, R.D. Gopalan, R.D. Singh and A.V. Prasad. 2024. Physiological performance and fruit quality of noni (*Morinda citrifolia* L.) cultivated in different agroclimatic zones of Fiji. *Adv. Hortic. Sci.* 38(2): 177–188.

Praveen, K. and B. Akash. 2025. *Harvesting technologies and practices in agriculture*. 1st ed. Palwal: Chyren Publication, pp. 5–60.

Pulidindi, K. and K. Ahuja. 2025. Specialty fruits and berries market size by product type, form, application, distribution channel, cultivation channel, growth forecast, 2025–2034. Global Market Insights, Report ID: GMI14522. https://www.gminsights.com/industry-analysis/specialty-fruits-and-berries-market (accessed 1 November 2025).

Raimondo, M.L., F. Lops, A. Tarantino, N. Bellantuono, A. Carlucci and F. Bimbo. 2025. The current state of Italian pomegranate production: agronomic, crop protection, economic, and managerial perspectives. *Agronomy* 15: 826.

Rangare, N.R., N. Lal, G. Shiurkar, K. Banjare, P. Singh, N. Jangre et al. 2025. Impact of climate change on guava production: challenges and adaptive strategies. *Appl. Fruit Sci.* 67: 404.

Rausch, H., S. Schröder, T. Friedemann, S. Cameron, K. Kuchta and M. Konrad. 2024. History and present of European traditional herbal medicine (phytotherapy). In: *History, present and prospect of world traditional medicine*, pp. 131–234.

Raza, A., T. Khare, X. Zhang, M.M. Rahman, M. Hussain, S.S. Gill et al. 2025. Novel strategies for designing climate-smart crops to ensure sustainable agriculture and future food security. *J. Sustain. Agric. Environ.* 4: e70048.

Rehman, F., S. Zeng, Y. Zhao, J. Zhao, K. Qin, C. Yang et al. 2024. Preservation and innovation of goji berry germplasm resources. *Med. Plant Biol.* 3: e022.

Rohela, G.K., P. Saini, S. Syam, P. Roy, B. Gonal, D. Aziz et al. 2025. Mulberry trees: a sustainable solution for urban forestry and improved air quality. In: J.A. Parray, N. Shameem and A.K. Haghi (eds.), *Sustainable urban environment and waste management*. Singapore: Springer.

Sagandyk, A.T., Â. Liberal, T.F.F. da Silveira, M.J. Alves, I.C.F.R. Ferreira, G.N. Zhakupova et al. 2024. Nutritional, phytochemical, and bioactive prospects of black chokeberry (*Aronia melanocarpa*) and saskatoon berry (*Amelanchier ovalis*) grown in the Republic of Kazakhstan. *Appl. Food Res.* 4(2): 100564.

Saini, R., N. Sharma, O.S. Oladeji, A. Sourirajan, K. Dev, G. Zengin et al. 2022. Traditional uses, bioactive composition, pharmacology, and toxicology of *Phyllanthus emblica* fruits: a comprehensive review. *J. Ethnopharmacol.* 282: 114570.

Salaria, N., I. Kumari and R. Kumar. 2025. Tissue culture technology: an emerging viable technology to protect biodiversity in the face of biodiversity loss. In: *Plant tissue culture technology*. 1st ed. Palm Bay, Fl: Apple Academic Press, p. 26.

Salinas, I., A. Carmona, J.J. Hueso and V. Pinillos. 2024. Optimal harvest maturity changes depending on the season throughout the year in papaya grown in Mediterranean climate- improved greenhouses. *Horticulturae* 10(12): 1360.

Sana, T., M. Aftab, M.M.A. Khan and M. Naeem. 2025. Overexploitation and conservation strategies for medicinal and aromatic plants. In: M. Naeem and M.M.A. Khan (eds.), *Essential oil-bearing plants*. Academic Press, pp. 95–105.

Shan, F., L. Liu, L. Li, W. Wang, Y. Bi and M. Li. 2025. Management, safety, and efficacy evaluation of nutraceutical and functional food: a global perspective. *Compr. Rev. Food Sci. Food Saf.* 24: e70222.

Sharawat, V. and Shilpa. 2025. Influence of integrated nutrient management on sustaining legume crop productivity and reducing environmental impact: a review. *J. Food Legumes* 38(1): 14–25.

Sharma, I. and S. Birman. 2024. Biodiversity loss, ecosystem services, and their role in promoting sustainable health. In: P. Singh and N. Yadav (eds.), *The climate-health-sustainability nexus*. Cham, Switzerland: Springer.

Sharma, K., S. Jarial and N.D.C. Patil. 2025. Integrating indigenous technical knowledge for sustainable agriculture and horticulture: insights from India. In: T. Mapotse, E. Tetteh and T. Matsekoleng (eds.),

Enabling indigenous knowledge systems in action research and action learning. Hershey, PA: IGI Global Scientific Publishing, pp. 207–230.

Shi, X., X. Wang, Y. Zheng and L. Fu. 2025. Advances in the study of bioactive compounds and nutraceutical properties of goji berry (*Lycium barbarum* L.). *Appl. Sci.* 15(1): 262.

Shinwari, A.K., M. Ahmad, A.A. Alsahli, A. Israr, Z.K. Shinwari, S. Shinwari et al. 2024. Himalayan medicinal plants wealth, threats and conservation for securing the future of biodiversity. *Pak. J. Bot.* 56(6): 2327–2350.

Shiwangee, N., N.S. Kaler, M. Sharma, S. Sharma and A. Sharma. 2025. Pomegranate-based agroforestry systems for sustainable agriculture, achieving food security and mitigating climate change through carbon sequestration in Northwest Himalayas. *Agrofor. Syst.* 99: 73.

Simsek, O. 2025. Harvesting sustainability: innovations and practices in modern agriculture. *Green Technol. Sustain.* 3(3): 100192.

Singh, N.K., K. Sachan, R. Ganjikunte, C.S., M.B.P., N. Panotra and D. Katiyar. 2024. Building soil health and fertility through organic amendments and practices: a review. *Asian J. Soil Sci. Plant Nutr.* 10: 175–197.

Song, Z.K., Z.X. Liu, C.R. Deng, G.Z. Duan, G.H. Fan and J.L. Li. 2025. Predicting suitable habitats of high-yield and quality *Lycium barbarum* based on climate characteristics in production area of Qinghai, China. *Ying Yong Sheng Tai Xue Bao* 36(4): 1118–1126.

Sota, V., H. Wilms, B. Yücesan, Y.Y. Mendi, B. Christie and J. Nisler et al. 2025. Challenges in the micropropagation of economically important fruit species in Europe. *Plant Cell Tiss. Organ Cult.* 162: 53.

Srinivasu, P., S. Parkavi, P. Ragul, N. Panotra, D. Thrilekha and L. Upadhyay. 2024. A critical review on fostering community involvement in sustainable horticulture initiatives. *J. Sci. Res. Rep.* 30: 394–404.

Supreetha, B.G., V.B. Mhetre, W.H. Raja, M.K. Verma, R.E. Rashmi, K.R. et al. 2025. Agronomic practices affecting the yield and quality of kiwifruit: a comprehensive review. *Appl. Fruit Sci.* 67: 239.

Swathi, S., S. Narayani, K.P. Vedasree, V. Pavitraa and V. Vadivel. 2024. Wound healing potential of an Indian underutilized fruit *Garcinia gummi-gutta*: evidences from in vitro and in vivo experiments. *Food Hum.* 2: 100271.

Szot, I. and G.P. Łysiak. 2025. Factors influencing the formation, development of buds, and flowering of temperate fruit trees. *Agriculture* 15(12): 1304.

Teixeira, B., G. Teixeira, E. Darnet, H. Schaller, H. Rogez and S. Darnet. 2025. A review of the genus *Euterpe*: botanical and genetic aspects of açai, the purple gold of the Amazon. *Bot. J. Linn. Soc.* 208: 1–13.

Teng, P. 2025. Asian food security issues and opportunities. In: *Asian food security issues and opportunities*. Springer, Singapore.

Tengsetasak, P., K. Tongkoom, J. Yomkerd, C. Susawaengsup, N. Khongdee, T. Chatsungnoen et al. 2024. Sustainable strategies for fresh mangosteen: adapting to climate challenges. *Earth Syst. Environ.* 8: 1829–1847.

Tian, Z., X. Guo, W. Ma and X. Xue. 2025. Research on kiwifruit harvesting robot worldwide: a solution for sustainable development of kiwifruit industry. *Smart Agric. Technol.* 10: 100792.

Torrance, A.W. and B. Tomlinson. 2025. Overharvesting: the why of biodiversity loss. *Tex. A&M J. Prop. L.* 11: 615.

Ubaidillah, M.A.M., D. Erwanto, A. Baidowi, M.T.N. Ngulya, M.R. Gymnastiar et al. 2025. Assistance to business actors through the cultivation of grapefruit seeds to improve the village economy. *Taawun* 5(2): 280–294.

Uhl, K. and A.E. Mitchell. 2024. Elderberry, an ancient remedy: a comprehensive study of the bioactive compounds in three *Sambucus nigra* L. subspecies. *Annu. Rev. Food Sci. Technol.* 15: 27–51.

Uyak, C., E. Aglar, B. Ozturk, A. Doğan, O. Tekin et al. 2024. Biochemical characterization of mulberry (*Morus* spp.) genotypes from Türkiye (Hizan, Bitlis): a comprehensive analysis of fruit properties and bioactive compounds. *Food Sci. Nutr.* 12: 6425–6441.

Vadivel, R., K.S. Reddy, Y. Singh and D.D. Nangare. 2024. Effect of pit and soil types on growth and development, nutrient content and fruit quality of pomegranate in the Central Deccan Plateau Region, India. *Sustainability* 16(18): 8099.

Vukelić-Nikolić, M., A.V. Dragićević, P. Vasiljević, N.M. Stojanović and D.R. Pavlović. 2025. Beyond the traditional applications of raspberry (*Rubus idaeus*): an in vitro, in vivo and in silico study. *Rec. Nat. Prod.* 19(4): 518–s.

Wang, J., M. Sun, J. Yu, J. Wang and Q. Cui. 2024. Pomegranate seeds: a comprehensive review of traditional uses, chemical composition, and pharmacological properties. *Front. Pharmacol.* 15: 1401826.

WHO. 2023. Increasing fruit and vegetable consumption to reduce the risk of noncommunicable diseases. https://www.who.int/tools/elena/interventions/fruit-vegetables-ncds (accessed 15 October 2025).

WHO. 2025. Noncommunicable diseases. https://www.who.int/news-room/fact-sheets/detail/noncommunicable-diseases (accessed 15 October 2025).

Woldegebriel, A.M. 2025. Fruit crops and their role in enhancing food security in Ethiopia. *Discov. Food* 5: 293.

Wu, L., X. Yan, J. Huang, X. Xu, S. Zhang, S. Cui et al. 2024. Towards sustainable fruit production: identifying challenges and optimization strategies. *Agric. Syst.* 221: 104132.

Xiao, T., H. Zhang and H. Li. 2025. Research progress on hawthorn (*Crataegus pinnatifida*) cultivation techniques and landscape application. *Proc. Int. Conf. Biol. Environ., Mater. Sci. Intell. Comput. (BEMSIC 2025)*, 14(1): 322–327.

Xu, J., C. Li, W. Wu, L. Kong, L. Xiao, W. Ma et al. 2025. Cranberry research progress: a systematic review of chemical composition, pharmacological mechanisms, clinical applications, and nutritional significance. *Int. J. Mol. Sci.* 26(19): 9707.

Yadav, V., D. Pal and A.K. Poonia. 2024. A study on genetically engineered foods: need, benefits, risk, and current knowledge. *Cell Biochem. Biophys.* 82: 1931–1946.

Yang, J.-L. and S.-Y. Lin. 2025. Evaluating sustainable media mixes using local agricultural and forestry wastes for enhancing the early establishment of container-grown southern highbush blueberry in subtropical climates. *Sci. Hortic.* 339: 113845.

Yang, W., Q. Gao, K. Ran, M. Jiao, X. Fu, Z. Han et al. 2025. The nutritional and bioactive components, potential health function and comprehensive utilization of hawthorn: a review. *Food Front.* 6: 2108–2128.

Yasseen, B.T. and R.F. Al-Thani. 2025. Adaptive strategies to biotic stress in Qatar's native flora. *Life* 15: 1645.

Young, K.J. 2024. Mimicking nature: a review of successional agroforestry systems as an analogue to natural regeneration of secondary forest stands. In: F. Montagnini (ed.), *Integrating landscapes: agroforestry for biodiversity conservation and food sovereignty*. Cham: Springer.

Zagórska-Dziok, M., A. Ziemlewska, M. Wójciak, I. Sowa, E. Wąsik-Szczepanek and Z. Nizioł-Łukaszewska. 2025. Comparison of cytotoxicity and antioxidant, antibacterial, and anti-inflammatory activity of aqueous and ethanolic extracts from *Malus domestica*, *Prunus armeniaca*, and *Prunus cerasus* leaves. *Molecules* 30(10): 2085.

Zejak, D., V. Spalevic and B. Dudic. 2025. Sustainability of raspberry (*Rubus idaeus* L.) cultivation in Montenegro with a focus on production systems and market dynamics. In: I. Karabegović, A. Kovačević and S. Mandžuka (eds.), *New technologies, development and application VIII*. Springer.

Zeneli, F., V. Ventura and D.G. Frisio. 2024. Sustainable fresh strawberry consumption: environmental, genetically modified food, and climate concerns in Europe and North Africa. *Front. Sustain. Food Syst.* 8: 1442074.

Zhang, Q., L. Wang, H. Li, F. Wu, Z. Tang, H. Li et al. 2024. Gartanin as an isoprenylated xanthone from mangosteen (*Garcinia mangostana*): insights into its synthesis, extraction, determination, and bioactivity. *Food Rev. Int.* 41(1): 218–235.

Zhang, Y., R. Wan, Z. Shi, L. Yang and S. Fang. 2025. Phenotypic and phytochemical variations in wolfberry varieties and their harvest times. *Horticulturae* 11(9): 1138.

Zhao, J., X. Liang, X. Kang, Y. Li and W. An. 2024. Estimation of goji berry (*Lycium barbarum* L.) canopy water content based on optimal spectral indices. *Sci. Hortic.* 337: 113589.

Zhou, W., Y. Arcot, R.F. Medina, J. Bernal, L. Cisneros-Zevallos and M.E. Akbulut. 2024. Integrated pest management: an update on the sustainability approach to crop protection. *ACS Omega* 9: 41130–41147.

Zou, Y., Z. Liu, Y. Chen, Y. Wang and S. Feng. 2024. Crop rotation and diversification in China: enhancing sustainable agriculture and resilience. *Agriculture* 14: 1465.

Index

Pages in *italics* refer to figures and pages in **bold** refer to tables.

B

C

D

E

F

G

M

N

O

P

For Product Safety Concerns and Information please contact our EU
representative GPSR@taylorandfrancis.com
Taylor & Francis Verlag GmbH, Kaufingerstraße 24, 80331 München, Germany

www.ingramcontent.com/pod-product-compliance
Ingram Content Group UK Ltd.
Pitfield, Milton Keynes, MK11 3LW, UK
UKHW052051230926
475702UK00025B/1092
* 9 7 8 1 0 4 1 1 6 3 8 4 8 *